Natural History and Ecology of Suriname

Natural History and Ecology of Suriname

LM Publishers
Parallelweg 37
1131 DM Volendam
The Netherlands

info@lmpublishers.nl
www.lmpublishers.nl

LM PUBLISHERS

Editor
Bart De Dijn, Environmental Services & Support NV
www.ess-environment.com

ESS environment

Graphic design
Brigitte Küchler, René Bosshard, Aisthesis.ch

aisthesis.ch

Production Hightrade bv, Zwolle

Artwork on cover by / © Isabella Kirkland

ISBN 978-94-6022-438-6

WWF

This publication has been made possible with the support of:

WWF Guianas

The Embassy of the Kingdom of the Netherlands in Suriname

Kingdom of the Netherlands

Contents

Preface

'Natural History and Ecology of Suriname' takes you on a journey through the natural world of what still is one of the greenest countries on earth. Indeed, Suriname maintains one of the highest forest covers of any nation worldwide, a treasure trove of natural richness. In addition, its savannah landscapes, swamps, rivers and mountains, all contribute to the colorfu diversity of plant and animal species to be found. To a wide range of nature lovers, naturalists and professionals, this book aims at unraveling some of the secrets of Suriname's natural world.

As it reflects on the history and status of our nature to date, this book offers a reference for those who wish to know more about and explore Suriname's natural world. Suriname's unique and little explored natural world deserves more of international attention and recognition. This book offers readers a wide knowledge base to help motivate and implement nature conservation for the years to come and to increase the awareness and appreciation of the values of our natural world.

WWFs' mission is to create a future in which humans live in harmony with nature. Knowing our natural history through the information in this book, we at WWF Guianas as a part of WWF, can actively work towards that goal.

We are grateful to Bart De Dijn and all contributors for all their hard work in producing this book. We want to thank the Embassy of the Kingdom of the Netherlands for providing the financial support to realize this unique publication.

We know that you will enjoy it.

Laurens Gomes
WWF Guianas Regional Representative

Acknowledgements

The editor extends thanks to WWF, in the first place the former director of the Guianas office, Dominique Plouvier, who initated the project to make this book. Also thanked are all others at WWF Guianas involved in the project, especially the current director Laurens Gomes for continued support, and Mark Wright, Sofie Ruysschaert and Jerrel Pinas who supervised, helped and gave valuable inputs. Barney Jeffries is thanked for language-checking semi-final chapters (his advice we mostly took, but remaining errors in the final product are not Barney's). The Government of The Netherlands is thanked for funding the entire project via the Royal Dutch Embassy in Paramaribo. It was implemented by a grant to WWF, which assigned project execution to Environmental Services & Support N.V. (ESS).

For their dedication and hard work, the authors are of course recognisized, as well as all who supplied images. Special thanks to Jan Mol, who not only wrote a chapter on his beloved fishes, but at the editor's request also co-wrote two more chapters. Junior author Chantal Landburg is recognized for taking charge of writing the book's closing chapter. Pieter Teunissen should probably get a statue for collaborating on this project while at some point facing life-threatening illnesses (and overcoming them). Brigitte Küchler and René Bosshard of Aisthesis spent many crazy hours on the layout of the book, and implementing final text corrections. Their hard work and flexibility was greatly appreciated. Peter Sanches and Ron Smit of LM Publishers are thanked for taking this project to the next level: final layout, printing and distribution.

For help with tracking down artwork by Gerrit Schouten, John Henri Lance, Maria Sibylla Merian and Louise van Panhuys for the introductory chapter, Bart De Dijn wishes to thank Clazien Medendorp. She and Pieter Teunissen are also thanked for a quick review of the chapter. For access to artwork, thanks go to Charlotte Brooks of the Lindley Library of the Royal Horticultural Society in London, also Jip Binsbergen of the Artis Library in Amsterdam, and Cornelia Gilb of the Zentralbibliothek of the Johann Christian Senckenberg Museum in Frankfurt am Main. Susanne Kridlo and Johannes Lerp of the Museum Wiesbaden are thanked for allowing access to insect specimens likely collected by Maria Sibylla Merian. Many thanks to Guido and Rosine De Dijn for hospitality abroad, and Afi for patience and a lot of breaks.

Salomon Kroonenberg and Dirk Noordam thank professor Theo Wong of the Anton de Kom University of Suriname for critically reading the geology part of this second chapter. Ton Markus of Kartomedia is thanked for professionally drawing two figures (Figures 6 and 15) for this chapter.

Riad Nurmohamed, Koos Groen and Sieuwnath Naipal gratefully acknowledge the Meteorological Service Suriname, the Hydraulic Research Division, the Maritime Authority Suriname and the Suriname Water Supply Company for data provided.

For support in his fieldwork in the region and study of the Guiana Shield, Brice Noonan would like to thank CNRS, NSF, the Centre for the Study of Biodiversity (Guyana), and the Nature Conservation Division (Suriname). Nathan Lujan, Brian O'Shea and Andrew Snyder kindly shared their thoughts and insights during the writing of the chapter on biogeography. Brice's wife Danielle and two children (Alivia and Caden) are thanked for having kindly tolerated his absence during many extended trips to the Guianas.

Bart De Dijn, Pieter Teunissen and Jan Mol wish to thank Salomon Kroonenberg, Dirk Noordam and Sofie Ruysschaert for helpful review of a draft version of the chapter on ecosystems. Dirk is also thanked for feedback on the deep history of Suriname's landscapes and the landscape-soil-vegetation relationship. Verginia Wortel and her employer CELOS are thanked for images of estuarine lakes.

Bruce Hoffman, Sofie Ruysschaert and Sabitrie Jairam-Doerga thank the following collaborators who contributed significantly to the completion of the chapter on plants: Pieter and Siela Teunissen (photography), Chantal van den Bergh-Lodeweyckx (background research, photography), Eric Gouda (Vriesea splendens species account), Scott Mori (Lecythidaceae species accounts), Reinaldo Aguilar (photography; bejucososa.blogspot.com/), Olivier Gaubert (photography; floredeguyane.piwigo.com/), Omar Kasijo (contributing artist; Figures 55, 56, 58, 59), Sara Fuste and Brigitte Küchler (image editors). Credits for many additional contributers, no less appreciated, are provided in image captions.

Bart De Dijn and Hélène Hiwat wish to thank those who contributed species accounts and photographs to the chapter on insects & relatives: Menno Reemer, Alies van Sauers-Muller, Hajo Gernaat, Simon Clavier, Atilano Contreras-Ramos, Nicolas Dedieu, Jocelia Grazia, Cristiano Schwertner, David Sillam-Dussès, Virginie Roy, Dominic Evangelista, Julio Rivera, Antonio Agudelo, Marcel Wasscher, Johan van't Bosch, František Kovařík and Volker Mahnert. Also thanked are Anil Gangadin, Yanouk Epelboin, Auke Hielkema and Vanessa Kadosoe for providing additional photographs. Maurice Leponce is thanked for a quick review of part of the chapter.

Paul Ouboter wishes to thank all scientists and technicians that assisted him in the field, especially Rawien Jairam, Vanessa Kadosoe, Gaitrie Satnarain and Gwen Landburg.

Brian O'Shea wishes to thank Otte Ottema, Serano Ramcharan, and Arie Spaans for their contributions to the species accounts, and Sean Dilrosun for sharing his extensive knowledge of indigenous names and stories of birds in the Interior of Suriname. Johan Ingels and Thomas Valqui assisted in contacting some of the photographers who kindly shared images to illustrate the species accounts: Carlos Alcantara, Nick Athanas, Hervé Breton, Marc Chrétien, Foek Chin-Joe, Kester Clarke, Ian Davies, Mathias Fernandez, Harold Greeney, Audrey Hogenboom, Sean McCann, Leon Moore, Meshach Pierre, Jean-Louis Rousselle, Frédéric Royer and Lloyd Spitalnik.

Burton Lim wishes to thank the Nature Conservation Division of the Suriname Forest Service for research and export permits. His fieldwork in Suriname has been funded and supported by Conservation International, SRK Consulting, ESS, University of Suriname, and Royal Ontario Museum Governors. Marilyn Norconk is grateful to the many students and colleagues in Suriname and from the US, UK and the Netherlands with whom she has worked at Raleighvallen-Voltzberg and at Brownsberg Nature Park.

Chantal Landburg and Jan Mol wish to thank the Foundation for Forest Management and Production Control (SBB) for providing land-use and deforestation maps through their online portal; Sarah Crabbe of SBB customized these maps for the final chapter of the book. They also recognize Aad Versteeg of Stichting Surinaams Museum who made images available of prehistoric man-made raised fields near Galibi and mounds near Wageningen.

General Introduction / History

Bart De Dijn

This book is about natural history and ecology, a subject currently within the domain of the biological sciences. It is also about Suriname, located in the American tropics (Neotropics), and part of the larger Amazon region (Amazonia). So it may be fitting that an introductory chapter that offers a historical perspective is told by a biologist who has been living and working in Suriname for the last 27 years. This chapter reflects upon the study of the country's natural history and ecology within the context of the 'Western' scientific tradition. This tradition is rooted in European culture, and is distinct from the Indigenous ('Amerindian'), African, Hindustani ('Indian'), Indonesian and other non-European traditions that are represented in Suriname. Yet it is informed by these traditions, often building on them: like Suriname itself, Western science is a bit of a melting pot, and no island. It is all about observing, collecting specimens, hypothesizing, testing, record keeping, and ultimately publishing findings based upon all this. The biological specimens and the scientific publications represent physical evidence that helps to reconstruct the history of the study of Suriname's nature in the Western tradition. The evidence is presented below, in its larger historical context.

Europeans came to the Americas as explorers from the late 16th till well into the 19th century. But exploring was more of a means than an end. The European goal at the time seems to have been to get richer and more powerful by any available means, be it trade and plunder, the appropriation of land and other resources, or the production of valuable commodities by a workforce of subjugated or even enslaved people (see for example Hemming 2009). Many practitioners of Western science at the very least piggybacked on what amounted to a sinister project, and the early stages of Western scientific study of Suriname are somewhat soiled by the link to this chapter in history.

Small steps, late 16th – early 18th century

Renaissance upheaval. What happened in Suriname from the late 16th till early 18th century should be understood in the context of the Renaissance. During the Renaissance there was upheaval in many aspects of society, such as on how knowledge was created and transmitted. Recent historical accounts of the Renaissance, such as by Cameron (2001) and Wiesner-Hanks (2006), explain this in some detail; below, a summary is provided based on these accounts. As the Renaissance proceeded from the 15th century onward, the full scope of the classical Greek and Roman arts and sciences was rediscovered in the Western European, Christian world. Europeans followed in the footsteps of Near-Eastern and North-African Muslim scholars. The latter had translated and drawn from the scientific works of, especially, the ancient Greek 'natural philosophers', and added new discoveries of their own to this body of work. The Western Europeans started to shift from scavenging the actual and proverbial ruins of classical civilization to wanting to fully understand, recreate and surpass the achievements of this civilization. Instead of uncritically accepting the dogmatic 'truths' found in the Bible and decreed by the Roman Catholic church, Western academics started to turn more to the empirical methods developed by ancient Greeks: observing the natural world, developing theories based on these observations, testing the theories in the real world, and thus making new discoveries.

The Renaissance was also a time of great financial and technological development in Western Europe. The first European banks date back to this time, as well as the first personal firearms, ships better suited to open ocean travel, and advanced astrolabes to aid in navigation. This enabled voyages into the unknown and the conquest of new territories, such as Suriname. In addition, this was a time of religious upheaval in Western Europe. Various reformed and protestant churches and nations were established, including in the Low Countries and parts of what we now call Germany. The colonies of the Guianas, such as Suriname, were established by exponents of protestant European nations; the rest of South America was the dominion of Iberian Roman Catholics. When Roman Catholic rulers expelled the 'heretic' Jews from the Iberian Peninsula and the Portuguese colony of Brazil, they were accommodated in protestant enclaves, including in the New World.

Another significant Renaissance development, given the importance of publishing in science, was the shift from manual copying of books to the technology to print books with text and black-and-white illustrations. This meant that ideas and knowledge could be shared faster, wider, and cheaper. Like translations of the Bible and protestant religious ideas, scientific ideas and accounts could thus reach a broader public, more widely sowing the seeds of change.

The new world discoverd and mapped. The year 1492 AD marks the 'discovery' of the New World by Columbus during his first expedition to find a western passage to the spice islands of

Figure 1. Hondius' 1599 'Guiana' map showing the coastline and main rivers of the Guiana Shield, with the Suriname coast located above the leopard image (source: Bijzondere Collecties van de Universiteit van Amsterdam, UBM: Kaartenzl : 104.05.04).

Southeast Asia ('the Indies'). Maps produced during the Renaissance confirm that shortly after 1492 the coast of the Guianas was explored by Iberian mariners, subjects of the Spanish and Portuguese kings (see Helman 1982, Bakker *et al.* 1998). One of the captains in Columbus' fleet during his second expedition to the New World in 1493 was Alonso de Ojeda. He sailed straight to the Guianas during a first expedition of his own in 1499, to explore and map the coast. Juan de la Cosa participated as a master navigator in that expedition, and made a crude map of the world ('Mappa Mundi') in 1500, on which the coastline of the Guianas can be recognized. These Iberian expeditions and the maps resulting from them were treated as state secrets. An example is the secret 1529 world map (the 'Padrón Real') of Diogo Ribeiro which shows 13 rivers between the 'Rio Dulce' (clearly the Essequibo), east of the island of Trinidad, and the 'Marañon' (the lower Amazon). No local names of these rivers are mentioned on this early map.

A century later, after the Englishman Sir Walter Raleigh's fleet explored the Orinoco and the coast of 'Guiana' in 1596, the secret was out. Based on this expedition, detailed maps were published in Dutch and other languages by Hondius and De Bry in 1599. Between the very recognizable Orinoco and Amazon, the maps show rivers with names quite similar to those on present-day maps of Suriname: 'Coreteny', 'Mickhery', 'Copanamo', 'Saramo', 'Saranano', 'Comojowini' and 'Cimarawini') (Figure 1). These maps were illustrated with an armadillo, jaguar, river turtle and peccary, alongside out-of-place illustrations of animals from the Old World. Apart from maps, narratives about the Amazon, Orinoco and the Guianas by for example Amerigo Vespucci (Markham 2010) and Walter Raleigh (Whitehead 1997) were published too. The narratives tell of mariners going upstream into the rivers and ashore, contacting the natives to extract information from them, as well as to acquire goods, including animals and plants.

During the early 17th century, maritime merchants from protestant nations found the Guianas poorly defended by the Spanish and Portuguese, and thus available for first establishing trading posts, later colonies. The Dutch from Zeeland established trading posts at Cayenne and on the Corantijn (Helman 1982). They successfully established colonies along the Essequibo in 1616, and the Demerara in 1627; these were the 'Dutch Guiana' colonies, in present-day Guyana. The first successful colonization of Suriname wasn't initiated until around 1650, by Englishmen hailing from Barbados, under the leadership of Francis Willoughby. They established sugar plantations, using the advanced 'factory' technology and practices – such as sugar mills and slave labor – that were introduced by displaced Jewish plantation owners from Brazil. This Jewish-protestant connection was a consequence of the religious persecution in Portuguese Brazil. In protestant territories like Suriname the authorities opportunistically welcomed the know-how and capital of the religious refugees.

Initially, the colony in Suriname had just one small town, Torarica, some 40 km up the Suriname River. By 1665, a Jewish town, Jodensavanna, had developed some 10 km further upstream. In 1667, the thriving English colony was taken over by an expeditionary force from Dutch Zeeland. At the time, it already had 175 plantations along the Commewijne River, the lower Suriname River, and their tributaries (Figure 2). By 1686 a town called Paramaribo had become Suriname's main settlement, located close to the mouth of the Suriname River, in the shadow of pre-existing fort Zeelandia (locally called *foto*, meaning fort; this word in present usage also refers to Suriname's capital Paramaribo). Aside from the English interregnum of 1804-1816, the colony essentially remained under control of the Dutch until Suriname's independence in 1975.

Figure 2. Anonymous 1667 map of 'Surranam', present-day Suriname (source: https://coursewikis.fas.harvard.edu/aiu18/File:Surinam.jpg).

Naturalia. The Renaissance's focus on discovering and observing the natural world led to new developments in the arts and sciences – which, at that time, were not really separated. The realistic paintings of still life scenes became fashionable: vases with flowers, insects and other animals added as embellishments, as well as baskets with fruits, fishes and game catches. There was a definite taste for depicting the exotic, such as the New World. Drawings and paintings were commissioned by noblemen, as well as merchants, whose wealth and power had been increasing. The members of this elite acquired a variety of 'naturalia': rocks, minerals, fossils, dried plants, stuffed animals, and their depictions (van Gelder 1993, Swan 2002, 2004; Reitsma 2008, Davenne & Fleurent 2011). Besides these naturalia, exotic ethnographic objects were collected. They were combined into a 'cabinet of curiosities' or 'cabinet of wonders': typically a single room, or even a single piece of furniture, with all objects on display (Figure 3). The cabinets were mostly private spaces in which the rich and powerful could retreat and enhance their knowledge by actively studying the collection of objects or simply passively contemplating their esthetics and meaning. Some cabinets comprised considerable collections of exotic objects, and were famous and more public, such as the Rumphius cabinet. The fascination with naturalia and exotic places at the time led to the publication of catalogs (for example Rumphius 1705) and natural history books. Brazil's environment, flora, fauna and people featured extensively for the first time in an illustrated book by Piso *et al.* (1648). Suriname and its flora and fauna were introduced to the world in a beautifully illustrated book by Maria Sibylla Merian (first edition 1705; second, posthumous expanded edition 1719) (see Box 1).

Figure 3. 'The Cabinet of a Collector' painting of part of a cabinet of wonder, by Frans Francken the Younger, 1617 (source: Royal Collection Trust / © HM Queen Elizabeth II 2017, RCIN 405781).

Living exotic plants and animals were shipped to Europe and kept in-house or in private 'gardens' (precursors of public botanical gardens and zoos). Plants were of interest for use as ornamentals, food and medicines. Records indicate that seeds from Suriname were shipped to the Netherlands from the mid 17th century onward (van Andel *et al.* 2012c). At that time, several botanical gardens ('hortus botanicus') already existed, such as the one of the University of Leiden (established 1590) and the city of Amsterdam (established 1638). Early, published accounts of useful plants and animals from Suriname can be found in Merian's work (see above and Box 1), and in early descriptions of the colony (van Berkel 1695, Harcourt 1707, Herlein 1718).

Adolescent leaps, mid 18th – late 19th century

LINNAEUS' APOSTLES. From the 1730s onward, the Swedish naturalist Linnaeus named and classified naturalia (see Box 2). He described what we now consider biological species based on specimens obtained from curiosity cabinet owners, as well as traders and correspondents in distant territories. He inspired others to collect, name and classify species, including two botanists that went to Suriname but alas died not long after their arrival in the country (Lanjouw & Uittien 1935, van Andel 2010). He sent out 17 students – his 'apostles' – to the planet's far corners to collect biological specimens (see **www.ikfoundation.org/downloads/LACompleteCatalogueCopyrightIK2.pdf**). Several of his apostles died on the road, or not long after returning to Sweden. Daniel Rolander was an apostle that survived. He was sent to Suriname in 1754 and left in 1763. Rolander collected mainly plants and insects in Suriname, where he was a guest of plantation owner Carl Gustav Dahlberg (van Andel *et al.* 2012a, de Moraes *et al.* 2014). Dahlberg was a friend of Linnaeus and sent him plant specimens for description, including the *kwasibita* plant (*Quassia amara*) with fever-suppressing properties (Linnaeus 1763, 1775; see also Oudschants Dentz 1941, Dragtenstein 2004).

BOX 1 Maria Sibylla Merian: naturalist-artist in a strange land

Maria Sibylla was the daughter of Matthäus Merian, a Swiss-German copper plate engraver and publisher. He was famous for his books with copper plate prints of landscapes (Wettengl 1998, Reitsma 2011). She followed in her father's footsteps, making watercolors and engravings of flowers and insects. She developed an interest in the metamorphosis of butterflies, a novel subject at the time. Exploring nature, looking for caterpillars and their food plants, made Maria Sibylla an early ecologist.

Maria Sibylla Merian married, but in the 1680s separated and left with her two daughters to join the Labadists, a reformed religious sect with a secluded congregation in the north of the Netherlands. This was located on lands of the van Sommelsdijck family. At the time, Cornelis van Aerssen van Sommelsdijck was co-owner and governor of Suriname. The Labadists also had a small beachhead in Suriname, on a plantation owned by governor van Sommelsdijk. Maria Sibylla and her daughters left the Labadist community to start an engraving and publishing business in Amsterdam in 1691. Yet the Labadist connection remained strong, at least via the oldest daughter's husband Jacob Hendrik Herolt, a devout Labadist with business interests in Suriname. In 1699, Maria Sibylla traveled to Suriname with her youngest daughter. She stayed there at the Labadist outpost for less than two years. After returning to Europe, she published her observations and illustrations in a book on the insects and plants of Suriname. It was originally published in a limited edition in 1705 (Figure 4), but became a bestseller, well after Maria Sibylla's death in 1717. Her hand-colored prints remain popular to this day, and are still of considerable scientific interest.

Both daughters continued to make watercolors under their mother's name after her death. Maria Sibylla's elder daughter moved to Suriname in the 1710s with her husband; she was buried in Paramaribo in 1728. The younger daughter married and moved to St. Petersburg (Russia) with her husband, where she died in 1743.

Figure 4. Watercolor of the butterfly *Morpho deidamia* (adult and caterpillar) on Suriname cherry, *Eugenia surinamensis*, by Maria Sibylla Merian, 1902-3 (source: Royal Collection Trust / © HM Queen Elizabeth II 2017, RCIN 921160).

BOX 2 The Linnaean system for naming and classifying species

Carl Linnaeus (1707-1778) – the 'Father of Taxonomy' – formalized the system for naming and classifying species that we still use today. Linnaeus referred to what we now call species with universal two-part ('binomial') names. These were unique, short names that expressed key characteristics of the species. Latin, the universal language of science in Linnaeus' time, was used to form the names. Before Linnaeus' innovation, different names – in Latin or local languages – were often assigned to the same species by different authors, which caused much confusion. Some authors used descriptive sentences to name species; these were hard to memorize. The usefulness of Linnaeus' innovation is evident from the fact that it has remained the basis for current scientific species names. The scientific name for the jaguar, for instance, is *Panthera onca*. It is written in italics in the modern literature. The first part of the name is the generic part, a single word with the first letter always capitalized in modern usage. This part is typically shared by several rather narrowly related species, which are then referred to as belonging to the same genus. It is also known as the genus name, and for example *Panthera* spp. is used as shorthand for all species of the genus *Panthera*. The second part of the name is the specific one, a single word that is never capitalized in modern usage. It makes a unique combination with the generic name, distinguishing, for example, *Panthera onca* from *Panthera tigris* (tiger) and *Panthera leo* (lion).

In 1753 Linnaeus published the first edition of 'Species Plantarum', in which all plants known to him were named and classified. The classification was hierarchical, and based on structural characteristics, such as the position and shape of leaves, flowers and fruits. Earlier, plants were often classified based on their usage. Linnaeus assigned structurally very similar plants to the same genus. Genera (plural of genus) with broad similarities were grouped into families. Families were grouped into higher categories, such as orders. The same principles were applied to other naturalia. By 1758, the tenth, extended edition of his publication 'Systema Naturae' was printed, in which all naturalia known to him were classified, including animals. Linnaeus' publications include descriptions of plants and insects that were based on the drawings and notes of Maria Sibylla Merian (1705). By the end of the 18th century, Linnaeus' works had become authoritative, and the basis of universal scientific names. Many scientific species names that are still in use today can be traced back to Linnaeus' 1753 and 1758 publications.

In the end, Linnaeus did not describe any of the plants Rolander brought back, because Rolander refused to hand over his collections, and tried to independently embark on an academic career. Rolander failed miserably at this, and ended up selling his collection to make ends meet; in this way at least some of the material he collected led to published descriptions of species (mainly by Rottbøll 1778). Rolander did process his notes and species descriptions into a manuscript that remained unpublished until it was translated 250 years later (Rolander 2008; also known as 'Diarium Surinamicum').

Natural history books and illustrations galore. In the mid 18th century there seems to have been an increasing appetite in Europe for natural history accounts of exotic territories. Several books on the subject as it related to Suriname were published in a variety of languages: by Fermin in French (1765), by Bancroft (1769) in English, and by Hartsinck (1770) in Dutch. A late 18th century bestseller that included much information on the country's natural history was John Gabriel Stedman's 1790 'Narrative' (reprinted repeatedly, such as in 1971). Stedman was a professional soldier, part of a contingent sent to Suriname to protect the colony against runaway slaves who had escaped from the plantations and taken refuge in the forest to establish independent Maroon communities there. Stedman's narrative contains a lot of information on natural history, but its prime purpose was to expose the cruel treatment of slaves in Suriname.

From the mid 18th till late 19th century many collecting activities took place in Suriname by amateur naturalists, often expatriates who were part of the country's colonial elite. Although these amateurs did not publish their findings, they showed their scientific interest by sending specimens to museums in Europe for further study. Examples include Dahlberg (already mentioned above), Dieperink, Hering, Hostmann, Lammens, Kegel, Wullschlägel and Voltz (Teunissen 1979; see also Medendorp 1999). Definitely not part of the elite was August Kappler, a low-level public servant stationed at the periphery of the colony. He published a narrative (Kappler 1854), as well as an extensive description of Suriname and its natural history (Kappler 1887).

The English interregnum (1804-1816) seems to have been a period of great synergy for the production of natural history illustrations in Suriname. At least four people were involved at the time: Adriaan François Lammens, a Dutch judge stationed in Paramaribo, John Lance, his English counterpart in the mixed Dutch-English court at Paramaribo, Louise van Panhuys, the wife of a plantation owner that was also briefly governor of Suriname, and Gerrit Schouten, a local artist (Burkhardt *et al.* 1991, Medendorp 1999, 2008) (see Box 3). Subsequently, some naturalists with more professional ambitions were active in Suriname. One was Hendrik Charles Focke; born in Suriname, he had studied law in The Netherlands, and was undoubtedly part of Paramaribo's elite. He published several works on the orchids of Suriname (such as Focke 1849, 1855). Another publishing amateur botanist in Suriname was the Dutchman Splitgerber (for example 1940, 1942).

Mature strides, early 20th century onward

Recent history. It is challenging to present the highlights of natural history studies in Suriname from the early 20th century onward. So much biological exploration and research activities took place in this period, and relatively little time has passed to digest what has transpired. Below an overview is provided of the kind of efforts that were undertaken, the key institutions and people involved, and the historical context. As in earlier times, scientific developments reflected the social and political ones. The historical backgrounds are outlined in what follows largely based on van Traa (1946), van Lier (1971), Helman (1982), Bakker *et al.* (1998), and the author's personal observations in the period 1990 till present. Biological research in Suriname till the late 1970s has been reviewed by Teunissen (1979), botanical expeditions by van 't Kloster *et al.* (2003). Below a limited number of references are mentioned; the other chapters in the present book contain more extensive references to recent research in the country by researchers mentioned ones.

BOX 3 Gerrit Schouten, home-grown artistic genius

Gerrit Schouten was a multi-talented professional artist: skilled in taxidermy, drawing and the making of dioramas (Medendorp 1999, 2008). He was a mixed-race Surinamer, and is assumed to have been largely self-taught. He likely absorbed the skills of people like Louise van Panhuys, who drew and painted plants and landscapes (Görner & Dobat 1991), and John Henri Lance, an amateur naturalist and draughtsman. Schouten sold his beautiful watercolors of plants and animals (Figure 5) to Lance and to the judge Adriaan François Lammens. These watercolors rival those of Maria Sibylla Merian.

Figure 5. Watercolor of the lizard *Iguana iguana*, with line drawing added to show details of the plates and scales on the head, by Gerrit Schouten, 1800-1839 (source: Artis Bibliotheek, Bijzondere Collecties van de Universiteit van Amsterdam, IZAA100171).

Schouten's drawings are currently in the Surinaams Museum in Paramaribo, in the Artis Library in Amsterdam (recently integrated in the 'Bijzondere Collecties' of the University of Amsterdam), in the Naturalis Museum in Leiden, in the Lindley Library of the Royal Horticultural Society in London, and in the Fine Arts Museum in San Francisco. The Lindley Library has the largest known collection of Schouten's watercolors, mixed with those of Lance in two large bound volumes. Many of Lance's and Schouten's drawings are near-identical in subject and composition, but not in quality: Schouten's drawings were executed with superior artistic skill. They were intended for use by naturalists, as is obvious from the fact that no backgrounds were drawn, and that the subjects are positioned to show all essential anatomical features to identify the species depicted.

Schouten's most amazing artistic product can be seen in the Boerhaave Museum in Leiden: a box filled with Surinamese butterflies. They look like regular pinned butterflies, but they are entirely artificial. The likeness to real butterflies is uncanny. This appears to be the only surviving box of three that were made by Schouten (Clazien Medendorp, personal communication) (see 'vlinderkast' video at **www.youtube.com/watch?v=WgqVia55CVg**).

A PERIOD OF CHANGE. In the early 20th century and the period leading up it, Suriname went through dramatic changes. While the Dutch banned the transatlantic slave trade in the early 19th century, slavery itself had not been abolished in Suriname. As the supply of slaves to the country gradually diminished, the management of plantations became difficult to impossible. Suriname's economy was entirely based on these plantations. As July 1st 1863, the date of the abolition of slavery by the Dutch, approached, the pace of abandonment of plantations increased. On that date, only 163 of the 500 previously active plantations remained. This number decreased further to 45 at the start of the 20th century. Some freed slaves stayed on the plantations, but many went to the capital city to settle in a part of Paramaribo aptly called *frimangron* (literally: free man's land). The colonial government managed to somewhat delay the demise of the plantations by recruiting an alternative workforce in the form of impoverished Dutch farmers and 'East Indian' laborers recruited in what is currently China, India and Indonesia.

For a few decades, the economic decline in the colony due to failing plantations was partly compensated by development of the gold mining and *balata* plant gum extraction industries (see chapter on Use, Threats and Conservation). Gold production peaked around the turn of the 19th-20th century, *balata* production some 20 years later. In the period 1901-1911, a railroad was built from Paramaribo deep into the Interior (Wicherts & Veltkamp 2012). The railroad track opened up previously inaccessible parts of the country, which led to social and economic development, as well as scientific exploration. The colonial government was eager to discover more potential sources of wealth in the Interior. It started a program to demarcate the southern borders of Suriname and map the geological riches in Suriname's entire territory. From 1901 till 1911, the Royal Dutch Geographical Society (KNAG) organized a total of seven expeditions to the far reaches of all the main rivers of Suriname. Typically one of the participants in these expeditions was an amateur naturalist, often the medical doctor that was part of the team to keep it healthy (Pieter Teunissen, personal communication). Between 1920 and 1944 at least ten expeditions took place; professional biologists and foresters led these expeditions or participated in them, such as Stahel, Gonggrijp, Geijskes and Maguire. Zoologists and botanists, such as Pulle and Lanjouw, also undertook several collecting trips in Interior during this period.

Pulle, a botanist at the herbarium of Utrecht University, initiated an important, high quality series of publications in 1928: the 'Flora of Suriname'. Pulle's initiative was continued by other Utrecht University botanists: Lanjouw, Stoffers and Lindeman. At present, the Flora of Suriname series encompasses over 2000 pages of species descriptions, including identification keys. During 1908-1910, two Surinamese brothers, Frederik Paul and Arthur Philip Penard, compiled 'De Vogels van Guyana', the first monograph on the birds of the three Guianas (present-day Guyana, Suriname and French Guiana).

During this period, applied biological research was also undertaken in Suriname. Forestry research took place under the auspices of Gonggrijp. Research on agricultural pests was undertaken at the newly (1904) created Agricultural Experiment Station (Landbouwproefstation) in Paramaribo by such researchers as entomologist (insect specialist) Geijskes and botanist and mycologist (fungi specialist) Stahel. Research on human and livestock parasites was undertaken by researchers such as Flu and Frickers. Medical interest in the transmission of malaria led to a monograph on the mosquitos of Suriname by Bonne and Bonne-Wepster (1925).

Second world war until independence. The Second World War (WWII) created major economic opportunities for Suriname. Bauxite was used to make aluminum for the war industry in the USA, and Suriname was a major source of it. Airstrips were built as part of 'Operation Grasshopper' in remote corners of the Interior to facilitate mineral exploration. The bauxite industry in Suriname boomed and led to the construction of the Brokopondo hydropower reservoir in the 1960s in the Suriname river basin (see also chapters Climate and Hydrology, and Use, Threats and Conservation). A road was built all the way around the reservoir to Pokigron, on the middle Suriname River. Road development increased in the decades after WWII: an east-west road in the swampy Coastal Plain, one in the Zanderij Belt with its savannas and forests on white sand, and a number of connections between the two, as well as extensions to the south, into the densely forested Interior. The coastal road served to connect population centers and new agricultural projects, such as wet rice farming in the northwest of Suriname. The other roads provided access to timber and mining concessions, and allowed for the creation of palm oil plantations in parts of the Interior.

This airstrip and road development opened up parts of Suriname previously inaccessible for researchers. It made biological exploration much easier than before, and offered opportunities for Dutch researchers, who were no longer welcome in the former colony 'Nederlands Oost Indië', which became independent shortly after WWII as the Republic of Indonesia. The work of Dutch researchers was mainly funded by the newly (1954) created research funding agency in The Netherlands, Wosuna, which became Wotro in 1961. Their publications were included in the Flora of Suriname series, in the newly created serial 'Studies on the Fauna of Suriname and other Guyanas' (started 1945), in pre-existing zoological serials of the Museum of Natural History in Leiden, and in a number of separately published studies. At that time, more than 30 extensive collecting trips were undertaken to many corners of Suriname by researchers such as the botanists and vegetation specialists Lindeman, Jonker, Kramer, Hekking, Wessels-Boer, Florschütz, Maas, Moolenaar, Schulz, Heyligers, Van Donselaar and Oldenburger, entomologist Geijskes, fish specialists Boeseman and Nijssen, amphibian and reptile specialist Hoogmoed, bird specialist Mees, and mammal specialist Husson. Important ecological research was also done in relation to the Brokopondo reservoir area, such as by Leentvaar on the aquatic environment (see Ouboter 1993), and by Genoways on small mammals. Spaans undertook ecological studies on shorebirds in the context of nature conservation. Publications of these researchers are referred to in other chapters of this book. Inventories of mollusks, such as snails and mussels, in the Estuarine Zone and elsewhere in Suriname were undertaken by van Regteren Altena (for example 1960, 1975).

A number of publications deserve specific attention here, because they are still commonly used. Lindeman and Mennega (1963) published a Suriname tree guide; a fully revised, English-language edition has recently been published (Bhikhi *et al.* 2016). Mennega (1976) edited *'Fa joe kan tak mi no moi'*, a pocket flora of mostly common plants in the Coastal Zone. Husson (1978) published a monograph on the mammals of Suriname, Haverschmidt (1968) on the birds, Schulz (1975; Dutch edition in 1980) on the marine turtles, and Hoogmoed (1973) on the lizards. In the domain of applied science, there were important publications on agricultural pest insects by van Dinther (1960), and on useful plants by Ostendorf (1962).

Immediately after independence. During the first few years after Suriname's independence in 1975, biological research flourished as never before. New research and development initiatives financed by The Netherlands created new opportunities. There was a strong focus on ecology, a reflection of the global trend at the time. An agreement was made between the University of Wageningen in The Netherlands and the University of Suriname (AdeKUS, with its research institute CELOS), to strengthen forestry and forest ecology research. This led to research on the so-called CELOS management system (CMS), a timber harvest and sylvicultural system described in a multi-author volume edited by Werger (2011).

During the late 1970s to mid 1980s, a number of important studies were published that were based on fieldwork done in the preceding years: on primates (monkeys) by Mittermeier (1977) and van Roosmalen (1985b), on river otters by Duplaix (for example 1980), on wild fruits by van Roosmalen (1985a), on vegetation and ecosystems by Teunissen (see references in Ecosystems chapter), and on the traditional use of plants (ethnobotany) by Plotkin (1986). What drew a lot of attention internationally was a popular publication by Trail (1983) on the behavior of one of Suriname's most charismatic birds, the Guianan Cock-of-the-rock. Much of the fieldwork of these authors took place in protected areas or focused on protected species. This was facilitated by several governmental institutions: the Nature Conservation Division of the State Forest Service (LBB), and the Foundation for Nature Preservation in Suriname (Stinasu). The role of the so-called 'Teacher's College' (IOL) merits mentioning here, because IOL offered the only advanced courses in biological sciences in Suriname at the time. IOL students and graduates were involved in some of the research mentioned above.

Trying times. The period from 1982 till the mid 1990s was politically tumultuous. An eight-year civil war led to the destruction, or at least sharp decline, of many development projects initiated in the previous decades. It also ruined Suriname's economy, and negatively affected governmental institutions, such as the Central Bureau for Aerial Survey (CBL), the State Forest Service (LBB), the Geological Mining Service (GMD), and the AdeKUS university. From the mid 1990s onward, the political and economic situation improved. Biological research activities in Suriname followed this trend of decline and subsequent improvement.

Just ahead of this period, in 1979, two new institutions had been created at AdeKUS: the National Herbarium of Suriname (BBS), and the National Zoological Collection of Suriname (NZCS). BBS was headed till the early 2000s by botanist Marga Werkhoven, who published a popular book on the orchids of Suriname (Werkhoven 1986). NZCS was and still is headed by herpetologist (amphibia and reptiles specialist) and ecologist Paul Ouboter, who in 1993 edited a multi-author volume on the 'Freshwater Ecosystems of Suriname' and in 1996 published his thesis on the ecology of caimans. At CELOS fish specialist Jan Mol did research on fish ecology, and published parts of his thesis work on armored catfish (for example Mol 1995). The NZCS and BBS undertook collecting expeditions to the remote corners of Suriname (see van 't Klooster *et al.* 2003, Ouboter & Jairam 2012). This has led to publications on Suriname's freshwater fishes (Mol 2012b) and amphibians (Ouboter & Jairam 2012), part of a series on the fauna of Suriname, which also includes a field guide to the country's birds (Spaans *et al.* 2016). Several international herbaria have collaborated with BBS since its establishment, resulting in reports and publications on biodiversity and ethnobotany by Raghoenandan, van Andel, Ruysschaert, Hoffman, ter Steege, Banki and collaborators (see references in other chapters). Much of the fieldwork of these researchers, as well as researchers of the previous generation, has been facilitated by Frits van Troon, a Surinamer with a remarkable, hybrid scientific-local knowledge. He remains the most knowledgeable person on rainforest plants in Suriname.

BOX 4 Taxa and molecular taxonomy

A taxon (plural: taxa) is a coherent group of species or populations that are part of a single species, ideally representing a complete lineage, meaning all descendants of a closest common ancestor (de Quieroz & Gauthier 1990). Taxa are organized hierarchically; species are very basal in the hierarchy, yet may be further subdivided in subspecies (distinct, geographically isolated populations). The species *Homo sapiens* (modern humans), for instance, belongs to the human taxon (genus *Homo*), which is a group of species most closely related to and including *Homo sapiens* (the only surviving species of the group); the genus *Homo* in turn belongs to the taxon of Hominidae ('great apes'), which also includes closely related chimps, gorillas and orangutans (McNulty 2016).

Taxonomy is the naming, description and classification of taxa, such as species, based on shared and distinguishing characteristics (de Quieroz & Gauthier 1990). It is traditionally based on morphology: the presence, size, shape and structure of body parts, taking into account what is understood about their evolution. Contemporary taxonomy often makes use of a wide range of data: morphological, ecological and behavioral, as well as 'molecular'. The latter refers to data encoded in biological molecules, particularly RNA and DNA which contains the genetic code. In this genetic material lies a detailed record of ancestry.

But what is a species? Based on the 'biological' species concept, individual organisms are considered part of the same species if they can interbreed (mate and produce fertile progeny) (discussion of this and other species concepts: Mallet 1995, de Queiroz 2007). Most extant (currently living) species were never proven by breeding experiments, but assumed not to be able to interbreed with other species based on obvious morphological differences between the species, and the absence of extant intermediary forms in nature.

Currently, species can be delimited based on molecular data (see for example Rannala 2015). No interbreeding between lineages, means that they do not physically exchange genetic material (no 'gene flow'), and that specific changes in the genetic code due to random mutation that occur over time in one lineage are very unlikely to be shared with other lineages. The accumulation of changes in the genetic code ultimately results in lineages whose representatives lose the ability to interbreed with those of other lineages, producing what may be considered different species (de Queiroz 2007).

Over the last three centuries, taxonomists (practitioners of taxonomy) have named and described taxa based on data that was available to them at the time and considered relevant, and based on different concepts of what should be considered a taxon, especially a species (see de Quieroz & Gauthier 1990, Mallet 1995, de Queiroz 2007). As a result, species and other taxon names, as well as taxonomic hierarchies, have often changed dramatically over time. In recent decades, better delimitations of species have been achieved based on molecular evidence of the interruption of gene flow; the classification of taxa was also improved based on genetic evidence of ancestral relations (see for example Heled & Drummond 2010).

In the late 1990s and the 2000s, Stinasu facilitated research by a number of visiting specialists and their associates, such as marine turtle specialists Hilterman and Goverse (supported by WWF), primatologists Boinski and Norconk, small mammals specialist Lim, ornithologist O'Shea, herpetologist Noonan, and seed dispersal expert Forget; Conservation International, in collaboration with local and international scientists, undertook several so-called rapid biological assessments (RAP; reports edited by Alonso and others) (see references in other chapters). Their work has produced important insights into biological diversity, ecology and nature conservation. Comprehensive biological studies were also undertaken in Suriname in the context of environmental and social impact assessment (ESIA): at bauxite mines and oil prospection and drilling sites in the Coastal Zone, and at bauxite and gold concessions in the northern part of the Interior. These studies were implemented by researchers such as Noordam, Teunissen, Bordenave, ter Steege, Ouboter, Mol, Lim, O'Shea, and De Dijn (see reports referenced in other chapters, but not all are in the public domain).

As part of some of these recent studies, tissue samples have been taken of birds, small mammals and amphibians. These samples have been used to generate so-called molecular data that are of critical importance in contemporary taxonomic research (see Box 4) (for example Velazco & Lim 2014), as well as biogeographical research (for example Noonan & Wray 2005) (see Biogeography chapter). This gives Suriname a taste of the new direction biological research is taking in the 21st century.

Geology, Landforms and Soils

Salomon Kroonenberg, Dirk Noordam

Whoever flies to Suriname for the first time cannot but wonder at the seemingly pristine landscapes of the country seen from the air. Upon approaching the coast from the north, you first see a half-submerged mangrove-clad coastal plain with narrow ridges. Then comes a somewhat dissected, but also partly drowned landscape, followed by white sand savannas just before you land. Further away to the south you see endless expanses of tropical forest disappearing into the hills. Before you touch down you have already seen the four main landscapes of Suriname: the Holocene Young Coastal Plain, the Pleistocene Old Coastal Plain, the Pliocene Zanderij Belt, and, in the distance, the Precambrian basement of the Interior. On the geological map of the country (Figure 6) they all come out clearly. The landscapes, rock units and their ages, and the sequence of events are given in Table 1.

The Precambrian basement occupies about 80 percent of the land surface of Suriname, and is a part of the Guiana Shield. The Guiana Shield is the northern half of the Amazonian Craton, which is divided into two parts by the river basin of the Amazon and its tributary the Solimões. Shields are the oldest cores of the continents, and the Precambrian spans the first 80 percent of the history of the earth, from its origin 4.6 billion years ago to 541 million years ago.

The Guiana Shield has a long geological history, spanning the time between 3.5 billion and 600 million years ago (Gibbs & Barron 1993, De Vletter *et al.* 1998, Delor *et al.* 2003a,b; Fraga *et al.* 2009, Kroonenberg & De Roever 2010). However, almost all rocks in Suriname are about 2 billion years old, and were formed mainly as a result of a pervasive period of mountain building that affected large parts of eastern South America: the Trans-Amazonian Orogeny (Priem *et al.* 1971, GMD 1977, Bosma *et al.* 1983, De Vletter 1984, Kroonenberg *et al.* 2016) (see Table 1). The major rock types that were formed during that period are distinguished on the simplified geological map in Figure 6.

Geology of the Precambrian basement

Marowijne Greenstone Belt. The northeastern part of the country is occupied by a greenstone belt, so called because of the predominance of green rocks. It consists of a series of volcanic and sedimentary rocks that have been folded and metamorphosed (recrystallized under high pressure and temperature) deep in the earth's crust. This greenstone belt is also the main area with gold mineralization (Daoust *et al.* 2011). Along its southern border large ellipsoidal bodies of granite-like rocks have intruded, also called TTG (tonalite-trondhjemite-granodiorite) bodies.

Older granites and gneisses. The area south of the greenstone-TTG belt, more or less coinciding with the Marowijne and Tapanahony drainage basins, is occupied by rather inhomogeneous granites, often showing slight banding or parallel orientation. Close to the greenstone belt, older gneisses (strongly banded and oriented metamorphic rocks) predominate.

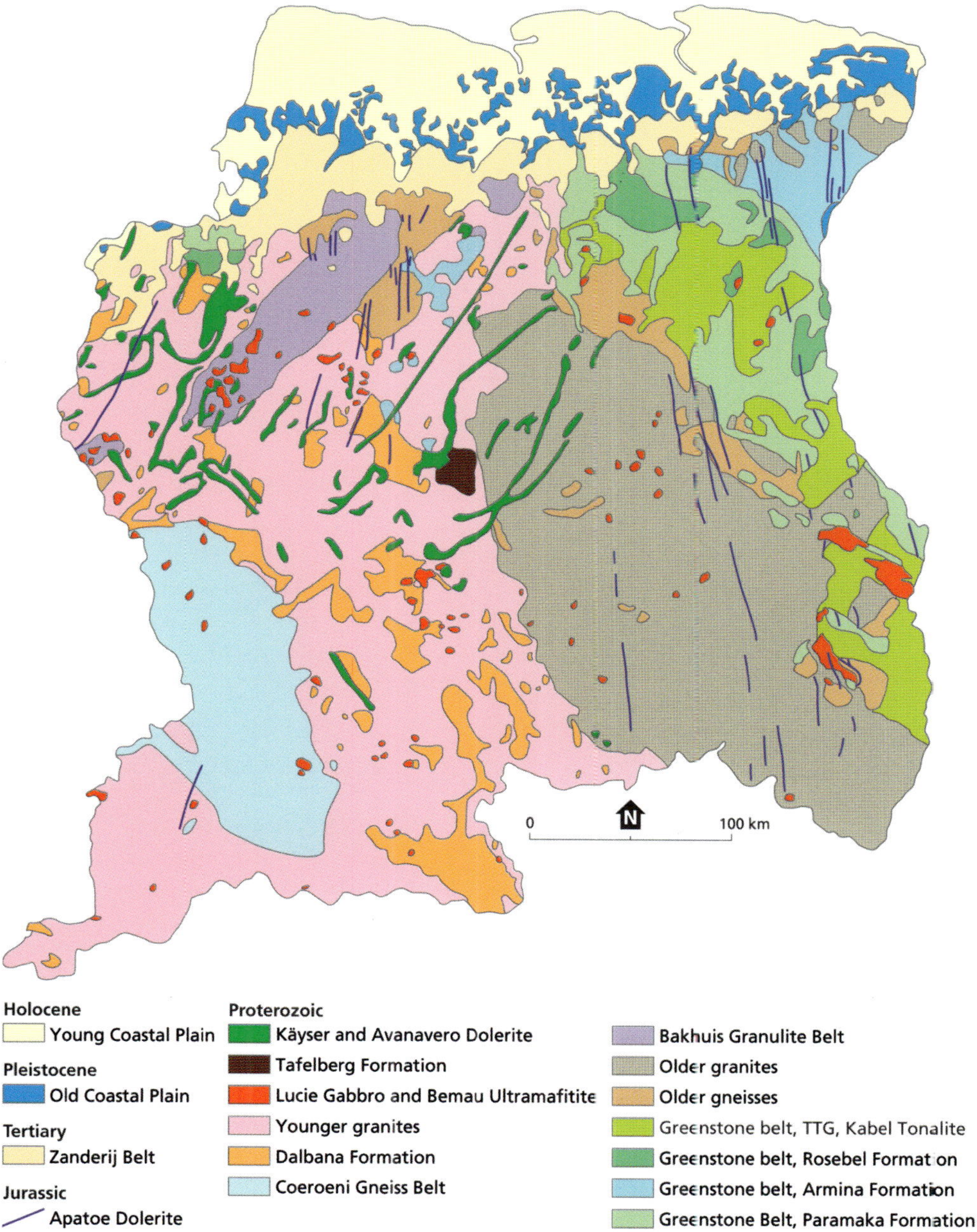

Figure 6. Geological sketch map of Suriname, simplified after Kroonenberg et al. (2016) (image © Kartomedia).

Bakhuis Granulite Belt. A striking unit on the map of Figure 6 is the Bakhuis horst in the NW part of the Precambrian basement. This area consists mainly of granulites, dark grey metamorphic rocks that have been subjected to much higher pressures and temperatures during the Trans-Amazonian Orogeny than the greenstone belt. They show features of early stage melting, and may once have been as deep in the earth's crust as 30 km (De Roever *et al.* 2003a). Here too there are intrusions, mainly of charnockites and anorthosites, equally dark gray intrusive rocks (Klaver *et al.* 2015).

Landscape	Formation [in map color]	Epoch	Age	Sediment/rock type	Geological event	
Coastal Zone						
Young Coastal Plain	**Coronie** Comowine	Holocene	1000-0	clay, sand, shells	deposition Amazon mud	
	Moleson		2500-1300	clay, sand, shells	deposition Amazon mud	
	Wanica		6000-3000	clay, sand, shells	deposition Amazon mud	
	Mara		10,000-6000	clay, peat	sea level rise, gully fills	
		Pleistocene		erosion gap	Ice Age, low sea level, dissection Old Coastal Plain	
Old Coastal Plain	**Coropina** Lelydorp		~120.000	sand, clay	Amazon mud deposition, high sea level	
	Para		~700.000?	clay, sand	Amazon mud deposition, high sea level	
				erosion gap	Ice Ages, low sea level	
Zanderij Belt	**Zanderij**	Pliocene	5-2 Ma	white & brown sands	deposition sands by rivers from weathered basement	
		Miocene	11-5 Ma	offshore sediments	birth Amazon River, initation westward mud transport	
Bauxite Belt	**Onverdacht**	Eocene-Oligocene		bauxite (duricrust)	deep weathering during major regression	
		Paleocene		oil, offshore sediments	offshore and onshore sedimentation	
		Jurassic-Cretaceous	180-65 Ma	offshore sediments	uplift Bakhuis horst, fill Takutu graben	
				erosion gap		
Precambrian basement						
duricrust-capped ridges	**Apatoe Dolerite**	Jurassic	196 Ma	dolerites	separation Africa-South America	
				erosion gap		
fault lineaments	**Nickerie Mylonite**	Meso-proterozoic	~1200 Ma	mylonite	ENE faulting, mylonitisation, collision Amazonia-Laurentia	
duricrust-capped ridges	**Avanavero Dolerite, Käyser Dolerite**	Paleo-proterozoic	1787/1500 Ma	dolerites	crustal extension, continental separation?	
Tafelberg Mountain	**Tafelberg Formation**		1873 Ma	sandstones, conglomerate	deposition by rivers from basement, acid volcanism	
				erosion gap		
duricrust-capped hills	**Lucie Gabbro, Bemau Utramafitite**		1980/2100 Ma	gabbros, ultramafic rocks	small mafic intrusions	Trans-Amazonian Orogeny
W granite landscape	**Younger granites**		~1980 Ma	granites, charnockites	intrusion at shallow to deep crustal level	
hills and ridges	**Dalbana Formation**			(meta)rhyolites, dacites	pyroclastic flows, lavas	
hilly lowlands	**Coeroeni Gneiss Belt**		~2050 Ma	gneisses, amphibolites	sedimentation, metamorphism	
Bakhuis Mountains	**Bakhuis Granulite Belt**			granulites	sedimentation, metamorphism	
E granite landscape	**Older granites & gneisses**		~2100 Ma	deep granites, gneisses	melting and deep-level intrusion	
hilly lowlands	**Kabel Tonalite, TTG bodies**			tonalites, trond-hjemites, etc.	syntectonic intrusions	
sandy plains, ridges	**Marowijne Greenstone Belt, Rosebel Formation**			quartz sandstones, conglomerates	deposition sands by rivers	
dissected hills	**Marowijne Greenstone Belt, Armina Formation**			greywackes	submarine mudflows, turbidites	
duricrust-capped hills	**Marowijne Greenstone Belt, Paramaka Formation**			mafic lavas, intrusions, etc.	ocean floor and island arc volcanism, sedimentation	

Table 1. Timeline of geological events (approximate ages after Kroonenberg et al. 2016).

Coeroeni Gneiss Belt. In southwestern Suriname a third unit of metamorphic rocks occurs, mainly gneisses with large amounts of micas. Some of them represent older sediments, others possibly older granitoid rocks. These rocks also have been subjected to intense pressures, high temperatures and incipient melting deep in the earth during the Trans-Amazonian Orogeny, though not as deep as the Bakhuis rocks (Kroonenberg 1976). In contrast to the Bakhuis Mountains, the Coeroeni Gneiss Belt occupies a lowland area, in which different types of gneisses give different landforms (Kroonenberg & Melitz 1983).

Younger granites, gabbros and volcanic rocks. The western half of central Suriname, especially in the Wilhelmina Mountains (Verhofstad 1971) and in the Corantijn and Sipaliwini drainage basins, consists of very fine grained volcanic rocks (Dalbana Formation) and granite bodies intruded at a shallow level in the earth's crust. Both rock types stem from the same magma sources. The volcanic rocks have been deposited mainly by pyroclastic flows (glowing ash clouds), and have been slightly recrystallized in contact with hot intruding granite magma. Small, round gabbroic intrusions of similar age (Lucie Gabbro) are widespread throughout the basement. Gabbros are dark-colored iron-rich rocks.

Tafelberg Formation. The Tafelberg is an eastern outlier of the Roraima Supergroup that forms impressive sandstone plateaus in Guyana, Venezuela and Brazil (Bisschops 1969, Bosma *et al.* 1983, Santos *et al.* 2003b).The Tafelberg Formation consists of at least 700 meters of indurated pinkish sandstones and quartz conglomerates. Sedimentary structures as trough bedding and ripple marks suggest deposition by rivers. The contact with the underlying granites is marked by a Precambrian fossil soil (Kroonenberg 1983). The age of the formation was debated in the past, until Priem *et al.* (1973) established its Proterozoic age from red volcanic ash layers between the sandstones. Since then, more accurate data have been obtained in Brazil by Santos *et al.* (2003), giving an age of 1873 million years.

Dolerite dykes. There are three generations of dolerite dykes, kilometers-long fractures filled by basaltic magma during periods of extension in the Guiana Shield. The oldest one, the Avanavero Dolerite suite (Reis *et al.* 2013), forms conspicuous mountain ridges such as the Van Asch van Wijckgebergte. The Käyser Dolerite suite in SW Suriname with predominant N-W orientation is slightly younger (De Roever *et al.* 2003b). The youngest one, the Jurassic Apatoe suite, with predominantly N-S orientation, is the only rock unit in the basement younger than the Precambrian; it marks the extension where Wegener's supercontinent Pangea started to break up and South America and Africa started to drift apart (Deckart *et al.* 2005).

Landforms and landform formation in the Precambrian basement

Physiography. The Guiana Shield area is dominated by an endless mosaic of low to moderately high convex hills. The northern part of the shield is characterized by undulating to moderately steep-sided, low hills with elevations of up to 100 m. Further south, large areas show a rolling relief with very steep-sided, low to high hills with elevations between 100 m and 250 m. Above this hilly terrain, mountains, plateaus and residual hills form prominent features up to 750 m. The higher summits are often covered with laterite or bauxite duricrusts (see below and Figure 10). Conspicuous mountainous terrain and high plateaus occur locally within almost all of the above zones, in some

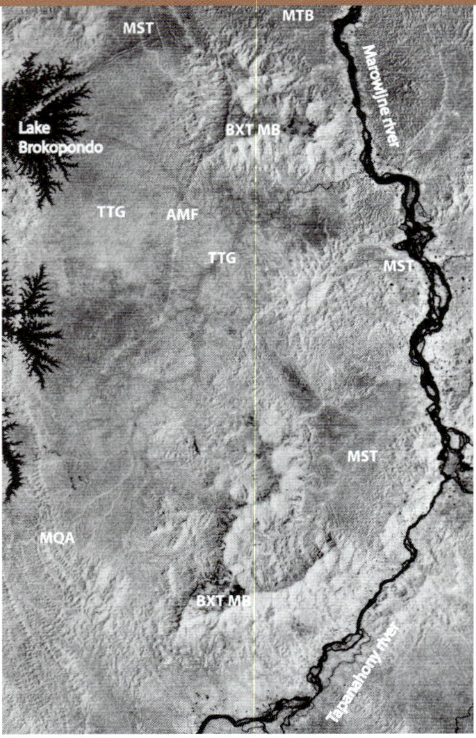

Figure 7. Landforms in the Marowijne Greenstone Belt strongly influenced by rock type (image © INPE (Brazil); Landsat image #2763050-130119-7 (band 7, infrared), October 31st, 1976).

BXT-MB: Bauxite plateau on mafic (iron-rich) rocks (Paramaka Formation, see Table 1); MTB: metamorphosed turbidites (Armina Formation); MST: metasandstone (Rosebel Formation); AMF: Amphibolite; MQA: Meta-quartzites; TTG: Tonalite-Trondhemite-Granodiorite plutons. Note also cataracts (sula) in Marowijne and Tapanahony rivers.

Figure 8. Devil's egg, granite inselberg (bare mountain top) and tor (the loose core stone) in the Wilhelmina Mountains, Central Suriname (source www.pinterest.com/pin/336573772124194603/; author unknown)

Figure 9. Tafelberg, view from subtop to the west (image © Bart De Dijn, 2008).

cases extending to elevations in excess of 750 m. The Juliana Peak at 1230 m is the highest point in Suriname.

Each rock type of the Precambrian basement produces its own landforms and soils (O'Herne 1969, Melitz 1975, Kroonenberg & Melitz 1983). The basaltic rocks in the greenstone belt (Paramaka Formation) form the typical bauxite plateaus of Brownsberg, Nassau and Lely mountains and smaller occurrences. The TTG rocks, on the other hand, form low areas: the Brokopondo storage lake (Brokopondo Reservoir) occupies such a low area (Figure 7). Even narrow ridges of more resistant rock types are directly visible in the LANDSAT satellite images. In granitic terrains, domed inselbergs (bornhardts) crop out (Figure 8); rock pavements (ruwares) and boulder inselbergs (tors) are also present here, in general covered by forest.

In the Sipaliwini Savanna and elsewhere, outcrops of fine-grained volcanic rocks of the Dalbana Formation (see above) may form long ridges in the landscape. Gabbro intrusions form

Figure 10. Plinthite horizon at Mosqueiro, near Belém, Brazil (image © Salomon Kroonenberg, 2013).

A plinthite horizon in a soil profile is the result of changing ground water levels in the soil. When plinthite is exposed to the air, for instance if the overlying soil is eroded away, or, as is the case here, the profile is exposed in a cliff, plinthite hardens into a duricrust, i.e. a laterite cap on mafic (iron-rich) basement rocks, bauxite on aluminous basement rocks as in the Bakhuis Mountains, or in aluminous sediments in the Suriname Coastal Plain.

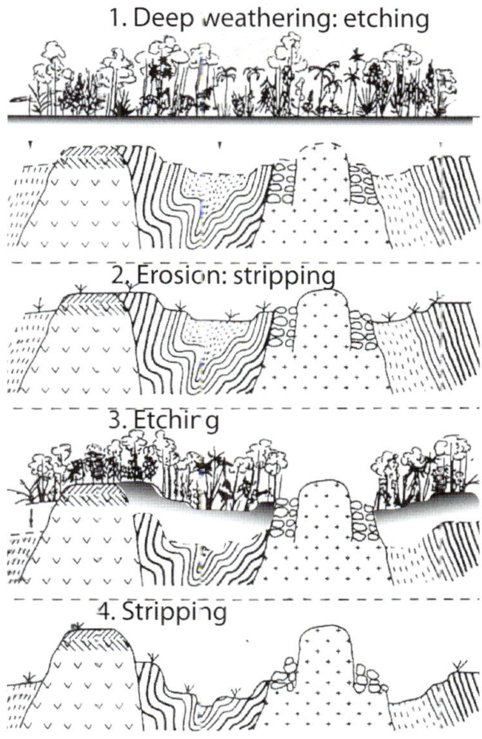

Figure 11. Stages in the development of an etchplain (Kroonenberg & Melitz 1983).

Stages 1) Humid climate: deep weathering, formation of plinthite over mafic rocks, shallow weathering in slightly jointed granite, deep weathering in gneisses; 2) Arid climate: stripping of the weathered rock, plinthite hardens to laterite or bauxite, inselbergs and tors are exposed; 3) Humid climate and 4) Arid climate: the processes are repeated, height differences increase.

conspicuous hills due to laterite duricrusts. Some of the rocks from the Bakhuis Granulite Belt contain so much aluminum that bauxite plateaus have been formed upon them by chemical weathering. The prominent horizontal sandstone plateau of Tafelberg in central Suriname reaches an elevation of 1026 m (Figure 9). A plateau at a lower level around the Tafelberg itself, including the Kappel Savanna, consists of the lowermost part of the same formation. Dolerite dyke suites form laterite-capped ridges protruding above the surrounding basement landscapes.

Differential etching and stripping. Rocks in humid tropical environments experience intense chemical weathering. Rainwater mixed with acids from rotting leaves and trees penetrates through cracks in the rock, and changes the most susceptible minerals such as feldspars and micas into clay and iron oxide, so that only a red loamy soil, often with residual quartz, remains. Open pit mining operations and drill holes show that sometimes you have to go down 100 m through soil and weathered rock before you touch the fresh rock. But not all rock types are equally susceptible to chemical weathering. Mica-rich rocks like gneisses are often deeply weathered, whereas granitic rocks usually weather less deeply. Iron-rich and aluminum-rich rocks may develop plinthite horizons (layers within the soil) rich in iron oxides which may harden upon exposure, and then develop into laterite or bauxite duricrusts (see explanation in Figure 10). This differential chemical weathering of rocks is called etching (Thomas 1975, Rabassa 2014).

Now imagine that a drier or at least more seasonal climatic episode sets in, in which the tropical forest is replaced by savanna vegetation. Savanna protects the soil much less efficiently than the tropical forest, and soon soil erosion leads to stripping of the weathering mantle and planation of the land. The deepest weathering mantles get most affected, and become deeply eroded terrains. Granites with widely spaced vertical joints (fissures), in the order of tens to hundreds of meters, have shallow weathering mantles and become exposed as fresh rock outcrops such as inselbergs and tors (Figure 8). Granites with more closely spaced joints may weather more deeply. Plinthite horizons developed from iron-rich rocks become exposed, harden, and form laterite or bauxite caps that prevent further stripping. When in a subsequent climate cycle moister conditions return, the tropical forest comes back again, but less luxuriously on laterite plateaus and not at all on exposed granite outcrops, and deep weathering is resumed. After a few of such cycles of etching and stripping, the height differences gradually increase between the laterite caps and the granite outcrops on the one hand and the gneiss landscapes on the other. Thus the well-known basement topography of plateaus and inselbergs and tors appears. The sequence of events is depicted in Figure 11. These climate cycles are usually correlated with supposedly drier climates during the Pleistocene ice ages (Van der Hammen & Absy 1994), although recent research shows little support for dry climate during the ice ages in Amazonia (Hoorn 1997, Hoorn *et al.* 2010).

Tectonics. Erosion is only capable of removing the soil mantle if there is potential energy in the landscape, i.e. a slope from the headwaters of the rivers towards the sea. What causes that elevation? The only way to achieve it is tectonic uplift. Without uplift it is impossible to imagine that rocks which once were 30 km deep in the interior of the earth are now exposed at the surface. It is also impossible to imagine the origin of the Tafelberg Mountain otherwise: in the Precambrian this must have been the lowest basin in the landscape to be able to receive river sediments. Now it is a high top: a clear case of relief inversion as a result of tectonic uplift. The rate at which the surface of the land is being lowered by the combination of uplift on the one hand and etching and stripping on the other is called denudation rate. The denudation rate in the Guiana Shield was calculated from the water chemistry of the Orinoco River at 15-20 m per million years in highlands and 5 m per million years in the dissected lowlands (Edmond *et al.* 1995).

The clearest example of uplift is in the Bakhuis Mountains. This entire area is a horst, meaning that it is uplifted with respect to the surrounding land. The resulting mountainous character is further enhanced by the presence of prominent bauxite caps. Although the SW-NE direction of the border faults is inherited from Precambrian events, the Bakhuis horst is still active under the sediments of the coastal plain, and has had an impact on the location of the oil fields in the Saramacca area (Bosma *et al.* 1983, Wong 2014, Nelson, 2016).

Other areas have subsided, such as the 6 km deep Jurassic Takutu graben in Guyana. The fact that the coastal area in western Suriname is much broader than in the east is also related to the subsidence of that graben. Suriname is tilting northwestwards. These up and down movements of the Guiana Shield are probably an echo of the separation of South America and Africa since the Jurassic.

Planation levels. In many areas underlain by iron-rich rocks such as the Paramaka Formation, several laterite and bauxite plateaus at different heights above each other have been found, separated by steeper steps in the terrain, for instance at Brownsberg Mountain. As denudation results in lowering of the surface, higher plateaus are considered older than lower ones; the elevation of a laterite plateau is thus regarded as a measure of its age. The Tertiary age (Eocene-Oligocene, see Table 1) of the oldest laterite plateau in the basement is confirmed by the occurrence of bauxites of the same age in the coastal plain (Wong 1989), and by absolute dating of the laterites themselves (Théveniaut *et al.* 2002).

The level of the highest laterite plateaus in Suriname slopes towards the north from Lely Mountain (700 m), Nassau Mountain (600 m) to Brownsberg Mountain (500 m). It is therefore tempting to connect them as evidence of a countrywide ancient surface sloping to the sea: a planation surface. The concept has been extended to surfaces without laterite caps as well: the whole continental surface was thought to be subdivided in planation surfaces separated by steps in the topography, formed by parallel slope retreat during periods with dry climate independent of rock type. The highest level was called the Early Tertiary Level, followed by the Late Tertiary I and II Level (Zonneveld 1993, Zonneveld & Krook 1998). But the steps between levels outside the lateritized areas appear to coincide with boundaries between rock types, suggesting they are formed by etching and stripping, not by parallel slope retreat (Kroonenberg & Melitz 1983). Planation by backwearing of slopes as envisaged for examples by Kips and Snel (1979) is not a viable process in crystalline rocks. The lithological control of unlateritized surfaces means that their altitude should not be taken as a measure of age. The landforms of the basement outside the laterite and bauxite plateaus are essentially an etchplain, not a succession of planation levels (Kroonenberg & Melitz 1983).

Rivers and creeks. A striking feature of the rivers in the basement is the occurrence of cataracts or rapids (known as *sula* in Suriname): sudden changes from a single channel pattern to a multi-channel pattern, not unlike that of braided rivers (Figure 12). These cataracts develop at sites where

Figure 12. Bonnidoro Falls, Marowijne district, typical sula complex in Armina Formation meta-greywackes (image © Google Earth, 2015).

Figure 13.
Creeks following fault patterns in dissected granite landscape, W of Oelemari river, at about 2°N55'30" N , 54°W 40'30" W (image © Centraal Bureau Luchtkartering, Suriname; aerial photograph run 13ZO, 6739, Z=4500 m f= 152.9 mm, November 1977).

rivers, while incising into the weathering mantle, suddenly encounter hard rock. Chemical weathering is so intense in the drainage basin that no gravel is formed, and therefore the rivers do not carry bedload coarser than sand. As a result they do not have enough erosive power to incise into the bedrock, and instead they try to negotiate any hard rock obstacle by passing it sideways. In this way the river splits in different channels around hard rock outcrops in sometimes very complex ways (as in Figure 12). River courses are also strongly influenced by linear features in the bedrock. They tend to follow faults, fractures and shear zones, and to avoid or be deflected by dolerite dykes.

Along the major rivers, staircases of river terraces are found: these are abandoned floodplains at a certain elevation, often at about 6 m and 15 m above the present floodplain (De Boer 1972). Some terraces contain rounded gravel, and some are gold bearing, whereas at present no gravel is being formed. This indicates that weathering conditions in the basement might have been less severe during the deposition of the terrace sediments than at present. Terraces might have been deposited by braided rivers in Pleistocene periods with more arid climate and less vegetation cover, and become dissected at the return of the tropical forest. During the next more arid climatic event a new terrace would have been deposited at a lower level, because tectonic uplift would have left the previous terrace at too high an elevation above the river to be flooded again. River terraces are especially well developed along the Marowijne, as that is where uplift is more vigorous (De Boer 1972). Absolute ages from these deposits are not available.

The smaller creeks that dissect the relief in numerous small hills are deeply incised into the weathered rock, and rock outcrops are often found at the creek bottoms; elsewhere, the valley bottoms are swampy and filled with young sediments. The creeks often follow rectilinear lineaments in the basement (Figure 13). Tectonic uplift is probably the main reason for this dissection. Sea-level change, as in the Old Coastal Plain (see below), cannot play a role, because dissection occurs everywhere in the Interior, not only close to the sea. Moreover, the cataracts form local base levels that prevent gradient changes in the coastal area from moving upstream.

Figure 14. Sipaliwini Savanna, with boundary savanna-forest maintained by burning (image © Salomon Kroonenberg, October 2013).

Savannas in the Interior. There are two larger savannas within the Interior, the Sipaliwini and the Gros savannas. These savannas cover approximately 0.5 percent of the Interior, but their combined area exceeds the combined area of the savannas in the Zanderij Belt (see below). The Sipaliwini Savanna (Figure 14) is underlain by exactly the same rocks as the surrounding forested area. Climatic conditions are also the same, as is the general morphology of low dissected hills. However, due to more active soil erosion, gully formation and slope movements, valley fills in the savanna are more prominent, leading to adverse soil factors (soil-water relations; Riezebos 1979). The Gros Savanna is underlain by sandy soils derived from quartz arenites of the Rosebel Formation. This might represent a model for river and slope processes under more arid conditions, though the origin or at least the persistence of the savanna is due to human activity, namely burning, rather than due to the prevailing climate (see Ecosystems chapter).

Soils of the Precambrian basement. The soils of the Precambrian basement are predominantly developed in the loose regolith that overlies the fresh, unweathered rock. Regolith is *in-situ* weathering material, colluvium (slope deposits) local alluvium (river sediments) or a combination thereof. The often deeply weathered parent material and its colluvium or alluvium are already strongly leached and poor in nutrients, and thus the soils that develop from it are mostly chemically poor as well. Kaolinite is the main clay mineral, which has only a limited capacity for retaining plant nutrients. There is a fair proportion of iron oxides. In spite of the high amount of litter production above the ground, organic matter is low in the soil, because of the high mineralization rate under the humid and warm conditions in the tropical forest. Also, nutrient content is low, because most nutrients from the litter are already scavenged by the standing vegetation before they can be incorporated into the soil. Texturally soils in the basement differ widely, mainly depending on the parent materials. Soils derived from coarse-grained granitoid rocks are usually rather sandy because of the presence of residual quartz, while rocks from gneisses are more clayey due to the high amount of micas weathered to clay and far less residual quartz. Soils on iron-rich rocks such as the meta-basalts in the greenstone belt, the gabbros and dolerite intrusions are deep red due to the abundance of ferric iron , and often have developed lateritic duricrusts. Soils on slopes over mafic rocks can be quite stony due to ferric iron nodules (resembling brown-red rounded gravel).

The physical characteristics of these soils are generally favorable, due to actively burrowing soil fauna and the non-swelling type of clay minerals. Drainage depends on position on the slope and soil texture: soils on laterite caps are stony and excessively drained, while the sandy-clayey soils at the footslopes and valley bottoms are imperfectly to poorly drained, and are grey and mottled as a result of seasonally changing groundwater levels. Soils on other slope sections are usually well drained, but some water stagnation may occur in the top of clay soils on gentle slopes. Imperfectly drained soils locally occur on plateaus in gently undulating landscapes.

Geology, landforms and soils in the Coastal Zone

Geological outline. In the Coastal Zone (i.e. the Zanderij Belt and the Coastal Plain), the Precambrian basement is overlain by younger sediments, ranging in age from Jurassic to Holocene and thickening towards the sea (Figure 15). Apart from a few bauxitized outcrops of the Paleocene Onverdacht Formation (Monsels 2016) (Figure 16) only the Pliocene, Pleistocene and Holocene crop out at the surface in the Coastal Zone. The older sediments have only been retrieved from onshore and offshore drilling for hydrocarbon (oil) and groundwater exploration. Sedimentation reflects variations in tectonic subsidence, global sea level, ocean currents and sediment supply.

Zanderij Belt. Along its northern boundary, the basement dips away beneath a cover of Pliocene sands, the Zanderij Formation (Wong 1989; Wong *et al.* 1998, 2009), locally characterized by white sands. The Zanderij Belt, also known as the Savanna Belt or Cover Landscape (in Dutch: *Savanne-gordel* or *Deklandschap*) is a gently rolling plateau landscape dipping to the north. Its elevation varies between 10 and 50 m in the eastern and central part of the country, while in West

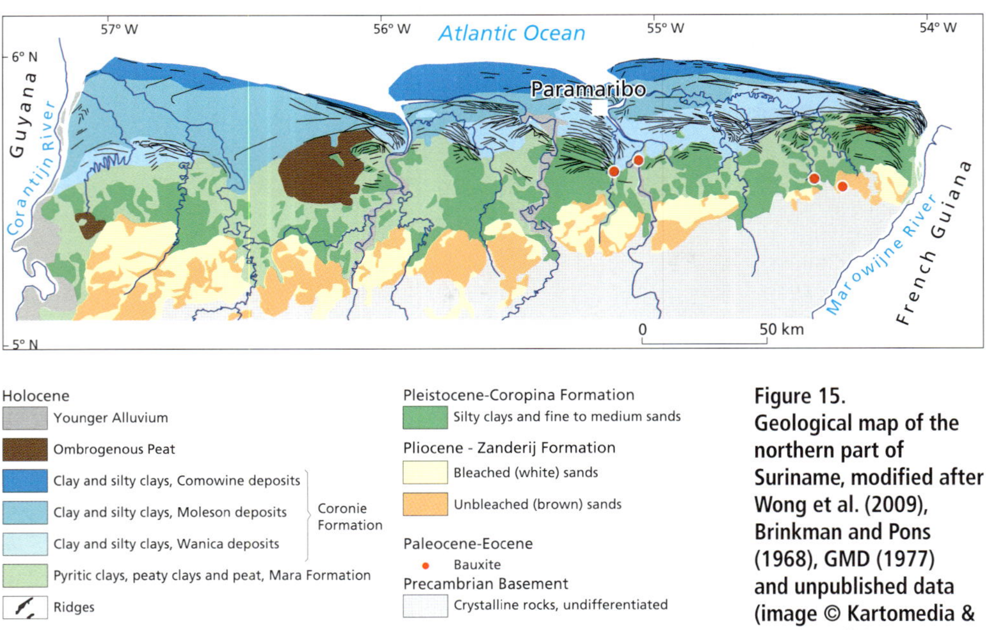

Figure 15. Geological map of the northern part of Suriname, modified after Wong et al. (2009), Brinkman and Pons (1968), GMD (1977) and unpublished data (image © Kartomedia & Bart De Dijn).

Figure 16.
Bauxite on top of Lobato Hill, Moengo (partly disturbed by mining), developed in strongly mottled, horizontally stratified Paleocene Onverdacht Formation sediments, which is in turn overlying saprolite of Paramaka Formation meta-andesite (image © Salomon Kroonenberg, 2016).

Suriname higher elevations (50 to 100 m) occur. In the west, the landscape also has a stronger relief, giving it a hilly character. Where the sediment cover is thin, mainly in the south, locally residual hills crop out. While about 40 percent of the landscape is occupied by white sands, only 7 percent is covered with savannas. Yet the name Savanna Landscape or Belt is still widely used, probably because most people only know the section south of Paramaribo, where savannas dominate the landscape. The sands are well layered (Figure 17), with locally gravelly beds. In deeper parts kaolinitic clayey layers also occur. They were deposited in the Pliocene in an arid climate by braided rivers draining the Precambrian basement towards the north (Krook & Mulders 1971, Wijmstra 1971, Wong 1989; Wong *et al.* 1998, 2009). Their thickness in the Zanderij area is about 20 meters, and increases coastwards. Part of the sands were deposited with their white color, as appears from sections in the bauxite mines and at the White Sand Cliffs in Orealla along the Guyanese bank of the Corantijn River. Elsewhere, however, and especially on top of the dissected plateaus in the Zanderij Belt, they owe their color to bleaching and podzolization (see below), while on the slopes unbleached brown soils predominate (Heyligers 1963).

Figure 17.
Cross-bedding in Zanderij Formation white sands, indicating fluvial deposition, at Lelydorp III mine (image © Salomon Kroonenberg, 2002).

The white sands in the Zanderij Belt are formed as giant podzols on medium to coarse sandy sediments. A podzol is a soil type in which organic compounds, such as humic acids, as well as iron have been washed out from the upper level of the soil to deeper levels. The result of this podzolization is a light grey or white 'leached' horizon in the soil below the uppermost darker humus horizon, and a brown or black horizon at deeper levels in which organic compounds and iron have been deposited. The dark, deeper horizon may consolidate, and then becomes a so-called humic hardpan. In part of the Zanderij Belt, such a humic hardpan occurs at 2-4 m depth, leading to waterlogging at the surface during the rainy season. Groundwater in these podzols is often very acid, down to a pH of 4, and is colored dark with humic substances (Krook & Mulders 1971). The creeks draining areas with podzols have black water (see Ecosystems chapter). Giant podzols are widespread in Amazonia. They are called giant because compared to podzols in temperate-climate regions, the grey to white soil horizon is very thick (Klinge 1966). Apart from the white sands, well drained non-bleached (usually yellowish brown to orange) sand to sandy clay soils cover 60 percent of the surface of the Zanderij Belt. The transition between bleached and unbleached soils in the terrain is often sharp (Lucas *et al.* 1982). The Savanna Belt is crossed by creek valleys with imperfectly to very poorly drained soils. Overall, the soils of the Zanderij Belt have a very low (podzols) to low fertility, and fair to good physical properties (Janssen & Wienk 1990).

Old Coastal Plain. The Zanderij Formation is overlain to the north by the Pleistocene sediments of the Old Coastal Plain, called the Coropina Formation. The Coropina Formation reflects the effect of several global sea-level cycles. In older publications, deposition of the Lelydorp Member was thought to have taken place partly during the last interglacial (Eemian) sea-level highstand, which was globally about 6 m above present sea levels. The Para Member would partly correspond to the preceding interglacial highstand (Veen 1970). Modern paleomagnetic research on extensive outcrops in the Lelydorp III mine, now closed, shows that the whole sequence might be considerably older (Wong *et al.* 2009). The Para Member is in any case older than 700,000 years – that is, at least six sea-level cycles back – or even more than 1 million years old. The Lelydorp Member was deposited on top of the Para Member, but it is not clear whether this was indeed during the last interglacial highstand 120,000 years ago or a much earlier one, maybe also 1 million years ago.

When sea level fell during the last ice age, it reached a level 120 m below the present one. A large part of the continental shelf was exposed, and in the clear shallow water near the coast corals could develop (Nota 1958). That was because at low sea levels, the Amazon River sheds its mud directly in the deep sea, instead of having it transported along the Guiana coast by ocean currents (Damuth & Kumar 1975). Another effect of the lowered sea level was that the rivers and creeks in the coastal plain started to incise into the Coropina Formation deposits (Brinkman & Pons 1968). When sea level rose again in the early Holocene, the gullies became filled with the peat and organic clays of the Mara Formation. Carbon-14 dating of peat at the bottom of these gullies shows ages around 10,000 years, the start of the Holocene. Pollen data show that at that time they were largely mangrove (*Rhizophora* spp.) swamps. Apart from the incision by creeks and rivers, considerable parts of the coastal plain at the time experienced surface erosion.

The Old Coastal Plain forms a discontinuous belt of erosion remnants of a formerly continuous coastal plain at 4-11 m elevation, separated by deep swampy gullies filled with early Holocene peat and clay (Veen 1970). The Old Coastal Plain consists of three different landscapes. The northern, younger part of the Old Coastal Plain, the Lelydorp Landscape, is underlain by sand ridges of the Lelydorp Member. The southern, older part, the Para Landscape, is underlain mainly by clays of the Para Member of the Coropina Formation, deposited by westwards migrating mudbanks originating at the mouth of the Amazon River, just as in the Young Coastal Plain (see next paragraph). The Mara Swamps Landscape ('schollen' landscape) refers to the swampy gu lies between the erosional remnants and the extensive deep swamps formed in the eroded sections in the belt of the Old Coastal Plain (Figure 18). These swamps are filled with peat and clay of the Mara Formation.

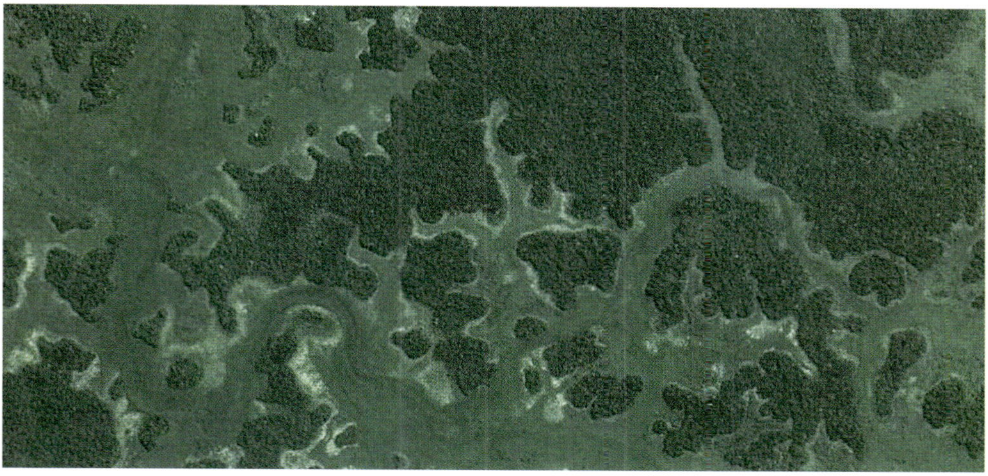

Figure 18. Old Coastal Plain 'schollen' landscape, dissected by gullies filled with Mara Formation clayey-peaty deposits; Commewijne district (image © Google Earth, 2015).

The Lelydorp or Old Ridges Landscape consists of a complex of eroded and flattened ridges, also known as cheniers (local Dutch name: *ritsen*), between 5 and 11 m elevation. Although the original orientation of the ridges and accompanying inter-ridge depressions can still be seen in places, the ridge pattern is usually rather indistinct. As a rule the higher ridges, with fine sandy to loamy soils, are well drained, while the low flats comprise imperfectly to poorly drained fine sands to fine sandy clay-loams. Podzol formation is common on the low flats. The inter-ridge depressions have poorly to very poorly drained fine sandy loam to fine sandy clay soils, but little peat is formed. Soil fertility is low for the sandy soils and slightly higher for the loamy and clayey soils. Poor drainage with water stagnation in the upper soil forms the main physical constraint of the flats and depressions (Noordam & Schellekens 1983).

The Para or Old Sea Clay Landscape is an almost level landscape at 4 to 7 m elevation. It is intersected by filled-in erosion gullies with swamps filled with Early Holocene peat and clay of the Mara Formation. The landscape is characterized by almost flat, low plateaus (local Dutch name: *schollen*) bounded by gullies or swamps. The soils of the plateaus comprise imperfectly to poorly drained silts to compact silty clays. Soil fertility is moderate to low, but the compact nature of the silty clays forms a major constraint and only soils with a silty to loamy topsoil have some potential for agricultural use (Edelman & Kartoredjo 1978).

Figure 19. Recent chenier formation at Braamspunt (image © Salomon Kroonenberg, 2004).

The Mara Swamps Landscape is formed on deposits of the Holocene Mara Formation. The swamps are predominantly found in the same zone as the Old Coastal Plain, at locations where most of this once coherent coastal plain has been eroded away during glacial periods. Scattered small outcrops of the Old Coastal Plain have remained as islands within the Mara Swamps Landscape (Figure 18). The elevation of the swamps is at about mean sea level or slightly above. The swamp soils comprise soft clays and peats, often in alternating layers. The clays contain pyrite (Brinkman & Pons 1968) and turn very acid upon drainage; they are indicated as potential acid sulfate soils (PASS; Dost 1973). The Mara Swamps are inundated throughout the year. Peat is still being actively formed on top of the deposits at some locations (such as the Coronie Swamp and the Nanni Swamp), thus forming domed low mounds of thick (>4 m) peat layers that extend above sea level. Such swamps are indicated as ombrogenous peat swamps (Brinkman & Pons 1968).

Young Coastal Plain. Since 6000 years ago, when postglacial sea-level rise started to slow down, the Amazon River has been shedding about 20 percent of its huge sediment load towards the coast of the Guianas. Westwards-flowing ocean currents transport the sediment in huge mud banks over 1500 km along the coast towards the Orinoco delta, which as a consequence itself consists of more than half Amazon mud. The mud banks are about 20 to 35 km long and move westwards at a velocity of up to 5 km per year (Augustinus 1978, Eisma *et al.* 1991). Part of the mud settles at the coast and is stabilized by mangrove colonization, especially by *Avicennia* and *Rhizophora* trees. Between the mud banks, ocean waves erode part of the newly deposited mud away, but at the same time they transport sand from the shelf, and in the east of Suriname also from local rivers, to the land. The waves are generated by the northeastern trade wind, and hit the east-west trending coast at an angle. This leads to westwards alongshore transport of the sand. When the surf throws part of the sand onshore, narrow ridges (cheniers) of marine sand are formed that protect the accreted mud banks from further erosion (Figure 19). Once a mud flat is isolated from the sea by a chenier, the abundant rainfall turns it into a freshwater swamp. The strong westwards coastal dynamics force smaller rivers such as the Saramacca and Commewijne to bend westwards until they reach a large river mouth capable of bypassing the moving mudbanks. In this way the coast has been growing seaward during the later parts of the Holocene, forming the Young Coastal Plain of Suriname. These Holocene sediments are called the Coronie Formation.

Figure 20. Wanica phase and Moleson phase cheniers fan out westwards in the Saramacca-Coesewijne area (image © Theo Wong, 2014).

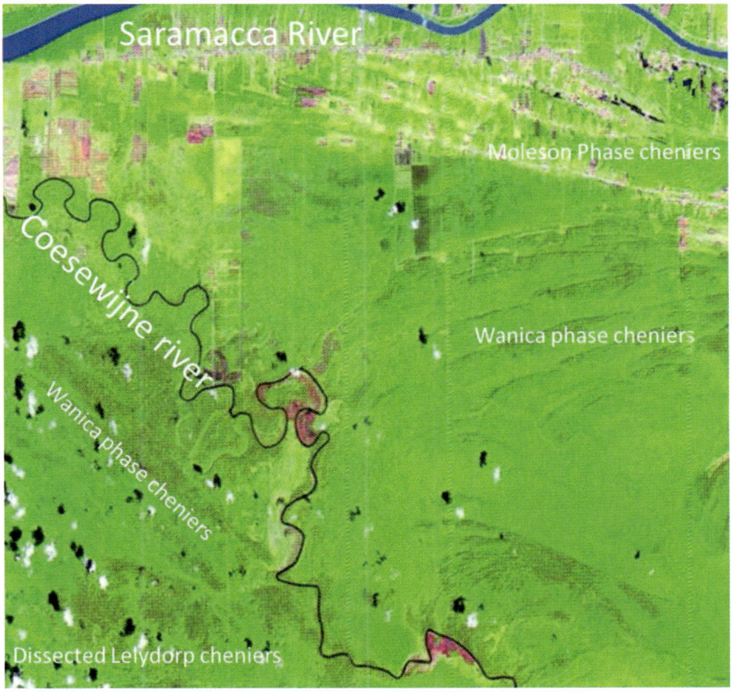

Three generations of ridge bundles with their respective mudflats can be discerned, representing three Holocene phases of deposition within the Coronie Formation separated by intervals of non-deposition or erosion (O'Herne 1966, Brinkman & Pons 1968, Roeleveld & Van Loon 1979, Versteeg 1985, Augustinus *et al.* 1989). The Wanica Phase lasted from 6000 to 3000 years ago, and ended with a short period with very little deposition. The Moleson Phase lasted from about 2500 to 1300 years ago. Versteeg (1985) showed from detailed archaeological studies of mounds in the coastal plain that the Moleson Phase becomes younger from east to west: Bucklebury 1800, Hertenrits 1300, and Prins Bernard Polder (Nickerie) 930 years ago. The Comowine Phase is the most recent phase, which started about 1000 years ago and is continuing to the present day. The two gaps between the three main phases may record reductions in the amount of mud being supplied by the Amazon River, leading to more erosion and less deposition along the coast. Other explanations are small changes in sea level, or changes in the direction of trade winds (Augustinus 2004). Also in modern times, there are variations in coastal dynamics, showing a net loss of sediment, for instance in the period 1947 to 1956, and a net gain, for instance from 1966 to 1981 (Augustinus *et al.* 1989). This reflects natural coastal processes.

The landscape of the Young Coastal Plain features extensive, flat and low-lying formations of heavy marine clays, usually overlain by a peat layer (locally known as *pegasse*). The clay flats are locally interspersed by roughly east-west striking ridges, consisting of sand, with admixture of shells locally. The ridges rise 1 to 3 m above the surrounding clay flats, are either individual or occur in bundles and are between 10 and 400 m wide. On any aerial photograph or satellite image of the Young Coastal Plain bundles of such cheniers strike the eye, starting from a left river bank and fanning out westwards (Figure 20). One bundle may partly cut a previous bundle at an angle. The most extensive ridge complexes were formed during the Wanica Phase and the Moleson Phase, but in recent time ridge formation has been limited to generally narrow ridges. During all phases ridge formation has been most active in the eastern and middle section of Suriname, with few ridges in western Suriname. This is due to differences in sediment load: rivers in the east of the country transport more sand to the sea than those in the west. Along the main rivers, the landscape consists of a mix of river and marine sediments. Already in Pre-Columbian times, landscape and soil were influenced by man. In the coastal swamps, for example, clay mounds were built for settlement, and raised fields for agriculture. Later, starting in the 15th century AD, land along rivers was reclaimed to form plantations, many of which are now abandoned. Land reclamation in recent decades has created polders for rice cultivation and dry cropping.

Several landscapes can be distinguished in the Young Coastal Plain. The first is the Young Salt- and Brackish Water Sea Clay Landscape, often referred to as the Estuarine Zone (Werkgroep Estuariene Kuststrook 1976). It comprises a zone several kilometers wide along the ocean and a narrow strip along the lower rivers. It covers the northern, most recent part of the Comowine Phase. The landscape along the ocean is characterized by extensive mudflats and coastal swamps, either bare (mudflats) or covered by mangroves and brackish to saline vegetation. 'Lagoons' (local name: *pan*), shallow ponds, are typical of this landscape. Extensive sand beaches occur along the ocean in eastern Suriname, but west of the Suriname River, sand and shell deposits are limited to a few small, narrow and shallow overwash bars, which are deposited on top of clay. South of the coastline only a few ridges are found within the Comowine Phase deposits. These ridges are usually short, narrow and relatively low-lying (up to 1.5 to 2 m elevation). The northern part of the unit is subjected to tidal flooding; the water level ranges from some decimeters during flood to slightly below the surface during low tide. The southern part is inundated by rainwater during the rainy season.

The landscape along the lower rivers is limited to low tidal flats covered by mangroves. The clay soils in this landscape are soft, and saline to brackish. Peat thickness in the coastal swamps is usually less than 20-30 cm, while peat is virtually absent in recent coastal mangroves. Because in the mangroves the land falls dry twice a day, peat does not form; peat only accumulates under prolonged wet conditions, such as those prevailing in brackish and freshwater swamps.

The Young Freshwater Sea Clay Landscape is found south of the Brackish Water Sea Clay Landscape, where it comprises the clay soils of the southern Comowine Phase and the Moleson and Wanica Phases (Van der Eyk 1957). The clay flats have a low elevation with the major part having elevations around 1.0 (± 0.5) m elevation. These areas are flooded in the rainy seasons, and often also part of the dry seasons, thus forming extensive swamps with peat formation. The (potential) thickness of the peat layer is positively correlated with the water depth in the swamp. Most clay flats of the freshwater swamps have a peat thickness of 20 to 60 cm. Soil formation is limited to ripening, which is a process initiated by the gradual and irreversible water loss from the soft and saline clay sediments. Ripening results in an increased firmness, a decreasing salinity and an increased amount and intensity of mottling of the clay. The northernmost (younger) Comowine clays show only some superficial ripening and desalinization, while the southern (older) Wanica clays are deeply ripened, with strong mottling and deep desalinization. The clay soils of the Young Coastal Plain are very fertile, with a slight decrease in fertility with age, but they have moderate to poor physical properties, such as low permeability to water. Narrow, individual sand, shell sand or sand ridges with better physical properties occur within this landscape.

Also within the freshwater zone, landscapes can be distinguished that are dominated by ridges (Van der Eyk 1957). This Young Ridges Landscape comprises areas with ridge bundles, which are mostly found to the west of main rivers (Figure 20), especially in the Moleson and Wanica Phases. The Moleson ridges comprise fine to coarse sand frequently mixed with shell fragments, but locally pure shell ridges are also found. The Wanica ridges contain fine sand and fine sandy loam. In between ridges, inter-ridge depressions are present, sometimes referred to on older soil maps in Dutch as *zwinnen*. Although marine clay is an important component of the soils in depressions, various admixtures with silts, sands and/or shells are present here as well, forming a complex pattern of clay, loam and sand soils. Usually the inter-ridge depressions are slightly inundated in the rainy season, but they fall dry in the dry season. This condition is indicated in Dutch as *dras*, and it is characterized by the presence of so-called *kawfutu*, a hummock or hog wallow micro-relief tied to biological activity (Scheltema 1964). Prolonged inundation in shallow gullies may lead to swamp conditions and peat formation. The soils in this Ridges Landscape are well drained to imperfectly drained at the ridges, and poorly to very poorly drained in the inter-ridge depressions. Soil fertility is low to moderate; it is highest on shell soils and loamy soils, and lowest on medium to coarse sands.

The River Landscape forms strips along all lowland rivers. The width may vary from only some tens of meters to over one kilometer from the river. Silting up has occurred till approximately 2 m elevation. Indistinct and narrow silty clay levees are found, gradually merging into low-lying backswamps with clay soils. Drainage of the river-bound zones is relatively good, due to the relatively high elevation and the proximity of the river. The levees are only inundated during spring tide. The backswamps are inundated during part of the rainy season, and after spring tide at high river discharge levels. The backswamps are characterized by a thin peat layer and *kawfutu*. At certain locations, the levees are cut by short and relatively straight creeks that drain the water from the backswamps and from the swamps deeper inland (Noordam 1986). The properties of the soils of this landscape are similar to those of the marine clays.

The Polder Landscape comprises the old plantation areas (dating back to the 17th-19th century AD) as well as the more recent polders (starting early 20th century) that have been developed on clay soils of the Young Coastal Plain. They differ from the natural clay flats by the presence of water management structures comprising sluices or pumps, canals, ditches and a system of cambered beds and trenches in case of dryland cropping. Land development has improved drainage conditions within the empoldered areas and it has accelerated soil formation. Plantation clays are usually completely ripened. In active polders hardly any inundation takes place and no peat or *kawfutu* are found. But most of the old plantations have been abandoned and drainage structures have deteriorated, so that inundation is commonly observed. In wetland rice polders typical paddy soils occur.

Climate and Hydrology

Riad Nurmohamed, Jacobus Groen, Sieuwnath Naipal

Suriname is located north of the equator, in the northern part of South America (1-6° N, 54-58° W). In this tropical part of South America, the climate is influenced by oceanic-atmospheric processes, most of which are indicated in Figure 21. These processes are caused by the variability in sea surface temperatures (SST) in the Atlantic and Pacific oceans (Rajagopolan *et al.* 1997, Giannini *et al.* 2000, Marshall *et al.* 2001, Wang 2001, Martis *et al.* 2002, Ambrizzi *et al.* 2005, Wang 2005, Peel *et al.* 2007).

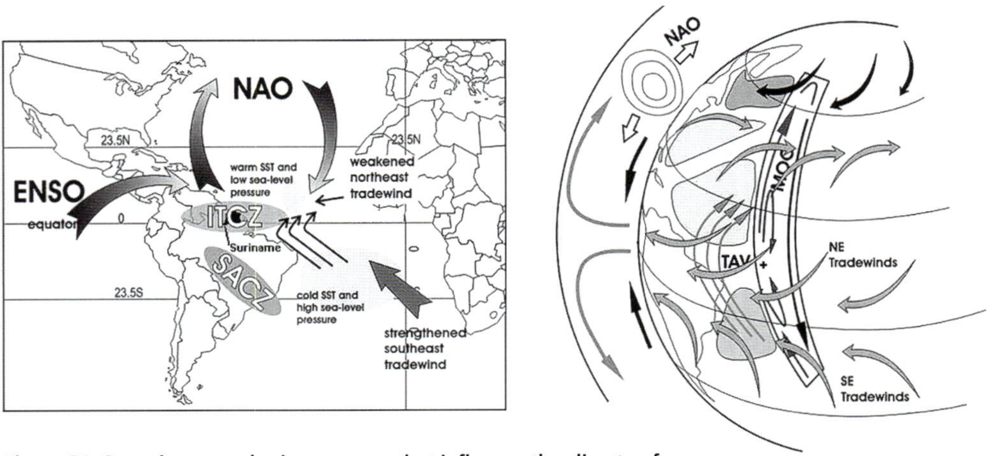

Figure 21. Oceanic-atmospheric processes that influence the climate of Suriname (schematic, after Marshall *et al.* 2001, Wang 2001, Robertson & Mechoso 2002).

Left side: The sea surface temperature (SST) in the Atlantic Ocean north of Suriname is influenced by the NAO and ENSO. The north–south SST anomaly is responsible for shifting the position of the ITCZ, reducing and enhancing NE en SE trade winds. The Pacific ENSO may influence the tropical Atlantic Ocean SST (north of Suriname).

Right side: The NAO is the primary source of the climate variability in the North Atlantic Ocean, and also an important source of variability for TAV and the MOC.

NOA = North Atlantic Oscillation; TAV = Tropical Atlantic Variability; ITCZ = Intertropical Convergence Zone; SACZ = South Atlantic Convergence Zone; ENSO = Pacific El Niño-Southern Oscillation; MOC = Atlantic Meridional Overturning Circulation; SST = Sea Surface Temperature; NE = north-eastern, SE = south-eastern.

The tropical part of South America is characterized by a large variability in rainfall between years and decades, high average rainfall (annually more than 2000 mm), high temperatures year-round (26-28°C), and high relative atmospheric humidity (generally more than 80%). The position of the Intertropical Convergence Zone (ITCZ) largely controls Suriname's climate. The ITCZ is a low-pressure region that produces much rain, located where the northeast and southeast trade winds converge. Due to the annual northward displacement of the ITCZ starting around April and southward displacement starting around October (between 10° N and 0° S), Suriname has a tropical climate with two dry and two wet seasons (Hammond 2005; Figure 22). Increase in the SST with low-pressure systems in the Atlantic Ocean brings increased precipitation, while high-pressure systems cause lower precipitation.

Other processes that also influence Suriname's climate, especially rainfall, include (based on Rasmussen & Wallace 1983, Gable & Aubrey 1990, Acevedo *et al*. 1999, Robertson & Mechoso 2002):

- sea-land interactions – for instance, ocean water stores a lot of solar radiation, and is constantly evaporating, increasing the temperature and humidity of the surrounding air to form rain or storms that are then carried by the northeast trade winds to Suriname's coastal lands;

- land processes – for instance, forest loss reduces evapotranspiration and as a result reduces cloud formation and rainfall (see below);

- topography – for instance, mountainous areas can act as barriers to wind and rainfall, with much of the rain falling on the windward side and top of a mountain (unless the top is located above the clouds, as in much of the Andes), and less on the leeward side;

- the El Niño southern oscillation (ENSO) – a periodical, global phenomenon characterized by changes in SST and a shift in upper atmosphere wind currents. The shifts are such that during some years ("El Niño years") moisture-bearing storms are directed away from Suriname, and thus cause drought in the country, while during other years ("La Niña years"), they direct more intense storms to Suriname, causing heavy rains and flooding. Like the NAO, TA and MOC (see below), this is an oceanic-atmospheric process caused by Earth's non-uniform solar energy input;

- the North Atlantic oscillation (NAO) – for instance, when NOA becomes strong, northeast trade winds become stronger and bring more moisture in the ITCZ, which results in more rainfall in Suriname;

- the Tropical Atlantic (TA) SST anomalies – for instance, a positive SST anomaly in the tropical North Atlantic brings warm and humid air to Suriname via northeast trade winds, resulting in more rainfall; and

- the Atlantic Meridional Overturning Circulation (MOC) – these deep-ocean currents are driven by differences in the water's density, which is controlled by temperature and salinity, and may play a role in abrupt climate change.

The temporal and spatial variability of rainfall in Suriname translates into variability in freshwater resources on land. Freshwater resources are vital for human life and the economy – in the form of drinking and irrigation water, for instance – and determine the nature and location of wetland ecosystems. Freshwater is generally defined as water with a chloride concentration lower than 250 mg chloride ions per liter (Cl-/l) (see for example Haripersad-Makhanlal & Ouboter 1993, Groen 2002). Suriname is characterized by high runoff (the amount of water that flows over the land surface), a dense river system and, along the Atlantic coast, extensive swamps. The major streams that dissect the Coastal Plain are estuarine, meaning that they are influenced by tides, and that saline water surges into them from the Atlantic Ocean. Groundwater is an important source for the industrial and domestic water supply in the Coastal Zone (the Coastal Plain and the Savanna Belt), in an area that covers only 18 percent of Suriname but holds about 85 percent of the population. In a humid country like Suriname groundwater is omnipresent and usually found within a few meters below the surface. In valleys in the Interior and in the low-lying Coastal Plain it is even closer to the surface and sustains freshwater and brackish water ecosystems.

Climate

Current climate characteristics. The average annual precipitation in Suriname varies from 1443 mm in the northwest margin of the Coastal Plain (district of Coronie) to 2852 mm in central Suriname (near Tafelberg Mountain) (Figure 23). The average annual evaporation in the country varies from about 1800 mm in the southwest to 2100 mm in north, southeast and northwest Suriname (Lenselink & van der Weert 1970). The rainforest in Suriname generates about half of its own rainfall, through evapotranspiration (Schlesinger & Bernhardt 2013, Heinrichs 2010). The forest thus plays an important role in regulating the local climate and sustaining the local water cycle (Morales & Vinicius 2003). The other half of the rainfall in Suriname is caused by the trade winds and the ITCZ. The forest releases water vapor into the atmosphere and this moisture returns as rainfall. It has been estimated that in Suriname, about two-thirds of all rainfall returns to the atmosphere by evaporation, and about a third flows towards the Atlantic Ocean (Schlesinger & Bernhardt 2013, Nurmohamed 2008). The average annual temperature varies between 25.5°C in the central part of Suriname and 27.2°C in the west and north; the annual national average is 26.8°C (Nurmohamed 2008). Atmospheric humidity averages annually about 65-85 percent in the Young Coastal Plain (Naipal & Amatali 1999a).

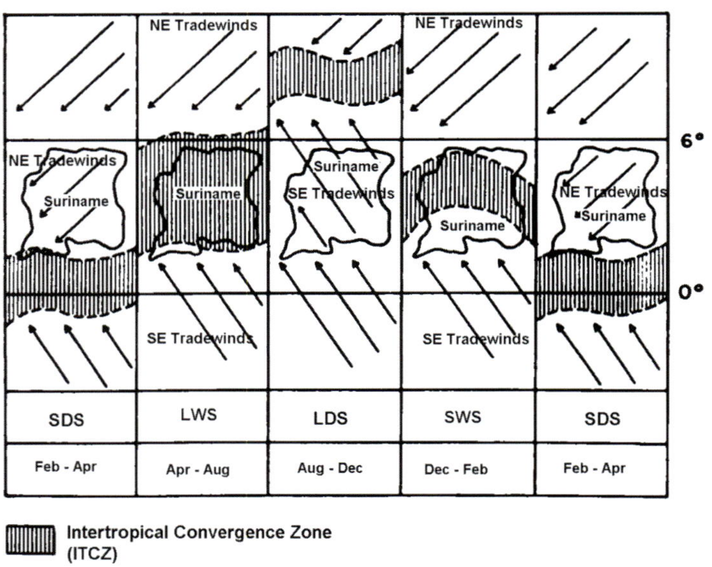

Figure 22. Movement of the Intertropical Convergence Zone (ITCZ) above Suriname and the resulting seasonality of the climate (SPS/OAS 1988).

SDS = short dry season; LWS = long wet season; LDS = long dry season; SWS = short wet season; NE = north-eastern; SE = south-eastern.

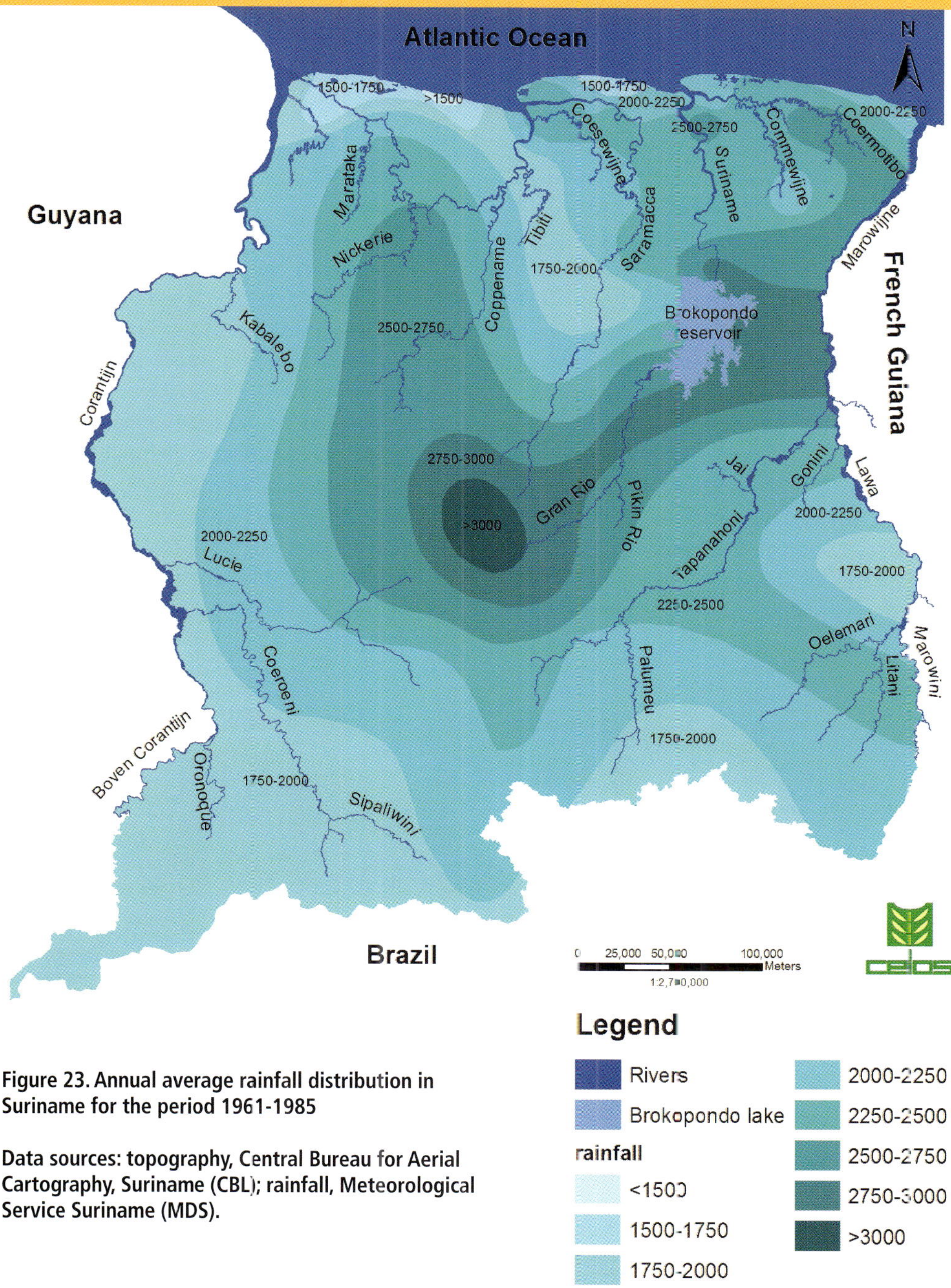

Figure 23. Annual average rainfall distribution in Suriname for the period 1961-1985

Data sources: topography, Central Bureau for Aerial Cartography, Suriname (CBL); rainfall, Meteorological Service Suriname (MDS).

As already mentioned, the seasonality of rainfall in the country is largely explained by the annual, cyclical ITCZ movements; the year can be divided into the following four seasons (see also Figure 22):

- Short wet season (SWS; mid December – early February). The SWS contributes 17 percent (on average) of the annual rainfall. During this period the ITCZ manifests itself as a narrow rainfall belt that moves southward across Suriname, resulting in an increase in precipitation.

• Short dry season (SDS; February – early April). The SDS contributes 14 percent of the annual rainfall. During this period the ITCZ is displaced further south, away from Suriname, resulting in less precipitation.

• Long wet season (LWS; April – mid-August). The LWS contributes 53 percent of the annual rainfall. During this period the ITCZ moves back north over Suriname. The rainfall belt is about twice as wide as during the SWS (about the width of the whole of Suriname), as south-easterly trade winds bring hot and humid air to Suriname from the Amazon, where massive amounts of water are evaporated by the luxurious vegetation. The long duration of the LWS compared to the SWS is due to the fact that during May-July, the northward moving equatorial air mass of the Amazon clashes with the high relief of the Guianas (as in the geographical center of Suriname), as well as with the northeast trade winds; this causes the ITCZ movement above Suriname to slow down. This also explains the high rainfall in central Suriname (see Figure 23).

• Long dry season (LDS; mid-August – mid December). The LDS contributes 16 percent of Suriname's annual rainfall. During this period, the ITCZ moves further north, away from Suriname, leading to reduced precipitation across the country.

However, in southwest Suriname (Kabalebo, Coeroeni and Sipaliwini meteo stations), only one wet and one dry season is observed (Emanuels 1968). Throughout the country, the months with the

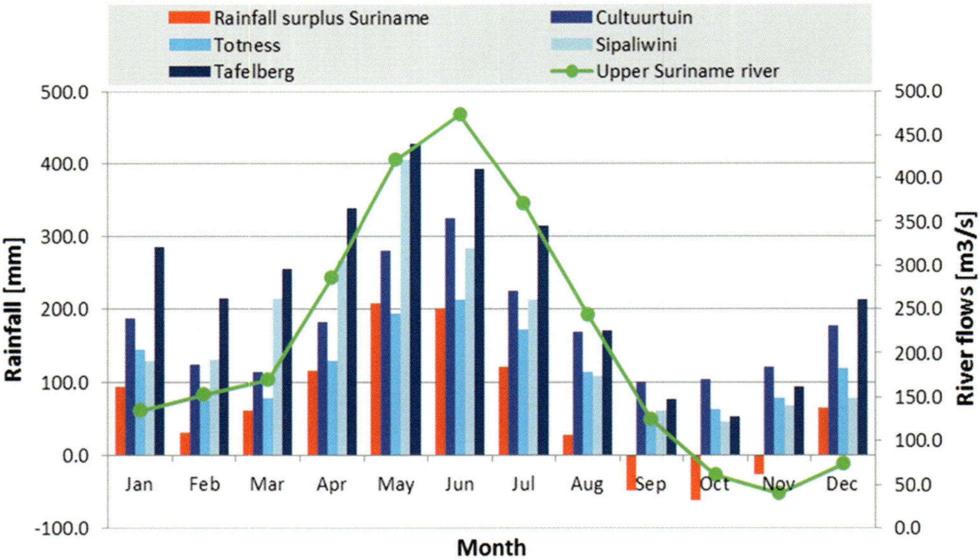

Figure 24. Average rainfall surplus 1960-1990 for entire Suriname (precipitation minus evapotranspiration; red bar), average monthly rainfall at 4 meteorological stations in Suriname (other colored bars), and monthly river flow of Upper Suriname River 1960-1985 (green line) (based on Nurmohamed 2008).

Meteorological stations: Cultuurtuin (Paramaribo, central part of N Suriname), Totness (Coronie district, NW Suriname), Sipaliwini (Sipaliwini district, SW Suriname), Tafelberg (Kappel Savanna, Sipaliwini District, geographical center of Suriname). Data sources: Meteorological Service Suriname (MDS) and the Hydraulic Research Division (WLA).

highest precipitation typically are May or June; those with the lowest precipitation typically are September or October (Figure 24) (Nurmohamed 2008). During the LDS, the weather may also be influenced by tropical storms or hurricanes in the Caribbean. Hurricane-related weather disturbances that reach Suriname may cause days with high or even torrential rainfall that punctuate the LDS (Evans *et al.* 2011).

PREHISTORIC CLIMATE. Suriname's climate has changed significantly during the past thousands of years. Various studies (NOTA 1958, van der Hammen 1963, Wijmstra & van der Hammen 1966, Nota 1967, Wijmstra 1969, Krook 1970, Nota 1971, Hoorn 1994a. Grimm *et al.* 2001, Clark *et al.* 2009) show that during the last glacial maximum, the coldest period of the last Ice Age, 30,000 - 15,000 years ago, the climate in Suriname was drier than it presently is. Grass savannas were probably much more extensive in the northern part of Suriname and Guyana (van Donselaar 1965, van der Hammen 1966). At the time, the sea level was some 100 to 130 m lower than today (Wong *et al.* 2009). During the early Holocene (starting 10,000 years ago), the sea level started to rise; it reached 25 m below the present level 9000 years ago, and by 6500 years ago reached a level close to the present one. During this period of rising sea level, the average rainfall and temperature were much lower than at present. Van der Hammen and Absy (1994) concluded that at this time rainfall must have been about 40 percent lower than currently (annually about 850 mm lower). A dryer and colder climate in Suriname at the time would help explain the presence and extent of savannas, such as the Sipaliwini Savanna, which likely were larger then than they currently are (Roeleveld 1969, Wijmstra 1969, Hoogmoed 1973, Kips & Snel 1979). The first indigenous hunter-gatherers must have also played a role then by burning the savannas to expand them and push back the forest (Van der Hammen & Hooghiemstra 1996).

HISTORIC CLIMATE (IN)STABILITY. The meteorological record shows that the climate in Suriname has been more or less stable during the last century. Studies by Nurmohamed and Naipal (2005) have shown that in Suriname annual rainfall has decreased by seven percent – about 200 mm – since the year 1900. The annual average temperature has increased by about 0.75°C since 1971 (few reliable measurements are available from the period before 1971).

Extreme climate events in Suriname have been recorded in historic times. Since 1821, at least 15 exceptionally dry periods (droughts) have been recorded, and at least 14 exceptionally wet periods (floods) (Geijskes 1943, Houben & Molenaar undated, Mol *et al.* 2000. Webster & Roebuck 2001). Droughts are longer periods with a rainfall deficit, meaning that (monthly) evaporation exceeds rainfall. Prolonged droughts, due to the influence of ENSO, were experienced in northwest Suriname in 1925/1926, 1939/1940, and 1982/1983. One of the more recent and well-documented droughts was that of August 1997 till February 1998. This caused the Bigi Pan wetlands in northwest Suriname to dry up, a high mortality of fish in the area, low rice production in the district of Nickerie, as well as severe shrinkage cracks in the asphalt of the nearby Henar bridge. This event was also experienced near Moengo (B. De Dijn, personal observation), and in neighboring Guyana and northeastern Brazil. The most recent drought was from September 2004 till January 2005, which resulted in food shortages in many villages in the Interior of Suriname, and the shutdown of several turbines at the Afobaka hydropower plant due to the extremely low level of the water in the Brokopondo Reservoir (see also Box 5).

BOX 5 The Brokopondo Reservoir (mainly based on Mol *et al.* 2007a)

The Brokopondo Lake or Reservoir – officially known in Dutch as the Prof. Dr. Ir. W. J. Van Blommesteinmeer – is a manmade lake situated at 4°45′ N, 55°05′ W in the Suriname river basin (Figure 23). Dating back to 1964, it is one of the oldest large hydroelectric reservoirs and the first large reservoir in a tropical rainforest area (Goodland 1979). The Afobaka hydroelectric dam was built in order to provide electricity for the processing of bauxite into alumina and aluminum metal by the Suriname Aluminum Company. About 75 percent of the dam's electricity was used to power this; the rest was used to power Suriname's capital city, Paramaribo. The reservoir has a surface area of about 1560 km^2, which is one percent of Suriname's land surface. The dam is 54 m high and is located 194 km upstream of the mouth of the Suriname River, a medium-sized river with a length of 500 km, a catchment of 16,500 km^2 and a discharge 440 m^3/s at the outlet. The catchment of the reservoir itself is about 12,550 km^2, or three-quarters of the Suriname river basin. Due to the great area of the reservoir, villages where around 5000 people in total lived had to be abandoned. The reservoir was put into service to generate electricity in 1965, but did not reach its optimal water level until 1971. The volume of water in the reservoir is 22.7 km^3 at its maximum water level of 48.2 m. The yearly discharge is 9.7 km^3. Water stays in the reservoir for around 28 months (van der Heide 1982). Prior to the closure of the dam, the Suriname River in the area of the current Brokopondo Reservoir was a low gradient river (0.64 m per km), with all the characteristics of a clear-water river (see also Ecosystems chapter): low annual sediment transport (7000 m^3/yr at Brokopondo village), high transparency (1 to 2 m when measured with a Secchi disk), oligotrophic (poor in nutrients; NO3, NO2, PO4 and SO4 < 0.1 mg/L), low conductivity (15 to 23 µS/cm), and mild acidity (pH 5.3 to 6.3); the water temperature used to vary between 27° and 31°C.

Most of these events are caused by SST anomalies in the Atlantic Ocean, tied to extreme SST anomalies in the Pacific Ocean. Rainfall deficits in Suriname may have been caused by warmer than normal SST in the tropical North Atlantic, which result in the ITCZ staying longer southwards of Suriname, in particular during December-January (see Figures 21 & 22), but also by El Niño events (for example in 1965/1966). Rainfall excess leading to floods in Suriname may have been caused by warmer than normal SST in the tropical South Atlantic, which result in the ITCZ staying longer above Suriname, in particular during March-April, but also by La Niña events (for example in 1970/1971; Ambrizzi *et al.* 2005, Nurmohamed 2008). Global climate change conditions will increase the variability in SST increases: it is expected that not only will El Niño and La Niña events change, but also the behavior of the ITCZ. A lot of progress has been made in the past 20 years in modeling and predicting extreme events like ENSO (see for example McPhaden 2004).

Hydrology

Surface water resources. The average annual river flow of the seven main rivers of Suriname combined is approximately 4800 m^3/s (Naipal & Amatali 1999a); this is the equivalent of 2.3 percent of the total annual flow of the Amazon River. Key characteristics of Suriname's large rivers are

River Catchment	Total Catchment area * (km2)	Annual Average Discharge (km2)	Maximum Discharge (m3/s)	Minimum Discharge (m3/s)	Min SIL** (km)	Max SIL** (km)
Corantijn	67,600	1,570	7,070	41	40	82
Nickerie	10,100	178	880	2	28	105
Coppename	21,700	500	2,200	6	31	83-95
Saramacca	9,000	225	1,260	5	37	89
Suriname	16,500	426	1,800	220***	54	90
Commewijne	6,600	120	215	28	5	150
Marowijne	68,700	1,780	6,160	48	37	59
Total	200,200	4,799				

* estimated value, till river mouth
** minimum (wet season) and maximum (dry season) saltwater intrusion limit (Min and Max SIL) measured along the river (river mouth = 0 km); saltwater is defined as > 250 mg Cl/l
*** river discharge is impacted by the hydropower dam at Afobaka

Table 2. Key characteristics of Suriname's main river catchments (based on Naipal & Amatali 1999a; 1999b).

detailed in Table 2; the main rivers and swamps are shown in Figure 23 and Figure 25 respectively. All Suriname's rivers ultimately discharge in the Atlantic Ocean, and near the coast they are subject to tidal movement and hence salt intrusion. The tidal cycle is approximately 12.5 hours. During the dry season, the tidal effect of seawater intrusion is observed far up these rivers. The extent of salt intrusion varies and is shown in Table 2. For example, salt water intrusion – a salt wedge of 200 mg Cl-/l – in the Corantijn River can reach between 40 and 82 km from the river mouth. Swamps contain brackish to saline water or even hypersaline water, often with a salt content several times higher than the sea. Brackish water contains 0.05-3 grams of salt per liter and saline water contains 3-5 grams of salt per liter (Stuyfzand 1989). Freshwater in large swamp complexes such as those of Nanni, Bigi pan, Coronie, Coesewijne, Surnau, Perica, and Buku originates directly from rainfall (for location, see Figure 25). The Brokopondo Lake or Reservoir (for location, see Figure 23), often simply referred to in Suriname as *het stuwmeer*, was constructed in the Suriname river basin for hydroelectricity purposes (see Box 5). Abandoned, exhausted bauxite mines in plains have turned into small lakes with stagnant water (Webster & Roebuck 2001); most are located near Lelydorp and Paranam (about 20 km south of Paramaribo).

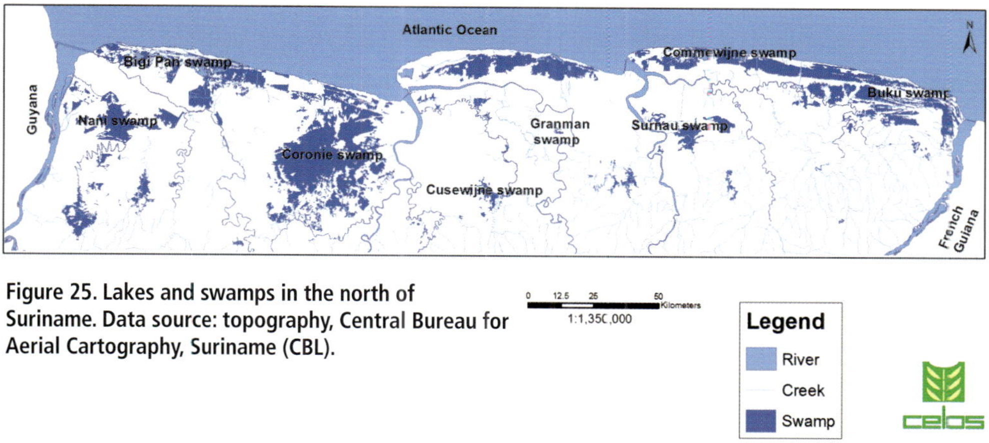

Figure 25. Lakes and swamps in the north of Suriname. Data source: topography, Central Bureau for Aerial Cartography, Suriname (CBL).

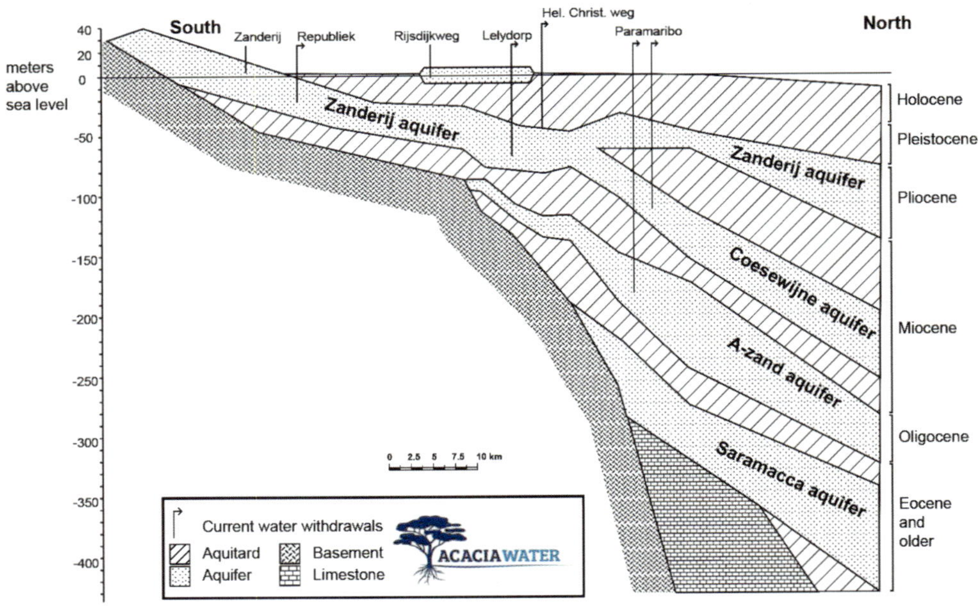

Figure 26. Schematic hydrogeological north-south section through the north of Suriname at the latitude of Paramaribo (Groen 2002).

Groundwater resources. More than half of the fresh exploitable groundwater resources in Suriname is found in the sedimentary deposits of the relatively small Coastal Zone (50 billion m^3 of water; HACAS 2016) (Figure 26). Below these sediments, in the crystalline and metamorphic rocks, there is an additional 40 billion m^3 of groundwater in the weathering layer and rock fractures. Though the Coastal Zone sediments are rich in groundwater resources, only 10 percent of these are renewable. The renewable groundwater is found in the Zanderij Belt where the outcrops of the Zanderij sands are recharged directly by rainfall (HACAS 2016, Groen 2002). Rainfall surplus – precipitation minus evaporation – which is directly related to recharge displays pronounced variations (Figure 24). Recharge may vary from 200 mm/yr near the coastline, where annual rainfall is low, to 500 mm/yr in the open savanna (Groen 1998, 2002). Recharge primarily takes place from December till August, and causes groundwater levels to rise during this period. During the long dry season groundwater levels fall (correlated to negative rainfall surplus values in Figure 24). The seasonal groundwater recharge pattern is mirrored by the flow pattern of rivers and creeks. During the long dry season rivers and creeks return to their base flows, which are sustained by the more or less permanent direct inflow of groundwater into streams, as well as via springs. Though renewable fresh groundwater also occurs in isolated sand ridges in the Coastal Plain, the major part of the plain has thick marine clay layers at the surface preventing deep percolation of rainfall.

The deeper sediments like the Zanderij, Coesewijne, Burnside (A-sand) and Saramacca formations contain non-renewable freshwater and saline groundwater (HACAS 2016). The age of the groundwater in these formations based on radiocarbon dating appears to vary from a few thousand to 21,000 years (Groen 2002). Figure 26 shows a typical section of the groundwater aquifers in the Coastal Plain. During the major part of the period during which these non-renewable aquifers filled,

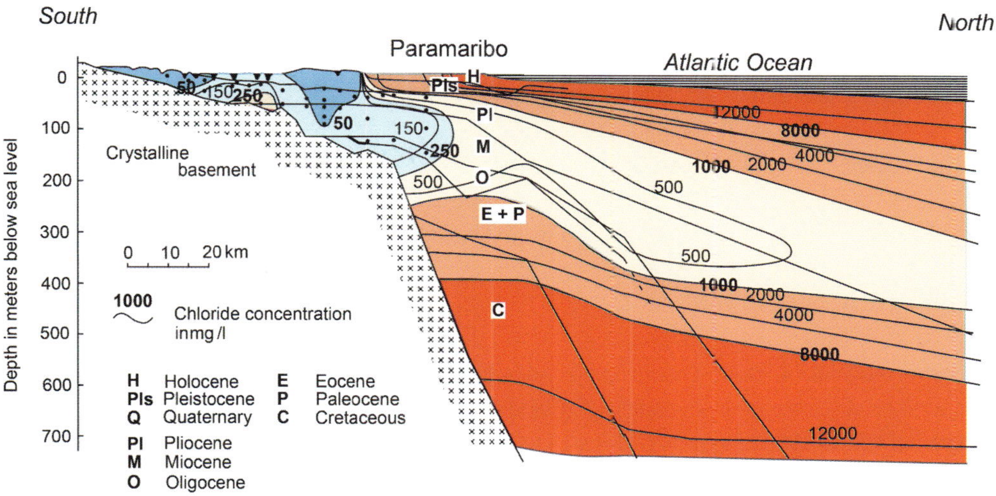

Figure 27. Hydrogeological north-south section through the north of Suriname and the offshore zone till 90 km north of the current coast, with indication of groundwater salinity (Groen 2002).

Dark red = chloride concentration above 8000 mg/l; medium red = between 1000 and 8000 mg/l; light red = 250 - 1000 mg/l; light blue = 50 - 250 mg/l; dark blue = below 50 mg/l. Dots indicate where this was recorded in water and oil wells.

sea level stood much lower than today, and the coastline was located far offshore. The major rivers, like the Suriname River, flowed through deeply incised river valleys (30 m near Paramaribo). These low river levels triggered recharge and deep groundwater circulation in the large, exposed Coastal Plain. After the Holocene rise in sea level, the deep groundwater flow system halted and left behind large bodies of fossil and non-renewable groundwater, including much low-salinity groundwater, below the current estuarine and marine environment. The glacial freshening and subsequent Holocene salinization has created an intricate pattern of groundwater salinities: shallow brackish groundwater can at present be found far inland, while low-salinity fossil groundwater extends 90 km offshore (Groen 2002) (see Figure 27).

Variability and trends in hydrology. The high temporal and spatial variability of rainfall in Suriname determines the hydrological regime of the rivers. The annual cycle of most of the river flows is characterized by a maximum in May-June and a minimum in October-November (Figure 24). The monthly discharge effectively follows almost the same seasonal pattern as the monthly rainfall. Reduced rainfall in the upper Suriname river basin has resulted in reduced inflows in the Brokopondo Reservoir in 1987/1988 (18 months), March 1999, January 2001 and September 2004 - January 2005, and has affected hydropower generation (Nurmohamed 2008). Statistical analysis of Suriname's historical river hydrology data shows no clear evidence of significant overall long-term changes in the mean annual river flows (Nurmohamed *et al.* 2007). In the Rio Negro, a main tributary of the Amazon River which drains a large part of the Western Guiana Shield, no significant change has been detected either (Gloor *et al.* 2013). However, a lack of continuous high-quality rainfall and river-flow data in our region has made the task of finding hydrological trends very difficult.

Seawater level observations (1960-2004) at the mouth of the Suriname River have shown that the mean water level has increased by 2.2-4.5 mm/yr since 1960 (Nurmohamed 2008). This is close to observations in the Caribbean region (Gable & Aubrey 1990, Mesquita 2000). An initial sea-level rise (SLR) estimation for Suriname by Nedeco in 1968 was lower, at approximately 1.1 mm/yr. If SLR increases linearly, it is expected that the coastal water levels will be about 22 to 45 cm higher by 2100 (with reference to 1960-2000 level). This value is close to the predictions of the Met Office Hadley Center (Rayner *et al.* 2003) and the IPCC (Solomon *et al.* 2007).

Future climate and hydrology

Climate change. Climate change projections for Suriname suggest that the annual average temperature will increase by between 1.7°C and 4.3°C by 2100, compared to a 25.8°C baseline in the period 1960-1990. These projections are based on scenarios called representative concentration pathways (RCPs), which refer to trends over time in the concentration of greenhouse gases - and indirectly emissions of air pollutants (see Taylor *et al.* 2012). The 1.7°C rise is based on RCP2.6, in which greenhouse gas emissions are reduced substantially, by 2100 than today's levels. The 4.3°C rise comes from RCP8.5, and lead to high greenhouse gas concentration levels by 2100 than today's levels.

The average annual rainfall is projected to decrease by about 37.9% (870 mm) by 2100, with reference to a 2292 mm baseline (1960-1990). (Taylor *et al.* 2012). In the near future, we can expect a longer period of rainfall deficit, from September till November; during the rest of the year, no rainfall deficits are projected for the country in general, but additional months with a deficit may occur in some parts of Suriname. River flows will reduce correspondingly during these dry months (see also below). A decrease in rainfall in the Coastal Zone may result in water shortages, a drop in agriculture output, high fish mortality in coastal wetlands, and loss of mangroves.

During the last decade, Suriname has been experiencing occasional periods of localized intense rainfall accompanied with strong wind gusts, causing local flooding and property damage (Buitelaar *et al.* 2007). Increasing temperature in the near future may increase the intensity or frequency of these events. There also seems to be a trend of a change in rainfall seasonality along the coastline: the amount of rainfall during the LDS has slightly increased, while the rainfall in the SDS seems to have decreased (Ministry of Labour, Technological Development and Environment 2013). The increased rainfall in Suriname's coastal area during the LDS is probably due to the increased number of storms and hurricanes in the Atlantic Ocean (Curry *et al.* 2006).

Hydrological changes. According to climate change predictions, rainfall will diminish and evapotranspiration will increase in Suriname. Consequently, groundwater recharge will decrease considerably in the future. For the Zanderij Belt, reductions are expected of 28% by 2040 and 86% by 2100 (HACAS 2016). This will be reflected in larger fluctuations and lower minimum groundwater levels, as well as lower base flows all over the country (Nurmohamed & Naipal 2007). Sea level is expected to rise by 0.45 m by 2100 (Nurmohamed, unpublished). As a result, brackish and saline water will further intrude upstream in the major rivers and in all connected creeks of the Coastal Plain, especially during the dry season. Some freshwater ecosystems in the northern part of the coastal plain will experience salinization due to sea level rise. The expected rise in surface water level will counteract groundwater level decline in the Coastal Plain. Changes in land cover because of deforestation, urbanization and infrastructure generally have a negative effect on groundwater recharge. If these developments continue, we expect they will locally contribute to falling groundwater levels and base flows.

In the Coastal Plain, saltwater intrusion is detrimental to agriculture and the persistence of the brackish and freshwater wetlands. Rainfall decreases will exacerbate this. In the hinterland, increased rainfall seasonality is a major threat for water resources in terms of both quantity and quality, especially when combined with mining activities. Polluted water from the mining areas containing fine sediments and mercury flows towards the north. This affects the health of all aquatic life in rivers and creeks, as well as the people who depend on these rivers as a source of protein (fish), and drinking and household water (see also chapter on Use, Threats and Conservation).

Biogeography

Brice Noonan

As part of the Guiana Shield, one of Earth's oldest exposed geological formations, Suriname has a long and complex biogeographic history. For starters, it may be useful to focus on the 'recent' history of the terrestrial biota of the Guiana Shield – the entirety of terrestrial plants, animals and other organisms of the region – and how it helps explain the high numbers of regional endemics. Guiana Shield endemic species, or endemics, are native species not found outside this region in the wild, unless introduced by man. For millions of years the biota of the Guiana Shield have been isolated from the rest of South America by the Atlantic Ocean and to some extent also by major rivers: the Orinoco and Negro to the west, and the Amazon to the south. This isolation would have been quite pronounced more than 20 million years ago, in the Miocene, due to the presence of shallow seaways covering much of Amazonia, as well as parts of current Venezuela and Colombia. Thus, the Guiana Shield terrestrial biota has a history of isolation from nearby regions, such as the Andes. The high proportion of Guiana Shield endemics – some 40 percent amongst the higher plants (Kelloff & Funk 2004) – may largely be attributed to species formation during such periods of isolation from neighboring lands that were, at other times, contiguous with the Guiana Shield.

The ancient history of the Guiana Shield, however, begins in the Precambrian, as obvious from rocky inselbergs, known as *lajas* in Latin America, that date back to at least 800 million years ago (Huber 1995). Much of the Precambrian basement is much older, and was covered with thick layers of sediment. After uplift of the Guiana Shield, the solidified sediments eroded away, except locally, where they remained as tepuis, steeply sloped table mountains of sandstone and conglomerate (Briceño et al. 1990). Tafelberg is the easternmost remaining tepui of the region, and the only one in Suriname. Over the millennia, the rest of the Guiana Shield has also continued eroding, resulting in more or less isolated high ranges in a lower, hilly matrix. Around 35 million years ago, thick ferrallitic duricrusts – crusts rich in iron and aluminum oxides – started forming locally in the soil. Many proved to be very resistant to erosion, and explain why at present we have isolated, flat-topped bauxite hills in Suriname. Subsequent erosion in the Miocene and Pliocene led to the downstream deposition of deep layers of sand, currently exposed in the Zanderij Belt in Suriname. Later, in the Pleistocene and Holocene, Amazonian river sediment was deposited along the Guiana Shield coastline, which in Suriname led to the formation of the Coastal Plain.

Precambrian geological events precede the origins of the terrestrial biota of the region. They set the stage for later biological diversification on land driven by the Mesozoic fragmentation (180 million years ago) of the Gondwana supercontinent, as well as by climatic cycles such as the recent (in geological time) Pleistocene Ice Ages. A few species found in Suriname clearly demonstrate biogeographic patterns that reflect the early Mesozoic connection of South America with Africa, such as pelomedusoid turtles which still occur on both continents (Noonan 2000). However, it is the changes to the South American landscape since the end of the Eocene (34 million years ago) that are responsible for most of the biogeographic patterns observed today in Suriname. Factors that drove continental-scale changes in biota occurred both near and far from Suriname – such as the distant formation of the Andes – yet their effects are evident in biogeographic patterns throughout the Guiana Shield.

To explore the biotic patterns of a region, we may consider the geographical distribution of a single species (phylogeography), of a group of closely related species (biogeography), or of unrelated biotic groups with a common distribution pattern. A clear understanding of the variety, distribution and continuity of various ecosystems is critical. Further, one must consider the full history of both the organisms and the ecosystems they inhabit. For instance, to understand the biogeographic pattern within a group of organisms whose most recent common ancestor lived 5 million years ago, one must consider not only the impact of recent glacial episodes up to 2 million years ago (Pleistocene) on these organisms and their habitat, but also the impact of earlier (Pliocene) climate changes. When populations of a single species become isolated from one another they may, over time, speciate, meaning that they differentiate and ultimately become distinct species (see also Box 4 in General Introduction / History chapter). Understanding, spatially, where one species stops and another begins, and relating that back to the past or present landscape in order to infer the factors driving differentiation is the heart of biogeographic study.

To understand the biogeography of Suriname, we must consider the past and present continuity of ecosystems through time. For example, the patchwork of savannas in Suriname today are separated from one another by a network of forests (Figure 28.a). Certain species exclusively inhabit these 'xeric' savanna patches, xeric meaning they are dry and typically much hotter than humid forest ('mesic' forest). If we were to explore the history of these species and find that isolated populations of, for example, the Neotropical Rattlesnake (*Crotalus durissus*; *sakasneki*) that inhabit them have only recently become isolated from one another, then there are two possible explanations. The first is that they moved through the forests to colonize distant savanna islands, and there has been little contact among populations since. The second is that at some point in the past these savannas were not isolated, but may have been more expansive and interconnected during 'dry' climatic cycles that favor xeric vegetation (Figure 28.b). The severing of these temporary connections between savanna patches would then result in the present occurrence of populations of savanna species, such as rattlesnakes, in distant, isolated patches. What remains to be explored is the correlation between the timing of evolutionary isolation of different populations of a species, and the independently determined history of abiotic events, such as climatic change. Only through such studies can we support or refute the hypothesis that certain events are at the heart of certain biotic distribution patterns.

While Suriname is not as geographically heterogeneous as some neighboring territories, its biodiversity exhibits complex patterns due to its geographic placement amidst these territories. Geographically, Suriname can be divided latitudinally into the relatively young Coastal Zone (Coastal Plain and Zanderij Belt), and the much older Precambrian Interior. Topographically, there are numerous, distinct mountainous regions throughout Suriname's Interior whose uplands exhibit varying degrees of isolation from their nearest neighbors. However, unlike the high elevation areas in the western part of the Guiana Shield, there are few true high-elevation areas in Suriname. Only a handful of peaks in the country exceed 1000 m elevation, and none exceeds 1300 m. These elevations appear to be largely insufficient to drive speciation of terrestrial animals and produce restricted-range, high-elevation endemics like those known from the tepuis of Guyana,

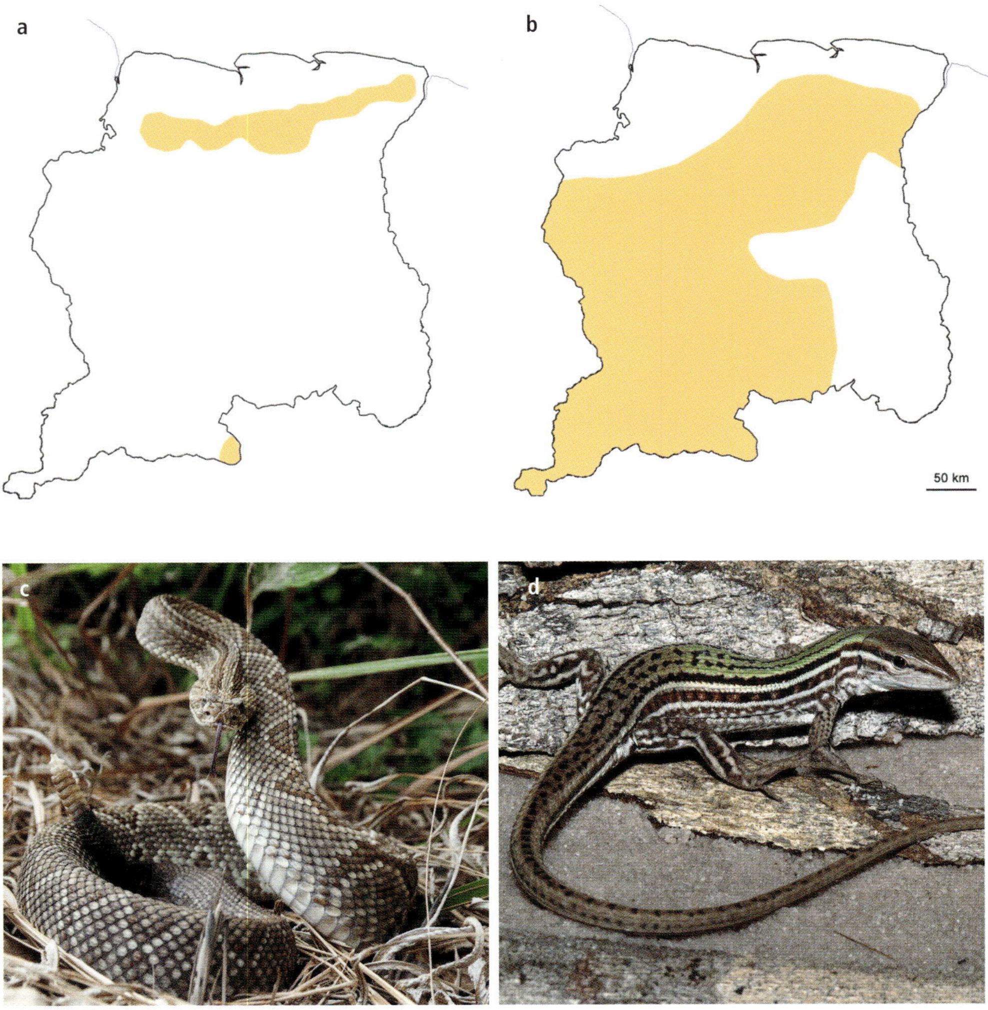

Figure 28. Schematic illustrating cyclic expansion and contraction of savanna biome (shaded) in Suriname, and illustrations of two exclusive savanna species:

a. present minimal extent and presumed extent during similar warm/moist periods,

b. maximal extent (hypothetical) during cool/dry periods.

c. Neotropical rattlesnake (*Crotalus durissus*; image © Andrew Snyder), and

d. Whiptail Lizard (*Kentropyx striatus*; image © Kenny Wray).

Venezuela and Brazil. Yet there are exceptions, such as the fish *Harttiella crassicauda* (see Freshwater Fishes chapter), which prove that isolated uplands in Suriname, which are known to contribute to biogeographic structure within certain species (see below), can also lead to speciation.

The large rivers of Suriname also contribute significantly to biogeographic patterns. For terrestrial biota, these rivers may serve as barriers, or at the very least filters that limit migration and gene flow – genetic exchange due to mating (interbreeding) – between populations on different sides of the rivers; this structures especially the fauna, more so than the flora. The extent to which rivers influence biotic patterns is, of course, species dependent. Though much of this chapter will focus on species that are part of Suriname's terrestrial biota, for which we know far more biogeographically, the aquatic species of Suriname exhibit some truly remarkable patterns of their own. This is, in large part, due to the near-complete isolation of Surinamese rivers from the Amazon drainage basin. By virtue of the mountainous ridge that runs east-west along the Brazilian border, nearly all rivers within Suriname's boundaries flow north, to spill in to the Atlantic. They thus have virtually no connection to the Amazon basin and its diverse aquatic biota. This isolation from Amazonia, while not complete, has led to a very diverse freshwater biota with a much greater proportion of endemics than among terrestrial species.

Biogeographic patterns due to connections between continents

Intercontinental-scale events. Events at a supercontinental scale have influenced biotic patterns in Suriname. They are largely limited to: i) those resulting from the Cretaceous (~110 million years ago) separation of Africa from South America as one of the earliest stages of the fragmentation of the Gondwanan supercontinent (Gondwana); ii) Cretaceous (~100 million years ago) migration of animals between North and South America; iii) migration through the continuous landmass consisting of present day South America, Antarctica and Australia through the early Eocene (~40 million years ago); and iv) those resulting from the recent (~3 million years ago) connection of North and South America by the rise of the Panamanian Isthmus (land bridge). Below, patterns resulting from these events, or at least presumed to result from them, are described and discussed.

Gondwanan patterns. Few plant or animal lineages with currently living representatives track biogeographic patterns to the Cretaceous (~110 million years ago) separation of South America from Africa (Sanmartín et al. 2004). Patterns in a number of groups have been suggested to reflect a very ancient presence on the conjoined Africa-South America portion of Gondwana. However, subsequent analyses have shown many of these to instead be the likely result of trans-Atlantic dispersal that took place much later, usually from Africa to South America. Bromeliads of the genus *Pitcairnia*, for example, are known from northern South America, but there is a lone species (*Pitcairnia feliciana*) in West Africa. Dating analyses based on examination of DNA differentiation among these species firmly place this diversification at less than 20 million years ago (Givnish *et al.* 2007), clearly ruling out the possibility of vicariance, which is the formation of species due to the isolation of populations by an insurmountable barrier, in this case the Atlantic Ocean when Gondwana fragmented.

There are, however, a few trans-Atlantic groups of organisms for which there appears to be sufficient support for their presence in Gondwana prior to late-Cretaceous fragmentation (Figure 29). An excellent example is the Strelitziaceae, a small plant family of giant herbs closely related to bananas (Musaceae) which is found in Madagascar (one species), southern Africa (five species) and northern South America, including Suriname (one species: *Phenakospermum guyannense*; *bigipalulu*). Rigorous testing by Sanmartín *et al.* (2004) of the relationships of these species reported by Kress et al. (2001) supports the possibility of divergence resulting from Gondwanan fragmentation.

Figure 29. Schematic illustrating relative positions of continents approximately 100 million years ago.

The present-day distribution of these species reflects the ancient Gondwanan connections of South America (Suriname in red), Africa, Australia, and Madagascar:
Strelitziaceae (blue; *Phenakospermum guiannense*; image © Pieter Teunissen),
boine snakes (white; *Boa constrictor*; image © Kenny Wray),
marsupials (orange; *Marmosa regina*; image © Kenny Wray), and
podocnemid turtles (yellow; *Podocnemis unifilis*; image © Kenny Wray).

However, stronger evidence of Gondwanan vicariance would be estimates of divergence times of the African and South American Strelitziaceae around 110 million years ago, at the time of the separation of South America and Africa (Madagascar split off from Africa far earlier).

Though the subject of great debate (Briggs 2003, Sparks & Smith 2005), both cichlids and aplocheiloid killifishes – both found in Suriname – have also been suggested to represent taxa whose present distribution – Africa and South America – is the product of the fragmentation of Gondwana. The killifishes in particular highlight another Southern Hemisphere connection, that of South America with Australia, via Antarctica, that persisted until the Eocene (roughly 55-35 million years ago). It is through this land connection that species such as marsupials, boid snakes, pelomedusoid turtles, and ratite birds are thought to have moved (see for example Noonan & Chippindale 2006).

Isthmian patterns. The Cretaceous exchange of biota among the Americas is clearly reflected in the geographic distribution of fossilized dinosaur lineages, as well as the present-day distribution of species such as marsupials in North America. Following a period of isolation of the South American continent, lasting for at least 30 million years after its isolation from Antarctica, another inter-continental connection was made possible by the rise of the Panamanian Isthmus roughly 3 million years ago (O'Dea et al. 2016). This land bridge reconnected North and South America, and the most recent great biotic interchange ensued. The southward migration of many North American plant and animal lineages into South America dramatically altered the composition of Neotropical communities, bringing such characteristic animals to Suriname as the Neotropical Rattlesnake (*Crotalus durissus*; *sakasneki*), White-tailed Deer (*Odocoileus virginianus*; *sabanadia*), and the raccoon (*Procyon cancrivorous*; *krabdagu*) (Helgen et al. 2013).

Biogeographic patterns due to changes within the continent

Intra-continental events. Dramatic changes have taken place within the South American continent, and these changes are thought to have contributed significantly to the evolution, diversification and current biogeographic patterns of the South American biota. More localized or intra-continental events have influenced biotic patterns within northern South America, and Suriname specifically, primarily through their alteration of habitat continuity. These forces can be generally classified as either fragmenting forces – such as barriers arising, isolating populations – or expanding forces – changes in the distribution of suitable habitat that may lead to a significant expansion or movement (see for example Figure 28). There is a diverse array of such forces that has contributed to South American diversification and biotic patterns; the timescale of these events influences whether they are likely to have affected patterns among species (interspecific) or within species (intraspecific). Furthermore, ecologically distinct organisms respond very differently to the same conditions, and thus patterns may not be reflected by many of the species subjected to a particular event. Below are detailed the primary events occurring within South America that are thought to have influenced biotic patterns; details are provided as to their timing, role in biotic diversification, and the influence on the biota of Suriname in particular.

Patterns due to climatic events. Perhaps the most pervasive of abiotic factors thought to have affected patterns of biodiversity through time is variation in climate. While physical changes to the landscape (see below) clearly impact the continuity and distribution of biota, the influence of climatic variation though time is widely recognized to be a significant driver of biotic patterns. This is primarily due to the fact that climate fluctuations, whether of temperature, precipitation, or both, are cyclical and can therefore represent alternating fragmenting and expanding forces.

To begin with, it should be noted that the history of climatic variation over the recent millennia has been well tracked and reconstructed from such diverse sources as palynology (fossil pollen; Bush 1994) and speleothem records (cave deposits; Wang *et al* 2008). From these, clear evidence is emerging to counter (Colinvaux *et al* 2000) the earliest hypotheses of glacial-cycle induced aridity and fragmentation of an otherwise uniform mesic forest throughout much of South America (Haffer 1969, Anhuf *et al*. 2006). Under Haffer's proposal, climatic variation would have led to reduction and fragmentation of the portions of the South American landscape covered in mesic forest, in particular during Ice Age (Pleistocene) glacial periods when both temperature and precipitation were assumed to have declined dramatically. For forest-dwelling species, this would have led to inhabitation of isolated patches of forest. Such isolation would have resulted in differentiation between populations and possibly speciation, resulting in 'new' sister species. A rise in both temperatures and precipitation, and a shift of the global climate to a warmer, wetter interglacial period – similar to the Holocene and the current climate – would have expanded these same forest patches, reconnecting them. This would allow formerly isolated populations to reconnect, and would offer any sister species opportunities to expand their range.

This simple pattern of cyclic expansion and fragmentation in concert with global cycles of glaciation was proposed as a, or even the, driver of speciation in the tropics. It was heralded by many as a 'species pump' of sorts, repeatedly leading to divergence and subsequent mixing of sister species, pushing up the total number of species. Overlap in the distribution of sister species was regarded as being the result of post-isolation range expansion. The hypothesis that this drives tropical speciation has, however, been shown to have a number of shortcomings. First, it seems clear that while temperature is likely to have changed throughout much of South America during these cycles, there is insufficient evidence to support a pattern of periodic aridity, which was proposed as the primary factor in the fragmentation of forests (Colinvaux *et al.* 2000, Cruz *et al.* 2005). In fact, analysis of the population history of Guiana Shield tree species has revealed a history of forest expansion over the last 3 million years (Barthe et al. 2016). Furthermore, the timing of diversification for most Neotropical species clearly indicates that their origins precede the onset of Pleistocene glaciation cycles (Rull 2008). Thus, while climatic fluctuations during Pleistocene glaciation surely impacted the distribution of species, and produced many of the intra-specific patterns observed in current biota (see below), these factors can explain neither the origins nor the distribution of most extant taxa (groups of organisms that have not gone extinct). The lone exception to this may be those species pairs or groups presently fragmented as a result of the current interglacial conditions, as these may be in the earliest stages of speciation. Hoogmoed (1973), for example, cites differentiation among populations of the savanna-inhabiting lizard species *Kentropyx striatus* and *Anolis auratus* of the coastal and Sipaliwini savannas (Figure 28) as reflecting the fragmentation of a previously continuous savanna across Suriname. During glacial periods, savannas are thought to be expansive and forests fragmented, during interglacials the converse. So, while climate has fluctuated dramatically during recent millennia, affecting species distributions, over short timescales these fluctuations appear insufficient to lead to the origin of new species.

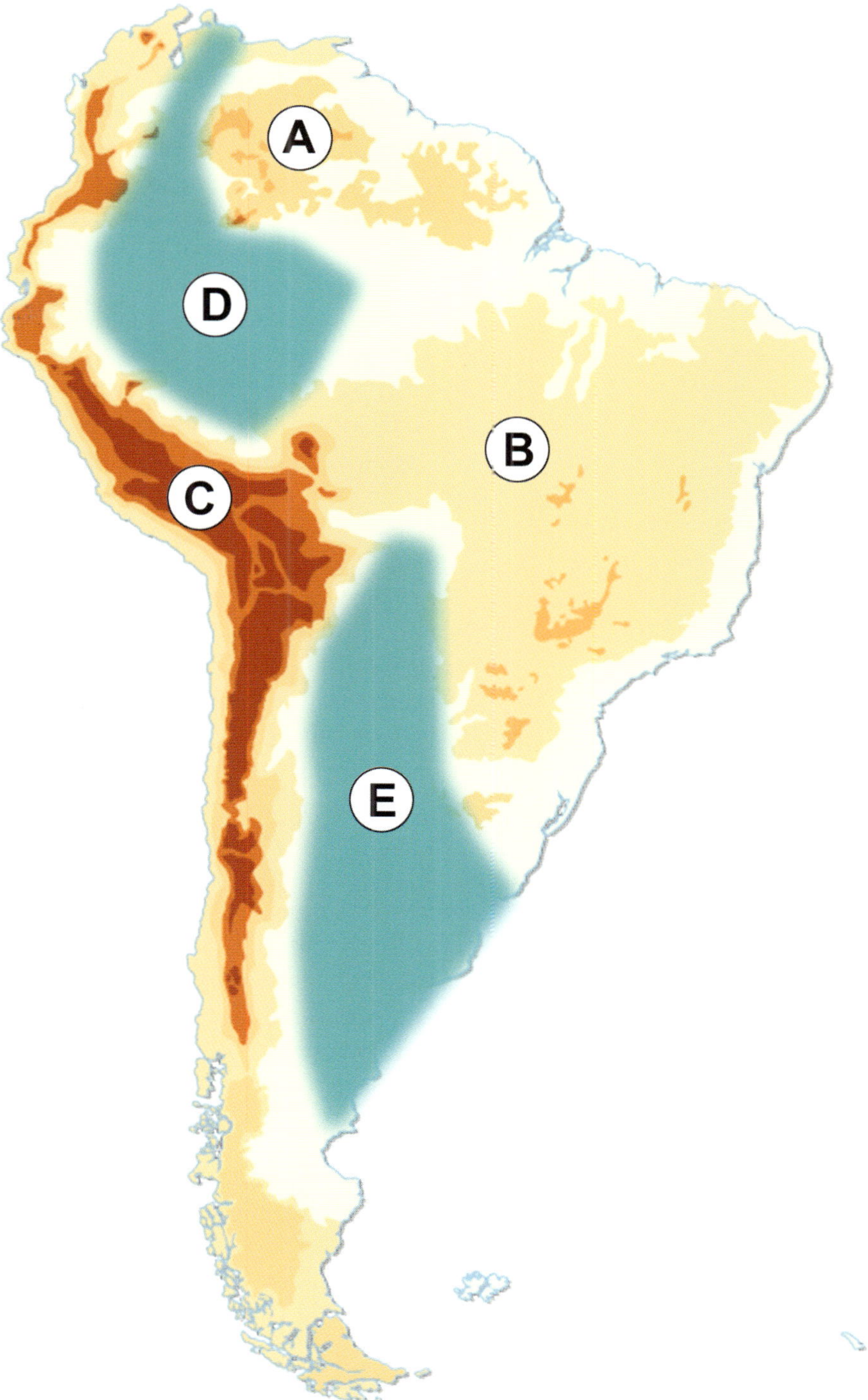

Figure 30. Illustration of major geological formations that have contributed to the biogeographic patterns of South America: Guiana Shield (A), Brazilian Shield (B), Andes (C), and the paleo-seas Pebas (D) and Paraense (E).

Patterns due to orogenic events. The orogeny or mountain building of the Guiana Shield is very ancient, and little activity has occurred in times recent enough to significantly impact biotic patterns. Yet recent mountain building and landscape deformation elsewhere in South America have altered the landscapes throughout the continent. The vicariant (isolating) effects are apparent in some Guiana Shield taxa. Among the primary contributors to these patterns are: i) the formation of the Andean mountain range, and ii) the topographic disruption of lowland Amazonia (Andean foreland basin, lower Amazon basin, eastern Guiana Shield), by paleo-ridges, a paleo-seaway, or both (Figure 30; see also below).

Due to the formation of the high Andes, South American biota were separated and isolated, and ultimately differentiated into biota east and west of the Andes, also referred to as cis- and trans-Andean. This separation was completed within the last 5 million years (Gregory-Wodzicki 2000). The resulting isolation, while punctuated by periodic crossings, was followed by differentiation of plant and animal lineages on either side of the Andes. In some cases, this has led to the repeated formation of cis- and trans-Andean species within genera. In other taxa, there has been significant speciation on either side, leading to surprising patterns whereby many a species that ranges widely across Amazonia, including Suriname, is most closely related to a species on the other side of the Andes.

In the Miocene, and as recently as the Pliocene, paleo-ridges – ancient uplifted or elevated areas – or marine incursions that produced paleo-seaways have resulted in discontinuities in the biota of South America, such as between eastern Amazonia and the Andean foreland, the area adjacent to the eastern slopes of the Andes. This has produced an Amazonian biota that can be clearly differentiated for many groups of organisms into Andean foreland, Guiana Shield, and Brazilian Shield sub-groups (Räsänen *et al.* 1995, Lougheed *et al.* 1999, Noonan & Wray 2006). Differentiation within the Guiana Shield has proceeded steadily for most groups studied (see for example Santos *et al* 2009), but current patterns are clearly derived from the effects of these Miocene and Pliocene events. It is becoming increasingly common to recover patterns similar to those recovered by Ribas *et al.* (2009) for parakeets of the genus *Brotogeris* (*kankantri-prakiki*): differentiation of taxa separated by the Andes, and of genetically distinct lineages restricted to the Andean foreland, Guiana Shield, and Brazilian Shield. Notably, the close relationship of Guiana Shield and Andean foreland sub-groups is not uncommon, as for instance with frogs of the genera *Atelopus* and *Ranitomeya* (Noonan & Gaucher 2005, Noonan & Wray 2006, Aleixo & Rossetti 2007).

The resulting biotic diversification within the Guiana Shield. There have been relatively few biogeographic studies of closely related Guiana Shield species (inter-specific studies), particularly in the eastern Guianan lowlands, which include all of Suriname (but see Fouquet *et al.* 2007a,b). Studies of patterns within species (intra-specific), on the other hand, are quite numerous. Noonan and Gaucher (2005) explored biogeographic patterns of toads of the genus *Atelopus*, finding the widespread *Atelopus spumarius* – occurring from French Guiana to Peru – to be distinguishable from the closely related *Atelopus franciscus* of northern French Guiana. The genetic differentiation among these species was observed to be low, and there was evidence of recent genetic exchange (interbreeding) among these regions. The pattern was similar to intra-specific patterns observed with other species, such as *Dendrobates tinctorius* (Noonan & Gaucher 2006).

As patterns recovered in the few inter-specific studies are largely coincident with phylogeographic (intra-specific) patterns, we will focus below on the latter since there is a much wider range of studies at this finer biogeographic scale.

To explore intra-specific patterns in the Guianan lowlands, it is useful to consider separately the histories of freshwater and terrestrial species. In the case of the latter, it helps to differentiate the forest-dwelling species – inhabiting what amounts to a relatively humid, mesic habitat – from those inhabiting savannas and the characteristic inselbergs of the region, which are xeric habitats, dry or hot, often both. As these mesic versus xeric terrestrial habitats would have alternated their periods of expansion and contraction or fragmentation, we shall consider reported patterns from each independently.

Diversification of aquatic species. The freshwater rivers of Suriname, like those in French Guiana, can be referred to as Atlantic; they are unidirectional in their drainage, from the southern part of the Interior toward the northern coastal region that borders the Atlantic Ocean. The 'Amazonian' rivers, on the other side of the political border in the south, drain into the Amazon River. Though hypothesized on the basis of biogeographic patterns (see below), there are no freshwater connections between the Surinamese drainage system and that of the Amazon River. This creates the potential for high levels of differentiation from the diverse freshwater Amazon biota, and is reflected by the large number of endemic fishes in Suriname (see Freshwater Fishes chapter). However, despite this lack of a direct connection, freshwater species frequently occur in unconnected Atlantic drainage systems, and some even occur in the Amazon basin. This similarity of adjacent, isolated drainages to one another is attributed to one or more of the following phenomena: i) migration out of the Amazon to and among Guiana Shield rivers via a near-shore area with low salinity wetlands that enables freshwater species to move from river mouth to river mouth along the coast, or ii) the merging of river mouths on the coast during periods of low sea level (note that at present many Guiana Shield rivers are merged at their mouths, such as the Suriname and Commewijne rivers in Suriname), or iii) the 'jumping' of species from one drainage to another by headwater stream capture (Lujan & Armbruster 2011). The latter involves a small headwater stream shifting its course from draining into, for example, the Coppename to the Corantijn River, taking with it all resident species.

Contrasting with this are the few reported examples of species believed to have crossed the southern border barrier between the Atlantic and Amazonian drainage. While such a connection has been well demonstrated in the Rupununi of Guyana, in Suriname it has only recently received empirical support. Species in the Atlantic drainage region that are notably absent from a significant portion of the eastern drainages nearest to the Amazon, such as *Corydoras bondi* (*seseiguse*), led some to suggest that headwater stream capture might have occurred between Amazonian rivers, like the Paru do Oeste, and Atlantic ones, such as the Sipaliwini (Nijssen 1970). As a similar pattern has been observed in the Rupununi Savanna, Nijssen suggested that the proximity of the headwaters of these rivers in a flooded Sipaliwini Savanna would provide ample opportunity for such species exchange. However, it is only with the recent work of Cardoso and Montoya-Burgos (2009) on *Pseudancistrus brevispinis* that we have strong evidence to suggest such connectivity, in this case between the Marowijne and the Jari, a tributary of the Amazon.

Unlike terrestrial vertebrates, for which the relatively low-elevation uplands of Suriname have apparently been insufficiently isolated, at least ecologically, there are a growing number of fish species known only from the Nassau and Lely mountains of eastern Suriname. *Harttiella crassicauda* (Nassau; Mol *et al.* 2007b) and two undescribed species of *Guyanancistrus* (one from Lely, the other from Nassau; see Freshwater Fishes chapter) have presumably evolved in the creeks originating on these mountains.

Diversification of savanna and inselberg species. As they share many species adapted to xeric environments, we shall consider together the biota of the savannas and of the rock savannas of inselbergs of the Guiana Shield. Notably, inselbergs are characterized by a number of species that at present do not occur in savanna-like ecosystems other than rock savanna, but may have been able to utilize open, xeric areas for migration between inselbergs during dry climatic periods with maximal savanna expansion. Most of the small and scattered savannas in the Guianan lowlands are concentrated in a narrow coastal belt between the Coastal Plain and Interior; this belt – referred to as Zanderij, Dek or Savanna Belt in Suriname – extends from Guyana into French Guiana. There are also large, isolated savanna expanses found in southwestern Guyana (Rupununi Savanna) which are connected to the Venezuelan Gran Sabana, as well as in the Suriname-Brazil border region, the Sipaliwini Savanna which is broadly connected to the Brazilian Paru Savanna. In Suriname, most of the coastal savannas are located on deposits that formed as recently as the Pliocene (Noordam 1993). The Sipaliwini-Paru savanna area, on the other hand, appears to be much older, a relict of what is assumed to have once been a more widespread savanna system (Hoogmoed 1973). Many species are shared among the three primary savanna regions of the Guianan lowlands, such as the Red-footed Tortoise *Chelonoidis carbonaria* (*redifutu sekrepatu*). Within Suriname, Hoogmoed (1973) interpreted distribution patterns of some savanna-inhabiting lizard species to reflect a history of Pleistocene savanna connectivity. This connectivity is hypothesized to be followed by the isolation of the Sipaliwini-Paru savanna area from the rest of the savannas, followed by separation of the coastal savanna belt from other central Suriname isolates, such as the tiny savanna near the Kayser River in central Suriname. As yet, this is simply based on observations of correlated distributions, and has not received rigorous testing. However, there have been studies of both fossil pollen (Van der Hammen 1974) and extant vertebrates (rattlesnakes; Wüster *et al.* 2005) that provide independent lines of evidence in support of Pleistocene expansion of open, savanna habitats.

The unique subset of biota, flora in particular, found on many inselbergs has recently received a great deal of attention (see for example Sarthou *et al* 2003). The ancient inselberg formations harbor dry and extremely hot environments – frequently reaching 45°C – that are largely devoid of actual soil. They are inhabited by a biota quite distinct ecologically from that of the surrounding forest matrix. One interesting example that spans coastal savannas to interior inselbergs is the plant *Manihot esculenta flabellifolia*. Duputié et al. (2009) explored the population structure of this species in French Guiana, finding coastal savanna populations to be highly differentiated from those of inselbergs in the interior. This suggests long-term isolation without gene flow. Among inselbergs, again of the French Guianan interior, gene flow has been shown to be very low for plants (Sarthou et al. 2001), and the long-term isolation of interior versus coastal savanna habitat has been further substantiated (Boisselier-Dubayle *et al* 2010). Although no such studies have taken place in Suriname, the French Guiana studies provide insights that likely also apply to Suriname.

Diversification of forest species. The exploration of patterns of diversity within species of the Guianan lowlands has recently exploded, with numerous studies exploring intra-specific patterns simultaneously in many species, in search of common patterns. The goal of these studies is identifying factors driving the process of diversification in the region. From these it can be concluded that the terrestrial Guianan lowland forest biota retains the signature of frequent and long-term exchange with greater Amazonia (Noonan & Wray 2006, Lim 2012a). Additionally, studies of fine-scale patterns within the region have been conducted that are more comprehensive than before, either in terms of the number of species or the number of localities sampled. These studies explore, primarily, biogeographic patterns within individual species of anuran amphibians (toads and frogs) (Noonan & Gaucher 2005, 2006; Fouquet et al. 2012). Results from these studies have strongly implicated features of the landscape, such as rivers, and intrinsic characteristics of the species in question, such as reproductive mode and dispersal capability, as the primary factors contributing to the generation of biodiversity. In particular, these multi-species studies have highlighted the importance of taxon-specific ecology as a driver of variation of biogeographic patterns among species.

Ecosystems

Bart De Dijn, Pieter Teunissen, Jan Mol

The edge of the continental shelf is located some 150 km north of the coast of Suriname, about 100 m below the surface of the sea. Here the sea is clear blue due to the scarcity of plankton. You cannot see the shelf edge when flying south over the Atlantic Ocean to Suriname, but about 100 km from the coast, you may notice the transition to green water, packed with photo-synthesizing plankton. Between 60 and 30 km from the coastline the sea becomes very shallow (40 to 10 m depth), and its water muddy brown (Lee 1958, Cadee 1975). The blue, green and brown water zones run more or less parallel to the coastline; their exact position shifts seasonally.

The mangrove forests and brackish herbaceous swamps along the shore, and the shallow, brackish near-shore sea (Table 3), form the bulk of the **Estuarine Zone.** This zone is the interface between land, rivers, and the sea with its saline water. The land part is a 1-15 km-wide coastal strip exposed to tidal brackish water with extensions upstream along the main rivers (Teunissen 1978g, 1984, 1987b, 1988c). Part of the Coastal Plain, it represents one percent of Suriname's land surface. Each year the sea creates and destroys stretches of land, but overall the coast is advancing (land gain).

The **Coastal Plain** (13% of Suriname's land surface) continues south for another 20-60 km, and up to twice as far along the main rivers. The northern half is the Young Coastal Plain (YCP; Mid to

Ecosystem	Temp. (C)	Conductivity (µS/cm)	pH (pH units)	Dissolved Oxygen (mg/L)	Biological Oxygen Demand (mg/L)	Tannin& Lignin (mg/L)	Secchi transparency (cn)	Turbidity (NTU)
Shallow inshore marine < 6 m depth	28 - 31	10000 - 55000	7.5 - 9.0	6.5 - 9.0	-	-	1 - 30	-
Brackish lakes / lagoons	24 - 34	12000 - 27000	7.0 - 10.0	1.0 - 9.0 usually < 3.0	-	-	5 - 10	20 - 100
Estuary of large river	26 - 31	1000* - 55000	7.0 - 8.5	6.5 - 9.0	0.1 - 0.5	-	5 - 30	20 - 43
Swamp Young Coastal Plain	24 - 29	90 - 4000	5.0 - 7.0	< 2.0	7.5 - 21.5	-	15 - 80	8 - 55
Swamp Old Coastal Plain	24 - 31	25 - 100	4.0 - 6.0	0.5 - 3.0	3.5 - 7.5	-	15 - 100	30 - 70
Black-water stream in the Zanderij Belt	25 - 26	30 - 60	4.0 - 6.0	1.0 - 3.5	2.0 - 22.5	3.0 - 4.0	50 - 130	2 - 9
Lower freshwater reach of large rivers (in Coastal Plain)	26 - 32	25 - 1000	6.0 - 7.0	4.5 - 8.0	0.1 - 3.5 usually < 2.0	-	15 - 100	1 – 15 usually < 10
Middle reach of rivers (in Interior)	25 - 31	10 - 25	6.0 - 7.0	5.0 - 8.5 rapids > 10.0	0.3 - 3.5 usually < 2.0	0.5 - 1.5	100 - 250	< 10 usually < 1
Headwaters of large rivers (in Interior)	22 - 25	10 - 20	5.0 - 6.0	6.0 - 8.0	-	2.0 - 5.0	-	5 - 10
Upland streams (> 500 m elevation)	22 - 24	20 - 30	4.5 - 6.5 usually 5.0 - 5.5	7.0 - 8.0	-	-	>100	0 - 4 usually < 1
Lowland terra firme rain forest creek	23 - 27	25 - 50	5.5 - 6.5	0.5** - 7.0 usually > 4.0	0.3 - 11.5	0.5 - 5.0	40 - 175	< 30

*Upstream limit of estuary set at the 300 mg Cl/L (0.5 ptt salinity or conductivity of ca. 1000 µS/cm) limit of saltwater intrusion
**Low DO values in the dry season when intermittent creeks split into series of pools with standing water

Table 3. Generalized water quality characteristics of undisturbed brackish and freshwater ecosystems in Suriname (data from Mohadin 1980, Crul & Reyrink 1980, Haripersad-Makhanlal & Ouboter 1993, Doerga 2013, and Jan Mol, unpublished data).

Late Holocene), the southern one the Old Coastal Plain (OCP; mostly Pleistocene to Early Holocene). The Coastal Plain includes shell and sand ridges that rarely if ever flood; in their natural state, they are forested, but much of the forest has been cleared for human land use. Most of this plain is low lying and poorly drained with freshwater swamps and marshes on clayey soils; it is partly natural, covered with forest or herbaceous vegetation, and partly drained and deforested (Teunissen 1978f, g; 1984, 1988c, 1993). During the wet season rainfall accumulates and floods most of the land, resulting in a near-continuous sheet of shallow, stagnant water. The lower fresh- to brackish-water reaches of the main rivers meander through the Coastal Plain towards the sea.

The **Zanderij Belt** or Savanna Belt (5% of Suriname's land surface) extends up to 50 km to the south of the Coastal Plain (Krook & Mulders 1971). It is well known for its white sand savannas, but it is in fact mostly forested (Teunissen 1978g, 1980a,b; 1984). The savannas and forests occur on more or less well-drained land with sandy soils that were deposited during the Pliocene epoch. Much of the forest, particularly on bleached sands, is low in stature and rich in xerophytes, and is referred to as savanna forest. Creeks and small rivers draining areas with bleached sands carry so-called 'black' water (the color of dark tea). Rivers and some large creeks that originate in the Interior cut broadly through the Zanderij Belt; their water is clear, not black.

Everything south of the Coastal Plain and Zanderij Belt is referred to as the **Interior** (82% of Suriname's land surface); here the weathered Precambrian Guiana Shield is not covered by more recent deposits. The land is mostly 'terra firme', land that does not flood. It is drained by numerous rain forest creeks, which merge to become clear-water rivers with rapids. The soil in the Interior is mostly lateritic (typically red-brown soil, rich in clay and iron), well drained, and covered with tall rain forest (Teunissen 1978f, 1984, 1988c). While rain forest dominates, the Interior also has swamp and marsh forests, and areas (mostly small) with savanna, savanna forest and related vegetation. Beyond 400-600 m elevation – where the climate is a few degrees cooler and typically more humid – upland ecosystems occur that, while often resembling those in the lowlands, have some unique flora and fauna elements.

This chapter focuses on the natural ecosystems that occur in the Estuarine Zone, the rest of the Coastal Plain, the Zanderij Belt, and the Interior. Ecosystems are defined and differentiated by a wide range of aspects, such as geography, climate, water, soil, vegetation, flora and fauna, and disturbance by flood or fire. Many of Suriname's ecosystems defy assignment to supposedly mutually exclusive categories such as terrestrial and aquatic: mangrove forests and other flooding forests along the lower reaches of the rivers, for example, have inundated and dry phases according to the seasonal and tidal cycles (Junk *et al.* 1989).

Brackish water ecosystems of the Estuarine Zone

Shallow sea and mudflats. Sediments from the Amazon River cause the band of muddy brown sea water and mudflats along the coast of the Guianas (Eisma *et al.* 1991, Hu *et al.* 2004) (Figure 31). The near-shore sea water, up to the 20 m depth contour and 30 km offshore, has relatively low salinity and is brackish in the rainy season (< 30 g salt/L; 'normal' sea water has 35 g/L); this has been referred to as 'estuarization of the inner shelf' (Lowe-McConnell 1962). The bottom substrate is soft mud. The tidal rise and fall is three meters at spring tides, and one at neap tides. Although

Figure 31. Mudflat along the Surinamese coast at low tide showing mangrove trees, Scarlet Ibis and other coastal birds (detail from panorama painting by Wim Eriks in Neotropical Butterfly Park, Lelydorp, Suriname).

nutrients derived from the breakdown of organic matter coming from the land abound, as do nutrients that well up from the sea bottom, low light penetration in the muddy water limits photosynthesis by algae and other primary producers. The most common planktonic producers here are floating colonies of algae of the genus *Phaeocystis* living in transparent gelatinous spheres 1 cm wide (known as *sowtu* in Suriname). Also at the base of the food web, living in the mud, are bacteria (Madrid *et al.* 2001), as well as diatoms that emerge during low tides and then color the mudflats brownish green (Artigas *et al.* 2003). Coastal birds, including millions of migrating North American waders, feed on invertebrate animals in the mud (Spaans 1978, 2003): fiddler crabs (*Uca* spp.; *odi-odi-botoman*) and large mangrove crabs (*Ucides cordatus*; *woyokrabu*) in the upper

River	Catchment (km^2)	Catchment area in Coastal Plain / Zanderij Belt (% of catchment)	Mean discharge at river mouth (m3/s)	Tidal volume (million m^3)	Maximum limit of salt intrusion (300 mg Cl/L or 0.5 ptt) during low flow	Minimum limit of salt intrusion during peak flow	Location of most downstream rapids (tidal limit)	Maximum depth along the thalweg (m) and location (km upstream along river)	Silt content (g/L) upstream of the minimum limit of salt intrusion	Sediment discharge (1,000,000 ton/year)
Corantijn	67,600*	5.4 / 4.6	1,572	300	km 82	km 40	km 235	30 (km 112)	0.02 - 4.00	1.20
Nickerie	10,100	34.1 / 23.6	178	10	km 110	km 28	km 240	17 (km 73)	0.03 - 6.00	0.10
Coppename	21,700	26.3 / 9.2	500	75	km 93	km 31	km 170	22 (km 82)	0.02 - 1.30	0.25
Saramacca	9,000	22.8 / 3.1	225	50	km 89	km 37	km 285	24 (km 145)	0.01 - 1.30	0.13
Suriname	16,500	10.7 / 4.5	426	125	km 90	km 54	km 194	17 (km 95)	0.01 - 1.50	0.25
Commewijne	6,600	52.1 / 11.5	120	40	km 150	km 55	NA	31 (km 71)	0.07 - 20.00	0.06
Marowijne	68,700*	0.5 / 0.1	1,785	200	km 59	km 37	km 115	21 (km 47)	0.01 - 0.08**	1.30**

*Catchment areas of these border rivers include areas in the territories bordering Suriname (French Guiana and Guyana)
**Values for silt content and sediment discharge of the Marowijne River are based on measurements before the modern gold rush that started around 1990; current values are probably higher due to ongoing extensive gold-mining activities upstream
NA = not applicable

Table 4. Hydrological characteristics of the seven large rivers of Suriname (data from Amatali 1993).

The Suriname River is regulated since the construction the Afobaka dam (at km 194) in 1964, and as a consequence discharge and salt intrusion depend on the discharge at the dam. Distances along the river course are measured in km along the thalweg (deep channel in the river bed) upward from the river mouth (km 0; at the 15-m depth contour at low water during spring tide).

Figure 32. River estuaries in Suriname:

a. Turbulent water in the large Corantijn River estuary (image © Jan Mol, Nov. 2007).

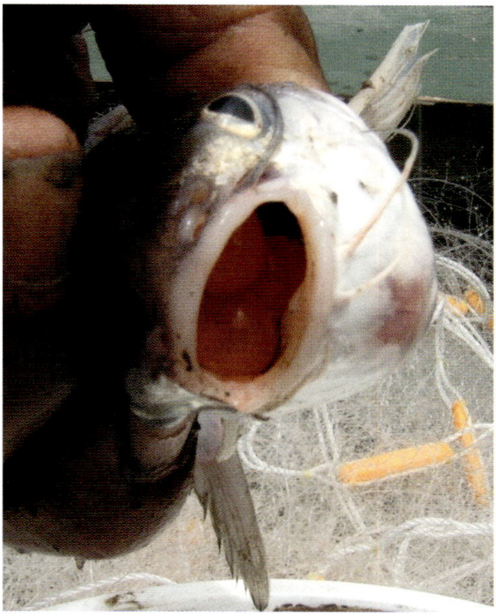

b. The sea catfish *Sciades herzbergii* (Ariidae) mouth-brooding its large eggs (image © Jan Mol, Jan. 2004).

tidal zone bordering the mangrove forest, and small, shrimp-like burying Tanaidacea lower on the mudflats (Swennen *et al.* 1982). Important predators include the Rufous Crab-Hawk (*Buteogallus aequinoctialis*; *krabu-aka*), the raccoon *Procyon cancrivorus* (*krabdagu*) and, entering from the sea, the swimming crab *Callinectes bocourti* (*srika*) and various shrimps (such as seabob *Xiphopenaeus kroyeri*) and fishes (Willems *et al.* 2015a,b).

Two shallow sea fish communities described from Guyana (Lowe-McConnell 1962) also occur in Suriname. The inshore 'brown fish' community includes ariid sea catfishes, such as *Sciades parkeri* (*yaribaka*) and *Amphiarus rugispinis* (*kodoku*), rays (*spari*), mullets (*Mugil* spp.; *kweriman*), toadfish (*Batrachoides surinamensis*; *lompu*) and snook (*Centropomus* spp.; *snuku*). The abundance of early life stages (larvae and juveniles) in the inshore environment indicates that it is an important spawning ground and nursery for marine fishes and shrimps. The 'golden fish' community in slightly deeper water is dominated by Sciaenidae, such as the important food fishes *Cynoscion acoupa* (*banban*) and *Cynoscion virescens* (*kandratiki*), and includes sharks, rays and tarpon (*Megalops atlanticus*; *trapun*).

River estuary – the brackish part of rivers. The estuary is the lowest reach of the river, between the river mouth and the upstream head where the seasonally shifting limit of salt water intrusion is located (0.5 g salt/L; Table 4). Current, water level, salinity, turbidity, pH and the amount of dissolved oxygen are highly variable and affected by the tide and river discharge. Tidal flows of seawater into and out of the estuary (the tidal volume) occur during the flood and ebb tides, respectively. The tidal flow diminishes further upstream, and beyond a certain location – the tidal limit – there is a continuous downstream flow and no tidal variation in water level (Table 4). In the major rivers that have rapids – only the Commewijne has none – the tidal limit will be the nearest rapids to the sea. River and tidal currents can be strong (>1 m/s); wind action can result in big waves up to 2 m high near the mouth of the estuary (Figure 32.a).

The water is brackish, varying from 0.5 g salt/L at the estuary head to 30 g/L at the mouth of the river (Table 3). At any one point, but especially mid-estuary, salinity varies both seasonally and daily (with the tide), making it a harsh environment for most organisms. The silt (brown 'mud') suspended in the estuarine water (Table 4) limits underwater visibility, and explains why the four-eyed fish (*Anableps anableps*; *kutai*) swims with the upper part of its eyes above the water surface. Silt

Figure 33. Coastal and riverine mangrove forest in the brackish-water Estuarine Zone:

a. *Avicennia germinans (parwa)* **trees dominating extensive forest near the ocean along the NDP creek, Commewijne District (image © Pieter Teunissen, Feb. 2008). A mat of upward growing parwa roots (pneumatophores) that are never fully submerged covers the soil; these pneumatophores, which ensure that the root system can breathe.**

b. *Rhizophora sp. (mangro)* **trees dominating the fringe of the bank of the Commewijne River at Alkmaar, Commewijne District (image © Bart De Dijn, Oct. 2016). The tall, branched stilt roots – many of which develop from branches – of** *mangro* **provide support in the muddy bank and also serve as pneumatophores.**

Figure 34. Early stage of colonization of a mud bank by *parwa* on the Atlantic coast, Coronie District (image © Pieter Teunissen, Feb. 2009).

can smother eggs and clog feeding and respiratory structures of aquatic animals. Estuarine fishes brood their eggs in the mouth (Figure 32.b) or attached to the body, like *Aspredo* spp. (*trompetfisi*), bear live young, like *kutai*, or spawn upstream in clear water, like *Pseudauchenipterus nodosus* (*botromanki*), or at sea, like *trapun* (see also chapter on Freshwater Fishes). Dissolved oxygen content is high in both the sea and rivers (Table 3), but can be low in the mid-estuary, especially near human settlements where biological oxygen demand is high due to pollution. Estuarine food webs are based on detritus (decaying organic matter) from marine, estuarine, freshwater and terrestrial sources. Seabob shrimp feed mainly on intertidal benthic (bottom-dwelling) diatoms (Willems *et al.* 2016b) and they, in turn, provide food for marine and estuarine fishes. Catches of juvenile marine fishes (*tri*) and shrimps (*sarasara*) near the river mouth (Engel 1981, Tito de Morais & Tito de Morais 1994) indicate important nurseries.

Marine species diversity is high near the mouth of the estuary and freshwater diversity is high near the head, but in the harsh mid-estuary, diversity is low (Whitfield *et al.* 2012). Estuarine fishes include anchovies, sea catfishes, aspredinid banjo catfishes (*trompetfisi*), goliath catfishes (for example *Brachyplatystoma filamentosum*; *lalaw*), *Hypostomus watwata* (*sekwikwi*) and puffers (*bosrokoman*) (Vari 1982). Two critically endangered marine fishes enter the estuaries: the Atlantic Goliath Grouper *Epinephelus itajara* (*granmorgu*) and the rare Largetooth Sawfish *Pristis pristis* (*safisi*). The charismatic Guiana Dolphins (*Sotalia guianensis*; *profosu*) feed, play and breed in the estuaries.

Mangroves and derived ecosystems. More than half of the Estuarine Zone is mangrove forest (89,000 ha in 2004; Parahoe *et al.* 2008). Along the ocean, there is a belt of **black mangrove forest** (*Avicennia germinans*; *parwa*) up to 10 km wide (Figure 33.a). It continues along the mouth of large rivers, and often also further upstream, quickly narrowing and transitioning into riverbank strips of **red mangrove forest** (*Rhizophora* spp.; *mangro*) (Figure 33.b). *Parwa* and *mangro* are pioneers with the rare ability amongst vascular plants to colonize and thrive in tidally flooded, brackish mud along the coast and rivers (Figure 34) (see for example Berrenstein 2012). Young *parwa* plants become established on coastal mud banks as these are taking shape; as *parwa*

Figure 35. Non-forest ecosystems on clay flats in the Estuarine Zone:

a. Hypersaline herbaceous marsh dominated by the herb ***Batis maritima*** (*krapegrasi*) at Matapica, Commewijne District (image © Bart De Dijn, Oct. 2001).

b. Hypersaline lake (*lagune, pan*) with the grass ***Sporobolus virginicus*** along the edge, at Bigi Pan, Nickerie District (image © CELOS, Sep. 2005).

c. Brackish lake (*lagune, pan*) with the cattail *Typha domingensis* (*langagrasi*) along the edge, north of Wageningen, Nickerie District (image © Bart De Dijn, Apr. 2012).
A breeding colony of Snail Kites (*pakro-aka*) is located in the mature *parwa* trees behind the lake (blackish specs in the tree tops are perching kites).

d. Brackish herbaceous swamp with the sedge *Eleocharis mutata* dominating along the NDP creek, Commewijne District (image © Pieter Teunissen, Feb. 2008).

mature, they help stabilize and grow the mud banks. As the banks age and the layer of mud thickens, the *parwa* are less exposed to tidal flooding, and more to rain. The mud gradually becomes clayey soil. With ageing the soil loses salt, organic matter and water, and shrinks, resulting in a progressive lowering of the land and accumulation of low-salinity water. This typically means that the soil becomes permanently or seasonally flooded; deep flooding causes the suffocation of the mangrove root system. The mangrove vegetation and other vegetation types of the Estuarine Zone are described below, based on Teunissen (1978g, 1980b, 1987b, 1988b) and earlier work by Lindeman (1953), Sterringa (1971), Pons (1972) and Julen (1974).

Extensive *parwa* groves may also be negatively affected by coastal changes that more abruptly modify the local hydrology and water quality, resulting in similar deep flooding or an increase of surface water salinity; the latter dehydrates the mangrove root system (Augustinus 1978). Very high salinities (hypersaline conditions) develop seasonally due to high net evaporation in poorly drained, low-lying areas near the coast that do not receive substantial amounts of fresh water from the southern hinterland, while regularly receiving salty sea water during high tide. An example of an area where this occurs is Bigi Pan (Parahoe *et al.* 2004), a stretch of land bordering Suriname's western coastline with a large, shallow lake (open water) at the center. It is separated from the bulk of the hinterland by the bank of the Nickerie River, and from the sea by a narrow ridge (see below). None of the creeks in the area effectively drain water away to the sea during the dry season,

Figure 36. Bigi Pan Lagoon, Nickerie District, northwest Suriname:

a. Fishermen's camps in the shallow lagoon (image © Jan Mol, Mar. 2005).

b. Catch of fisherman consisting of snook (*Centropomus* spp.), tarpon (*Megalops atlanticus*), mullet (*Mugil* spp), tilapia (*Oreochromis mossambicus*) and crabs (*Callinectes bocourti*) (image © Thomas Kerkhove, Feb. 2016).

and there is no continuous mud bank located north of Bigi Pan that, during high tide, prevents waves from overtopping parts of the coastline where the ridge is low or lacking.

All of these changes that negatively affect the mangrove root system typically lead to the same result: parwa trees and entire parwa groves die off ('mangrove graveyards'). The death of parwa is followed by the formation of: i) shallow **hypersaline herbaceous marshes** (Figure 35.a) dominated by succulent plants with a high salt tolerance, such as *Sesuvium portulacastrum* (*seposren*) and *Batis maritima* (*krapegrasi*), or the grass *Sporobolus virginicus*; and ii) slightly deeper **hypersaline lakes** (locally known as *lagune* or *pan*) (Figure 35.b) in the lowest-lying parts of the land with the above-mentioned plants along the edge, as well as the submerged aquatic herb *Ruppia maritima* (*sewar*). **Brackish lakes** (also known as *lagune* or *pan*) (Figure 35.c) seem to be a follow-up stage in these lowest-lying parts where the inflow of salt water has become less important compared to that of freshwater. Poor drainage combined with a limited influx of sea water, or a substantial influx of fresh water from the hinterland, results in **brackish herbaceous swamps** (Figure 35.d). These swamps, as well as the lake edges, are dominated by salt-tolerant aquatic plants, such as the grass *Paspalum vaginatum*, the sedge *Eleocharis mutata*, the fern

Figure 37. Shell and sand beach and ridge (*rits*) ecosystems along the coast:

a. Herbaceous beach vegetation dominated by the herb *Ipomoea pes-caprae* (*geitenhoefblad*) on a recently formed spit near Matapica, Commewijne District (image © Bart De Dijn, Oct. 2001).

b. Littoral scrub and thicket with the tall cactus *Cereus hexagonus* (right side of image) and the tall shrub *Hibiscus tiliaceus* (*maho*; left side and center) at Babunsanti beach, near Galibi, Marowijne District (image © Bart De Dijn, Apr. 2015).

Acrostichum aureum (*tabakatiki*), or the cattail *Typha domingensis* (*langagrasi*). The swamps and lakes may hold fresh water during the rainy season, but are brackish during the dry season. The brackish swamps can be quite distinct when separated from the hinterland by a ridge or dyke; in the absence of such a barrier, there is a gradual transition to freshwater conditions and freshwater swamp vegetation (see below). Since Suriname's coastline has been advancing north since the early Holocene, the influence of the receding sea on mangrove forests wanes as they age. The forest becomes progressively less brackish, and first is infused, later entirely taken over, by freshwater swamp tree species. This change is readily observable along the major rivers as one moves upstream.

Various animals, mostly invertebrates, live in and on the muddy substrate in which mangrove trees grow. The mangrove roots provide shelter and food for motile shrimps, crabs and fishes. Fishes mainly occur in tidal creeks and brackish-water lakes. The lakes (Figure 36.a) are typically shallow (< 1m deep), located within a few kilometers of the coast, and up to several square kilometers in size. During spring tides they receive sea water and early life stages of marine organisms via tidal creeks. The salinity of the water can increase through evaporation from 5 g/L in the rainy season up to 40 g/L at the end of the dry season (Table 3); even large lakes can dry up completely during extended dry spells in El Niño years (Mol *et al.* 2000). Mangrove litter (decaying leaves and wood) and animals that feed on it (detritivores) are important in the food web. With abundant food and shelter, and low predation pressure, mangroves function as a nursery that supports populations of crabs and shrimp, as well as offshore fisheries (Werkgroep Estuarine Kuststrook 1976; Nagelkerken *et al.* 2008). They are also important breeding and feeding areas for resident and migratory birds (Spaans 1978, 2003). Lagoon fisheries target snook, mullet, tarpon, juvenile Southern Brown Shrimp *Penaeus* (*Farfantepenaeus*) *subtilis*, *srika*, *woyokrabu* and, since its introduction in 1955 (Lijding 1958), tilapia (*Oreochromis mossambicus*) (Figure 36.b) (Geijskes 1943, Mol *et al.* 2000).

Beaches. Along parts of the coastline, extensive sand and shell beaches are found. They flood during spring tides, except the windswept highest parts. They represent incipient cheniers or ridges (*ritsen*) or remnants of pre-existing cheniers being eroded by the sea (Augustinus 1978). They are covered with species-poor herbaceous beach vegetation (Figure 37.a), typically with the creeping *Ipomoea pes-caprae* (*geitenhoefblad*) and *Canavalia rosea* (*zeeboon*). The oldest and tallest parts of these beaches are covered with littoral scrub and thicket (Figure 37.b), which is richer in plant species, and includes tall woody plants such as the shrubs *Hibiscus tilaceus* (*maho*) and *Chrysobalanus icaco* (*pruim*), and the columnar cactus *Cereus hexagonus.* The beaches tend to be broader and taller along Suriname's eastern coastline, where they are nesting places of international importance for four sea turtle species (Schulz 1975, 1980), including the leatherback turtle (*Dermochelys coriacea*; *aitkanti*) (Fossette *et al.* 2008).

Freshwater swamps

Defining characteristcs. Freshwater swamps are perennially flooded or at least waterlogged during most years; the water in these swamps is essentially stagnant. Flooding tends to subside only during long dry spells, which are most pronounced in El Niño years (Mol *et al.* 2000; Teunissen 1993; see also 'extremely dry years' in Bubberman 1973). Freshwater swamps dominate much of the low-lying Coastal Plain: 'YCP swamps' in the Young Coastal Plain, and 'OCP swamps' in the

Old Coastal Plain (on Early Holocene Mara deposits) (these are the 'young' and 'old' swamps of Teunissen 1978g, 1980b). Freshwater swamps also occur in the Zanderij Belt and the Interior, but are less extensive there. Swampy ecosystems associated with creeks and rivers south of the Coastal Plain are regarded as riparian (see below), not freshwater swamps in the strict sence. Freshwater swamp vegetation is herbaceous, (shrubby) low forest, tall forest, or a mosaic of these. In a swamp environment plant litter accumulates in the poorly oxygenated, stagnant water. The oxygen shortage slows the breakdown of plant litter, and causes peat (*pegasse*) to develop on top of the mineral soil; locally thick (> 2.5 m) peat layers may develop (Brinkman & Pons 1968, Bubberman 1973, Teunissen 1987a, Noordam & Teunissen 2008). Suriname's freshwater swamp ecosystems can be differentiated based on vegetation physiognomy (stature, layering, etc.) and composition (see below), and the geographical setting in which they occur. The vegetation of these ecosystems has been studied by Lindeman (1953), Lindeman and Moolenaar (1959), Van Donselaar *et al.* (1964-1975) and Teunissen (1976, 1978g, 1980b).

In shallow swamps, a thin peat layer develops that may catch fire during long dry spells (especially in El Niño years). As a dry spell proceeds, more of the water above the mineral soil evaporates, and more of the peat layer – which is highly combustible when dry – is exposed and desiccates. When fire rages in swamp forest, it destroys the exposed peat layer, kills the trees rooted in it, and initiates

Figure 38. Herbaceous freshwater swamp dominated by the sedge *Eleocharis interstincta* (*prapragrasi*; bottom of image), intermixed with the tall herb *Montrichardia arborescens* (*mokomoko*; middle), near Hannover, Para District (image © Bart De Dijn, Aug. 2016).

The forest (top) bordering the swamp is riparian forest on the bank of the black-water Para River; the presence of the tall palm *Mauritia flexuosa* (*morisi*) is suggestive of gallery forest, which would imply repeated burning of the swamp in historic times.

secondary vegetation succession (see below and Box 6). In deeper swamps, it takes much longer for the water to evaporate and the peat to desiccate; the thick peat layer that often forms in deep swamps may thus burn during extremely dry years only (typically when El Niño is particularly 'strong'). Intense peat fires in deep, forested swamps leave virtually no survivors, not even seeds in the soil. As a consequence, primary succession sets in. The initial colonization is by floating aquatic plants, which form floating meadows and some initial peat. As the peat layer continues to develop over centuries, the vegetation changes from mainly herbaceous, to woody, to forest (Teunissen 1993). However, the succession to swamp forest is slowed down or stopped if herbaceaous vegetation takes over that prevents establishment of shrubs and trees that are hydrophytic, adapted to growing in water or on soil that is flooded or waterlogged. For example, dense stands of the cattail *Typha domingensis* (*langagrasi*) or the invasive grass *Brachiaria mutica* (paragrass; *paragrasi*) in swamps seem to at least delay further succession. Swamps dominated by these plants quickly accumulate large amounts of dead biomass, and burn frequently and intensely. Repeatedly in the last centuries, both infrequent catastrophic fires and frequent more limited fires – all likely initated by people – have raged in swamps of various depths in the Coastal Plain (Bubberman 1973). The fires and subsequent vegetation succession explain much of the current varied aspect of the landscape in the Coastal Plain. The existence of groves of the tall palm *Mauritia flexuosa* (*morisi*) and tree *Erythrina fusca* (*kofimama*) can be attributed to their high fire resistance compared to other woody hydrophytes. Mature inviduals of these plants that survive a fire serve as on-site seed sources, and surviving seeds and young individuals likely benefit from being in a post-fire environment from which less fire resistant competitors have been eliminated (at least partially and/or temporarily).

Herbaceous freshwater swamps. Low vegetation dominated by herbs, often with scattered shrub individuals or groups of shrubs forming islets or groves, is a key feature of herbaceous freshwater swamps (Figure 38). They resemble savannas, but are perennially flooded and dominated by hydrophytes. In YCP swamps, the herb layer typically is a rich mixture of tall, emergent herbs, more or less dominated by *langagrasi*, *Cyperus giganteus* (*prasorograsi*) or *Montrichardia arborescens* (*mokomoko*). In OCP swamps, large sedges are often strongly dominant, such as *Eleocharis interstincta* (*prapragrasi*), *Lagenocarpus guianensis*, or *Rhynchospora gigantea* (the latter two known in Suriname as *twatwagrasi*).

Herbaceous YCP swamps are often shallow (<1 m deep), with a low drainage density (few creeks) and a slow flow (tenths of mm/s at most). Herbaceous OCP swamps have lower nutrient concentrations and can be several meters deep. At the end of the dry season, fishes concentrate in deeper pools, creeks and drainage canals. The often dense growth of emergent plants hides vast areas of water from view, and shields the water from sunlight and wind. Floating plants (such as *Azolla*, *Salvinia* and *Eichhornia* spp.) as well as submerged ones (*Cabomba* and *Utricularia* spp.; especially in OCP swamps) and those firmly rooted in the soil (such as *Nymphaea* and *Nymphoides* spp.) grow in unshaded areas (Werkhoven & Peeters 1993, Doerga 2013). Swamp invertebrates include air-breathing apple snails (Pomacea; *pakro*) that deposit their conspicuous pink (*Pomacea dolioides*) or green (*Pomacea glauca*) eggs in clusters above the waterline, the shrimp *Palaemonetes carteri*, water bugs, water beetles, and nymphs (immature life stages) of dragonflies and damselflies (Mol 1996b). The snail *Biomphalaria glabrata*, the intermediate host of the bilharzia parasite *Schistosoma mansoni*, occurs in swamps bordering shell ridges (Reyenga 1971).

Figure 39. Forested freshwater swamp ecosystems (perennially flooded):

a. Swamp wood (low swamp forest) dominated by the strongly buttressed tree *Pterocarpus officinalis* (*watrabebe*), near Lelydorp, Wanica District (image © Bart De Dijn, Jun. 2011).

b. Swamp forest (high or mixed swamp forest) dominated by the tall palm *Euterpe oleracea* (*pina*) in combination with taller *Virola surinamensis* (*babun-udu*) trees, along the highway south to Paranam, Wanica District (image © Bart De Dijn, Oct. 2016).

Swamp fishes obtain oxygen from the air, like Callichthyinae (*kwikwi*), *Hoplerythrinus unitaeniatus* (*walapa*) and the swamp eel *Synbranchus marmoratus (snekfisi)* (Carter & Beadle 1931), or from a thin layer below the water surface by 'aquatic surface respiration', like *Astyanax bimaculatus* (Winemiller 1989a). The swamp eel and *Callichthys callichthys* (*plata-ede-kwikwi*) construct burrows during unfavorable dry season conditions, and migrate overland to escape from drying pools; hence the local expression *lowe fisi*: fish that walk away. Most fishes spawn in the rainy season when swamps fill up. The mainly fish-eating *walapa* and *pataka* (*Hoplias malabaricus*) are the first to spawn in newly flooded swamps, thus avoiding low dissolved oxygen levels that develop later in the season. This may give their young an advantage in age and size over their prey. *Hoplosternum littorale* (*hei-ede kwikwi*) deposit their eggs above the water in a floating bubble nest under a cover of plant material, thus assuring oxygen supply to the eggs while preventing desiccation (Mol 1993; see chapter on Freshwater Fishes).

Most swamp fishes such as *pataka*, *walapa*, *Crenicichla saxatilis* (*datrafisi*), *Cichlasoma bimaculatum* (*krobia*), *Rhamdia quelen* (*dyaki*), *Trachelyopterus galeatus* (*noya*), *Gymnotus carapo* (*logologo*), *Hyphessobrycon simulatus*, and *Pristella maxillaris* (the latter two known locally as *reditere-sriba*) also occur in forest creeks and may enter swamp forest when water levels are high. But *hei-ede kwikwi*, *Loricariichthys maculatus* (*basyafisi*; apparently endemic to Suriname) and *Ctenobrycon spilurus* only occur in coastal herbaceous swamps. Other vertebrates of herbaceous swamps that are much less common in forested swamps include the Suriname Toad

Figure 40. Marsh forest (seasonally flooded):

a. *Hura crepitans* (*posentri*; with spiny bark) dominating forest on poorly drained margin of a low ridge in Paramaribo (image © Bart De Dijn, Oct. 2016).

b. Soil-level view of hogwallow (*kawfutu*) soil structure in *posentri*-dominated forest at Paramaribo (image © Bart De Dijn, Oct. 2016).

The shallow gullies between the small hummocks fill with water in the rainy season; in the dry season (when the photo was taken) they just contain forest litter. The trees and most other plants grow on the hummocks, not in the gullies.

(*Pipa pipa; pipatodo*), giant tadpoles of *Pseudis paradoxa* (*tododyaki*), Spectacled Caiman (*Caiman crococilus; wetberekaiman*), Green Anaconda (*Eunectes murinus; aboma*), Capybara (*Hydrochaeris hydrochaeris; kapuwa*), Wattled Jacana (*Jacana jacana; kepanki*) and Purple Gallinule (*Porphyrula martinica*; *blawkepanki*).

Forested freshwater swamps. Perennially flooded forest of low stature (10-25 m) is **swamp wood**, also known in Suriname as low swamp forest (Figure 39.a); it typically has a single mature tree layer. A few types are distinguished in the Coastal Plain, based on the dominating tree species (all hydrophytes): i) *Pterocarpus officinalis* – *Tabebuia insignis* (*watrabebe – swampupanta)*, ii) *Mauritia flexuosa* (*morisi*; a large palm), iii) *Triplaris weigeltiana* (*mira-udu*), iv) *Erythrina fusca* (*kofimama*), and v) *Chrysobalanus icaco* – *Annona glabra* (*zwamppruim – zwampzuurzak*). **Swamp forest**, also known in Suriname as high swamp forest (Figure 39.b), is perennially flooded forest of tall stature (25-35 m), typically with two mature tree layers. Throughout the Coastal Plain, in shallow swamps where the water level fluctuates some 30-80 cm, a type of mixed swamp forest occurs that is rich in tree species, and characterized by the presence of the trees *Virola surinamensis* (*babun-udu*) and *Symphonia globulifera* (*mataki*), and the tall palm *Euterpe oleracea* (*pina*). One or more of these characteristic species can be quite dominant, alongside *Hura crepitans* (*posentri*) at some locations in the Coastal Plain. In deep OCP swamps where water levels fluctuate up to 3 m, swamp forest is less rich in tree species. The above-mentioned species are lacking, and instead the characteristic trees are *Crudia glaberrima* (*watabiri*) and *Macrolobium acaciifolium* (*waratapa*). For more details and more images of various swamp woods and forests, see Teunissen (1993).

In both swamp wood and swamp forest epiphytic plants (plants growing on the stems, twigs or even leaves of other plants) are often abundant, alongside tree-inhabiting animals such as tree opossums, monkeys, sloths, anteaters, rats, porcupines, squirrels; also common are bats, birds, tree and water snakes, tree frogs and fishes (see Teunissen 1987a, and examples in the species account chapters of the present book).

Marsh forests

Defining characteristics. Marsh forest has a tall stature (>25 m) and is better known in Suriname as high marsh forest (Figure 40.a). It occurs on land that is seasonally shallowly inundated with fresh water or where the soil is at least seasonally waterlogged. During prolonged dry periods, the upper soil dries out. A seasonal change in the soil from very wet, but not deeply flooded, to dry is characteristic. Marsh forests are most extensive in the Coastal Plain on river levee formations (in Dutch: *oeverwallen*; Holocene to present-day river bank deposits in the tidal zone), and low clayey plateaus (*schollen*; remnants of Pleistocene mudflats in the Old Coastal Plain). They are less extensive in the Coastal Plain on low ridges (cheniers, *ritsen*): 'YCP ridges' (Holocene cheniers in the Young Coastal Plain), and 'OCP ridges' (Pleistocene cheniers in the Old Coastal Plain). They also occur in the Zanderij Belt and the Interior. Ecosystems with marsh forest vegetation associated with creeks and rivers south of the Coastal Plain are regarded as riparian forest (see below). In the Interior, marsh forest areas occur on poorly drained river terraces, as well as in broad valleys in some headwater regions (Bart De Dijn, personal observation; as in French Guiana, see De Granville 1994, p. 26). Typical marsh forest soil has a hogwallow relief (*kawfutu*), characterized by hummocks in a network of channels 10-50 cm deep (Figure 40.b). During rainy periods, the channels fill with water, and hummocks may flood briefly. Due to the shallowness and temporary nature of the flooding, peat is typically lacking in marsh forest. Teunissen (1978d,f; 1980b) differentiates between several types of marsh forest ecosystems on the basis the geographical setting in which they occur.

Figure 41. Riparian forest rich in *pina* palms and with a tall *Ceiba pentandra* (*kankantri*) tree leaning over the lower Nickerie River, Nickerie District (image © Bart De Dijn, Apr. 2011).

Flora and fauna. Marsh forest vegetation has been studied in Suriname by Lindeman (1953), Lindeman and Moolenaar (1959), Teunissen (1976, 1978g, 1980b), and Ruysschaert (2017). It is usually rich in, but not dominated by, large flood-tolerant palms such as *Attalea maripa* (*maripa*) and *Euterpe oleracea (pina)*, as well as typical marsh forest tree species such as *Carapa* spp. (*krapa*). *Eperua falcata* (*walaba*) can be one of the dominant tree species, but is largely restricted to the Zanderij Belt and Interior. The tree *Hura crepitans* (*posentri)* dominates locally near the coast, and *Mora excelsa* (*mora*) in northwest Suriname along the Corantijn River. The fauna associated with marsh forests is rich, and overlaps with that of swamp forest and rain forest because of the seasonal alternation of wet and dry soil conditions, accommodating both swamp and rain forest animals in different seasons. The fauna overlap is also due to similarities in vegetation structure and – to a lesser extent – composition. Proximity also plays a role, since marsh forest is often located where swamp land transitions to dry land with rain forest.

Riparian forests

Defining characteristics. Riparian forest (Figure 41) occurs along creeks and rivers where much of the land is affected by waterlogging and often also seasonal flooding. It occurs in both the Zanderij Belt and the Interior of Suriname. Its vegetation has been studied in the Zanderij Belt by Heyligers (1963) and Teunissen (1978g, 1980b), in the Interior by Lindeman and Moolenaar (1959) and Hoffman (2013). In the Zanderij Belt, it is often fairly homogeneous, with a vegetation that resembles mixed swamp forest (with *babun-udu* and *mataki* trees and *pina* palms; typically with numerous *walaba* trees). It is more heterogeneous in the Interior, with parts resembling mixed swamp forest, marsh forest, and dryland forest infused with hydrophytes. Riparian forest occurs on high river banks as well as low river banks, from the stream edge to the base of hill or plateau slopes, and often also up the base where the soil is wet due to downslope groundwater flow. *Kawfutu* relief or a thin peat layer is often present in the low-lying parts where hydrophytes are particularly well represented. Hydrophytes are much rarer to absent in nearby forests on better drained soil higher upslope. However, riparian forests are typically enriched by dryland plant species which do well in the drier parts in the swampy-marshy matrix. Riparian ecosystems are referred to in Suriname as creek or creek valley forest, and river valley forest. Although literally these terms could also apply, they are rarely if ever used for swampy or marshy streamside forests in the Coastal Plain (see for example Teunissen 1978g, 1980b). The latter forests are designated using other terms (see above), and are fundamentally different since they are not associated with adjoining hills and plateaus which can feed them with groundwater. De Granville (1986, 1994), writing about French Guiana, restricts the usage of the term riparian forest to forest along rivers and large creeks (not small creeks) in the interior.

Flora and fauna. Marsh forest trees such as *Eperua falcata* (*walaba*) are often quite abundant in riparian forest, but in the Interior *Eperua rubiginosa* (*libawalaba*) often occurs in stead of *walaba*, especially along major rivers and on river islands (Teunissen, personal observation). The vegetation composition is strongly influenced by seed dispersal via the flowing and seasonally rising waters (as in the case of both trees mentioned above), as well as by some fishes (Goulding 1980; see Freshwater Fishes chapter). The overlap in fauna with swamp, marsh and rain forest is great, yet riparian forest seems to be special in the sense that its vegetation suffers less from drought due to the influx of groundwater, and thus offers important feeding opportunities to many forest

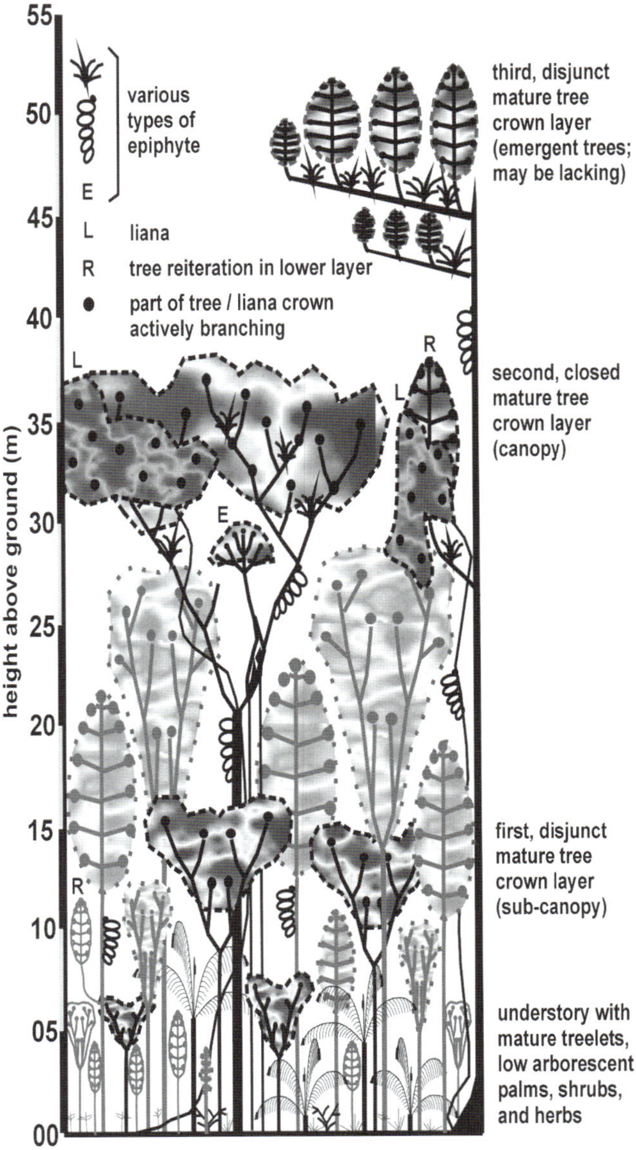

Figure 42. Rain forest (high dryland forest):

a. Huge trees of the genus *Couratari* (with high buttresses) pierce the forest canopy in the western foothills of Nassau Mountain (image © Bart De Dijn, Feb. 2008).

b. Model of the rain forest in which layers are distinguished based on the location of the crowns of mature trees (drawn by Bart De Dijn, based on original drawing in Oldeman 1974).

Mature trees are those that have essentially reached their final height. They typically have a single, finely branched crown that is at least as wide as tall, with a weakly convex top. As a result of, for example, injury, they may 'reiterate' below this crown, that is, develop one or more additional stems with separate crowns. Immature trees typically have a single weakly branched crown that is taller than it is wide, with a pointed or strongly convex top. They are in the process of extending their crowns upward and – less so when younger – sideways. This extension is often temporarily arrested when light availability is severely limited. Not shown here are treefalls, nor trees that break at the stem or main branches, and often reiterate before finally dying.

BOX 6 Vegetation succession in rain forest

When a storm topples a large tree in the forest, a natural clearing or gap is created. Seeds of pioneer tree species already present in the soil ('seed bank') or subsequently dispersing into the gap ('seed rain') germinate in response to direct sunlight reaching the soil. They produce pioneer or early secondary vegetation (*kapuweri*): a dense, single layer of low vegetation that forms an initial canopy. In the shade of the pioneers, seedlings and saplings of other tree species develop. As the short-lived pioneers die off, the fastest growing of the other tree species break through the initial canopy. What then develops is late secondary forest (*kapuweri busi*): a taller, typically more stratified and less dense forest. Over time, many of the slowest growing seedlings and saplings also reach the canopy, and the forest becomes mature or climax forest (*gran busi*). This is natural **secondary** vegetation succession (Figure 43), as opposed to an anthropogenic one (caused by man), which often follows a similar sequence, at least after small-scale slashing and burning of the forest.

Primary vegetation succession begins in the absence of a seed bank. It is rarely observed in Suriname, except along the coast (see Mangroves) and when it is anthropogenic, such as after intensive burning (see Freshwater swamps) or abandonment after the removal of the pre-existing vegetation and soil (such as mining). It often follows a course similar to secondary succession, but with the initial stages more strongly dominated by water or wind-dispersed plants.

Useful plants, such as palms of the genus *Astrocaryum* and bamboo of the genus *Guadua*, are indicators of anthropogenic succession. On some lands previously inhabited by indigenous people, the process of succession seems to have followed a different path, leading to **bamboo forest**, a dense bamboo-dominated thicket (Versteeg 2003, De Granville 1994). **Liana forest** may have a similar origin; it forms a dense, low thicket dominated by lianas and vines, but with scattered emergent trees.

Rain forests succession described by Hallé and Oldeman (1975) and Hallé *et al.* (1978), and well studied in French Guiana (see for example De Foresta & Lescure 1989, Riéra *et al.* 1989,1998). Pioneering research was done in Suriname by Schulz (1960), followed by research by Boerboom (1974).

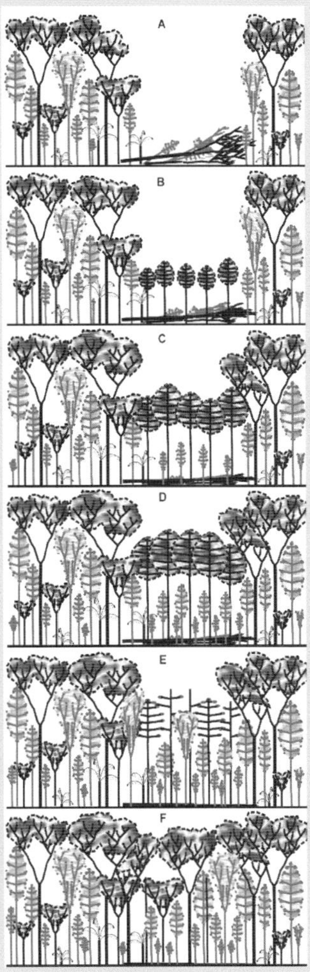

Figure 43. Model of secondary vegetation succession in rain forest after a major treefall (drawn by Bart De Dijn, based on original drawing in Hallé *et al.* 1978).

Profiles a) through f) shows the sequence of events: a) a large tree has fallen, damaging other trees and creating a gap in the forest canopy, meaning that more light is now available in and near that gap; b) softwood, pioneer trees germinate in the gap and grow much faster than pre-existing saplings of hardwood trees that survived the treefall; c) the pioneers form a low, closed canopy in the gap below which pre-existing hardwood trees slowly continue to develop, and new ones germinate; d) this low canopy increases in height, as the pioneer trees quickly reach maturity, e) the pioneers also die quickly, thus allowing immature hardwood trees in the gap to grow faster; f) ultimately, part of these hardwood trees reaches maturity and closes the gap.

animals during prolonged dry periods. Another special phenomenon is that when parts of the riparian forests flood, during rainy periods, much of the aquatic fauna moves from the streams into the flooded forests.

Rain forests and rain forest creeks

Rain forest. The term rain forest or rainforest is often used in a very broad sense, encompassing not just what is described in the present paragraph, but also other forest ecosystem and vegetation types (such as tall marsh and savanna forest described elsewhere in the present chapter). Here, it is interpreted rather narrowly, in approximate agreement with Beard's (1944, 1955) Rain Forest Vegetation Formation. Such rain forest is known as high dryland forest in Suriname, 'high' referring to tall (tree) stature (Figure 42.a). It occurs in the lowlands, below about 400-600 m elevation, depending on geographical position and local topography (see section below on upland ecosystems). It covers about 80 percent of Suriname's land territory (Teunissen 1988c), and is most extensive throughout the Interior, on hilly terra firme with lateritic soils, and on well-drained river terraces. It covers about half of the Zanderij Belt, at locations with well-drained, unbleached 'brown' sand (locally also loamy or clayey soils). It is less extensive in the Coastal Plain, where it is found on high cheniers (ridges) and on higher, better-drained parts of river levee formations and low clayey plateaus (*schollen*). Rain forest soil is deep, with the deeper parts always moist; it almost never floods, and if it does, or becomes waterlogged, this condition tends to be ephemeral (lasting hours, not days). Teunissen (1978d,f; 1980b) differentiates between rain forest ecosystems in the Coastal Zone based on the geographical setting in which they occur (see above). The Interior is, however, poorly studied compared to the Coastal Zone, and this may explain why the rain forests of the Interior are often regarded and mapped as a single, uniform ecosystem type (with the exception of rain forest in uplands; see final parts of this chapter). In French Guiana, research has been done in a wider range of forested sites in the interior of the country, and this has led to the differentiation of rain forests on the basis of the nature of the parent rock from which the soil is derived (such as greenschist, granite, quartz-rich subgreywacke), which seems to affect the stature and density of the forest (De Granville 1984). A recent study in French Guiana using both remote sensing data (obtained via satellites) and field survey data, confirms this and has led to further differentiation (Guitet *et al.* 2015a,b).

Rain forest is very rich in plant species, especially woody plants such as trees: up to 143 species of trees with stems of at least 10 cm diameter at breast height have been recorded per hectare of mature rain forest in the lowlands of Suriname (see Banki 2010, Table S.3.1.: 'lowland rain forest on ferralsol and ferralic arenosol'). The overlap in tree species composition, even between nearby plots with lowland rain forest in essentially the same setting and with the same set of dominant tree species, is typically low (for example, 34% tree species overlap between two lowland 1 ha plots in the foothills of Lely Mountain that were just 1 km apart; based on data in ter Steege *et al.* 2005). Trees are the main primary producers of the rain forest; they constitute the bulk of the biomass and provide the structure of the forest. Data from an undisturbed rain forest site in Suriname bears this out: the dry biomass of living plants was 539 ton/ha; trees represented 464 ton/ha, or 86 percent of total living plant biomass (Schmidt *et al.* 2011). In a sample of multiple sites in Suriname, trees represented on average 94 percent of living plant biomass in rain forest stands ('high dryland forest'; Crabbe 2012); the rather modest (289 ton/ha) average amount of total living biomass recorded suggests that these may be selectively logged stands.

The rain forest fauna is also very rich, but there is comprehensive data only on vertebrate animals in Suriname's rain forests (see species accounts in the present book). Most terrestrial animal groups reach their highest levels of diversity in the rain forest, and most of the animals featured in the present book are rain forest species. The rain forest fauna includes large species such as the South American Tapir (*Tapirus terrestris*, *bofru*) and the Jaguar (*Panthera onca*, *penitigri*), and eight species of monkey (*yapiyapi*) (Husson 1978). More than 500 bird species occur in the rain forests of the Interior (Spaans *et al.* 2016), such as the Harpy Eagle (*Harpia harpyja*; *gonini*), and a variety of macaws, parrots and parakeets (Psittacidae; *popokai*). Although data on invertebrate taxa in Suriname is extremely fragmentary, there is broad consensus that invertebrates, especially insects, are several orders of magnitude more species-rich than vertebrates, certainly in rain forests (based on global data, such as in Wilson 1988, Ponder & Lunney 1999, Mora *et al.* 2011). There is some regional data on fungi, but mainly in relation to peculiar forests in Guyana dominated by single tree species (for example Henkel *et al.* 2012, Smith *et al.* 2013). Apparently there is no regional data yet on microbial diversity, despite a recent global surge in the study of the 'microbiome' (see for example Gilbert *et al.* 2014). No comprehensive data on the animal biomass in the Suriname rain forest is available, but Fittkau and Klinge (1973) made the following estimates of proportions in relation to rain forest in nearby Central Amazonia: plants equal 83 percent of living biomass, herbivores 2 percent, carnivores 1 percent, and 'soil fauna' 14 percent, soil fauna refers to mostly tiny animals that live in the soil and in leaf litter such as springtails (Collembola) and mites (Acarina), many of which are fungus feeders and detritivores. However, they did not factor in fungi, nor microbes, and may have underestimated animal biomass, especially that of social insects whose biomass is hard to assess (see for example Roubik 1993). Rain forest animals in our region are more than just herbivores, carnivores or detritivores; many of them also interact in complex ways with plants as pollinators and seed dispersers (De Dijn 2005, Forget & Hammond 2005), and with plants or each other as parasites and disease vectors (examples can be found in the species account chapters of the present book).

A typical feature of lowland tropical rain forest is the lack of a thick layer of fine litter or humus on top of the mineral soil, and the low fertility of the soil, despite the luxuriance of the vegetation (see global review in Mabberley 1992). Rain forest in an undisturbed setting in Suriname had 35 tons of litter (dry weight) per hectare, only 35 percent of which was fine litter, mainly dead leaves and twigs, the rest being dead branches and stems (Poels 1987, Schmidt *et al.* 2011). This means that fine litter represented 44 times less biomass than living plant biomass. Data from a wider range of rain forest plots in Suriname (Crabbe 2012) yields rougly the same ratio. The combined amount of nutrients 'locked' in living plants in this forest, in the forest litter, and in organic particles incorprated in the soil (to 120 cm depth) was many times higher than the amount of nutrients in the soil that was readily available for plants. Soil minerals in Suriname's rain forest incorporate substantial amounts of nutrients, or at least their chemical precursors, in particular of phosphates, but in chemically bound form: these nutrients are released gradually over the long term, as the soil continues to weather chemically. To satisfy their short-term 'hunger' for nutrients, growing and reproducing rain forest plants must thus turn to extracting them from the forest litter and organic matter in the soil, both of which are supplied by the living biomass.

Studies undertaken elsewhere in our region help to understand how this extraction takes place: the litter layer and the soil immediately below it are rich in small tree roots with symbiotic

Mycorrhizae (St. John 1980, Béreau *et al.* 1997). Mycorrhizae are fungi that help the trees with nitrogen and phosphorus acquisition, and likely also improve water and micro-nutrient uptake (Turner 2001). The relatively stable humid and warm rain forest micro-climate at the soil-litter interface (Schultz 1960) allows for fast breakdown and mineralization of litter. At this interface, the shallow and dense root-Mycorrhizae mat is ideally positioned for nutrient uptake. Additional small amounts of nutrients enter the rain forest ecosystem via the atmosphere (gas, dust, aerosol, dissolved in rain), but more or less equivalent amounts leave the system via runoff of rain water and via groundwater flow towards creeks (Poels 1987). Nitrogen compounds are added by nitrogen-fixing bacteria that occur in root nodules of certain tree species; such nodules have also been recorded in our region (in various species of Leguminosae; Perreijn 2002). The rain forest water cycle is not as tight as the nutrient cycle, nevertheless, most of the water evaporated by the vegetation – via local cloud formation that frequently results in precipitation – quickly returns to the soil and vegetation (see chapter on Climate and Hydrology).

Rain forest vegetation. In Suriname, rain forest vegetation has been studied by Lindeman (1953), Lindeman and Moolenaar (1959), Schulz (1960), Maas (1971), Teunissen (1980b), Ter Steege *et al.* (2007), Banki (2010), Bordenave *et al.* (2011), Banki and Bhikhi (2011, 2013) and Hoffman (2013) (see also Werger 2011). Mature rain forest is climax vegetation – the final stage of vegetation succession (see Box 6) – which typically has two to three mature tree layers: a diffuse one below the canopy, a compact one which forms the canopy at 25-45 m, and often also a discontinuous one formed by emergent trees up to 50-60 m tall (Figure 42.b). 'Mature' trees are those that are or have been reproducing sexually (flowering and fruiting). The layering of mature trees is often hard to appreciate, especially during a brief field assessment, because flowering and fruiting are intermittent, seasonal events for most tree species in our region, and because individual trees may 'skip' multiple years (Sabatier & Puig 1985, Norden *et al.* 2007). Immature trees of various stages of development are often more numerous than mature trees, and tend to obscure the layering. A rule of thumb is that mature trees, unlike immature ones, typically have finely branched crowns that are at least as wide as tall, and have weakly convex, umbrella-like tops (Oldeman 1974). Although tree composition in rain forest is typically mixed, certain species may dominate locally, such as *Mora gonggrijpii* (*morabukeya*) in parts of western Suriname and *Voucapoua americana* (*broin-ati*) in other parts of the country.

Rain forest on cheniers (ridges) is often referred to as **ridge forest** (this is the 'littoral woodland' of Lindeman & Moolenaar 1959). It lacks emergent trees, except where *Ceiba pentandra* (*kankantri*) has invaded gaps caused by fire. It is poorer in tree species than rain forest elsewhere in Suriname; species richness is the lowest close to the coast on younger ridges. Near the windswept coastline, where rainfall is less than further inland and evaporation higher, it may represents a distinct, drier ecosystem type. The columnar cactus *Cereus hexagonus*, which is present in many ridge forests, is considered an indicator of a physiologically dry environment, meaning one with little water in the soil, or with lots of sodium and calcium ions in the water – due to lots of salt and dissolving shells in the soil – which makes the uptake of water more difficult for plants.

Rain forest creeks. Shielded from direct sunlight under a closed canopy of trees, rain forest creeks (Figure 44.a) have temperatures of 23-27 °C (Table 3), that are lower than those of larger, sunlit rivers, and vary very little during the day or through the year (±1 °C; Geijskes 1942). Water in these creeks is low in electrolytes, and clear to somewhat brownish from tannins.

Dissolved oxygen content is high in the rainy season, but low in the dry season when creeks split into a series of disconnected pools (Figure 44.b). Natural fish kills have been observed in such pools (see Beebe 1945). The outer margins of creek meanders are erosion zones with deep water, steep undercut banks and overhanging root mats – refuges for caimans (*kaiman*) and electric eels (*Electrophorus* spp.; *stroomfisi*) – whereas the shallow, inner bends build up sand and leaf litter, each with its own characteristic fish fauna.

Although light reaching the forest floor and creeks is just one percent or less of the light reaching the forest canopy, it does energize some primary producers that are present in the creeks: such as dense stands of the emergent herb *Thurnia sphaerocephala*, red algae (*Batrachospermum* spp.) and diatoms (Fittkau 1967, Walker 1995, Crul & Reyrink 1980, Mol *et al.* 2007b). However, the bulk of the energy and matter supporting the food web is derived from leaf litter (4200 kg/ha in central Amazonia) and woody debris (1500 kg/ha) that does not originate from plants growing in the creeks, but from the forest above (Vannote *et al.* 1980, Walker 1995). Terrestrial invertebrates that fall into creeks are an important source of nitrogen and phosphorus (Small *et al.* 2013). The food chain is cyclic: nutrients absorbed by the forest vegetation enter the creek in the form of forest litter, fruits and small animals, and are incorporated in the aquatic food web (see Knöppel 1970), from which nutrients return to the forest in the form of adult insects and frogs leaving the water, and aquatic animals taken by terrestrial predators such as otters, Fishing Bats (*Noctilio leporinus*) and birds.

Chironomid midge larvae, shrimps, crabs, fishes and tadpoles live in the submerged litter (Henderson & Walker 1990). Several fishes resemble dead leaves (Leaf Fish *Polycentrus schomburgkii*, *Helogenes marmoratus*, *Bunocephalus* spp.) and are thus camouflaged

Figure 44. *Terra firme* rain forest creeks in Suriname:

a. The canopy closes above a small clear-water rain forest creek in southern Suriname (upper Corantijn River catchment (image © Jan Mol).

b. A small rain forest stream has been reduced to a series of interconnected pools in the dry season; Maykabu Creek, Mindrineti River, Saramacca river system (image © Rosebel Gold Mines Environmental Department, Nov. 2015).

Figure 45. Ephemeral aquatic habitats in the tropical rain forest of Suriname:

a. Rain puddle on a trail in the forest, upper Nickerie River catchment; type locality of the killifish *Anaplepsoides fransvermeuleni* (image © Frans Vermeulen, Oct. 2010).

b. A pool associated with a flooded forest creek in the lower Paloemeu River catchment; type locality of the killifish *Kryptolebias sepia* (image © Frans Vermeulen, Jul. 2003).

c. Plant-held aquatic habitat in Bromeliacea on branches high in the canopy of the Surinamese rain forest (painting by Wim Eriks in Neotropical Butterfly Park, Lelydorp, Suriname).

to deceive visually oriented predators (Sazima *et al.* 2006) or prey. Fish eaters in the creeks include *pataka*, *Hoplias aimara* (*anyumara*), *Acestrorhynchus* spp. (*dagufisi*), *Crenicichla* spp. (*datrafisi*) and electric eels. Creek fishes partition resources spatially – living near the surface, on the creek bed, or mid-water – and by time of activity, gymnotiform fishes and catfishes being active at night, characoids and cichlids during the day.

Ephemeral aquatic habitats. Such habitats are an integral part of forest ecosystems, especially in tropical rain forests. In Suriname, they include rain puddles (Figure 45.a), pools associated with flooded creeks (Figure 45.b) and small plant-held water bodies (phytotelmata). Phytotelmata occur in dead leaves and fruit husks, in the leaf axils of live Bromeliaceae (Figure 45.c), as well as in bracts – robust leafy structures that envelop single or multiple flowers or fruits – such as the living floral bracts of Heliconiaceae and Strelitziaceae, and the dead, fallen bracts of palms (Greeney 2001). These patchily distributed microhabitats are utilized in a variety of ways by aquatic algae, micro-organisms, annelid worms, mites, zooplankton, insects (such as mosquito larvae), crabs and frogs (Zahl 1975). Small-sized fish species colonize temporary pools during flood events (Pazin *et al.* 2006), and killifishes (Rivulidae) flip across humid leaf litter to colonize areas beyond the reach of the flood (Figure 45.a) (Ouboter & Mol 1993). These small predatory fishes have dramatic effects on invertebrates and tadpoles in pools and puddles, in the sense that they may eat them all (Hero *et al.* 2001).

Clear-water rivers

Lower freshwater reaches. The lower freshwater reaches of rivers are situated in the flat Coastal Plain, from the head of the estuary to the limit of tidal changes in water level at the most seaward rapids (Table 4). Here the rivers meander extensively, rise and fall with the tides, and seasonally expand into large areas of forested floodplain. Large river islands are also important features here, at least in Suriname's largest rivers, the Marowijne and Corantijn. In the Suriname River, stands of submerged herbs such as *Cabomba* and *Mayaca* spp. (Doerga 2013) form important spawning habitat for fishes (Mol 2012b). Dissolved oxygen content and transparency are high (Table 3), but conductivity, pH and amounts of dissolved oxygen decrease in the rainy season (Crul & Reyrink 1980). The river bottom changes from muddy downstream, to silty, then sandy upstream. Plankton – mainly diatoms, desmids and other green algae – is usually not abundant, but blooms can occur when water levels are low (individuals of the diatom *Melosira granulata* can occur at densities >100,000/L; Crul & Reyrink 1980).

In the rainy season, river levels rise substantially, which allows fishes to move away from the river channel into the inundated floodplain to spawn and feed (Lowe-McConnell 1964, Vari 1982, Junk

Figure 46. Capybara (*Hydrochoerus hydrochaeris*) and Hoatzin birds (*Opisthocomus hoazin*) in riparian vegetation along the lower Corantijn River (detail from panorama painting by Wim Eriks in Neotropical Butterfly Park, Lelydorp, Suriname).

et al. 1989). Vegetarian fishes can ingest large numbers of seeds and thus assist in their dispersal (Goulding 1980, Vari 1982). Large-scale spawning migrations up- or downriver by fishes are known for prochilodontids (*kwimata*) and goliath catfishes (*Brachyplatystoma* spp. such as *lalaw*) (Carolsfeld *et al.* 2003), but have not been observed in the Corantijn River (Vari 1982). These fish migrations represent landscape-scale flows of energy, materials and organisms, out of one system into another and back again (Flecker *et al.* 2010). Fisheries in the lower freshwater part of the rivers mainly target *Plagioscion* spp. (*kubi*), tarpon, goliath catfishes, and the filter-feeding pelagic catfish *Hypophthalmus marginatus* (*kwasimama*) (Vari 1982, Boujard & Rojas-Beltrán 1988).

Giant Otters *Pteronura brasiliensis* (*bigiwatradagu*; Duplaix 1980), West Indian Manatees *Trichechus manatus* (*seku*), Capybaras (*Hydrochoerus hydrochaeris*) and piranhas (*Serrasalmus rhombeus*; *piren*) occur in most, if not all, lower freshwater reaches, but Hoatzins (*Opisthocomus hoazin*) are only know from the Corantijn River (Figure 46). The Corantijn River also has an endemic freshwater stingray (*Potamotrygon boesemani*; *libaspari*) and a miniature freshwater pipefish (*Pseudophallus brasiliensis*). Common fish-eating birds include kingfishers (Alcedinidae), skimmers (Rynchopidae) and egrets (Ardeidae; *sabaku*), as well as the Large-billed Tern (*Phaetusa simplex*), Neotropical Cormorant (*Phalacrocorax brasilianus*), and Anhinga or snakebird (*Anhinga anhinga*) (the latter two known in Suriname as *doiklari*).

Clear-water rivers of the Interior. The rivers in the Interior show a relative lack of both color and suspended sediment (Table 4), and thus their transparency is high (Table 3; 'clear water' rivers of Sioli 1950). During intense rainstorms the rivers can have high sediment loads, but this is of short duration, as the sediments flow downriver once the rains stop. The water level of smaller rivers can

Figure 47. Mabuka Sula, rapids with flowering Podostemaceae (*kumarunyanyan*) herbs attached to the rocks (image © Pieter Teunissen, Oct. 2013).

Figure 48. Rapids and other fast-water habitats in Suriname:

a. The large Blanche Marie Falls of the upper Nickerie River (image © Jan Mol, Jan. 2006).

b. Submersed Podostemaceae (*kumarunyanyan*) in the strong current of a rapid in the upper Suriname River (image © Willem Kolvoort).

c. Aquatic insect larvae (Diptera) on the underside of a Podostemaceae leaf, upper Corantijn River (image © Jan Mol, Oct. 2007).

d. Juvenile *Pseudancistrus kwinti* in the Raleighvallen rapids, Coppename River (image © Willem Kolvoort).

rise quickly during rain storms, such as 6 m overnight on the Mindrineti River (Jan Mol pers. obs.). The river courses follow troughs that coincide with joints and faults in the bedrock. Rapids (*sula*) and straight stretches ending in irregular and angular bends (Zonneveld 1972) replace the meandering and substantial floodplains of lower reaches in the Coastal Plain. Between the rapids, the river is relatively 'flat', deep and sluggish; it resembles a staircase with mostly low risers (the rapids) and very long treads (the flat parts). In the flat parts, large aquatic plants and phytoplankton are sparse. Dissloved oxygen content is high, but conductivity is even lower than in creeks (Table 3). Primary production in the river itself is largely restricted to Podostemaceae (*kumarunyanyan*; Werkhoven & Peeters 1993) and diatoms (like *Cymbella* and *Navicula* spp.) in the rapids (see below). Plankton is most abundant in the dry season and consists of diatoms, desmids, Sarcodina, rotifers and micro-crustaceans (Leentvaar 1975, Crul & Reyrink 1980). Detritus-eating and migratory fishes (Flecker 1996, Carolsfeld *et al.* 2003), as well as caimans (Fittkau 1970) are important in the cycling of nutrients and their flow back and forth between ecosystems (Vanni 2002). Fishes not specifically associated with rapids include large pimelodid catfishes, such as *lalaw* and *Pseudoplatystoma* spp. (*spigrikati*), *Brycon falcatus* (*moroko*), *Prochilodus rubrotaeniatus* (*kwimata*) and *Semaprochilodus varii* (endemic to the Marowijne River), *anyumara*, *Cichla* spp. (*tukunari*) and freshwater stingrays (*Potamotrygon* spp.; *libaspari*). Shoals of small *Corydoras* catfishes (*seseiguse*; many species, most endemic) are conspicuous in shallow water.

Rapids. In the middle reaches of the rivers, between the lower freshwater reaches and the headwaters, rapids are a common feature and represent a distinct ecosystem (Figure 47). In these rapids, extensive 'meadows' of herbaceaous Podostemaceae (*kumarunyanyan*) cover bedrock and boulders that are submerged throughout the year with the exception of the long dry season, when *kumarunyanyan* flowers. Scattered xeromorphic shrubs (see below) such as *Psidium persoonii* (*sula-guyaba*) may also be present at the margins (De Granville 1994). The vegetation is severely stressed by flooding during most of the year, and when not flooded, it quickly becomes drought-stressed. The force of the water currents in the riverbed prevents soil formation and succession towards closed scrub or forest. Moving away from the river bed, into a zone with a rocky substrate that is never submerged, the ecosystem transitions to a kind of rock savanna or mountain savanna forest (see below).

From an aquatic perspective, rapids are high-energy systems with flow velocity > 0.5 m/s. They are characterized by *kumarunyanyan* (Figure 48.abc), aquatic invertebrates (Figure 48.c) and fastwater fishes (Figure 48.d) (Lujan & Conway 2015). Big waves and hissing waters in rapids suggest strong

Figure 49. Typical savanna landscape of the Zanderij Belt with Giant Anteater (*Myrmecophaga tridactyla*) and a black-water stream in the foreground (detail from panorama painting by Wim Eriks in Neotropical Butterfly Park, Lelydorp, Suriname).

currents, but can be misleading because slow currents in shallow water look as impressive as swift currents in deep water. Currents are up to 2.8 m/s in the rapids, compared to 0.7 m/s elsewhere in the river (Marrenga & Ruleman 2008). To maintain their position and avoid downstream displacement, organisms must avoid or counteract shear forces and turbulence which are both high in shallow water over an irregular bottom (Hoover & Ackerman 2004). Small, flat-bodied larvae of aquatic insects can live in a thin (5 mm) boundary layer above the substrate where current velocities approach zero (Statzner & Holm 1982). Larger organisms are too big to fit entirely within this boundary layer; they may instead find refuge from shear forces within, behind or beneath rocks and wood, or in *kumarunyanyan* vegetation. Predator avoidance may be an important motive to live in rapids (Schlosser 1987). Food – *kumarunyanyan*, algae, invertebrates – also tends to be more abundant in rapids than in slow- and deep-water habitats. Vegetarian serrasalmines, such as *Acnodon oligacanthus*, *Myloplus* spp. and *Tometes lebaili* (*pakusi* and *kumaru*), feed on *kumarunyanyan* leaves, while other fishes feed on invertebrates associated with these leaves (Horeau *et al.* 1998; Figure 48.c). Yet others, such as the electric eel and *anyumara*, feed on fishes. Many fishes have bodies that are spindle-shaped, such as *Parodon guyanensis*, flattened, such as *Lithoxus*, *Harttia* and *Pseudancistrus* spp, or eel-shaped (Trichomycteridae, *Heptapterus* spp.). They may also have organs to adhere to substrates, and thus prevent displacement by currents. Loricariid catfishes (Figure 48.d) have a broad sucking mouth that is directed downward to attach themselves to rock, while simultaneously taking in water for breathing and scraping algae from the substrate (Geerinckx *et al.* 2011). Snails of the family Ampullariidae and the genus *Aylacostoma* – often with an eroded tip – feed on algae they scrape from the rocks (Massemin *et al.* 2009).

Savannas, related xeric ecosystems, and black-water streams.

General characteristics. Savannas are ecosystems with savanna vegetation (see below; as in Figure 49). 'Rock savanna', 'savanna forest' and 'mountain savanna forest' (see below) are terms used in Suriname for related ecosystems. What they have in common is that the majority of the woody plants that are part of the vegetation are xerophytes, plants adapted to dry conditions. These woody xerophytes have leaves that are hard or leathery (xeromorphic), to allow the plants to survive recurring periods of severe water shortage. In savannas they are shrubs or treelets, but in forests a number of them are tall trees. They are a diverse group of species; some of the more widespread in Suriname are *Clusia* spp. (*sabanamangro*), *Bombax flaviflorum* (*sabanakatun*), *Licania incana* (*sabanafungu*), and *Humiria balsamifera* (*meri* or *blakaberi*). Exclusively associated with some of the savannas and related ecosystems are black-water streams, which are nutrient poor, acidic, and tannin-rich.

Savanna vegetation in Suriname is described by Heyligers (1962), Van Donselaar (1965, 1968, 1969), Van Donselaar-Ten Bokkel Huinink (1966), Kramer and Van Donselaar (1968), Teunissen and Wildschut (1970), Norde and Oldenburger (1975), Teunissen (1980a) and Jansma (1994) (see also remarks in Riezebos 1979). These authors define savanna – in approximate agreement with Beard (1953) – as vegetation dominated by terrestrial grasses (Poaceae), sedges (Cyperaceae) and other herbs that form an often dense low layer, with shrubs or trees either absent or occuring as scattered individuals or groups (shrub or forest islets), or as linear features (such as gallery forest along creeks;

Figure 50. Sipaliwini Savanna near Vier Gebroeders Mountains, Sipaliwini District (image © Bart De Dijn, May 2003).

On the tall hill (top part of image) some forest is left, but on lower hills (bottom part) repeated burning has resulted in orchard savanna with very few trees; in the valley in between (middle part) gallery forest remains, dominated by fire-resistant *morisi* palms.

Figure 51. Savanna ecosystems:

a. Wet shrub savanna on white sand with abundant *Bactris campestris* palms at Blakawatra, Para District (image © Pieter Teunissen, Nov. 2009).

b. Orchard savanna with gnarled *Curatella americana* (*sabanakasyu*) treelets along the dirt road from Witsanti to Hannover, Para District (image © Bart De Dijn, Aug. 2016).

c. Dry scrub savanna on white sand with abundant *Ternstroemia punctata* shrubs and scattered tufts of the sedge *Bulbostylis conifera* at Blakawatra, Para District (image © Pieter Teunissen, Jan. 2010).

d. Rock savanna with the cactus *Cereus hexagonus* and shrubs of the genus *Clusia* (at center of image) on granitic substrate at the foot of the Voltzberg inselberg (its rock dome is visible at the horizon), Sipaliwini District (image © Bart De Dijn, Jan. 2008).

see below). Rock savanna is described by Lindeman and Moolenaar (1959), Teunissen (1978c), Raghoenandan (2000), Oldenburger and Norde (2009) (see also unpublished reports: Jansen-Jacobs *et al.* 2002, Van Donselaar & Schulz, undated). Savanna forest is dealt with by Lindeman (1953), Lindeman and Moolenaar (1959), Heyligers 1962, and Teunissen 1980a; mountain savanna forest by Lindeman and Moolenaar (1959), Schulz (1960), Ter Steege *et al.* (2007), Banki (2010), Bordenave *et al.* (2011) and Ruysschaert (2017) (see also unpublished reports by De Dijn et al. 2005 and De Dijn 2012, and remarks in De Dijn *et al.* 2007).

Classification and distribution. Cohen and Van der Eijk (1953) recognize nine types of natural savanna landscapes that occur in Suriname, and Van Donselaar (1965) and Teunissen (1978g, 1980b) distinguish different types of savanna vegetation and ecosystems along the same lines. These authors, along with Lindeman and Moolenaar (1959), regard a range of forest vegetation types as successional stages that develop from a savanna stage. Many, but not all, of these successional stages are rich in woody xerophytes. The successional series depend on the type of substrate on which the vegetation occurs (type of soil, subsoil, underlying or exposed bedrock or duricrust), and its hydrology ('wet' or 'dry') (see sources mentioned above).

Just seven percent (700 km²) of the Zanderij Belt has savanna vegetation. Most of it occurs as small savannas on bleached 'white' sand (for example, near Zanderij and Jodensavanne), in a matrix of more extensive savanna forest and riparian forest. A handful of savannas in the Zanderij Belt occur on unbleached 'brown' sand, such as the Lobin Savanna, north of Zanderij, and the Coesewijne Savanna, in a matrix of rain forest and riparian forest on the same substrate. The Coastal Plain has a few small savannas and associated savanna forest patches on a variety of substrates, such as the Welgelegen Savanna, on clay soil. The northernmost part of the Interior has some savanna and savanna forest clusters, mainly on quartz-rich weathering material, such as the Sabanpasi and Rosebel savannas. A large savanna is located in the southernmost part of the Interior: the Sipaliwini Savanna (Figure 50), covering 650 km², and extending another 5,000 km² into Brazil, where it is known as the Paru Savanna. Scattered across the rest of the Interior are numerous small rock savannas, savanna forests, and mountain savanna forests. This includes the savannas and forests on sandstone at Tafelberg.

Wet shrub savanna. This is grassland with scattered xeromorphic shrubs (see set of species mentioned earlier) and/or shrub islets (Figure 51.a). Typical species are sedges of the genera *Lagenocarpus* and *Rhynchospora*, and herbs of the families Eriocaulaceae and Xyridaceae. This is considered true savanna, known as *campo sujo* in Brazil, or *campo limpo* if shrubs are scarce, which may be due to burning. In Suriname, they qualify as 'wet' since they occur on a variety of soils subject to prolonged seasonal flooding or waterlogging. Wet shrub savannas are widespread on the bleached sands of the Zanderij Belt ('Zanderij-type' savannas), and to a lesser extent on sandy ridges in the Old Coastal Plain ('Watamaleo-type'). Wet soil conditions in these sandy savannas are typically due to an impermeable hardpan near the soil surface. Wet shrub savannas are also found on subgreywacke rock, and soils derived from such rock that have an impermeable layer of loam ('wet Sabanpasi-type'; occurs in the Brinckheuvel nature reserve). They are even found at Tafelberg on poorly drained shallow soil over sandstone. During the rainy season the thin permeable soil layer (above the impermeable layer) is quickly saturated with rain water; during the dry season the water is quickly depleted. If the vegetation is not burnt, succession sets in:

xeromorphic shrubs spread and develop into shrub islets, which coalesce, and ultimately become wet savanna forest (see below). Fire seems essential to maintain wet shrub savanna.

Orchard savanna. This is open grassland with scattered xeromorphic treelets and/or forest islets (Figure 51.b). Typical herbaceous species are grasses such as *Trachypogon plumosus*, and sedges of the genus *Bulbostylis*; typical xeromorphic woody plants are gnarled treelets including *Curatella americana* (*sabanakasyu*) and treelets of the genus *Byrsonima* (*lontukasi*). This is also considered true savanna, known as *campo coberto* or *campo cerrado* in Brazil. It is found on a variety of soils that may only be waterlogged briefly after heavy rains. Drought stress is more prolonged than in wet shrub savanna. In the absence of fire, the ground shaded by xeromorphic treelets is colonized by secondary tree species with rain forest affinities, thus giving rise to secondary forest islands, which coalesce, and ultimately become rain forest. Fire thus seems essential to maintain orchard savannas. The frequency of fires matters, as is obvious from observations that intermediate burning frequency leads to a low treelet density. This is likely due to the following chain of causation: longer periods between fires lead to more flammable biomass accumulation, which burns more intensively or longer when set alight, thus causing high tree mortality (especially of young trees). Examples of orchard savannas in Suriname are clay savannas on low plateaus (*schollen*) in the Coastal Plain ('Welgelegen-type'), on brown sand in the Zanderij Belt ('Coesewijne-type'), on quartz-rich sandstone and conglomerate ('dry Sabanpasi-type') and (mostly) on lateritic soil in the south of Suriname (such as Sipaliwini Savanna; 'Paru-type').

Dry shrub savanna. This ecosystem type varies between two extremes: dense xerophytic shrubland with occasional near-barren grassy spots, and a near-barren grassy matrix with more or less isolated groups of shrubs (Figure 51.c). Typical herbaceous species are the grasses *Axonopus flabelliformis* and *Trachypogon plumosus*, as well as sedges such as *Bulbostylis conifera*. Typical woody plants are xeromorphic shrubs such as *Ternstroemia punctata* and *Matayba opaca* (*sabanagawetri*). It occurs on well-drained, deep bleached sands of the Zanderij Belt, and is referred to as 'Cassipora-type' savanna. It is not universally regarded as savanna, and in Brazil it is called (Amazonian) *campina* or *caatinga* (Anderson 1981). It may develop into a grassy *campo limpo* when burnt. There is at most occasional waterlogging stress, but drought stress is prolonged, as the water table is located very deep in the soil, beyond the reach of most plants. Plants cope by using water that remains after showers in soil pores and bound to humus. Dry shrub savanna ultimately becomes dry savanna forest (see below) in the absence of fire. To maintain dry shrub savanna in Suriname, fire is required.

Rock savanna. This is a scattering of individuals or mats of succulent plants on bare rock, as well as patches of grass, xeromorphic shrub and trees (Figure 51.d). In Brazil, it is known as *campo rupestre*. It is well known from granitic inselbergs in Suriname, where typical species are *Pitcairnia* spp. bromeliads, *Cyrtopodium andersonii* and *Catasetum planiceps* orchids, the cacti *Cereus hexagonus* and *Melocactus neryi*, and the agave *Furcrea foetida* (*ingisopo*). Sparse vegetation at sites with exposed duricrust or sandstone bedrock also qualifies as a type of rock savanna (see De Granville 1994). Shrubby, rocky areas along rivers, typically bordering extensive rapids, may also be considered rock savanna; here one may encounter the shrub *Calliandra surinamensis* (*officierskwast*; planted as an ornamental in Paramaribo). Flooding stress may occur very locally or occasionally, but severe drought stress is what governs most of the vegetation most of the time.

Figure 52. Forested ecosystems related to savanna:

a. Savanna forest on white sand dominated by *Dimorphandra conjugata* trees (*dakama*; with strongly fluted trunk and flaky bark) and the tall, terrestrial *Bromelia alta* (*singrasi*), near Witsanti, Para District (image © Bart De Dijn, Aug. 2016).

b. Low, stunted mountain savanna forest rich in Myrtaceae shrubs and treelets (*busiguyaba*), on lateritic soil with duricrust, on the slope of Brownsberg Mountain, near the Weti Creek, Sipaliwini District (image © Bart De Dijn, Mar. 2002).

To cope with this, plants access water remaining (briefly) after showers in scattered humus patches and rock depressions. Soil formation and succession to forest may occur locally, but is often set back when rainstorms soak and wash away soil and topple trees. Fire may affect rock savanna, but does not seem to be required to maintain it.

Savanna forest. This is thicket, woodland or (multi-layered) forest, dominated by woody xerophytes (Figure 52.a), including xerophytic trees, which are much rarer in rain forest. 'High' savanna forest has an upper layer of mature trees forming the canopy at 25-35 m; structurally, it resembles rain forest, but tree stem density is typically higher, and it lacks very tall emergent trees. 'Low' savanna forest, which is thicket to woodland, has a high density of mature trees typically no more than 5-20 m tall; it is structurally even less complex, typically having just a single mature tree layer. It typically consists mainly of thin-stemmed treelets – *tikitiki* forest, which is very dense – and may have gnarled trees, like the trees in an orchard savanna. There are indications – for example, charcoal in the soil, and observations of succession in progress – that in Suriname part of the savanna forest represents climax vegetation, while the other part represents early successional stages initiated by burning. The fire regime seems to have a great impact on vegetation composition: for example, forest dominated by *Dimorphandra conjugata* (*dakama*) is considered indicative of frequent burning in the not too distant past. Wet and dry savanna forest occur under soil and hydrological conditions roughly identical to those in wet and dry shrub savanna, respectively. In wet savanna forest, palms such as *Mauritia flexuosa* (*morisi*) and *Bactris campestris (keskesmaka)* are common; *Eperua falcata* (*walaba)* trees can locally dominate. Wet savanna forest along creeks at Tafelberg is very different: it is dominated by *Micrandra glabra*, a tall tree in the Euphorbiaceae family. In dry savanna forest these species tend to be absent. Along small creeks, the forest suffers less from seasonal drought and fire damage, and often remains as gallery forest, which is a strip of riparian forest in an otherwise non-forest setting. All trees except the *morisi* palm can be eliminated from this forest due to burning. Creek-side and gallery forests are typical of savanna landscapes and are often considered an integral part of them.

Mountain savanna forest. Mountain savanna forest refers to any type of ecosystem with a patchy or shallow soil over an otherwise rocky substrate or duricrust, covered by vegetation with a savanna forest physiognomy (Figure 52.b). It is often associated with rock savannas, which

Figure 53. Black-water streams in Suriname:

a. A black-water stream meets a clear-water stream (image © Willem Kolvoort).

b. Small school of *Hemigrammus bellottii* in a black-water tributary of Para River (image © Willem Kolvoort).

essentially occur where the soil is even shallower and patchier. In French Guiana, it has been studied extensively (see De Granville 1994). It is associated with a variety of rocky substrates, such as granite and sandstone bedrock, and ferralitic duricrust; it can be either dry or wet. Of all mountain savanna forests in Suriname, those associated with ferralitic duricrusts have been the best investigated. More so than in savanna forest, Myrtaceae trees are abundant or at least conspicuous in low-stature, species-poor mountain savanna forest, whereas Burseraceae trees are numerous in high-stature, more species-rich forest. Trees of both families also appear to be abundant in forest on shallow soil over granite (inselbergs) and sandstone (Tafelberg).

Terrestrial fauna. Large animals are not common in Suriname's savannas, except for the Giant Anteater (*Myrmecophaga tridactyla*), and, locally, the White-tailed Deer (*Odocoileus virginianus*; *sabanadia*; this species may currently be extinct in many savanna areas). The Savanna Fox (*Cerdocyon thous*) is very rare. Contrary to the strictly forest-inhabiting Yellow-footed Tortoise (*Chelonoidis denticulata*; *busisekrepatu*), the Red-footed Tortoise (*Chelonoidis carbonaria*; *sabanasekrepatu*) is a typical savanna species. A few additional examples of typical savanna species can be found in the species accounts of the present book. Savanna forest and mountain savanna forest ecosystems have a fauna similar to that of rain forest, but less diverse, especially in low forest. A spectacular species associated with mountain savanna forest is the Guianan Cock-of-the-rock (*Rupicola rupicola*; *rotshaan*); it is almost exclusively known from areas around the bases of granitic inselbergs and sandstone tepuis.

Black-water streams. Most creeks and small rivers that originate in the Zanderij Belt carry 'black water' (terminology of Sioli 1950) with a striking reddish-brown color (like Assam tea; appearing black in deep water), low pH, low conductivity, low turbidity (high transparency) and low amounts of dissolved oxygen (Figure 53.a; Table 3). Aptly named after the Rio Negro, black water originates in white sand savanna and savanna forest ecosystems where plant litter on the soil decomposes much more slowly and incompletely than in rain forest (Klinge 1967), forming soluble humic acids. Black-water creeks are also known from the Tafelberg sandstone mountain. The black-water Para River in Suriname is known for its many water plants, including *Cabomba aquatica* and *Marsilea polycarpa* (Doerga 2013). In Amazonia, some black-water rivers are considered 'hunger rivers' because they have few large edible fishes; however, black-water streams are also known for having many small-sized colorful fishes that are popular with aquarium hobbyists (Goulding *et al.* 1988) (Figure 53.b). In Suriname, these small fishes include pencilfishes (*Nannostomus* spp.), hatchetfishes (Gasteropelecidae), Head-and-taillight Tetra (*Hemigrammus ocellifer*), Splashing Tetra (*Copella arnoldi*), the killifish *Laimosemion agilae*, freshwater angelfish (*Pterophyllum scalare*), dwarf cichlids (*Nannacara anomala*, *Apistogramma*; *pikinkrobia*), *Acanthodoras cataphractus* (*merkikwikwi*), small knifefishes (*Hypopygus lepturus*, *Gymnotus coropinae*) and, in the Coropina Creek, the introduced Cardinal Tetra, *Paracheirodon axelrodi*, native to the Rio Negro. Cuvier's Dwarf Caiman *Paleosuchus palpebrosus* (*redikaiman*) is restricted to black-water streams (Ouboter 1996).

Upland ecosystems

In the Andes, the climate in the high mountains ultimately becomes too cold for trees to grow (von Humboldt 1814-1825; Lujan *et al.* 2013). The scattered tall hills and plateaus in the Interior of

Suriname – referred to locally as mountains – are far too low (no more than 1250 m elevation) for this to occur, but the climate and ecosystems do change as you climb up these hills. While the soil conditions and vegetation physiognomy may remain quite similar to those in the lowlands, the vegetation becomes more often shrouded in clouds, the atmospheric humidity increases, and the ambient temperature lowers; there are also greater changes in temperature and humidy in the course of 24 hours (see De Dijn *et al.* 2007 for Brownsberg). This transition to a cooler and more humid upland environment (known as premontane or submontane) seems to occur at around 400-600 m elevation in Suriname.

Upland terrestrial vegetation. Upland vegetation has been studied in Suriname by Maguire (1945), Maguire *et al.* (1948), Maguire (1970), Molgo and De Dijn (2007), Ter Steege *et al.* (2007), Banki (2010), Bordenave *et al.* (2011), and Banki and Bhikhi (2013) (see also unpublished report by De Dijn *et al.* 2005, and remarks in De Dijn *et al.* 2007). An obvious change in the vegetation as

Figure 54. Upland ecosystems:

a. Moss-covered tree stems in upland mountain savanna forest, on the high duricrust plateau of Lely Mountain, Sipaliwini District (image © Bart De Dijn, Jun. 2003).

b. Upland riparian forest dominated by *Micrandra glabra*, on the sandstone plateau of Tafelberg Mountain, Sipaliwini District (image © Bart De Dijn, Sep. 2002).

c. A rocky upland creek cascading down the duricrust plateau of Nassau Mountain, with the cyclanth herb *Dicranopygium pygmaeum* (resembling palm seedling) very abundant on the rocks, Sipaliwini District (image © Bart De Dijn, Mar. 2010).

one moves into the uplands is the increase in the bulk of epiphytic mosses, ferns, orchids and bromeliads, which is most dramatic on the windward side of mountains (Figure 54.a). This should not be confused with changes in vegetation physiognomy and composition due to changes in the substrate (soil, drainage, etc.) that happen to coincide with elevational changes in climate. Where the bedrock or duricrust is fully exposed in an upland setting, the vegetation looks similar to rock savanna in a lowland setting. Where it is covered with a thin or patchy layer of soil, it resembles lowland mountain savanna forest. Where the soil is deeper but still stressed due to wet/dry cycles, it may resemble anything from lowland savanna forest to savanna. Where the drainage is impeded, it resembles lowland swamp or marsh vegetation, and along creeks, it resembles lowland riparian forest.

Despite these similarities in the vegetation, the differences in climate, hydrology and amount of epiphytes between lowland and upland ecosystems are real, and are reflected in unique upland fauna and flora elements. The upland ecosystems of Suriname are like islands in a sea of lowland ecosystems. One of the largest, and arguably the most idiosyncratic, upland areas of Suriname is the Tafelberg (Figure 54.b). Guianan highland plant species – such as the grass *Thrasyas axillaris*, the climbing herb *Stenospermation maguirei*, the shrub *Myrcia tafelbergica*, the treelet *Clusia tabulamontana*, and the tree *Micrandra glabra* – occur here that are otherwise only found 350 or more kilometers to the west in Guyana or Venezuela (plants listed in Mittermeijer *et al.* 1990, whose distribution was double checked by the first author of the present chapter using the **www.tropicos.org** database). Suriname's tall lateritic hills with duricrust plateaus, such as the Lely, Nassau and Brownsberg mountains, have many plant species that are restricted to high elevations in the Guianas, for example the shrub *Cavendishia callista* and other Ericaceae. De Granville (1991) lists 31 plant species, most of them ferns, restricted to bauxitic and lateritic plateaus in the Guianas, including the cyclanth *Dicranopygium pygmaeum* which grows in abundance on rocks in creek beds (Figure 54.c). In addition, these lateritic hills are exceptionally rich in orchid and tree species.

Upland streams and headwaters. Upland streams in Suriname differ from lowland streams mainly in their higher velocities, more extensively rocky bottoms, lower temperatures, and the presence of small waterfalls (Figure 54.c). Stream size and fish diversity decrease with altitude. Upland fishes include *platkop kwikwi*, swamp eel, killifishes (Rivulidae), and small loricariid and trichomycterid catfishes (Mol *et al.* 2007b, Brosse *et al.* 2013). The Nassau and Brownsberg mountains each have a unique restricted-range endemic (RRE) *Guyanancistrus* catfish (Fisch-Muller *et al.* in prep.) and at least one undescribed RRE trichomycterid catfish. Nassau Mountain is the only place where the RRE catfish *Harttiella crassicauda* has been found (Mol *et al.* 2007b, Covain *et al.* 2012), alongside two RRE frogs, *Atelopus hoogmoedi nassaui* and *Ameerega trivittata nassaui* (Ouboter & Jairam 2012). Headwaters of Suriname's major rivers have several RRE fishes that are endemic to these headwaters, such as *Kryptolebias sepia* and *Hyphessobrycon georgettae*, or with a distribution extending over the border to the adjoining headwaters of northern tributaries of the Lower Amazon, such as *Parotocinclus halbothi* and *Stenolicmus ix* (see species accounts in Freshwater Fishes chapter).

Plants

Bruce Hoffman, Sofie Ruysschaert, Sabitrie Jairam-Doerga

Whether viewed in satellite imagery, from a small plane or while canoeing down a local river, a green mosaic of plant life is the most conspicuous feature of Suriname's ancient, weathered landscape. Indeed, Suriname is literally the greenest nation on Earth, with approximately 93 percent of the land cover classified as tropical forest and the largest per-capita forest area at 28 ha (SBB 2015). Furthermore, more than 5,100 vascular plant species have been documented within the country's borders (Funk *et al.* 2007, Feuillet 2009, Mota de Oliviera 2012, BDGS 2016) and up to 170 tree species in one forest hectare (Bankí 2010).

Such numbers highlight the abundance and diversity of plants in Suriname, yet fail to convey the dynamic, on-the-ground reality and interactions of these plants in nature. Details on the ecology and natural history of Suriname's flora may be published in scientific journals or embedded within forest people's traditional knowledge, yet little information is available to the general public. In this chapter, we shed light on this topic with a broad introduction to plants followed by 50 in-depth accounts of plant species occurring in Suriname's ecosystems, in alphabetical order by plant family. Most italicized plant names in this introduction refer to plants presented in the species accounts.

From Suriname's coastal mangroves to the vast, closed canopy forests of the Interior, plants support the development and maintenance of complex food webs as 'primary producers' or 'autotrophs'. Autotrophs are organisms that produce complex, energy-rich molecules such as carbohydrates from simple energy sources in the environment. At the core of primary production by plants is the process of photosynthesis, in which sunlight, water and atmospheric CO2 is converted into sugars and oxygen. By contrast, 'heterotrophs', such as animals, fungi, and bacteria, are organisms that cannot produce their own food. In most cases, their very existence depends upon the direct or indirect consumption of the organic compounds produced by plant life (Forget and Hammond 2005, Begon *et al.* 2006, Evert *et al.* 2013).

Plant life in Suriname also plays an invaluable role in hydrological cycle regulation, nutrient cycling (including carbon sequestration), soil formation, ecological succession, providing animal shelter, and as human food, fiber and medicinal resources, to name a few (Schulz 1960, Hammond 2005, Schulze *et al.* 2005, van Andel and Ruysschaert 2011, Hoffman 2012).

Plant groups and forms

Evolutionary lineages. A lineage is an evolutionary concept; it is a grouping of all species descended from a closest common ancestor (see also Box 4 in General Introduction / History chapter). The taxonomy or subdivision of plants in groups has been increasingly aligned with this concept, meaning that nowadays most groups such as the different plant families contain only species of common descent, and none that are more distantly related and part of another lineage altogether. The related groups or units recognized, named and ranked by taxonomists, such as family, genus and species, are known as taxa (singular: taxon). Some familiarity with major plant lineages in Suriname is highly relevant here, because the potential ecological roles of plants are strongly influenced by their evolutionary heritage (Mori *et al.* 1997, 2002; Coelho de Souza *et al.* 2016, Judd *et al.* 2016).

The most ancient plant lineages include aquatic green algae and bryophytes, which are terrestrial mosses and other moss-like plants. Both these lineages lack a vascular system – transport tubes analogous to veins and arteries in animals – and reproduce via water and spores rather than seeds. Species of another major lineage, the ferns, possess a vascular system, but still reproduce via spores. Collectively, the lineages mentioned above are referred to as the 'lower' plants.

In contrast, gymnosperms and angiosperms are vascular plant lineages referred to as the 'higher' plants. Ancestral gymnosperms were the first plants to produce seeds, which are embryos within a protective seed coat. In Suriname, examples of contemporary gymnosperms include introduced pine trees (*Pinus caribaea)* and native lianas of the genus *Gnetum*. Ancestral angiosperms, better known as flowering plants, emerged and rapidly diversified between 130 and 90 million years ago, in the Early Cretaceous Era.Their eventual dominance on Earth was largely due to the evolution of a sophisticated breeding system (see below) involving biochemical defense, pollination and seed dispersal (Herrera and Pelimyr 2009, Bongers *et al.* 2013). The success of angiosperms is evident in Suriname, where they comprise more than 93 percent of documented vascular plant species (Funk *et al.* 2007, BDGS 2016). The higher plants, in particular the angiosperms, are the main focus of the species accounts presented later in this chapter.

Angiosperms are traditionally divided into two main lineages: monocotyledons (monocots) and dicotyledons (dicots) (Judd *et al.* 2016). As seedlings, monocot species have one seed leaf (cotyledon), while dicot species possess two seed leaves. Past the seedling stage, the leaves of dicots typically mature with oval to oblong blades and feather-like (pinnate) venation (Figure 55). Examples include *Ocotea guianensis* (Lauraceae family) and *Piper marginatum* (Piperaceae family). By contrast, the leaves of most monocots are oblong to narrow and pointed with several semi-parallel main veins. This pattern is typical in well-known groups such as the grasses and bamboos (Poaceae family), bromeliads (Bromeliaceae family) and orchids (Orchidaceae family). Among both monocots and dicots, many species have evolved with atypical leaf vein patterns. *Montrichardia arborescens* (Araceae family), for example, is a monocot species with broad dicot-like leaves and pinnate venation. The leaves of a dicot species, *Clusia grandiflora* (Clusiaceae family), are fleshy and have straight, parallel and barely visible secondary veins.

Dicot species include many growth forms and textures, ranging from small, delicate or fleshy herbs to densely woody lianas, shrubs and giant trees. Tropical forest canopies are formed primarily of dicot tree species. By constrast, monocots are mostly herbaceous or fleshy and, with the exception of some tree-like palms, do not attain great heights. The trunks of some monocot palm species, such as *Astrocaryum sciophilum* (*bugrumaka*), resemble the hard-woody texture of dicot tree trunks. The two lineages also differ in characteristics of their flowers (Figures 56 and 59). Dicot species typically produce flowers with four to five petals or petal lobes, or multiples thereof. Most monocot species produce tubular flowers with three to six petal lobes, or multiples thereof. Typical and atypical examples of flowers for both groups are included in the species accounts.

Beyond the distinction of monocots and dicots, recognition of taxa becomes much more challenging due to the great morphological diversity of angiosperm plant families, genera and species – especially in the Neotropics (Milliken *et al.* 2009, Mota de Oliveira 2012). However, by reading the species accounts and directly observing plants in Suriname's ecosystems, a botanical knowledge base of common taxa and identification skills can be rapidly built up.

The three most species-rich and abundant dicot tree families in Suriname are Fabaceae, Chrysobalanaceae and Lecythidaceae (Funk *et al.*, 2007, ter Steege *et al.* 2016). Dicot families with the largest number of woody climbers (lianas) in Suriname include Apocynaceae, Bignoniaceae and Malpighiaceae. Examples of common Suriname palms (Arecaceae or Palmae family) include *Attalea maripa* (*maripa*), *Euterpe oleracea* (*pina, podosiri*) and *Mauritia flexuosa* (*morisi*). Smaller-stature shrub and herb species are common in dicot families such as Asteraceae, Melastomataceae and Rubiaceae, and in monocot families, such as Araceae, Heliconiaceae and Orchidaceae.

Growth / life forms. Perhaps the most conspicuous aspect of plant species is their occurrence as particular 'growth forms' or 'life forms', such as tree, shrub, herb, vine and liana species (Evert *et al.* 2013) (Box 7). Growth forms are highly significant in the natural history of plant species and populations, but are not reliable indicators of evolutionary relatedness. The coffee family (Rubiaceae), for example, includes a bewildering variety of growth forms, such as trees, shrubs, lianas, erect herbs, and tiny, flat herbs. And visitors hailing from non-tropical regions of the world might be surprised to learn that in tropical Suriname, the violet family (Violaceae) includes small trees, such as *Rinorea pubiflora* (*manaritiki*), as well as lianas.

Many unrelated plants in extreme ecosystems, such as deserts, have evolved with a similar 'look' due to similar physiological and reproductive challenges. This phenomenon is known as convergent evolution (Judd *et al.* 2016). A Suriname example is the shrubby, low-stature, thick-leaved appearance of many woody plant species growing on infertile white-sand soils (Lindeman and Molenaar 1959, Schulz 1960) (see also Ecosystems chapter).

Growth forms may even change during the life cycle of a given species. This can be observed in some lianas in Suriname, such as *Strychnos melinonii* (*dobrudua*). Such plants initially grow as erect shrubs, and develop tendrils or climbing hooks – typical attributes of many lianas – only when they encounter a tree trunk or other suitable climbing surface.

Plant structure and life cycle

Development and structure of plants. A generalized flowering plant life cycle is presented in Figure 57, including seed germination, growth and development, maintenance, and sexual and asexual reproduction. Basic morphological terms are also introduced here to assist the reader in understanding plant structures and growth and interpreting the information provided in species accounts.

Box 7. Plant Growth Forms (based on Evert et al. 2013)

Herbs. These typically include low-stature, short-lived, herbaceous plants. Some herbs are succulent (stems and leaves thickened and fleshy) or subwoody (stems hardened, yet green and without woody bark). Some robust herbs are not short-lived. Dicot herbs are common in open, disturbed areas or savannas in Suriname, with some adapted to aquatic habitats – such as *Mourera fluviatilis* (*kumarunyanyan*). Monocot herbs are common in the forest understory, including *Costus scaber* (*redisangrafu*) and *Hymenocallis tubiflora* (spider lily). *Phenakospermum guyannense* (*bigipalulu*) is a giant monocot herb that can attain 12 meters in height.

Rosette plants and cacti. Rosette plants possess round or pointed leaves clustered around a central inflorescence stalk. They are mostly monocots, and may occur on boulders, forest soils or tree branches. Species account examples include *Furcraea foetida* (*ingisopo, Agavaceae*) and *Vriesea splendens* (Bromeliaceae). Cacti (singular: cactus) are herbaceous to semi-woody dicots in the Cactaceae family. They typically possess green, fleshy stems and conspicuous spines. *Cereus hexagonus* is a columnar cactus species commonly observed in old Suriname plantations.

Grasses and sedges. These are herbaceous monocots with (usually) narrow, pointed leaves, jointed stems, spreading roots and rhizomes. Their tiny, wind-pollinated flowers are borne in erect, narrow inflorescences (spikes or spikelets) or as loose clusters. They occur mostly in open, disturbed areas in the Coastal Plain, herbaceous swamps,creeks and savannas.

Shrubs. These are relatively slender, woody dicots that often branch profusely near the base. To fit the definition of a shrub, the diameter of individual basal stems on a woody plant should be no greater than ten times the diameter of the uppermost branches (twigs). Shrubs are typical of savannas and other open areas, but are also relatively common in the forest understory. *Tibouchina aspera* (*wasduku*) is a shrub with showy blue-purple flowers commonly occurring in the white-sand savannas of Suriname.

Trees. These are woody dicots, usually with a single main trunk. Trees are the dominant structural elements of forests and mangroves. In tropical forests, the spatial distribution of tree crowns is typically not uniform, due to the occurrence of many different tree species with diverse heights and branching widths. Some ecologists recognize 1-3 canopy 'layers' at different vertical heights in tropical forests (Schulz 1960; see also Ecosystems chapter). A few examples of tree species include *Rinorea pubiflora* (*manaritiki*) – a treelet, *Carapa procera* (*redikrapa*) – a medium-height tree, and *Terminalia amazonia* (*swampu bus'amandra*) – a tall emergent tree.

Palms. These include single or multiple-stemmed, tree-like monocots of 1-20 meters height. In the Neotropics, they also include one genus of climbers, *Desmoncus* (*bambamaka*). There are numerous iconic, easily recognized palm species in Suriname, including *Euterpe oleracea* (*pina*) in marsh forest, *Mauritia flexuosa* (*morisi*) in savanna swamps, and *Astrocaryum sciophilum* (*bugrumaka*) in non-flooded, old-growth forests. Palms do not possess true wood, but palm trunks often have a texture that feels as hard as wood.

Vines and lianas. These are climbers rooted in the ground that require other plants for support. Vines are slender, non-woody climbers especially common in disturbed and open areas. Lianas are woody climbers common in old-growth and secondary forest. A climbing growth form has evolved in many unrelated plant families (van Donselaar 1970, Hoffman and Ruysschaert 2017).

Epiphytes. These plants colonize other plants, usually trees, and are not rooted in soil. Most epiphytic species are monocots in the orchid or bromeliad families. The higher levels of the tropical forest canopy can be a harsh, dessicating environment, due to a lack of soil holding nutrients, irregular water supply, and high exposure to sunlight. To increase the odds of survival in this environment, many epiphytic species possess unique adaptive traits. Two of the most important of these are specialized stems and roots for water and nutrient storage, and an alternative, water-conserving type of photosynthesis known as crassulacean acid metabolism.

Hemi-epiphytes. These are plants living in tree canopies like epiphytes, but with aerial roots connected to the forest floor and transporting water and nutrients upwards. 'Primary' hemi-epiphytes, such as *Ficus nymphaeifolia* (*abrasa*), germinate from bird- or bat-dispersed seeds in the canopy and send aerial roots down. 'Secondary' hemi-epiphytes, such as *Heteropsis (kamina)* and *Philodendron* spp., germinate in the soil, seek out and climb nearby host trees and eventually also extend aerial roots down to the forest floor (Hoffman 2015).

Parasites and saprophytes. Parasitic plants use modified roots (haustoria) to access the vascular system of other plants and obtain a portion of their nutrients and water. Some taxa, such as *Phorodendron* spp. (*fowrudoti*), have a dual strategy – they absorb food through parasitism and produce food with green, photosynthetic stems and leaves. Fully host-dependent parasites, such as *Helosis cayannensis* (*sneki-ede*), have non-photosynthetic (non-green) stems and leaves. Saprophytes are flowering plants that obtain their nutrients from dead or decaying organic matter. An easily overlooked yet common saprophyte in Suriname is *Voyria aphylla,* which appears as a tiny, single yellow flower on the forest floor.

When a seed germinates, food stored within the cotyledon(s) (seed leaves) – a key structure in the plant embryo – is activated and supports the transition into a seedling (Raghavan 2012, Evert *et al.* 2013). In seedlings and adult plants, primary growth is produced at the growing tips of roots and shoots (terminal meristems), where actively dividing cells give rise to a variety of tissues and structures.

The shoot apical meristem at the tip of shoots produces stems, leaves, axillary buds, and flowers. The above ground plant body is made up of repeating modular units, including nodes where leaves are attached to the stem, and, in between, internodes where the stem lengthens (Figure 55) (Harris and Harris 1994, Raghavan 2012). Vascular tissues throughout the plant body transport both nutrients and water absorbed from the soil, and sugars produced via photosynthesis by the leaves. In woody plants, wood and bark (secondary growth) is produced horizontally (radially) within stems through the development and division of specialized cells (cambium).

The root apical meristem produces roots. Many plants produce modified underground stems that colonize new areas or store nutrients. Examples include rhizomes [a horizontal stem that produces roots and shoots], bulbs [a short swollen stem surrounded by fleshy scale leaves or leaf bases] and tubers [a thick subterranean stem, such as the potato].

Leaves (Figure 55) are composed of two major parts: the leaf stalk (petiole) and the leaf blade. Leaf characteristics, such as size, shape and texture, are highly variable in different plant species and can change across the developmental stages of a single species. Grasses and many other monocots lack a true petiole. Instead, the lower part of the leaf forms a circular sheath that encloses the stem. Examples include *Costus scaber* (*redisangrafu*) and *Cyperus giganteus* (*piripiri*).

The type of leaf attachment at stem nodes is a critical vegetative characteristic for the recognition of major lineages and species. Leaf attachment may be 'alternate', with one leaf per node (Figure 55),

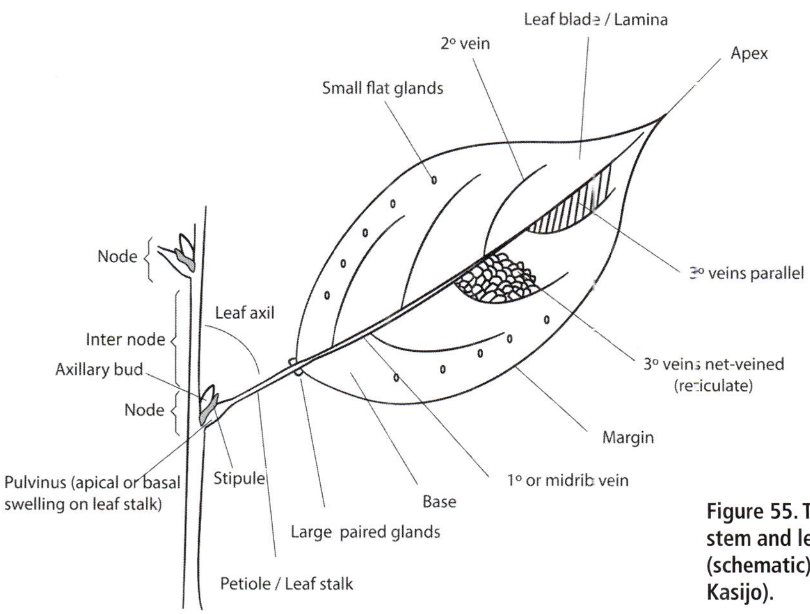

Figure 55. Typical dicot-type stem and leaf structure (schematic) (image by Omar Kasijo).

'opposite', with two leaves per node, or 'whorled', with three or more leaves per node. Leaf division and shape are also important characteristics. Most plants produce 'simple' leaves – with a single undivided blade and diverse shapes in outline. Other plants produce 'compound' leaves, with multiple leaflet subunits. Compound leaves are classified as 'pinnately compound', with the leaflets spread out along an axis, or 'palmately compound', with the leaflets radiating from a central point like fingers on a human hand. Some leaves are 'multiple compound' with leaflets divided along second or even third order axes. A compound leaf can usually be confirmed by the presence of an axillary bud or stipule at the petiole base and the absence of such structures at the leaflet bases.

The vein pattern (venation) within a leaf or leaflet blade is another important leaf characteristic. Venation can be 'pinnate' – with a conspicuous, central, primary vein and numerous smaller secondary veins branching off to both sides; 'palmate' – with veins radiating from a central point; or 'parallel' – with several semi-parallel main veins and no obvious secondary side veins.

Stipules (Figure 55) are narrow and pointed to broad, leafy tissues, attached at stem nodes. They occur in three distinct forms. The first form is the 'intrapetiolar' stipule, occurring within the leaf axil – where the leaf stalk (petiole) joins the stem. The second form, the 'interpetiolar' stipule, extends across the stem node between the petiole bases of opposite-leaved plants. A third form of stipule is the 'ochrea', which forms a cylinder encircling the node. In many plant species, stipules are absent or tiny and inconspicuous. The form, size and shape of stipules are often important characteristics for identifying plants lacking flowers or fruits. Many species in the legume family (Fabaceae or Leguminosae) have unique compound leaves, with 'stipels' (small stipule-like structures) at the base of the leaflets.

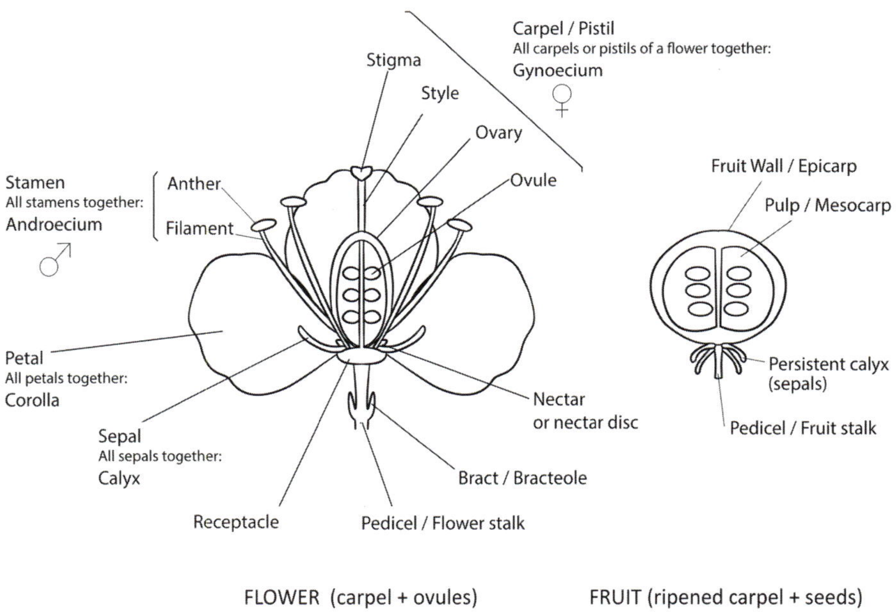

Figure 56. Flower and fruit structure (schematic cross-cut of angiosperm flower and fruit with reduced number of stamens and carpels) (image by Omar Kasijo)

Asexual reproduction. Many plant species, especially monocots, increase their population or the area they colonize through non-sexual (asexual) reproduction (Figure 57). A common colonization method is the clonal growth of rhizomes or sucker shoots, as with *Phenakospermum guyannense* (*bigipalulu*). The rhizomes advance just below ground, forming an extended network of genetically identical plants. Other methods include reproduction from underground bulbs or tubers, as with *Hymenocallis tubiflora* (*spider lily*), or the production of asexual plantlets (propagules) on above-ground plant parts, as with *Furcraea foetida* (*ingisopo*). Asexual reproduction has the advantage of very quickly reducing colonization space for competitors and being less energetically costly than sexual reproduction. A major disadvantage is the greater vulnerability to environmental stress, diseases and pests due to diminished genetic diversity (Schulze *et al.* 2005).

Sexual reproduction. The evolution of flowers as integrated reproductive structures led to the rapid diversification of the angiosperms from the Early Cretaceous onward. Understanding reproductive biology and learning to identify plants requires familiarity with basic flower morphology (Figure 56) (Evert *et al.* 2013, Judd *et al.* 2016). Bisexual flowers – with both male and female

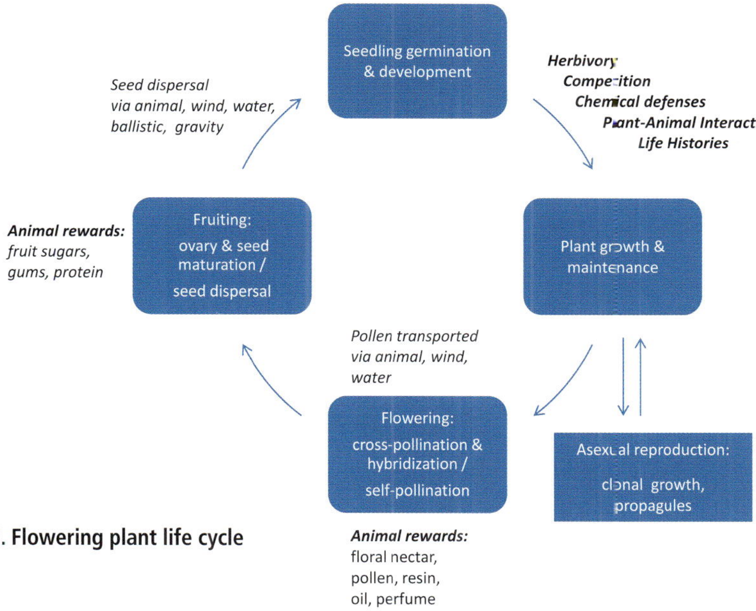

Figure 57. Flowering plant life cycle

parts, which is the norm – are typically composed of four whorls or organs attached to a shared flower base (the basal receptacle). These comprise the sepals which collectively form the calyx, the petals which collectively form the corolla, the stamens (androecium), which are the male parts that produce pollen, and the pistils or carpels (gynoecium), which are the female parts with ovaries that contain the eggs. In unisexual flowers, either the male or female organs are absent or reduced and non-functional. The position of the ovary in a flower (or fruit) is an important feature in plant classification. When it is on top of the receptacle, the ovary is termed 'superior'; when below, which is very common in monocots, it is termed 'inferior'. There are three basic forms of flower symmetry based on the relative position and size of the petals: radial (actinomorphic), bilateral (zygomorphic) and irregular (Figure 58). Some plant families have highly modified flowers that require unique descriptive terms, such as the Fabaceae (Papilionoideae) (Figure 59).

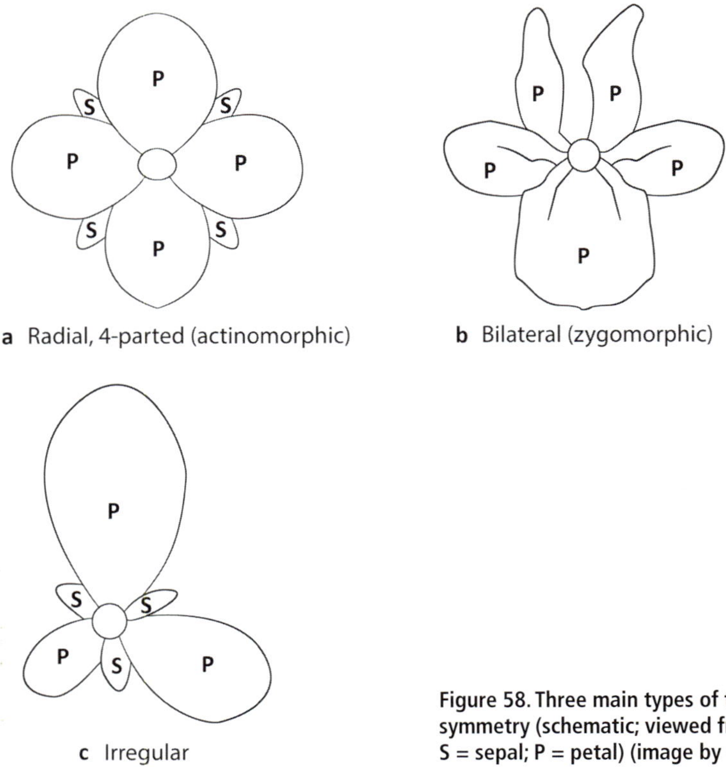

Figure 58. Three main types of flowers symmetry (schematic; viewed from top; S = sepal; P = petal) (image by Omar Kasijo)

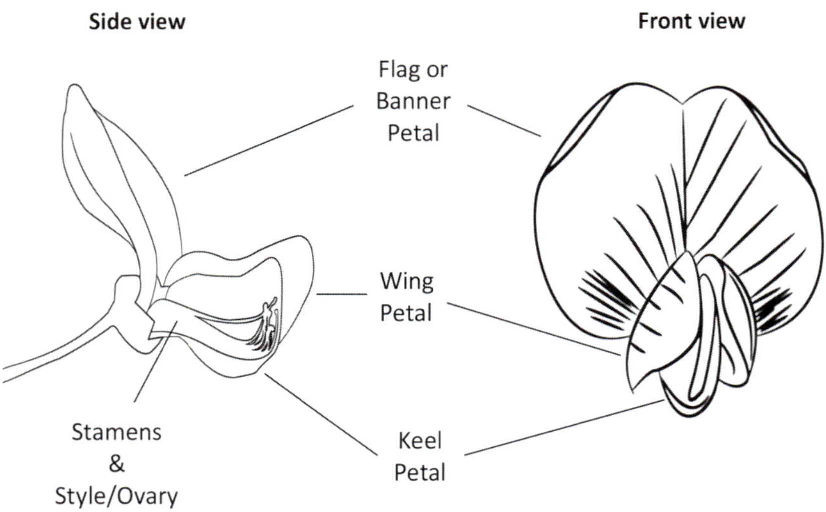

Figure 59. Legume flower (schematic; viewed from side and front) (image by Omar Kasijo)

Pollination. Pollination is the transfer of pollen from the (male) anthers to the stigma (female), the receptive part of a pistil or carpel (Figure 56; Evert *et al.* 2013). Mechanisms have evolved that engage wind, water or animals as agents of pollen transfer (see below) (Faegri and Van Der Pijl 2016). These mechanisms serve to promote cross-pollination, the transfer of pollen between the flowers of different individuals, thus achieving fertilization and genetic hybridization (sexual reproduction). This flow and mixing of genes leads to greater seed and seedling diversity, so cross-pollination is key to the adaptive capacity and survival of many plant species. In self-pollination, the pollen transfer occurs within a single flower or between several flowers on the same plant; note that individual plants that are the result of asexual reproduction are genetically identical, so are effectively the same plant for the purposes of pollination.

Abiotic pollination (without an animal agent) involves the transfer of pollen via wind, or by water for aquatic plants. Massive quantities of pollen must be produced, because fertilization depends upon the random drift of pollen, rather than the more targeted pollen delivery of biotic pollination. Wind-pollinated angiosperms often have small, inconspicuous flowers with no color, odor or reward to attract animal pollinators. This is an uncommon pollination strategy in the tropics.

Biotic pollination involves the participation of animals – insects, mammals, birds and bats, to name the most important ones – in moving pollen from anthers to stigmas; it is very common among angiosperms (Bawa *et al.* 1985, Faegri and Van Der Pijl 2016; see also review by De Dijn 2005 in regarding our region). Pollination 'syndromes' based upon combined flower traits – mainly color, shape, size, seasonality – allow for predictions of the type of pollination and, in the case of biotic pollination, the group(s) of animals involved. Bat-pollinated plants, for example, typically have large, white and strongly scented flowers with abundant nectar, such as those of *Parkia pendula* (*kwatakama*). Bird-pollinated flowers are commonly tubular, in the pink to red color spectrum and without a strong scent, such as those of *Costus scaber* (*sangrafu*). Bee-pollinated flowers vary, but are often yellow or blue and scented. In the species accounts of this chapter, several cases of beetle-pollination are highlighted, including *Montrichardia arborescens* (*mokomoko*), *Nymphaea ampla* (*pankukuwiwiri*) and *Xylopia aromatica* (*pedreku*). An example of an extremely specialized pollination mechanism occurs in the *Ficus* (*abrasa*) genus: as a rule, each species of *Ficus* has its own species of agaonid wasp to effect pollination (Heer *et al.* 2015). Overall, most plant-animal pollination relationships in our region are at most group-specific; with only a very few being species-specific. Flower visitors termed 'robbers' subvert normal pollinator-plant relationships. They do not pollinate, but instead take floral resources 'intended' for other visitors that do pollinate. (De Dijn 2005). A closer examination of the behavior of the different flower visitors is typically required to know which ones are potential pollinators, and experiments during which certain visitors are excluded (or allowed exclusive access) are often required to confirm species as actual ('effective') pollinators (Kearns and Inouye 1993).

The importance of fruits, and fruit and seed dispersal. An important factor in the success of flowering plants is that the ovaries develop into seed and fruit walls after pollination, thus protecting plant embryos from predators, desiccation and other environmental threats (Figure 56). Furthermore, fruits and seeds are essential for dispersal – the movement away from the mother plant into potentially suitable environments. Close seedling-to-mother proximity is unfavorable for plants due to competition with the mother and adjoining seedlings for space, nutrients and light (Levin *et al.* 2003, Schulze *et al.* 2005). Close proximity also increases vulnerability to seed predators, parasites and browsers, attracted to the many fruits, seeds and seedlings of the mother plant (Forget and Cuipers 2008, Bongers *et al.* 2013). The fruit and seed distribution patterns that result from dispersal events are significant not only for the plant populations involved, but also because they determine the structure and species composition of plant communities for years to come (Gentry 1982, Redford 1992, Levin *et al.* 2003, Forget and Hammond 2005, Stropp *et al.* 2011). For this reason, over-hunting of important seed-dispersing mammals and birds can have unforseen side-effects upon seed dispersal patterns and plant communities (Redford 1992).

Plant species have successfully evolved with both active and passive mechanisms of seed dispersal (see next section). As with pollination syndromes, dispersal syndromes based upon combined fruit traits – mainly texture, size, color, shape, seasonal availability – allows for the prediction of dispersal type and, in the case of biotic pollination, the group(s) of animals involved (Forget and Hammond 2005, Horn *et al.* 2011, Bongers *et al.* 2013).

Animal dispersal. This is the most common dispersal type among tropical plants by far, with diverse examples in the species accounts. *Pouteria guianensis* (*rode jamboka*) trees and many *Strychnos* spp. (*dobrudua*) lianas produce large, round, pulpy, orange fruits. This type of fruit is especially attractive to large monkeys, which swallow and defecate the large seeds whole. *Carapa spp.* (*krapa*) and *Vouacapoua americana* (*bruinhart*) trees produce dry capsules with large seeds. When the fruits fall to the forest floor, agoutis and acouchies quickly gather and bury the seeds. Many plant species produce seeds with a fleshy, energy-rich appendage known as an 'aril', which is attractive to many birds and smaller monkeys. An alternative seed dispersal strategy in the genus *Pisonia* is the production of fruits with tiny hooks or adhesives that stick to animal bodies. Even ant colonies may be seed dispersers for plant species that produce very small seeds.

Wind dispersal. Many fast-growing secondary forest trees, herbs and canopy lianas occurring in open areas are adapted for wind dispersal with winged seeds. Examples include *Stigmaphyllon sinuatum* (*konkonikasaba*) and *Triplaris weigeltiana* (*mira-udu*). Some plants produce tiny, lightweight seeds that are dispersed on wind currents, for example *Mourera fluviatilis* (*kumarunyanyan*), and orchids such as *Catasetum discolor*.

Water dispersal. Seed dispersal by water is the only option for many aquatic plants, including *Avicennia germinans* (*parwa*) and *Nymphaea ampla* (*pankukuwiwiri*). The seeds of trees in seasonally flooded forests, such as *Carapa spp.* (*krapa*), may be water-dispersed in the rainy season. Water-dispersed fruits and seeds often have air chambers, to facilitate flotation.

Ballistic dispersal. Some plants exhibit a 'do-it-yourself' style of seed dispersal. The fruit capsules of *Dalechampia tiliifolia* slowly twist and tighten as they develop and then release tension in a quick burst that sends seeds flying (Armbruster 1982). A similar mechanism occurs in *Eperua falcata* (*walaba*) and other legume (Fabaceae) trees and lianas.

Passive dispersal. Some fruits simply fall due to gravity and decompose on the ground to release seeds. *Couroupita guianensis (cannon-ball tree)* and *Crescentia cujete* (*krabasi*) fruits appear to lack dispersal agents. However, Janzen (1985) hypothesized that such species were originally adapted for dispersal by now extinct giant animals (megafauna) of the Pleistocene Epoch.

Life-history traits, survival strategies and ecosystem functions

Trade-offs. Plant species in Suriname exhibit a continuum of life-history traits or strategies and associated ecosystem functions. Some species are clearly adapted for and strongly influence particular environments, while others tolerate a wide range of conditions and vegetation zones (Schulz 1960, Schulze *et al.* 2005). The life-history traits of each species reflects its particular evolutionary path and energetic trade-offs (see below) to achieve successful growth and reproduction.

Pioneer gap species. These tree species, such as *Cecropia sciadophylla* (*manbusipapaya*), invest in the widespread and random dispersal of many small seeds, each with few stored resources for survival (Lobova *et al.* 2003). When a forest gap or other disturbance occurs, *Cecropia* spp. seeds germinate and grow rapidly in the sunlight with lightweight, hollow stems. They are 'shade intolerant' and must reach maturity and produce seeds before other trees begin to shade them out. The shade and leaf litter subsequently formed by pioneer saplings and trees creates favorable microhabitat conditions for later stages of forest succession to advance (Maury-Lechon 1991).

Early and late successional species. These tree species require light for seed germination, grow rapidly in gaps and have relatively lightweight wood. However, they can also tolerate some shade and are often found as large trees in old growth forests in Suriname (Schulz 1960). Examples include *Goupia glabra* (*kopi*), *Simarouba amara* (*sumaruba*), and *Laetia procera* (*pintokopi*).

Old-growth species. At the other end of the spectrum, old-growth tree species invest in fewer, larger seeds, and benefit from targeted dispersal by animals. Each seed has a greater chance of survival due to an abundance of stored food. Examples include *Carapa procera* (*redikrapa*), *Eperua* spp. (*walaba*), and *Vouacapoua americana* (*bruinhart*) (Jansen *et al.* 2004, Jansen and Forget 2005). The seedlings and saplings are 'shade tolerant' and can persist over long periods in the dim light of the understory, eventually outgrowing pioneer and late successional species.

Keystone tree species. Any pioneer, successional or old-growth species might qualify as a keystone species. Such species play a disproportionate, yet often inconspicuous role in supporting the integrity of ecosystems and food webs (Peres 2000). Several tree taxa in Suriname are considered keystones because they are reliable food sources for mammals and birds during the dry 'starvation' season. Examples include *Ficus nymphaefolia* (*abrasa*), *Virola surinamensis* (*laagland babun-udu*), and *Parkia pendula* (*kwatakama*). *Mauritia flexuosa* (*morisi*) palm groves form miniature ecosystems that provide year-round food and shelter to many animals, including rodents, monkeys, macaws, tapirs, deer, peccaries and fish (de Granville 1999, Ellison *et al.* 2005, Brightsmith and Bravo 2006).

Survival strategies As rooted organisms, plants are highly vulnerable to herbivorous insects and microorganisms, and most plant species have developed strategic deterrents (Herrera and Pelimyr 2009). One survival strategy is the production of biochemical defense compounds. Sticky latex and poisonous plant tissues protect against chewing insects, while chemicals such as tannins are effective against microorganisms (Coley 1986, Courtois *et al.* 2009 Dussourd and Hoyle 2000). Many plants also produce physical defense structures, such as thick bark, hairs or spines (Leuenberger 1997). An alternative strategy is to attract insect allies, especially ants, as active defenders against herbivores. To accomplish this, plants such as *Passiflora coccinea* (*snekimarkusa*) produce nectar-secreting glands (extra-floral nectaries) on leaf blades, petioles or flowers (Leal *et al.* 2006). In other cases, plant species produce hollow structures known as 'domiciles' or 'domatia' that provide shelter and often food to attract ants (Dyer *et al.* 1999, de Jean *et al.* 2004). Examples of plants often bearing domatia include *Cordia nodosa* (*knopotafrabon*) and *Piper marginatum* (*aneisiwiwiri*).

Tree communities in Suriname

The vast green canopy blanketing Suriname and adjoining territories appears homogenous from a distance, but a closer look inside reveals a complex mosaic of soils and landforms, and tree and palm species compositions that quickly change from place to place (see Ecosystems chapter). As ecologists have tried to document and explain forest diversity patterns and change, two major theories have continued to be debated. 'Niche theory' focuses upon species adaptation or specialization to local environments (niches), including soils, climate, and competition with other plants (Clements 1916, Futuyma and Moreno 1988). Strong proponents of niche theory believe that plant communities transit over time through 'inevitable' ecological succession phases towards a stable 'climax' state such as 'primary forest'. By contrast, 'neutral theory' focuses upon random chance, individual plant characteristics and limitations, and disturbance (Gleason 1926, Chave 2004). Neutral theory predicts that forests are in a state of constant, dynamic change, and rejects rigidly-defined forest types and 'lock-step' transitions. Both theories hold some validity. There is no doubt that forest 'type' categories traditionally used in Suriname – such as high dryland forest, mountain savanna forest, and swamp forest (Lindeman and Molenaar 1959) – are recognizable on the ground and can be predicted by environmental gradients such as elevation, soil type and drainage. However, it is also clear that natural forests are regularly disturbed by treefalls or local storms and any particular patch of forest rarely, if ever, reaches a stable 'climax' state.

Over the past century, understanding of tree diversity patterns in the Guiana Shield region has increased significantly (Davis and Richards 1933, Fanshawe 1952, Lindeman and Molenaar 1959, Schulz 1960, ter Steege *et al.* 1993, ter Steege *et al.* 2003, ter Steege *et al.* 2006, Bánki 2010, Stropp *et al.* 2011, ter Steege *et al.* 2013, Guitet *et al.* 2015). Compared to the western Amazon, Guiana Shield forests have lower tree species diversity, poorer soils and a higher percentage of endemic species. The region's forests are distinguished floristically by the co-dominance of tree species of the Fabaceae, Lecythidaceae, and Chrysobalanaceae families. At many forest sites in Suriname, at least one species from each of these families will occur.

Fabaceae tree species, of the sub-family Caesalpinioideae, occur in exceptionally high abundance within the forests of Suriname and the region. Of particular importance is *walaba* (*Eperua falcata*, Caesalpinioideae), which tolerates a wide range of ecological conditions and is the single most common tree species in the greater region (ter Steege *et al.* 2006). Other common Caesalpinioideae trees include *Dicorynia guianensis* (*basralokus*), *Vouacapoua americana* (*bruinhart*), *and Hymenaea courbaril* (*rode lokus*). Cases of extreme legume dominance – mono-dominance if a single tree species is involved – and, thus, extremely low species richness, are mostly found in swamps or on white sand (Lindeman and Molenaar 1959).

The relatively low tree species diversity in Suriname, at least in plots of limited size (say, one hectare). appears to result, in part, from the great stem abundance of a relatively few legume species. In ter Steege *et al.* (2006) and Hammond and Brown (1995), it is hypothesized that the geological stability, small area, and isolation (low disturbance) of the Guiana Shield has provided ecological conditions favorable to large-seeded, shade-tolerant legume (Caesalpinioideae) trees, such as *walaba*. In addition to large seeds, trees in this sub-family are widely known for root adaptations – symbiotic fungi (Mycorrhizae) and N-fixating bacteria – that make them nutritionally self-sufficient and able to survive and exclude other trees on very poor soils.

Moving southwards from central Suriname, forests become increasingly diverse and have a higher percentage of late successional tree species (see page 109) (ter Steege *et al.* 2001). Examples include *Sclerolobium melinonii (dyadidya), Goupia glabra (kopi), Jacaranda spp., Schefflera morototoni (morototo), and Tachigali spp.(gedu)*. Such species are characterized by light wood and small seeds dispersed via wind or birds. A higher forest disturbance regime - which would favor late successional species over large-seeded (and less species-rich) legume species – is a likely influence that could lead to higher diversity. Also, indigenous peoples may have played an ecological role in centuries past, by increasing the regularity and extent of forest disturbance and sparing or planting useful, late-successional tree species whose populations persist today (ter Steege *et al.* 2001, Levin *et al.* 2017).

Avicennia germinans
Black mangrove
Parwa
Acanthaceae (formerly Verbenaceae)

Black mangrove (*parwa*) is a fundamental component of mangrove ecosystems in the Guianas. Homogenous stands of this coastal pioneer tree species are common along the front line of Suriname's constantly shifting mudbanks and ocean shores (Figure 60). Further inland, it occurs intermixed with red mangrove, *Rhizophora mangle* (*mangro*), and other species (Proisy *et al.* 2009). Intact mangrove communities provide critical ecosystem services in Suriname: they stabilize the mudflats that buffer the hinterland from flooding and storms, trap nutrients and provide feeding and breeding grounds for wildlife, shellfish and fish populations (Fromard *et al.* 1998, Erftemeijer and Teunissen 2009). In Suriname, *parwa* is a key pollen provider for the coastal stingless-bee species *Melipona favosa* (Engel and Dingemans-Bakels 1980), and for the honeybee, *Apis mellifera*.

Parwa exhibits remarkable adaptations for survival and reproduction in brackish water and rapidly changing tidal environments. Excess salt absorbed by the roots is excreted through the leaves. Pneumatophores, upward extensions of the root system that reach into the air, allow for submerged roots to breath. *Parwa* seeds germinate and develop roots while still attached to the mother tree. This reproductive strategy, known as vivipary, primes seedlings (propagules) for rapid anchoring on muddy banks (Rabinowitz 1978). However, to ensure dispersal away from parent populations, newly released propagules temporarily suspend metabolic activity and growth. In this condition, they can float and survive in salt water for more than three months. Such attributes largely explain the pioneering and dominant character of *parwa* along the dynamic Suriname coast.

Figure 60. *Avicennia germinans (Parwa)*, mangrove tree stand with bed of pneumatophore roots in coastal Suriname (image © Pieter & Siela Teunissen).

Figure 61. Opposite, robust leaves and white tubular flowers of *parwa* (image © Pieter & Siela Teunissen).

Figure 62. Salt-water resistant fruits of *parwa* (image © Pieter & Siela Teunissen).

Description: Mature trees forming dense stands 15-35 m high; root pneumatophores conspicuous as a mat of dark, finger-like upward projections above the soil or water; leaves opposite, simple, oblong with pointed tip, leathery, shiny above, dull and with salt-secreting hair-glands below; flowers small, yellowish-white, five-lobed; sepals persistent in fruit; capsules green, egg-shaped, apex pointed, 2-2.5 cm diameter; one seed (Figures 61, 62).

Distribution: Widespread as a member of mangrove communities along the tropical and subtropical coasts of the Americas and western Africa.

Use: Firewood, charcoal, tannins, wood for fences and building; key in production of 'parwa honey'; stabilizes shorelines and protects the hinterland.

Similar species: The common red mangrove, *Rhizophora mangle (mangro)*, is distinguished by conspicuous, arching prop-roots and buttresses. The locally uncommon white mangrove, *Laguncularia racemosa*, lacks arching prop-roots and has conspicuous paired glands on the leaf stalk.

Hymenocallis tubiflora
Spider lily
Lelie (local name used for many Amaryllidaceae)
Amaryllidaceae

Hymenocallis tubiflora is a spectacular perennial lily, often found alongside creeks and rivers in Suriname (Figures 63, 64). Some Hymenocallis species pass through a physiologically inactive stage (dormancy) after flowering, but *Hymenocallis tubiflora* continues growing and producing leaves throughout the year. The unique, showy membrane of the flower likely serves as an indicator or flag for pollinators (Sealy 1954). The long, tubular flowers of the Hymenocallis genus appear to be specialized for hawkmoth (Sphingidae) pollination (Haber & Frankie 1989). The length of the corolla tube approximates the length of hawkmoth tongues and nectar is produced during evening hours when hawkmoths are active. The seeds of *Hymenocallis tubiflora* are green and fleshy at maturity and begin germinating while still on the parent plant (Martin *et al.* 2012). This reproductive trait, known as vivipary, was described above for the black mangrove (*parwa*). Like mangroves, *Hymenocallis tubiflora* seedlings apparently need a developmental headstart to ensure successful establishment in an unpredictable, watery environment.

Figure 63. ***Hymenocallis tubiflora* (Spider lily), (image © Olivier Gaubert).**

Figure 64. Close-up of *spider lily* flower with central, cup-like membrane (image © Olivier Gaubert).

Description: Perennial monocot herb with a large bulb and basal rosette of 4-7 spoon-shaped leaves; leaves pointed, 30-35 x 10-12 cm, leaf stalk to 11 cm long; flowers 5-10, each borne on a robust central stalk to 60 cm tall, corolla large, greenish-white, tubular, tube 12.5-20 cm long, with six narrow, curved petals attached to a shallow, thin, cup-like membrane (formed from fused stamens), filaments 4-6.5 cm long; ovary inferior (below); fruit capsules green with spongy seeds.

Distribution: Northeastern South America and Trinidad; occasional in coastal swamps, seasonally flooded forests and along rivers throughout Suriname.

Use: The bulb and leaves are used in Suriname for herbal baths and to cure fevers; alkaloids have been isolated from the bulbs in phytochemical research (Seaforth *et al.* 1998); the bulb is inedible, but not poisonous (van Andel and Ruysschaert 2011).

Similar species: *Crinum erubescens* is a perennial lily growing to 90 cm tall and commonly found in swamps; *Crinum* spp. possess narrow, strap-shaped leaves and lack the unique floral membrane of Hymenocallis species. *Hymenocallis tubiflora* is distinguished from other Hymenocallis species by conspicuous, long leaf stalks (petioles), broad leaves and large flowers.

Spondias mombin
Hogplum
Mope
Anacardiaceae

Mope is an early or late forest succession species that tolerates a wide range of ecological conditions in Suriname (Figure 65). It is known for seasonal 'mass' or 'mast' flowering and fruiting – the simultaneous production of flowers, and subsequently of fruits, by an entire plant population (Kelly 1994). In a good year, *mope* trees flower and fruit profusely in the rainy season and drop their leaves and orange fruits in the dry season (Figures 66, 67). One explanation for this phenomenon is that masses of flowers and fruits are likely to attract more insect pollinators and seed dispersers (Kelly 1994, Smythe 1970). As a result there is likely to be more effective pollination and fruiting, and ultimately better seed dispersal. Mass fruiting may also be a strategic way to 'over-stuff' or satiate seed predators, increasing the odds that at least some seeds are overlooked and survive intact.

Mope fruits are likely dispersed via water and by turtles, fish, toucans, monkeys, tapirs and humans (Henry *et al.* 2000). In Costa Rica, Janzen (1985) observed that most fallen *mope* fruits remained on the forest floor and were ineffectively dispersed. Janzen included *mope* among several tropical tree species he hypothesized were originally adapted for seed dispersal by large, now-extinct animals (megafauna) of the Pleistocene Epoch. If true, the widely hunted tapir (*Tapirus terrestris*) – the largest native mammal in Suriname's forests – may be the only remaining effective seed disperser of *mope* (Fragoso and Huffman 2000).

Figure 65. ***Spondias mombin (mope)*** **tree (image © Bruce Hoffman).**

Figure 66. Inflorescence of ***mope*** **(image © Olivier Gaubert).**

Figure 67. ***Mope*** **fruits (image © Andres Hernandez).**

Description: Tree to 30 m tall; bark greyish with large vertical grooves, exuding blackish sap; leaves clustered, 40-80 cm long, alternate, pinnately compound with terminal leaflet; inflorescences of many small (< 0.5 cm) yellow-white flowers; fruits orangish-yellow, oblong drupes (single-seeded fruits that do not naturally split apart at maturity), each 3-5 cm long, thin-skinned, and containing a large stone enveloped in a mesh of fibers.

Distribution: Widespread in tropical America; in non-flooded and seasonally flooded rainforest, secondary vegetation, village sites.

Use: Used for a wide range of ailments, including psychological illnesses (Ayoka *et al.* 2006); fruit pulp is made into a tasty, astringent juice; wood used in carpentry.

Similar species: Burseraceae species, such as *Protium heptaphyllum* (*tingimoni*), are distinguished from *mope* by a conspicuous swelling or flex at the apex of leaflet stalks and a bark slash with whitish, fragrant resin. *Mope* and most Anacardiaceae species have uniform leaflet stalks and exude a blackish, odorless sap from the bark slash.

Xylopia aromatica
(no English name)
Pegrekupisi (local name used for several *Xylopia* spp.)
Annonaceae

In the Brazilian Amazon, tiny insects known as *thrips* (Thysanoptera) were documented as primary pollinators of *Xylopia aromatica (pegrekupisi)* (Gottsberger 1994, 1999). It is likely that these insects are also the main pollinators in Suriname. Christianini and Oliveira (2010) found that both birds and ants disperse the seeds of *pegrekupisi*. These two animal agents are complementary rather than competitive, because they act at different spatial and temporal scales. Craig *et al.* (1991) documented very complex plant-insect and insect-insect interactions involving leaf bud galls on *pegrekupisi*. The galls first form due to the activity of weevils (Curculionidae), but are occupied over time by a diverse array of insects including parasitoid wasps and ants.

Figure 68. ***Xylopia aromatica (pegrekupisi)*** **fruiting branch (image © Reginaldo Baiao).**

Figure 69. Flowering shoot of *pegrekupisi* (image © Mauricio Mercadante).

Figure 70. Close-up of fruits and seeds of *pegrukupisi* (image © Reginaldo Baiao).

Description: Tree or treelet with round trunk and many horizontal branches; bark smooth, yellowish-brown; leaves alternate, simple, two-ranked (Figure 68); flowers 3-4 cm long, strongly fragrant, creamy white, and shaped like a star (Figure 69); fruits sickle-shaped, brownish-black, up to 2.5 long, splitting open to expose many seeds with arils (Figure 70).

Distribution: Widely distributed from Central America to the Guianas and southern Brazil; occurs in non-flooded forests and savannas.

Use: *Xylopia* spp are used medicinally and ritually by Maroon groups in Suriname (van Andel and Ruysschaert 2011); wood used locally to make furniture.

Similar species: The genus is easily distinguished from other Annonaceae genera due to the distinctive flowers. *Xylopia nitida* is distinguished by silver-green leaf undersides and *Xylopia longifolia* has longer (15-18 cm) leaves.

Geissospermum laevis, Geissospermum sericeum (no English name) *Bergibita* (local name used for all *Geissospermum* spp.) Apocynaceae

Bergibita is a common and characteristic element of Suriname's non-flooded forests, especially in hilly areas (Figures 71, 72). The common name, *bergibita*, refers to both the habitat ('bergi' means 'hill') and the bitter, medicinal quality of the bark ('bita' means 'bitter'). Bitter tastes in plant parts are due to the production of alkaloids and other bioactive compounds that help defend plants against herbivores and pathogens (Herrera and Pelimyr 2009). *Bergibita* also produce a white milky sap in young shoots and fruits that can gum up insect mouth parts. Interestingly, the white sap is not produced in the living bark, possibly because the bitter compounds stored there are already an effective deterrent.

Consumption of the large, juicy fruits of *bergibita* and dispersal of intact seeds has been documented for monkeys and tapirs (Figure 73) (van Roosmalen 1985b, Henry *et al.* 2000). Tapirs are important in the distribution patterns of *bergibita* and other trees because they defecate and deposit seeds repeatedly at non-random 'latrine' sites (Fragoso and Huffman 2000). In French Guiana, Hilbert *et al.* (2011) found that more that one-third of 53 tapir dung piles contained fragments or seeds of *Geissospermum laevis*. Thus, when tapirs are locally depleted through hunting, it can be expected that *bergibita* seedling dispersal and establishment is also negatively impacted.

Figure 71. ***Geissospermum laevis (bergibita)*** **tree trunk with characteristic vertical grooves and holes (image © Pieter and Siela Teunissen).**

Figure 72. Leafy branch of *bergibita* with silvery leaf undersides (image © Bruce Hoffman).

Figure 73. Fruits of *bergibita* (image © Bruce Hoffman).

Description: Canopy tree or a treelet in secondary forest; trunk irregular with deep vertical furrows; bark grayish-pink-brown; leaves alternate, simple, long-ovate, 7-10 cm long, often shiny below; inflorescences in dense clusters, flowers < 1 cm long, with silky-brown hairs; fruits two per flower, diverging horizontally, each long and pointed, to 14 x 5 cm, green but ripening brown with a softly velvety skin, containing flat seeds, exuding white milky sap.

Distribution: Venezuela and the Guianas; non-flooded lowland forest and secondary forest.

Use: Bitter bark used traditionally to treat malaria, fevers, indigestion, pain, liver ailments and constipation, among other ailments (Brandão *et al.* 1992); clinical trials indicate effectiveness against prostate cancer, HIV, Altzheimer's disease, malaria, and leishmaniasis or 'busi yasi' (Vital *et al.* 2015).

Similar species: *Minquartia guianensis* is a common tree in Suriname. Like *bergibita*, this species has deep vertical furrows and holes along the trunk. It differs by the non-bitter bark, clear, non-white sap, broadly oval leaves, and small, single, ovoid fruits.

Matelea denticulata
(no English name)
Konkonibonki
Apocynaceae, Asclepiadoideae subfamily

Most members of the Asclepioideae subfamily, such as *Asclepias curassavica*, possess unique, complicated flowers and pollination strategies (Ollerton and Liede 1997, Fishbein 2001). The pollen is produced in dense packages of grains (*pollinaria*) rather than loose grains as in most flowering plants. Orchidaceae species, including *Catasetum discolor,* also produce pollen in pollinaria. In most of the Asclepioideae subfamily, male and female flower parts are fused into a drum-like structure (*gynostegium*) with slits that requires a precise physical alignment of the pollinator with the flower. By contrast, *Matelea* spp. possesses a simple 'open pollination' system, specialized for flies (Figure 74). When flies land on *Matelea* spp. flowers, the pollinaria become stuck to their feet and are thereby transported to pollinate adjoining flowers. Fly pollination is rare in flowering plants, because most fly species are irregular pollinators and individual flies cannot transport large amounts of pollen.

Konkonibonki and most members of the family produce abundant white, sticky sap (*latex*) from damaged plant parts (Diego and Castro 2008). Latex is an important deterrent against herbivory. It is toxic to many generalist butterfly larvae and gums up insect mouth parts (Dussourd and Hoyle 2000). The latex seals plant wounds through coagulation and has antifungal and antibacterial properties. However, some herbivorous insects have evolved compensatory strategies such as cutting veins to drain the latex out of leaves before consumption or sequestering (absorbing) and compartmentalizing the toxins (Dussour 1990, Trigo 2000).

Figure 74. *Matelea denticulata*, slender vine with thin, greenish-yellow petals (image © Pieter & Siela Teunissen).

Description: Slender liana or vine; young branchlets densely hairy (villose); leaves opposite, simple, ovate, to 9 x 6 cm, heart-shaped; inflorescences few-flowered; corolla 3 cm in diameter, petals open and flat or gently sloping, greenish-yellow; gynostegium dark yellow, five-sided; anthers and pollinia horizontal; fruits single at maturity, spindle-shaped, 5 cm long x 1.5-2 cm wide, green, hairless, smooth or bumpy.

Distribution: Guianas, adjacent Brazil and Venezuela, northwards to Mexico; common in old-growth forest edges and secondary forest.

Use: (No known use)

Similar species: *Matelea surinamensis* has similar leaves, but is distinguished by a bell-shaped, yellow corolla of approximately 2.5 cm diameter; fruit unknown.

Montrichardia arborescens
(no English name)
Mokomoko
Araceae

In Suriname, herbaceous swamps are often dominated by *mokomoko* plants with their broad, arrowhead-shaped leaves (Figure 75). Dense *mokomoko* stands line the margins of many rivers and creeks in Suriname. This species is well adapted to tidal (daily) and seasonally changing water levels. The spike-like inflorescence of *mokomoko* has unisexual flowers, with female flowers crowded at the base and male flowers scattered above (Figures 76, 77). As documented by Gibernau *et al.* (2003), pollination is a two-day process. On the evening of the first day, the spathe partially opens and female flowers on the spadix become receptive. The spadix heats up to release volatile oil scents and attract potential pollinators. One of the most common and effective pollinators of *mokomoko* is the scarab beetle, *Cyclocephala colasi*. Before midnight, the spathe slowly closes, trapping some of the beetle visitors. The trapped beetles benefit from plant-provided 'feeding tissue' and mating opportunities with other trapped beetles. On the second day within the spathe, female flowers become inactive and male flowers release energy-rich pollen. The spadix slowly re-opens, allowing for the pollen-dusted, mated and well-fed beetles to escape. The released beetles follow the volatile oil scents of additional *mokomoko* flowers and, in the process, accomplish cross-pollination. Dispersal of *mokomoko* seeds occurs primarily through water transport and fruit-eating fish (Horn *et al.* 2011).

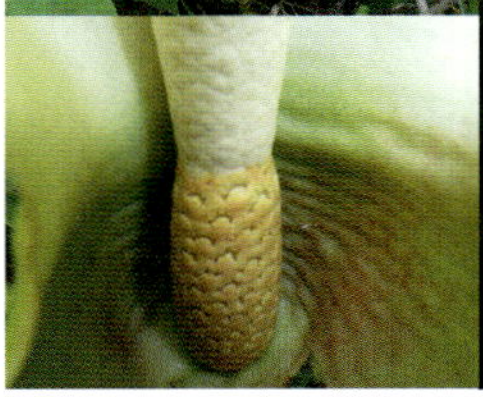

Figure 75. *Montrichardia arborescens* (*mokomoko*), riparian herb with arrowhead-shaped leaves (image © Pieter & Siela Teunissen).

Figure 76. *Mokomoko* spathe and spadix (image © Pieter & Siela Teunissen).

Figure 77. *Mokomoko*, closeup of spadix with male flowers above and female flowers below (image © Pieter & Siela Teunissen).

Figure 78. Mature, many-seeded fruit (infructescence) of *mokomoko* (image © Pieter & Siela Teunissen).

Description: Erect, perennial herb 1-4 meters tall; leaves alternate, simple, arrow-head shaped, blade to 35 x 30 cm; stem with conspicuous leaf-scars, armed with basal 'prickles'; inflorescence a cream-colored, many-flowered spike enclosed by a sheath, up to 15 cm long, greenish-yellow outside and white inside; infructescence to 12 x 8 cm, with many green, one-seeded, irregular-shaped fruits (Figure 78).

Distribution: Northern South America to southern Mexico; abundant in brackish and freshwater ecosystems in Suriname.

Uses: Sap used to stop bleeding; seeds edible when cooked or roasted; rope made from stem fibers (Ostendorf 1962).

Similar species: *Montrichardia linifera*, which is more common in the Interior of Suriname, has heart-shaped or rounded leaves.

Astrocaryum sciophilum
(no English name)
Bugrumaka
Arecaceae (or Palmae)

***B**ugrumaka* is a characteristic understory palm of old-growth, non-flooded rainforest throughout Suriname (Figures 79,. Charles-Dominique *et al.* (2003) studied the life cycle of *bugrumaka* in French Guiana and found that the species is unexpectedly slow growing and long lasting. Based upon empirical data and modeling, it was estimated that a single leaf can last for 10-12 years and that juveniles become adults at 170 ± 70 years of age. Thus, a *bugrumaka* palm with a 2 m tall stem may be as old as 270 years, and one with a 4 m stem may be much older than most of the neighboring trees that tower above it (Chambers *et al.* 1998).

Bugrumaka is a specialist in survival and growth in the dark forest understory. This species possesses an extensive underground rhizome that remains viable when above-ground parts are burned in forest fires or crushed in tree falls. Furthermore, *bugrumaka* architecture allows for leaves to be torn away in tree falls without much damage to the stem. Such traits may explain why it is such a common species in Suriname's forest understory.

The primary dispersers of *bugrumaka* seeds are scatter-hoarding rodents, including squirrels (*Sciurus aestuans*), agoutis (*Dasyprocta leporina)* and acouchis (*Myoprocta acouchy).* Based upon average rodent seed dispersal distances and bugrumaka seed production, Charles-Dominique *et al.* (2003) estimated that *Astrocaryum sciophilum* populations were colonizing new areas at an average rate of 2.3 meters per year.

Figure 79. ***Astrocaryum sciophilum (Bugrumaka)*****, common understory palm in old-growth, non-flooded forests of Suriname (image © Pieter & Siela Teunissen).**

Figure 80. ***Bugrumaka*** **spines (image © Bruce Hoffman).**

Figure 81. Spiny fruits and protective spathe of ***Bugrumaka*** **(image © Pieter & Siela Teunissen).**

Description: Solitary palms, growing to 10 m tall, stem spineless and 10-15 cm in diameter, younger plants without an above-ground stem; leaves to 8 m long with 50-90 leaflet pairs, leaf-base with many long black spines (Figure 80); inflorescences with a spine-covered spathe up to 70 cm long, enclosing a large, brush-like spadix with tiny yellow flowers; fruits brown, spiny, egg-shaped, to 6 x 3.5 cm (Figure 81) (Wessels Boer 1965).

Distribution: Guianas to just north of the Amazon river mouth; common in Suriname, especially in non-flooded forest south of the Coastal Plain.

Use: Maroon communities harvest the fruits to extract oil; indigenous groups use the sap of young shoots to staunch blood flow in wounds.

Similar species: *Astrocaryum paramaca* (*paramaka*) is a spiny, stemless palm in Suriname appearing similar to younger, stemless *bugrumaka* plants. The terminal leaf margins of the *paramaka* palm are free (vs. fused in *bugrumaka*) and the leaf base spines are winged (vs. wingless in *bugrumaka*).

Euterpe oleracea
Açaí (global trade name)
Pina, podosiri, prasara
Arecaceae (or Palmae)

The colonial, multi-stemmed habit of the graceful *pina* palm provides many advantages for survival, growth and reproduction. The multi-stem clumps allow individuals to expand laterally and capture more light, space and nutrients than single-stemmed palms (Figures 82, 83). Aerial roots are produced by basal clumps above the water line to facilitate gas exchange in swampy habitats. The multiple stems and basal sucker shoots of *pina* clumps greatly increase the odds of survival against threats such as tree falls and human harvest of stems or fruits.

Flowering and fruiting of *pina* populations occurs throughout the year. To avoid self-fertilization, the male flowers open and shed pollen before the female flowers become receptive. The main flower visitors are small bees, social wasps and flies, but it remains unknown if any of these insects are true pollinators (Wessels Boer 1965). Seed dispersal occurs primarily via birds (parrots and macaws), monkeys, fish, water and humans. Seeds germinate soon after dispersal, as they lack long-term dormancy.

Description: Colonial (multi-stemmed) palm up to 25 m tall with bright red aerial roots at the stem base; stems 7-18 cm diameter; leaves smooth, 3-4 meters long, with 50-80 leaflets per side; inflorescences have a white-haired, many-branched spadix protected by a green spathe to 1 m long, flowers tiny; fruits are purple-black, thin-skinned globes, 1-1.5 cm in diameter, each with one seed (Figure 84).

Figure 82. ***Euterpe oleracea (pina)*, multi-stemmed, colonial palm often forming dense stands in swamp and creek forests (image © Bruce Hoffman).**

Figure 83. *Pina* palm producing aerial roots above the water line (image © Bruce Hoffman).

Figure 84. *Pina* palm, widely branching, erect fruiting spadix producing hundreds of purplish-black fruits (image © Pieter & Siela Teunissen).

Distribution: Widespread in northern South America; occurs in Suriname in coastal swamps and along rivers and creeks of the Interior forests, often in mono-dominant stands; common around villages and camps due to planting or seeds discarded in refuse heaps.

Use: Popular among Aukaner Maroons and other groups as a liquid pulp made from the fruit pulp. Sugar is added for a sweet juice or the pulp is mixed with 'kwak', a granular form of cassava; the pulp is also sold in Paramaribo markets; the soft stem core is eaten (palm heart); the stem is also split and used locally to build floors and walls (Clay and Clement 1993, van Andel *et al.* 2003, van Andel and Ruysschaert 2011).

Similar species: *Euterpe precatoria* has similar leaves and fruits but is single-stemmed and grows mostly in non-flooded forest sites. *Oenocarpus bacaba* (*kumbu*) is a larger, single-stemmed, upland palm with similar, but larger fruits.

Mauritia flexuosa
Mauritius palm
Morisi
Arecaceae (or Palmae)

The *morisi* palm is the most widely distributed palm species in South America. Tall and beautiful stands of *morisi* are common in open, herbaceous swamps on the Coastal Plain of Suriname. *Morisi* is an ecosystem foundation species, meaning that it "creates and defines an entire ecological community or ecosystem" (de Granville 1999, Ellison *et al.* 2005). A variety of plants and animals live in *morisi* crowns and stems, including orchids, lizards and nesting Blue-and-yellow Macaws (Brightsmith and Bravo 2006). Mini-ecosystems of macro-invertebrates and amphibians occur within standing pools of water in *morisi* leaf bases and epiphytic bromeliads (Torreias *et al.* 2010). *Morisi* palm fruits are an important food for many forest animals, including mammals, reptiles, birds and fish.

In southern Suriname, Trio and Wai-wai indigenous peoples transported *morisi* seeds from savanna swamps and planted them at new, missionary-founded villages in the mid-20th century (Kamainja, Kwamalasamutu resident, pers. comm.). Today, conspicuous stands of *morisi* palms are present in swamps adjoining many local communities of the south. The modern distribution of *morisi* in Suriname is thus a result of both human and non-human influences.

Figure 85. *Mauritia flexuosa (morisi)*, robust, single-stemmed palm occurring in swamps; leaves palmately-compound (image © Bruce Hoffman).

Figure 86. *Morisi*, inflorescence of male flowers (image © Bruce Hoffman).

Description: Large single-stem palm up to 35 m tall (Figure 85); stem 30-40 cm diameter; leaves fan-shaped, leaf stalk long (to 3 m); leaf blade to 2.5 x 4.5 m, leaflets to 90 x 3 cm; inflorescences with a many-branched spadix, 2 m or longer; plants dioecious – each palm tree producing only male or female flowers (Figure 86); fruits round or oblong, ca. 5 x 4 cm, with reddish-brown scaly skin, orange pulp and one seed per fruit.

Distribution: Widespread across the Guianas and the Amazon Basin; in Suriname, locally abundant in herbaceous swamps and savannas, especially in the Coastal Plain, but also elsewhere along streams.

Use: Fruit pulp consumed in Suriname, rich in vitamins A and C; palm heart (young shoots) and palm grubs consumed; morisi roots used medicinally or ritually, especially by Maroon women, to avoid miscarriages or reduce menstruation pain (van Andel and Ruysschaert 2011); leaf fibers and stalk used to make clothes, toys, bags, soft hammocks, slippers, and tourist crafts.

Similar species: *Mauritiella armata* is a palm with similar leaves and fruit, but easily distinguished by multiple, slender stems, each < 15 cm diameter, with blunt, spine-like projections. This species occurs in the Sipaliwini Savanna of Suriname.

Cordia nodosa
(no English name)
Knopotafrabon
Boraginaceae

This easily recognized, very common shrub or treelet is one of the best known 'ant plants' in the Neotropics (Figures 87, 88, 89, 90). Characteristic swollen nodes ('domatia'; singular 'domatium') at branching points often house ant colonies. In this mutually beneficial relationship, occupying ant colonies protect the host plant against herbivores and, in exchange, receive food and a thick-walled shelter. *Knopotafrabon* plants passively control access to new domatia by producing sealed entrances of a precise diameter and thickness. Only those ant species with a small enough body size and sufficient jaw strength to cut open the domatia 'door' can enter.

The ant species residing in *knopotafrabon* domatia differ by region in the Neotropics and often change during the lifetime of an individual plant (Solano *et al.* 2003, DeJean *et al.* 2004). In French Guiana and Suriname, *Allomerus octoarticulatus* is the most common resident (Frederickson 2009). This ant species typically destroys *knopotafrabon* flowers before they mature and colonies remain for long periods. As a result, occupied plants are forced to prioritize vegetative growth and domatia production over sexual reproduction until the ant colony migrates elsewhere. Other common domatia residents are ant species of the *Azteca* genus (Frederickson 2009). These ants do not destroy flowers and occupied plants are free to produce flowers and seeds through sexual reproduction. However, a tradeoff is that *Azteca* spp. ant colonies lack fidelity. They often abandon *knopotafrabon* domatia after short periods of time, leaving the plant temporarily vulnerable to herbivores and seed predators.

Figure 87. *Cordia nodosa (knopotafrabon)*, understory shrub or treelet (image © Bruce Hoffman).

Figure 88. Swollen branch node (domatium) of *knopotafrabon*, typically hollow and housing ant colonies (© Pieter & Siela Teunissen).

Figure 89. Flowers of *knopotafrabon* (image © Bruce Hoffman).

Figure 90. Fruits of *knopotafrabon* (image © Pieter & Siela Teunissen).

Description: Understory tree, 2 to 5 m tall; branchlets growing horizontally and dividing evenly at branching points with swollen nodes; leaves ± opposite with a round, backwards-pointing leaf at each node; swollen nodes housing ants usually present; flowers white, radial, with five petals, five stamens, and four stigmas; fruits ovoid, green, one-seeded drupes.

Distribution: Guianas, greater Amazonia; common in understory of tropical forests, lowland to montane.

Use: Medicinal – cold, fever, headache, rheumatism, whooping cough (van Andel and Ruysschaert 2011).

Similar species: There are a few *Cordia* spp. with similar branching patterns but these lack the swollen nodes occupied by ants.

Vriesea splendens

Eric Gouda

Flaming sword, painted feather, zebra bromeliad
Businanasi (local name used for most Bromeliaceae)
Bromeliaceae

The diverse Bromeliaceae family is a widely cited example of adaptive radiation – the evolution of multiple forms from a basic ancestor through environmental specialization. Notable Bromeliaceae adaptations include the rosette-leaf water tanks that both hold water and capture nutrients, modified hairs that absorb nutrients, and the water-conserving photosynthetic pathway known as crassulacean acid metabolism. These traits allow tree-based, epiphytic bromeliads to survive on tree branches without soil and ground-based bromeliads to live on rocks (lithophytic) and very infertile soils.

In French Guiana, Brouard *et al.* (2012) observed that the pools of water held within epiphytic tank bromeliads (Bromeliaceae) form aquatic 'islands' within a terrestrial 'ocean' of forest. Miniature aquatic ecosystems and complex food webs occur within the tanks, which simultaneously provide a reliable nutrient and water supply to the bromeliad. This research showed that bromeliads at lower canopy levels, with less light, supported tank ecosystems dominated by filter-feeding insects and organic matter. By contrast, bromeliad water tanks at higher canopy levels, with greater light, were dominated by green algae, rotifers and predatory invertebrates.

Figure 91. ***Vriesea splendens*, a bromeliad with tall flowering stalks (image © Eric Gouda).**

Description: Plants with a broad rosette of mostly purple-red, horizontal-striped leaves (Figure 91); blades thin-to sub-leathery, 40-75 cm long, margins without spines; inflorescence simple, linear or lanceolate, rising to more than 1 meter above rosette (main stem 30-50 cm long, flowering portion of stem 20-55 cm long), strongly flattened, with two-ranked, tightly overlapping, bright red floral bracts; flowers 4-18, appearing at one side of the bracts; sepals yellowish, red at the apex; petals long and narrow, 8 cm long, yellow; fruit capsule round (in cross-section), 3.5-4 cm long (Gouda 1987).

Distribution and habitat: Venezuela and the Guianas; Sipaliwini and Marowijne districts in Suriname; mostly epiphytic on tree trunks in moist forests, also terrestrial along the edges of rocky areas.

Use: Ornamental – a stunning species that is easily cultivated and one of the first commercially grown Bromeliad species (Gouda 1987).

Similar species: There is a green-leaved variety of *Vriesea splendens* not yet observed in Suriname. *Vriesea splendens* belongs to the subfamily Tillandsioideae, with spineless leaves. Many other common bromeliad genera, such as *Aechmea* and *Ananas*, possess spiny leaves (Crayn *et al.* 2004).

Protium heptaphyllum
(no English name)
Sabana tingimoni (tingimoni is a local name used for many Burseraceae)
Burseraceae

As a family with high species diversity and abundance, Burseraceae has been used as a model to study biogeographic patterns across the Amazon Basin (Daly and Fine 2011). *Protium* is the largest and most variable genus in the family, with highest species diversity in the central Amazon rainforest and Guiana Shield (Daly 1987). *Protium heptaphyllum* is a common and morphologically highly variable species. As with most Burseraceae, *sabana tingimoni* flowers are pollinated by generalist insects and seeds are dispersed mostly by birds and monkeys (Daly 1987).

The strongly scented resin of *sabana tingimoni* is often visible as a whitish dried substance on trunk wounds. Resin production is important in defense against tropical forest pathogens, but it is also a valuable ecosystem resource for a variety of organisms. Orchid bees, for example, collect *tingimoni* resin to construct their nests, while weevils and some fly larvae populations develop within the resin lumps on the trunk (Langenheim 2003).

Figure 92. ***Protium heptaphyllum (sabana tingimoni)*** **tree, with pinnately compound leaves and small flowers (image © Pieter & Siela Teunissen).**

Figure 93. Close-up of *sabana tingimoni* flowers (image © Pieter & Siela Teunissen).

Figure 94. Immature *sabana tingimoni* fruits (image © Pieter & Siela Teunissen).

Figure 95. Mature fruits of *sabana tingimoni* on forest floor, seeds with white, fleshy and edible arils (image © Pieter & Siela Teunissen).

Description: Shrub or tree to 20 m, trunk often with aromatic resin drying white and hard; leaves alternate, pinnately-compound (5-7 leaflets) (Figure 92); flowers tiny, light yellow (Figure 93); fruits ovoid with pointed ends, reddish-green; seeds surrounded by white-edible pulp (Figures 94, 95).

Distribution: Widespread in the Neotropics, from Costa Rica to southern Brazil; in Suriname, common in open savannas, ridge forests, and river banks.

Use: Wood of poor durability but useful for interiors; resin widely used for light, incense, insect repellent, ritual, and medicine (anti-inflammatory, antioxidant) (Langenheim 2003; Rüdigera *et al.* 2007).

Similar species: Among Burseraceae genera, the leaflet stalks of Protium species are typically longer, thinner and more conspicuously swollen or flexed at the apex than *Tetragastris* or *Trattinnickia* spp. Anacardiaceae species, such as *Spondias mombin (mope*), lack a leaflet stalk swelling and exude a blackish, odorless sap in the bark slash.

Cereus hexagonus
(no English name)
Zuilcactus
Cactaceae

The conspicuous, columnar *zuilcactus* is an ecological generalist – having successfully colonized a wide range of wet, dry and disturbed ecosystems in Suriname (Figure 96). The photosynthetic trunks store water and utilize crassulean acid metabolism - an alternative type of photosynthesis that conserves water by taking up CO2 only at night. Such strategies are common in many epiphytic and rock-inhabiting plant species that lack a reliable supply of water and soil nutrients.

The beautiful, fragrant flowers of the *zuilcactus* bloom at night and produce copious amounts of nectar. Primary pollinators include night-flying animals such as bats and hawkmoths (Sphingidae). Beetles have been observed visiting *zuilcactus* flowers, but there is no evidence that they are pollinators. Seed dispersal is accomplished primarily by birds that consume the bright red fruit and defecate the tiny seeds (Figure 97) (Ruiz *et al.* 2000).

Figure 96. *Cereus hexagonus*, a columnar cactus species in flower (image © Willem Kolvoort).

Figure 97. *Cereus hexagonus* fruit with seeds in pulp (image © Willem Kolvoort).

Description: A columnar cactus growing as a branching tree, with 6-8 vertical ridges on trunks, 1-10 m high; regularly spaced clusters of 5-6 cm long spines present along the stem; flowers single to few, white and showy, 20-25 cm long, with many petals, stamens and stigmas; fruits egg-shaped, 5-13 cm long, bright red at maturity with white or pink flesh; many-seeded (Leuenberger 1997).

Distribution: Guyana, Suriname and Venezuela; uncommon but locally abundant; on sandy ridges near the coast, savanna, granite flats and outcrops, naturalized in coastal swamps from old plantation lands.

Use: Fruits edible; historically grown on plantations as a natural fence.

Similar species: There are no similar wild species in Suriname.

Parinari campestris
(no English name)
Redifungu, rode fungu
Chrysobalanaceae

This genus and species were first described in 1775 by the French botanist Aublet, in the *Histoire des Plantes de la Guiane Françoise* (Aublet 1775). *Rode fungu* trees often house large numbers of defensive ants. The ants collect wooly reddish-brown hairs from young *rode fungu* leaves to construct odd, spongy, hanging nests within tree branches (Prance 1992). The name 'fungu' refers to these nests. Two ant species known to form *Parinari* ant nests in Brazil are *Dolichoderus mucronifer* (Sefer 1957) and an unidentified species of *Hypoclinea* (Prance 1992). Very little is known about this phenomenon in Suriname, but it is likely a case of mutual benefit for both ant and plant. *Rode fungu* branchlets bear small nectar gland pairs at the base of leaf stalks (petioles) that attract ant visitors. This species is pollinated by bees and the large seeds are dispersed mostly by rodents.

Figure 98. *Parinari campestris (rode fungu)* tree (image © Pieter & Siela Teunissen).

Figure 99. Leafy branch of *rode fungu*, with large, spear-shaped stipules (image © Pieter & Siela Teunissen).

Figure 100. Flowers and fruit of rode *fungu* (image © Pieter & Siela Teunissen).

Description: Medium- to large-sized trees, to 35 m tall, with low buttresses (Figures 98); bark gray, scaly, bumpy; vegetation with a reddish-brown dusty appearance; branchlets densely hairy with 3 cm long spear-shaped stipules (Figure 99); leaves simple, alternate, broadly elliptic, 5-15 x 2.5-6 cm; always appearing withered, with hair-filled cavities below; inflorescences densely hairy; flowers five-parted, 0.5 cm long, white; fruits oval-irregular, 4-6 x 2-3 cm, dark green to brown, with white raised specks (lenticels) (Figure 100) (Prance 1972).

Distribution: Brazilian Amazon, Guianas, to Venezuela, Trinidad; common on non-flooded rainforest hills, old ridges, seasonally flooded forest and savanna, common in coastal Suriname (Maas 1971, Banki *et al.* 2003).

Use: Fruit pulp and seeds edible; ant nests as a 'magical medicine' for some Maroon groups; timber, firewood, ant-nest fiber as fire starter (van Andel and Ruysschaert 2011).

Similar species: *Exellodendron* species are closely related to *Parinari* spp.; they are distinguished by the absence of hair-filled cavities on leaf undersides and a smooth fruit wall.

Clusia grandiflora
(no English name)
Abrasa (local name used for many Clusia and Ficus spp.)
Clusiaceae

Clusia is an easily recognized and common genus of hemi-epiphytic climbers in Suriname (Figure 101; Box 7). *Clusia grandiflora* plants typically begin life as sticky, tiny seeds dispersed high in the forest canopy by birds or monkeys. As juvenile plants develop, they send aerial roots to the forest floor to collect nutrients and water. The woody, robust, hanging aerial roots of *Clusia* spp. are a common sight in Suriname's old-growth forests. Clusia spp. also commonly grow as small, shrubby trees in savannas and on exposed mountaintops.

Individual *Clusia grandiflora* plants are either male or female (dioecious). Male flowers resemble sea anemones, with outer rings of fertile and infertile stamens and a thick pool of sticky resin in the center. Female flowers have a stocky central pistil with fertile patches (stigmas) and infertile resin-producing stamens (Figures 102, 103). The main pollinators of *Clusia grandiflora* flowers are small to medium-sized bees – including Meliponinae, Euglossinae and Megachilid species (Armbruster 1984, Kaminski and Absy 2006).

Clusia is one of the only plant genera in the world to produce resin as a floral reward (Armbruster 1984). *Clusia* spp. resins and latexes are commonly used by native bee species in nest construction. Lokvam and Braddock (1999) found that *Clusia* spp. resins were also highly effective at battling the most problematic bacterial pathogens of bee brood.

Figure 101. *Clusia grandiflora (abrasa)* established in a tree crown, with hanging aerial roots (image © Bruce Hoffman).

Figure 102 (below, left). Female flower of *abrasa* with green, fertile pistil (image © Pieter & Siela Teunissen).

Figure 103. Male flower of *abrasa* (image © Pieter & Siela Teunissen).

Figure 104. Fleshy leaves and immature fruit of *abrasa* (image © Bruce Hoffman).

Description: Hemi-epiphytic trees; plant parts with a sticky, white to tan sap, oxidizing orange; leaves opposite and thick-leathery, oval, 10-35 cm long; male and female flowers large and showy, ca. 15 cm diameter, with snow-white petals and yellow-brown resin, often found on the forest floor; fruits sub-globose capsules to 8.5 cm diameter, resembling an onion (Figure 104), 10-15 chambered, resembling a many-pointed star upon opening, seeds tiny, with bright red-orange arils.

Distribution: Amazon, Guianas; widespread in non-flooded and flooded forests in Suriname.

Use: Aerial roots used as ingredient in a general strength and aphrodisiac tonics (van Andel and Ruysschaert 2011), roots used in Guyana in wicker-like furniture production.

Similar species: Hemi-epiphytic habit and all plant parts distinctive; may be confused with other *Clusia* spp., but *Clusia grandiflora* has by far the largest flowers among Suriname Clusia spp.

Terminalia amazonia
(no English name)
Swampu bus'amandra
Combretaceae

The small, ornamental almond or *amandra* tree (*Terminalia catappa*) is well known in urban coastal areas of Suriname. By contrast, the *Swampu bus'amandra* presented here is a tall to giant forest tree. In the forests of Suriname's Interior, extended dry seasons can cause evergreen trees to drop their leaves to conserve energy and avoid continued water loss through the leaves (evapotranspiration). When the rains return, drought-stricken *Swampu bus'amandra* trees often produce new leaves and flowers simultaneously. *Swampu bus'amandra* trees typically fruit during the long rainy season in Suriname. Medium-sized bees are the primary pollinators, but wind pollination has also been observed (Bawa *et al.* 1985). Most *Terminalia* spp. are specialized for wind dispersal, with dry, winged fruits (Flores-Vindas 1994).

Figure 105. *Terminalia amazonia (swampu bus'amandra)* tree, with distinctive branching pattern, clustered leaves, and inflorescences (image © Olivier Gaubert).

Figure 106. Papery, winged fruits of *swampu bus'amandra* (image © Dominiek Plouvier).

Description: Large tree with a flattened canopy, 25-50 m tall, trunk straight, with distinctive pagoda-like branching (horizontal whorls), large buttresses present in swampy areas; bark thin, cream- to brown-colored, with vertical cracks; leaves tightly clustered in whorls (Figure 105), blades oval, wider towards tip, 5-10 x 3-5 cm; inflorescences spikes of tiny yellow-green to white flowers borne below leaf clusters; fruits butterfly-like, 1.2 cm x 0.5 cm, with two large, papery horizontal wings and three tiny vertical or lateral wings (Figure 106).

Distribution: Amazonian Brazil to Mexico – the most widely distributed Neotropical species of the genus; common in swampy lowland rainforest, also occurs in upland ridge forest, deciduous forest and savannas; prefers clay soils.

Use: High quality wood for construction; a favored tree in native plant reforestation projects (Nicols 1994).

Similar species: Distinct from other *Terminalia* species, the finest (3°) veins of *Terminalia amazonia* leaves are parallel and at a 90° angle to the midvein. The fruits of a similar species, *Terminalia dichotoma*, are thick, woody and lack papery wings.

Ipomoea aquatica (formerly *Ipomoea reptans*)
Swamp cabbage, water spinach
Dagublat
Convolvulaceae

***D**agublat* is a common, fast-growing vine found mostly in canals and natural waterways in coastal Suriname (Figures 107, 108). Populations spread primarily through the vegetative breaking-off and transport of plant parts through water. Sexual reproduction also occurs, but seeds do not germinate well under water. A single *dagublat* stem can grow up to 10 cm per day and extend for many meters in length. Under some conditions, *dagublat* forms dense vegetation mats that clog waterways, reduce oxygen and shade out native aquatic plants (Langeland and Burks 2000). On the beneficial side, *dagublat* has proven effective at purifying eutrophic (nutrient-rich) waters polluted by industrial activities and farm run-off. Significant reductions in nitrogen, phosphorus, total suspended solids and heavy metals were found in *dagublat*-filtered water (Hua *et al.* 2008, Ghosh 2010).

Figure 107. *Ipomoea aquatica (dagublat)* vine, in a Paramaribo canal (image © Pieter & Siela Teunissen).

Figure 108. Close-up of flowering *dagublat* colony (image © Pieter & Siela Teunissen).

Description: Perennial or annual, non-native herb, often found floating or emergent in fresh water; leaves alternate; blade shape variable, often heart-shaped, 4-15 x 2-8 cm; inflorescence one- to few-flowered, flowers funnel-shaped, to 6 cm long, light pink-purple or white, often with a dark purple throat; fruit a brown capsule, four-parted or splitting irregularly; seeds 1-4.

Distribution: Pan-tropical, extending into subtropical zones, introduced in the Americas; common; freshwater rivers and coastal swamps, cultivated and naturalized in Suriname.

Use: Edible stems and leaves, rich in minerals and vitamins A, C and E; livestock feed; remedy for gastric and intestinal disorders, heart problems, blood purification, stress, psychological problems (Austin 2007, van Andel and Ruysschaert 2011).

Similar species: There are two widely recognized *dagublat* varieties. Red (*redi*) *dagublat* has reddish-purple stems, dark green leaves and light pink to purple flowers. White (*weti*) *dagublat* has green stems, green leaves, greenish-white petioles, and white flowers. An upland cultivar, sometimes known as *Ipomoea reptans,* has green leaves and white flowers.

Costus scaber
Wild cane, spiral ginger (English names used for many Costus spp.)
Redisangrafu
Costaceae

Costus spp. are common understory plants in Suriname's forests and in urban gardens. They produce showy flowers and fruits throughout the year. Most *Costus* spp. are pollinated by either hummingbirds or euglossine bees (Maas 1972, Kay and Schemske 2003, Kay *et al.* 2005). The bright red-orange inflorescence and long, tubular flowers of *redisangrafu* (*Costus scaber*) are specialized for hummingbird pollination. Bee-pollinated *Costus* spp. flowers are shorter, folded, and mostly white or yellow.

A pollination study in the Colombian Amazon found that *Costus scaber* was visited by five different hermit hummingbird species (Rodríguez-Flores *et al.* 2012). Four of these species occur in Suriname, including the Rufous-breasted Hermit (*Glaucis hirsutus*), Great-billed Hermit (*Phaethornis malaris*), Straight-billed Hermit (*Phaethornis bourcieri)* and Reddish Hermit (*Phaethornis ruber*). The last species is considered a specialist pollinator of *Costus* spp.

Mutually beneficial interactions have been documented between ants and *Costus* species (Schemske 1980, 1982). Upon close observation, ants can be found clustered on most *Costus scaber* inflorescences. The ants are attracted by 'extra-floral nectaries' on the overlapping floral bracts (leaf-like, colored appendages) of the inflorescence that produce sugars. In some cases, ant colonies also nest within the floral bracts. Sheltered and fed ants typically function to protect host plants from herbivores and may also assist in seed dispersal (Kay *et al.* 2005).

Figure 109. *Costus scaber (redisangrafu)* herb, spiral stem growth and inflorescence spike (image © Pieter & Siela Teunissen).

Figure 110. Ants occupying a *redisangrafu* inflorescence spike (image © Pieter & Siela Teunissen).

Figure 111. *Redisangrafu fruits* (image © Pieter & Siela Teunissen).

Description: Large perennial monocot herb to 3 m tall, with spiral stem growth and large fleshy, spirally arranged leaves (Figure 109); inflorescences terminal (occurring only at the ultimate end of the plant stem and not produced laterally along the stem), each one forming a tight cluster of bright red to orange-red leathery bracts, to 22 cm long x 4.5 cm wide (Figure 110); three sepals, united into narrow red tube; three petals, orangish-yellow, to 4 cm long. Fruit a white, ovoid capsule with many black seeds bearing white arils (Figure 111).

Distribution: Mexico, Central America, the West Indies, and northern South America; common along small streams, forest openings and clearings throughout Suriname.

Use: Primary herb for the traditional Afro-Surinamese religion (van Andel *et al.* 2013); medicine for colds, skin problems, postpartum vaginal baths (van Andel and Ruysschaert 2011).

Similar species: *Costus arabicus* is similar to *Costus scaber* but has white flowers.

Cyperus giganteus
(no English name)
Piripiri, prasorograsi
Cyperaceae

Piripiri is a giant, grass-like sedge that is common in Suriname's coastal swamps (Figures 112, 113). The ecological dominance of this species at some local sites is likely due to its large size and production of compounds toxic to other plants (allelopathy). Vilhena *et al.* (2009) found that the essential oils (primarily cyperotundone) of *piripiri* rhizomes inhibited the growth of three common Amazonian weed species – *Mimosa pudica, Senna obtusifolia* and *Pueraria pliaseoloides*. In the Brazilian Amazon, large vegetation mats formed mostly of *piripiri* sometimes break away and form floating islands on the Amazon River. Rocha *et al.* (2015) found that a large fire on one of these islands did not reduce the dominating capacity of the sedge over other plants in cover and abundance.

Figure 112. ***Cyperus giganteus (piripiri)*** **sedge, typical habitat and growth form (image © Pieter & Siela Teunissen).**

Figure 113. ***Piripiri*** **inflorescence, close-up (image © Pieter & Siela Teunissen).**

Description: Giant perennial sedge with leafless stem (culm), 1.5-3.5 m tall; Inflorescence a spoke-like array of horizontal branchlets (umbel) at the stem apex, approximately 60 cm across; plants bisexual, with clusters of spikelets that are not easily broken apart; flowers tiny; fruits three-sided achenes (single seeded) (Tucker and Ghandi 2013).

Distribution: Widespread in Central and South America, from Uruguay to central Mexico and the Caribbean; locally abundant in marshes in coastal Suriname and in the Interior.

Use: Pest and disease control: the essential oil of *Cyperus giganteus* rhizomes was found effective at killing eggs and larvae of the tiger mosquito *(Aedes albopictus)* (Kempraj and Bhat 2008)

Similar species: This species has a similar appearance and is closely related to the Old World papyrus (*Cyperus papyrus*), a source of fiber from which paper was made in the Nile and Mediterranean region in classical times.

Doliocarpus dentatus
Water liana (name used for many Dilleniaceae liana species)
Diatitei, watratitei (used for many Dilleniaceae liana species)
Dilleniaceae

In the old-growth forests of Suriname, the most impressive, giant woody climbers (lianas) are often Doliocarpus species. The genus includes nine species in Suriname (Aymard 1998). They are easily recognized by the reddish-brown, flaky bark, soft wood and alternate, robust leaves with canal-like petioles (Figures 114, 115). *Watratitei* leaves often go through conspicuous transitions during development – with large, spine-tipped or toothed leaf blades as juveniles and smaller, untoothed leaf blades in older plants.

Watratitei species are 'big bang' reproductive strategists. Masses of showy, short-lived flowers are produced simultaneously, in the hopes of quickly attracting masses of pollinators (Figure 116). By contrast, other plant species are 'steady state' reproductive strategists. Their approach is to produce a few flowers per day over long periods and secure the long-term fidelity of a few pollinators. The open flower of *watratitei* makes pollen and nectar available to a wide variety of insects, but bees seem to be the main effective pollinators (Rech *et al.* 2011). The red fruits of *watratitei* are typical for bird-dispersed fruits, with bright white, fleshy sweet arils surrounding black seeds.

Figure 114. *Doliocarpus dentatus (watratitei)*, woody canopy liana (image © Bruce Hoffman, standing in photo: Keeng Kumu of Kwamalasamutu village, Suriname).

Figure 115. Young leaves of *Doliocarpus sp.* in open area (image © Pieter & Siela Teunissen).

Figure 116. Flowering branch of *Doliocarpus spraguei, flowers similar to Doliocarpus dentatus* (image © Pieter & Siela Teunissen).

Figure 117. Robust, toothed leaves and red fruit capsules of *Doliocarpus dentatus* (image © Pieter & Siela Teunissen).

Description: Robust liana, 5-35 cm diameter, with flaky, red bark; leaves alternate, simple, leathery, often with teeth and a winged leaf stalk; flowers clustered on the stem, 0.5-2 cm wide, petals white, showy, with many stamens; berries round, red, 1-2 cm diameter (Figure 117); 1-2 seeds, shiny with fleshy white aril.

Distribution: Widespread in the Neotropics, from southern Mexico to Paraguay; common in Suriname, in lowland rainforest and seasonally dry forest, especially along creeks and rivers.

Use: Medicinal – more than 130 compounds isolated, many with proven biochemical activity (Lima et al. 2014), aphrodisiac (tonic), leishmaniasis and eye complaints; robust stems contain abundant, pure water (Jagessar *et al.* 2013).

Similar species: The leaves and young shoots of most Dilleniaceae species are conspicuously rough to the touch, but *Doliocarpus dentatus* vegetation is smooth.

Dalechampia tiliifolia
(no English name)
Bronwiwiri
Euphorbiaceae

Dalechampia spp. are slender, inconspicuous vines with highly complex flowers. What appears as a 'flower' at a distance is actually two large 'bracts' enclosing several complex groups of unisexual flowers and small 'bractlets' (Figures 118, 119). Female (pistillate) flowers are present in three-flowered 'cymule' clusters and male (staminate) flowers occur in clusters with bracts and glands. This basic arrangement changes little across the genus. *Dalechampia* and *Clusia* spp. are the only plant genera in the world known to produce resin as a pollinator reward (Armbruster 1984).

In *bronwiwiri,* the male bractlets secrete a yellow resin that is collected and used by pollinating female euglossine bees in nest building *(Eulaema, Euglossa)* (Armbruster and Herzig 1984). Flowering of this species occurs year-round. *Dalechampia brownsbergiensis* is a rare species discovered in Suriname in 1979 (Armbruster and Webster 1982). This species produces essential oil fragrances as a pollinator reward rather than resin. Male euglossine pollinator bees (*Euglossa, Eulaema, Eufriesea* spp.) may be using the 'perfume' to produce sex pheromones (Armbruster 1993).

Seed dispersal in *Dalechampia* species is largely a do-it-yourself activity. Fruit capsules develop a twisted tension as they mature, eventually bursting and flinging seeds many meters away (ballistic dispersal). The showy, paired inflorescence bracts wither or bend to the side, just before the capsules begin exploding. The seeds lack arils or any other animal reward and Armbruster (1982) documented a lack of interest in the seeds by ants and agoutis. A common, dangerous leaf pest of *Dalechampia* species are larvae of the butterfly genus *Hamadyras* (Lepidopteran). The butterfly lays eggs on the leaves and the larvae quickly defoliate individuals.

Figure 118. *Dalechampia tiliifolia* vine, leafy branch with large white bracts enclosing inflorescence (pseudanthium) (image © Olivier Gaubert).

Figure 119. Close-up of a *Dalechampia tiliiflolia* inflorescence (image © Olivier Gaubert).

Description: Slender twining vine covered with stinging hairs; stipules present; leaves simple, alternate, ovate or narrow, three-lobed and unlobed (both present during flowering), 6-15 cm long, velvety below; large bracts cream-white, almost round, tips three-lobed; floral resin yellow; sepals with dense, stiff hairs; style tip unevenly dilated; fruit a three-parted capsule with bristly hairs; seeds mottled brown and gray, smooth (Gillespie and Armbruster 1997).

Distribution: Widespread from Mexico to southern Brazil; open, disturbed areas, secondary scrub, and forest edges.

Use: (No known use)

Similar species: In the similar *Dalechampia affinis,* leaves and showy floral bracts are usually rounded (vs. three-lobed) and the style forms a symmetrical umbrella shape at the apex (vs. unevenly expanding).

Eperua falcata
Walaba, soft walaba
Walaba, babunwalaba
Fabaceae (subfamily Caesalpinioideae)

Eperua falcata is one of the 20 most common canopy tree species in the entire Amazon Basin, with highest density in the Guiana Shield region (ter Steege *et al*. 2013). This species occurs in almost all forest types in Suriname and often dominates in seasonally flooded and white-sand forests. *Walaba* occurs in especially high abundance in west Suriname (Schulz 1960, Bánki 2010). Several hypotheses have been offered to explain the considerable ecological success of *walaba* (ter Steege 2006). First, legume trees of the Cesalpinioid sub-family are widely known for root adaptations (ectomycorrhizae, root node nitrogen fixation) that make them nutritionally self-sufficent and able to survive on very poor soils. The second hypothesis is based upon the relatively low forest disturbance in the Guiana Shield region compared to the Amazon Basin on a geological timescale. Under such conditions, large-seeded, shade-tolerant species such as *walaba* likely have a competitive advantage over small-seeded, shade-intolerant species.

Walaba trees flower and fruit during dry seasons in Suriname. The long hanging 'strings' that display flowers and fruits below the tree canopy suggest that bats are important pollinators (Figure 120). Ballistic (exploding) pods disperse seeds at short distances from mother trees (Cowan 1975, Hammond and Brown 1995). Short-range dispersal, in combination with the special capacity to colonize poor soils, could explain the clumped, yet widespread range of this species (Brousseau 2013).

Figure 120. *Eperua falcata (walaba)* tree, with suspended fruit pods (image © Pieter & Siela Teunissen).

Figure 121. Open flowers in hanging inflorescence of walaba (image © Pieter & Siela Teunissen).

Description: Tree to 40 m tall, and up to 80 cm diameter, trunk base fluted; bark brownish-gray with large lenticels; leaves alternate, compound, even-pinnate with sickle-shaped leaflets; flowers red, pink to whitish on long, hanging, brown-hairy inflorescence stalks to 2 m long (Figure 121); fruit pods flat, curved, reddish-brown, twisting tightly after releasing seeds; seeds oval, compressed.

Distribution: In the Guiana Shield region and adjacent areas; frequent to dominant in forest on a wide range of soils (often pure white sand) at elevations up to 700 m.

Use: Important timber for construction – resistant to insects and fungi (Amusant *et al.* 2007, Vink 1965); medicinal and ritual – resin, bark and leaves (van Andel and Ruysschaert 2011).

Similar species: *Eperua falcata* is the most widely distributed of the four *Eperua* species in Suriname. *Eperua rubiginosa,* with which it is sometimes confused, differs by its non-sickle-shaped leaflets.

Parkia pendula
(no English name)
Kwatakama
Fabaceae (subfamily Mimosoideae)

Kwatakama is one of the iconic giant tree species of Suriname (Figures 122, 123, 124). It is recognizable even from a great distance due to the broad, horizontally extended crown and branches. Flowering and fruiting occurs mostly in the dry season, when flower heads or fruit pods hang conspicuously on long stalks below the crown. *Kwatakama* is pollinated by large, unspecialized bats in the *Phyllostomus* genus, primarily *Phyllostomus discolor*, that can easily access the hanging flowers. The complex flower heads (capitula) of *kwatakama* include approximately 250 sterile nectar-producing flowers and 1000 fertile flowers. The sterile and fertile flowers open in sequence and nectar production begins at dusk. Each flower head is available to pollinator bats for only one night (Piechowski *et al.* 2010).

At maturity, the fruit pod walls of *kwatakama* split along one side, abundant clear, sticky gum is produced, and small seeds within the pod become stuck in the gum. The gummy fruits remain attached to the tree for several months and are an important food resource for many primates. Peres (2000) considered the gums a 'keystone resource' because they are nutritional, highly desired and available during seasonal food scarcity. In central Suriname, van Roosmalen (1985b) observed Golden-handed Tamarins (*Saguinus midas midas*) and spider monkeys (*Ateles paniscus*) consuming the gums and dispersing the seeds.

Figure 122. *Parkia pendula (kwatakama)* tree with typical flattened crown (image © Pieter & Siela Teunissen).

Figure 123. *Kwatakama* tree trunk (image © Reinaldo Aguilar).

Figure 124. Hanging clusters of *kwatakama* fruit pods (image © Pieter & Siela Teunissen).

Description: Common emergent tree of rainforest, to 40 m height and with a horizontal, flat crown, often with 1 m high buttresses; bark reddish-brown, rough, peeling in large rectangular flakes; leaves alternate, compound, bi-pinnate with many small leaflets; main leaf stalk with small flat gland on upper side; compound inflorescences (heads) hang down like strings on long stalks; mature flower heads bright red, dusted with yellow pollen, turning purplish; fruit a slightly curved, flat pod, 7.5-20 x 2 cm, with thick edges; seeds covered in sticky gum (Hopkins 1986).

Distribution: Most widely distributed of Neotropical *Parkia* species, from Bolivia and southeast coastal Brazil to Honduras, up to 500 m elevation. Common in non-flooded old-growth forest.

Use: Medicinal – multi-use; cultural – trees believed to be spirit-inhabited; gum used for catching birds (van Andel and Ruysschaert 2011).

Similar species: *Parkia pendula* is distinguished from similar *Parkia* species by very fine leaflets, red bark and smooth fruit pods. *Parkia nitida* has yellowish bark and *Parkia ulei* has hairy fruit pods.

Vouacapoua americana
Partridgewood, brownheart
Broin-ati, bruinhart
Fabaceae (subfamily Caesalpinioideae)

B*ruinhart* is a slow-growing, shade-tolerant canopy tree species growing most commonly in non-flooded, old-growth forests (Figure 125). Its tiny, fragrant flowers are pollinated by tiny insects, such as thrips (Figure 126). Production of a high volume, simultaneous fruit fall (mast-fruiting) occurs every 3-4 years in the rainy season, from April to June. Scatter-hoarding rodents, such as agoutis and acouchies, disperse the large seeds (Figure 127) mostly within a short distance (< 5 m) of the parent tree (Forget 1994). Buried seeds have a higher survival rate due to protection from pathogens, fungi and insects. Relatively short-distance dispersal explains the often clustered distribution of *bruinhart* individuals.

Synchronization between seed fall timing, rodent activity and soil moisture is needed for effective regeneration of *bruinhart*. Populations are thus highly vulnerable to logging and hunting, which negatively affect both soils and rodent populations. Intense logging more than 100 years ago likely led to the absence of *bruinhart* in some older secondary forests (Forget 1997). This species is still reasonably abundant in Suriname, but populations have been severely impacted in Brazil due to logging. As a result, *bruinhart* has been listed as 'critically endangered' on the IUCN Red List (IUCN 2015). *Bruinhart* has been the subject of extensive research on population dynamics and genetic diversity in Neotropical forests, especially in French Guiana (see for example Dutech *et al.* 2004).

Figure 125. ***Vouacapoua americana (bruinhart)*** **tree, trunk irregular, with deep grooves and holes (image © Pieter & Siela Teunissen).**

Figure 126. Inflorescences of ***bruinhart*** **(image © Pieter & Siela Teunissen).**

Figure 127. Fruits and large seeds of ***bruinhart*** **(image © Pieter & Siela Teunissen).**

Description: Tree to 35 m, buttresses low, trunk irregular, with deep grooves and holes, bark brown; leaves pinnately-compound; inflorescence brown-hairy; small flowers yellow; fruit pod ovoid, yellowish-brown, hairy, splitting open along midrib,one-seeded; seed round, brown or black.

Distribution: Endemic to the (eastern) Guiana Shield region; locally abundant in non-flooded forest, especially on lateritic ridges in the (northern) savanna belt in Suriname (Schulz 1960).

Use: Timber – a highly valued commercial timber tree in the Guianas and Amazonian Brazil.

Similar species: Only one, easily distinguished species in the genus.

Codonanthe crassifolia
(no English name)
Brokobatrawiwiri
Gesneriaceae

Scientific reports of arboreal ant colonies associated with epiphytic plants began to emerge in the early 20th century. In Guyana, Wheeler (1921) discovered that these biotic assemblages – termed ant gardens – were commonly occupied by two or more unrelated ant species. The ant genera included *Crematogaster, Camponotus, Azteca* and *Anchoetus*. Working in Guyana, Venezuela and Suriname, Weber (1943) discovered three ant species in some nests along with ant-parasitizing flies and other insects. It soon became clear that they were not random assemblages but involved mutually beneficial, co-dependent relationships between a wide variety of ants and plants (Davidson 1988, Marini-Filho 1999).

Figure 128. ***Codonanthe crassifolia*, forming an epiphytic 'ant garden' (image © Olivier Gaubert).**

Figure 129. Flowering shoot of *Codonanthe crassifolia* (image © Pieter & Siela Teunissen).

In Suriname and elsewhere in the Neotropics, *brokobatrawiwiri* is one of the most commonly encountered ant-garden epiphytes (Figures128). Kleinfeldt (1978) studied ant gardens formed by *brokobatrawiwiri* and the ant species *Crematogaster longispina*. The primary benefits for ants were a structurally sound and dry nest and a reliable food supply from nectaries, fruit pulp and seed arils. The primary benefits for *Codonanthe* spp. were an increased growth rate due to the nest's concentrated, rich organic matter and ant-assisted seed dispersal. Likely additional benefits for ant-garden plants are protection against herbivores, and pollination.

Description: Epiphytic vine; stems hanging, to 2 m long, branches few; leaves opposite, simple; blade elliptic with smooth margin, 2-8 cm long, green or red, fleshy to leathery, underside often with nectaries; flowers one-few; calyx five-lobed, corolla tubular, white to yellow, ca. 2 cm long, nectary in basal spur; four stamens; ovary superior; berry fleshy, round, bright red, 1-1.5 cm diameter, many seeded.

Distribution: Neotropics, from Bolivia to Mexico; diverse habitats in Suriname, including old-growth forest and secondary forest.

Use: Medicinal, cultural – sap to heal eyes; magical medicine for walking on glass barefoot with no injury. The common name, *brokobatrawiwiri*, means 'broken bottle herb' (van Andel and Ruysschaert 2011).

Similar species: *Codonanthe calcarata* is similar but has a two-lobed calyx and leaves with teeth at the apex.

Ocotea guianensis
(no English name)
Sorfupisi, yoro-yoro-pisi, zilverpisi
Lauraceae

Figure 130. ***Ocotea guianensis (zilverpisi)*, leaves dark green above, bright silver to gold below (image © Olivier Gaubert).**

Figure 131. Flowering branch of *zilverpisi* (image © Olivier Gaubert).

Figure 132. Close-up of *zilverpisi* flowers (image © Olivier Gaubert).

Most Lauraceae in Suriname are medium-height tree species with spicy-scented wood, and tiny, complex flowers (van der Werff and Richter 1996). Identification of Lauraceae species is notoriously difficult. Tree spotters in Suriname distinguish the family ('pisi') and know local names for conspicuous species such as zilverpisi. Scientists recognize many more species based upon herbarium research, but this knowledge is rarely of practical use in the forest.

Lauraceae species populations are composed of separate male and female trees (dioecious) (Kubitzki and Kurz 1984). In *zilverpisi,* each fertile tree has reduced flowers with only stamens or only ovaries. The occurrence of separate sexes encourages cross-pollination when pollinators move from the flowers of one tree to another. Small bees are the main pollinators of *zilverpisi* and the family at large.

The medium-sized, acorn-like fruits of *zilverpisi* are specialized for bird dispersal. Lauraceae fruits are very important in the diet of parrots, toucans and bellbirds (Gibson and Wheelwright 1995, Tabarelli and Peres 2002).

Description: Tree to 25 m height (Figure 130) with straight, cylindrical stem and high buttresses; bark grayish, wood bright yellow-white and pleasantly aromatic (incense or spice); terminal leaves typically silvery-white, rolled up and pointing upwards; leaves alternate, blade narrow 11-16 x 2-3.5 cm (Figure 131); underside golden or silver, gland-dotted; plants unisexual; inflorescences 5-17 cm long; flowers tiny, yellow, to 0.35 cm long, six petals (Figure 132); fruits egg-shaped, 1-1.5 cm long, cupule 0.5 cm high; one seed.

Distribution: Widespread in the Amazon Basin and Guiana Shield, but only rarely in dense stands; lowland old-growth forest, savanna, secondary forest, and submontane forests on tepui summits.

Use: Essential oil, contains camphor and safrole; decay-resistant, commercial timber

Similar species: As an indicator of the identification challenges in Lauraceae, male and female individuals of a species are sometimes mistaken for two species, and two species are easily mistaken for one species. *Zilverpisi,* however, is easily recognized due to the conspicuous silver or gold leaf undersides.

Couroupita guianensis
Cannon-ball tree
Boskalebas, busikrabasi
Lecythidaceae

Scott Mori

The cannon-ball tree was first described by the French botanist Aublet in 1775 in his four-volume illustrated flora of French Guiana. It is one of the most unusual trees in the world because of the pleasantly smelling and beautiful flowers arising from its trunk, the cannon-ball-like fruits, and its interesting pollination and dispersal systems (Figure 133).

The most common pollinators of *boskalebas* are carpenter bees (*Xylocopa* spp.), attracted to the trees by their aromatic flowers (Figure 134). Female bees gather sterile pollen from the inner surface of the 'hood' to feed their larvae. While collecting sterile pollen, fertile pollen is rubbed off onto the bee's head and back and subsequently scraped onto the stigmas of new flowers. Studies have demonstrated that the two types of *Couroupita guianensis* pollen are morphologically and physiologically different (Mori *et al.* 1980).

The fruits fall to the ground at maturity and the seeds are dispersed by mammals, such as peccaries and pacas. These animals bite their way into the fruit, eat the pulp and expel the seeds after digestion. Janzen and Martin (1982) hypothesized that this species was originally dispersed by now extinct giant animals (megafauna) of the Pleistocene Epoch.

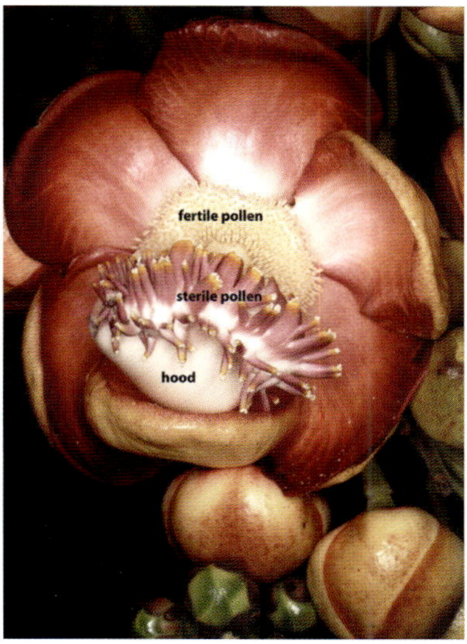

Figure 133. *Couroupita guianensis (busikrabasi)* tree, unique cannonball-like fruits borne on the trunk (image © L. E. Garnez Álvarez).

Figure 134. Flower of *busikrabasi*, with fertile pollen present at the base of the androecium (male part) and sterile pollen present in the 'hood' - the strangely formed top part of the androecium (image © Scott Mori).

Description: Trees to 35 m tall, with flowers and fruits borne on long stalks emerging from the trunk; flowers zygomorphic, 5-6 cm diameter, with six petals, pink to red, male part of flower (androecium) forming a flat hood on one side that arches over the ovary; fruits globose, 12-25 cm diameter, not breaking open; seeds embedded in pulp.

Distribution: Occurs across the northern and western side of the Amazon and in the Guianas; in central Amazonia replaced by the riverside *Couroupita subsessilis* (Mori *et al.* 2010); In Suriname, occurs in marshy areas of the Coastal Plain and, close to Paramaribo, in forest dominated by *possentri* (*Hura crepitans*).

Use: Ornamental – widely planted in tropical and subtropical botanical gardens; fruit pulp fed to domestic farm animals but not consumed by humans.

Similar species: *Couroupita guianensis* is highly distinctive and is the only *Couroupita* spp. occurring in Suriname (Mori and Prance 1993).

Gustavia augusta
(no English name)
Watramamabobi
Lecythidaceae

Scott Mori

The large, radially symmetric flowers of *watramamabobi* have anthers that open with pores at their apex (Figure 135a) – a feature absent in other genera of the Brazil Nut family. Mori and Boeke (1987) documented the pollination of this species in French Guiana. The flowers opened during the night and, before dawn, individuals of a large bee (*Megalopta genalis*) entered the flowers. This was the first time that nocturnal pollination had been recorded for this species. The bees produced a buzzing sound (vibration) that caused pollen to be actively shaken out of the anthers like salt from a salt shaker. Much of the pollen collected in this way is fed to bee larvae, but subsequently visited flowers and receptive stigmas certainly receive an unintentional share. Mori and Boeke (1987) also observed that *watramamabobi* flowers were visited in the daytime by *Trigona spp.* stingless bees. These bees are inefficient cross-pollinators because they typically collect pollen from a single flower on each foraging trip outside of the nest.

Figure 135. *Gustavia augusta (watramamabobi)*: a. flowers; b. fruit capsule; c. seeds with orange arils (images © Scott Mori).

The fruits of *watramamabobi* have an edible pulp and the seeds have an expanded, contorted yellow-orange aril that is eaten by animals (Figures 135b and 135c). There are reports that monkeys, other mammals such as agoutis, and even fish consume the fruits and disperse the seeds when they defecate.

Description: Understory tree; leaves simple, alternate, clustered at branch tips, blades longer than 20 cm; flowers radial, calyx lobes small or undeveloped, eight petals, pinkish-white, stamens arising from a tube, filaments often pink-tipped, anthers with terminal pores; fruit pods remain closed at maturity and do not naturally split along seams (indehiscent) (Figure 135b), ± round, smooth, pulp yellow, seeds black, aril yellow to orange (Figure 135c).

Distribution: Widespread in the Amazon Basin and Guiana Shield region; also the northernmost Atlantic coastal forests of Brazil; especially common in seasonally flooded (marsh) forest and along creeks in Suriname.

Uses: Medicinal – a remedy for vomiting in children; applied to lesions of leishmaniasis; a remedy for dizziness (Grenand *et al.* 2004); wood sometimes used in house construction.

Similar species: *Gustavia hexapetala* is the only other *Gustavia* spp. in Suriname. In contrast to *Gustavia augusta*, this species prefers well-drained soils; leaves are less than 20 cm long and not tightly clustered at branch ends; flowers are white with conspicuous calyx lobes; and fruits are winged (Mori *et al.* 2010).

Strychnos mitscherlichii
Devildoer, curare (English names use for many *Strychnos* spp.)
Dobrudua (local name used for most *Strychnos* spp.)
Loganiaceae

The *Strychnos* genus includes mostly woody canopy lianas that are important components of Neotropical forest ecosystems. Robust lianas form a vertical and lateral web of stems in forest canopies that can reduce tree fall from storms and facilitates cross-canopy movement of forest animals in search of food or shelter. Due to the thick and tough fruit wall of most *Strychnos* species, only specialized fruit-eaters, such as spider monkeys in Suriname and chimpanzees in Africa, are able to open the fruits. The fruits contain highly toxic seeds embedded in a nutritious pulp. In French Guiana, researchers found that the pulp of *Strychnos erichsonii* was sweet, but also contained almost 7 percent protein in dry weight (Julien-Laferrière 1999). Animals must lightly gum and swallow the toxic seeds, rather than biting them, in order to safely consume the pulp (McGrew *et al.* 1999, van Roosmalen 1985a). *Strychnos* spp. thus present an effective deterrent to destructive seed consumption and increase the odds of effective seed dispersal.

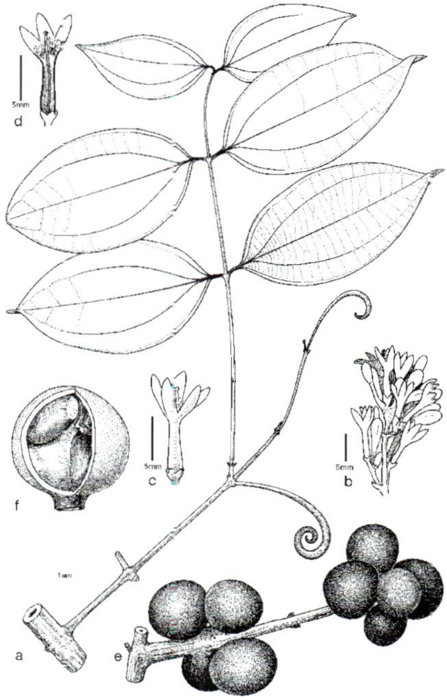

Description: Large, woody canopy liana to 30 cm in diameter (Figure 136); bark light-colored with large orange-brown raised specks; leaves opposite, simple, 8-15 x 4.5-7 cm, 3-5 palmately veined; inflorescence with dense flower clusters, corolla tubular, 0.3-0.4 cm long, whitish and woolly haired; berries round, to 3 x 2 cm diameter, maturing yellowish-orange, shell thin; two seeds, embedded in pulp; cut parts often have a disagreeable, 'poisonous' scent and bitter taste.

Distribution: Widespread in northern-eastern South America; common in flooded and non-flooded old-growth forests in Suriname.

Use: Medicine/poison – genus globally famous as an arrow and ordeal poison (curare); more than 300 *Strychnos* spp. alkaloids have been isolated (Quetin-Leclercq *et al.* 1990, Bisset 2002); aphrodisiac – *Strychnos mitscherlichii* and other *dobrudua* species are components of aphrodisiac tonics in Suriname (van Andel and Ruysschaert 2011, van Andel *et al.* 2012b).

Figure 136. ***Strychnos mitscherlichii (dobrudua)*** **woody canopy liana: a. leafy climbing branch with tendril-hooks; b. dense inflorescence of tubular flowers; c. single flower, corolla tube 0.5–1.1 cm long; d. corolla cut lengthwise – long style, short anthers, hairy throat; e. globose fruits on branch; f. fruit cut open to reveal disc-shaped seeds (image © Naturalis Biodiversity Center, Hendrik Rypkema).**

Similar species: Some *Bauhinia* spp. (Fabaceae) and *Uncaria* spp. (Rubiaceae) liana species have superficially similar hook-like tendrils, but no other genus or species has the combination of opposite leaves, palmate veins and hook-like tendrils.

Stigmaphyllon sinuatum
(no English name)
Konkonikasaba
Malpighiaceae

The Malpighiaceae family includes a diversity of growth forms and fruit types, but the flowers are remarkably similar across the family (Anderson 1997). The flowers are unique due to conspicuous, oil-bearing glands on the underside of each sepal and five claw-like petals (Figures 138, 139). Vogel (1990) recognized a unique pollination 'syndrome' in the family, based upon the presentation of high-energy oils rather than nectar as a floral reward.

Centridini bees are specialized pollinators of Neotropical Malpighiaceae and collect the oils as food and to waterproof nests for their bee brood (Buchman 1987). The structure of the flower – the claw of one petal is longer and thicker than the other petals – is designed to manipulate the pollinator. Bees hang on to the larger claw petal, using specialized combs to pierce and collect the oil while they also brush pollen off onto the stigmas (Sigrist and Sazima 2004).

Konkonikasaba and most Malpighiaceae lianas have winged seeds and are adapted for wind dispersal. However, some species adapted for water dispersal have reduced wings and air pockets to ensure flotation.

Figure 140. *Stigmaphyllon sinuatum* (*konkonikasaba*) leafy stem; inset: underground tuber (image © Pieter & Siela Teunissen).

Figure 138. Clawed-petal flowers and winged fruits of *konkonikasaba* (image © Pieter & Siela Teunissen).

Figure 139. Oil glands on underside of *konkonikasaba* flower, attached to sepals (image © Pieter & Siela Teunissen).

Description: Robust, woody liana; leaves opposite, simple, often heart-shaped (Figure 137), with silky, flattened hairs below; leaf stalk long (to 13 cm), with two flat, oval glands near apex; stipules present; flowers clustered on sparse branches, five-parted (Figures 138, 139); sepals each with two oil glands below; petals claw-shaped, reddish-yellow; seven stamens; three stigmas, each bearing a tiny 'leaf'; fruits three-winged, breaking into separate winged seeds, wing ca. 4.5 x 1.5 cm.

Distribution: Widespread in northern South America; occurs in lowland rainforest on white and brown sands in Suriname; also common as spreading, weedy liana in secondary forest and farms.

Use: Underground tubers of *konkonikasaba* inedible (Figure 140); medicinal – tuber used for eye pain, 'bosyaws' sores (leishmaniasis), stomach pain, menstruation problems; leaves used for rheumatism, trembling limbs, paralysis, genital steambaths, and malaria (van Andel and Ruysschaert 2011).

Similar species: *Stigmaphyllon convolvuliifolium* has smaller leaves with entire (untoothed) margins, lightly hairy leaf undersides and shorter leaf stalks.

Ceiba pentandra
Kapok, silk-cotton tree
Kankantri
Malvaceae (formerly Bombacaceae)

Kankantri is a giant canopy tree occurring on riverbanks and sand ridges throughout Suriname (Figures 141). Flowering, fruiting and leaf fall occur mostly during the dry season. *Kankantri* produces thousands of flowers in a short time and the strong scent attracts a wide range of nocturnal and diurnal animals. Only leaf-nosed bats (Phyllostomidae), however, play a serious role in pollination (Gribel *et al.* 1999; Rojas-Sandoval *et al.* 2008). Despite the impressive floral display on leafless branches, trees do not always bear abundant fruits. Many flower buds and young fruits are infested by weevils and consequently aborted (Fernandez *et al.* 2008). The *kankantri* has spread to tropical plant communities worldwide due to extreme long-distance dispersal of the light-weight, kapok-laden seeds (Figure 143) (Dick *et al.* 2007).

Figure 141. *Ceiba pentandra (kankantri)* tree, on river bank (image © Pieter & Siela Teunissen).

Figure 142. Alternate, palmately compound leaves of *kankantri* (image © Pieter & Siela Teunissen).

Figure 143. Fruit capsules of *kankantri*, seeds embedded in 'kapok' fibers (image © Pieter & Siela Teunissen).

Description: Tree to 55 m with straight, cylindrical trunk up to 2.5 m in diameter, often with large spines (especially in juveniles), buttresses high, crown broadly umbrella-shaped; leaves palmately compound, with 5-9 finger-like leaflets (Figure 142); flowers white; fruits brown capsules, opening with woody valves to expose numerous woolly-haired seeds (kapok) (Figure 143).

Distribution: Pantropical – including the full north-south extent of the Neotropics; occurs in well-drained and flooded old-growth and riparian forests.

Use: Often spared in Suriname due to poor quality wood, shading benefits, and to cultural taboos concerning tree spirits (van Andel and Ruysschaert 2011).

Similar species: Some tree species have similar leaves and produce kapok, but none attain the size of the *kankantri*. *Jacaratia spinosa* (Caricaceae) juveniles have similar leaves and spiny trunks but are distinguished by milky latex.

Theobroma subincanum
Wild cacao
Busikakaw (local name for all wild *Theobroma* spp.)
Malvaceae (formerly Sterculiaceae)

As the renowned source of chocolate and cacao butter, *Theobroma cacao (kakaw)* has long commanded the world's attention. However, some wild cacao relatives, including *Theobroma bicolor, Thebroma grandiflora* (cupuaçu) and *Theobroma speciosum*, also produce attractive, tasty fruits. Baleé (1989) viewed forests with a high abundance of *Theobroma speciosum* trees in Brazil as artifacts of ancient human cultivation.

Wild *Theobroma* populations have potential significance as gene reservoirs that contribute to pest and disease resistance of current cacao cultivars and as sources of new foods or condiments. The three non-cultivated cacao species occurring in Suriname are wild *Theobroma cacao*, *Theobroma subincanum* and *Theobroma velutina*. Unfortunately, crop improvement initiatives with rainforest trees are challenging due to the low genetic diversity often found in isolated populations (Rivas *et al.* 2013).

Theobroma subincanum is uncommon in Suriname, but widespread in the Amazon Basin. It is most abundant in lowland tropical forest with high annual rainfall. Pollinators include small nocturnal insects known as thrips, ants, midges and aphids (Young *et al.* 1987, Rodrigues and Venturieri 1997). The fruits do not naturally break open (indehiscent) and require the action of animals or water for seed dispersal. Monkeys chew open the capsules, swallow the seeds and pulp and later defecate the seeds.

Figure 144. Theobroma subincanum (*busikakaw*) leafy branch, fruit with edible, sweet seed pulp (image © Bruce Hoffman).

Figure 145. Flower clusters borne on the trunk of *busikakaw* (image © Marcus Vinicius Lameiras).

Description: Medium-sized tree to 12 m tall, 15-20 cm diameter; bark gray, flaky; vegetation with reddish-brown, stellate (star-shaped) hairs; leaves alternate, simple, large and leathery, elliptic-oblong, 16-40 x 5-20 cm (Figure 144); main trunk and branches bearing dense clusters of dark red, five-parted flowers each, 2 cm long (Figure 145); corolla complex, in two rows including thin 'hood' petals and fleshy 'lamina' petals; stamens also in two rows; fruits ovoid-rectangular, 10 x 6 cm, dull orange and smooth, with seeds immersed in sweet, scentless pulp (Cuatrecasas 1964).

Distribution: Widespread in the Amazon Basin, Orinoco watershed and the Guianas; non-flooded rainforest understory, also along creeks and rivers.

Use: Edible – Pulp around seeds consumed in Suriname; pulp used in Brazil to make sweet drinks.

Similar species: Distinguished from other *Theobroma* spp. in Suriname by densely flowering branches and a smooth fruit wall without ridges.

Tibouchina aspera
(no English name)
Wasduku
Melastomataceae

Melastomataceae is one of the largest plant families, with over 4300 species and ca. 170 genera native to (sub)tropical regions of the world (Clausing and Renners 2001). The genus *Tibouchina* has more than 400 species, with a center of diversity in Brazil. Melastomataceae species are common in open and disturbed places, because seeds often require high light conditions for effective germination (Ellison *et al.* 1993). In these open environments, the family is an important food source for many smaller fruit-eating birds, such as manakins, which are the main seed dispersers (Stiles and Rosselli 1993).

Melastomataceae species present mostly pollen, rather than nectar, as a floral reward and bees are the primary effective pollinators (Renner 1989). The showy stamens of *wasduku* are meant to attract pollinators, but often in a deceptive way. They mimic a pollen-rich flower, even when they are empty, and bees must directly visit the flowers to determine if they possess a reward (Buchmann 1983). As a further demand upon pollinators, the stamens are specialized for 'buzz pollination'. Bee visitors must vibrate their flight muscles at a high and specific frequency to shake pollen out of the anther pores (See also the *Gustavia augusta* (Lecythidaceae) species account). In some cases, wasduku flowers change color after pollination (Pereira *et al.* 2011). This is likely a signal to focus the attention of bees on finding any remaining unpollinated flowers.

Figure 146. Flowering *Tibouchina aspera* *(wasduku)* shrub in a white-sand savanna (image © Pieter & Siela Teunissen).

Description: Strongly branched shrub (Figure 146), to 3 m tall; stem and leaves rough and grayish-green, with scale-like hairs; leaves opposite and simple, rigid-leathery, 5-7 palmately veined, veins diverging at base and meeting again at apex; inflorescence many flowered, flowers ca. 2 cm diameter, rose-purple to pink; berry red, 0.7-1.2 cm diameter.

Distribution: Widespread in the Neotropics, from Central America to the Amazon Basin and the Guianas; common in savannas and open, weedy environments.

Use: Ornamental – popular garden plant in Suriname; medicinal/ritual – used in herbal steam baths and remedy for genitourinary problems and body pain (van Andel and Ruysschaert 2011).

Similar species: Many unrelated savanna plants have a similar, shrubby vegetative appearance, reflecting adaptation to a similar habitat. The Melastomataceae family is easy to distinguish with opposite, simple leaves, absence of stipules, and palmate main veins with conspicuous to faint ladder-like cross-veins. *Tibouchina* spp. have relatively large flowers in comparison with other species in the family.

Carapa procera (formerly *Carapa surinamensis*)
Crabwood (English name used for all *Carapa* spp.)
Redikrapa, krapa (latter local name used for all *Carapa* spp.)
Meliaceae

Redikrapa (*Carapa procera*) and *wetikrapa* (*Carapa guianensis*) are two very similar primary forest species that can grow in flooded or non-flooded forests (Figure 147). In a productive year, trees flower simultaneously and abundant fruit-fall occurs in the mid-to-late Long Rainy Season (May-July). In non-flooded areas, fallen pods split open upon impact and spill their angular, oil-rich seeds on the soil (Figures 148, 149). As discussed in the introduction, close proximity to a mother tree poses many threats to seeds and seedlings – including predator consumption and resource competition with both siblings and the mother tree.

A mutually beneficial relationship (mutualism) has developed between *krapa* species and forest rodents as seed dispersers (Forget 1996, Jansen and Forget 2001). The Red Acouchy (*Myoprocta acouchy)* and Red-rumped Agouti (*Dasyprocta leporina*) are very quick to discover, move and hide *krapa* seeds for later consumption, an activity known as 'scatter-hoarding'. Seeds thus escape immediate predation and may be placed in more favorable spots for germination. Many seeds survive because rodents forget the caches or are themselves eaten by hunters or forest predators. Rodents benefit from an abundant, energy-rich resource (mass fruiting, large seeds) that is timed perfectly – just before the rains subside and many animals face starvation (Jansen *et al.* 2004). One of the less obvious impacts of over-hunting is the loss of rodents as dispersal agents for *krapa* and other large-seeded tree species (Forget and Jansen 2007).

Figure 147. Flowering *Carapa procera (redikrapa)* tree (image © Pieter & Siela Teunissen).

Figure 148. Five-parted fruit capsules of *redikrapa* (image © Pieter & Siela Teunissen).

Figure 149. *Carapa guianensis (wetikrapa)* seeds with flat-angular surfaces (image © Bruce Hoffman).

Description: Tree to 35 m tall, with a straight trunk and often swollen or buttressed base, diameter to 95 cm; bark thick, light brown, wood yellow-white, bitter; leaves alternate compound, clustered at branch ends; leaflets 6-14, to 45 x 12 cm each, smooth and shiny; inflorescences large, terminal displays; flowers small, cream-colored, five-parted, unisexual; fruits brown, woody capsules, 7-10 cm diameter, five-chambered with five prominent ridges; seeds 4-16 per fruit, with conspicuous flat sides.

Distribution: Widely distributed in the Neotropics and tropical Africa; common in seasonally flooded and upland forests, highest densities in middle and eastern Suriname.

Use: High-quality oil from seeds as insect repellent (local market product); soap and candles; excellent insect-resistant timber.

Similar species: *Carapa guianensis* (*wetikrapa*) is distinguished from *redikrapa* by four-parted flowers and four-ridged fruit capsules. Also *wetikrapa* occurs more in western Suriname and *redikrapa* more to the east.

Ficus nymphaeifolia
Strangler fig (English name used for many *Ficus* spp.)
Abrasa (local name used for many *Ficus and Clusia* spp.)
Moraceae

The common name *abrasa* refers to the habit of *Ficus nymphaeifolia* and many other 'strangler' figs to overgrow and squeeze the life out of forest trees). *Abrasa* begins life as a small, sticky seed defecated on a tree branch by a bird or monkey high in the forest canopy. Young *abrasa* plants send roots downwards from their tree perches, seeking the nutrients and water of the forest floor soil below (Figure 150). The aerial roots of strangler figs thicken and fuse together where they touch to build a self-supporting structure (Figure 151). Eventually the host tree dies, becoming a rich source of nutrients for the strangler fig and the local ecosystem.

Strangler figs may seem cruel and opportunistic, but their way of life plays a vital ecological role in Suriname's forest ecosystems. The curves and crevices formed by the fused roots provide ideal living spaces for many smaller animals such as lizards, frogs, ants, bees and scorpions. The high-energy tropical fig is a favored food of birds, bats and tree-dwelling mammals (Vanderhoff and Grafton 1996). Bird and monkeys mostly pursue brightly colored, smaller figs, while bats prefer larger, dull-colored figs (Lomascolo *et al.* 2008). Figs are produced year-round, even during the dry season when food is extremely scarce and many animals die (Compton *et al.* 1996).

Figs have complex pollination systems involving co-evolutionary relationships (mutualisms) with tiny wasps (family Agaonidae). The wasps are poor flyers but may be blown by wind and inadvertently transport pollen across great distances (Heer *et al.* 2015). The unique (pseudo) fruit of figs is hollow and lined with tiny flowers inside. After a female pollinator wasp gains access, she uses some flowers as larval breeding chambers. When the larvae mature, males mate with females and pierce the fig wall, allowing pollen-laden females to escape and seek more figs.

Figure 150. Leafy shoot of *Ficus nymphaeifolia (abrasa)*, growing on host tree (image © Pieter & Siela Teunissen).

Figure 151 (right of Figure 150). Aerial roots of *abrasa*, encircling host tree (image © Pieter & Siela Teunissen).

Figure 152. Conical stipule and circular stipule scars of *Ficus nymphaeifolia* (image © Reinaldo Aguilar).

Figure 153. *Ficus nymphaeifolia* fig fruits (image © Reinaldo Aguilar).

Description: Hemi-epiphytic strangler tree; outer bark smooth, reddish-brown when young; milky latex present; leaves simple, alternate, ovate, 10-25 x 7-18 cm, smooth, fleshy; stipule conical or absent and leaving a circular scar at nodes (Figure 152); fruits round, paired, *ca.* 2.3 x 2.2 cm, green and purple (Figure 153).

Distribution: Widespread in the Neotropics and West Indies; Coastal Plain, savanna, old-growth and secondary forest in Suriname.

Use: Considered sacred and spirit-inhabited (Maroons); milk used as plaster or for relieving pain (Grenand *et al.* 2004, van Andel and Ruysschaert 2011).

Similar species: The genus is unmistakable due to the growth form and white milk in cut parts. Distinguishing among *Ficus* spp., however, is typically difficult.

Virola surinamensis
Wild nutmeg (English name used for all *Virola* spp.)
Laagland babun-udu (name *babun-udu* used for all *Virola* spp.)
Myristicaceae

Laagland babun-udu is common in Suriname and thrives under a wide range of ecological conditions. Individual trees are either male or female (dioecious). Male trees flower all year and females flower and fruit mostly in the drier months. *Babun-udu* species are considered 'keystone' or 'foundation elements' in Suriname's forest ecosystems because they provide critical dry-season food for animals that, in turn, positively affect seed distribution and germination, forest structure and diversity (Howe 1984). The ovoid seed, with its bright red, energy-rich aril, is dispersed primarily by birds, especially toucans. Agoutis play a secondary dispersal role by collecting and burying seeds that are dropped or discarded (Howe and Kerckhove 1980, Forget and Cujipers 2008).

The Brazilian government lists *laagland babun-udu* as 'threatened with extinction' in Brazil (IBAMA 2008). Ferreira *et al.* (2014) found that current logging practices in Brazil compromise the genetic health of *laagland babun-udu* populations and recommended protecting seed-producing female trees. In Suriname, it was once an important timber species, but there is currently less demand and populations are not greatly threatened.

Description: Canopy tree, with conspicuous horizontal branches, prop roots and buttresses to 1.5 m; bark gray, slash scent spicy, sap red and watery (Figures 154, 155); leaves simple, alternate, elongate, 10-22 cm long, twigs and petioles with stellate (star-shaped) hairs; male flower displays to 13 cm long, open; female flower displays to 9 cm long, narrow, flowers 3-8; corolla small, three lobed; fruit capsule ovoid, *ca.* 2 cm wide; seed 1.5 x 1 cm, black with bright red aril (Figures 156, 157).

Figure 154. ***Laagland babun-udu*, branches with narrow, two-ranked leaves (image © Pieter & Siela Teunissen).**

Figure 155. Tree trunk buttresses of *laagland babun-udu* tree (image © Bruce Hoffman).

Figure 156. Immature fruits of *laagland babun-udu* (image © Pieter & Siela Teunissen).

Figure 157. ***Laagland babun-udu*, fruit capsule and arillate seed (image © Bruce Hoffman).**

Distribution: Lowland Amazonia, Guiana Shield, to Costa Rica and West Indies; seasonally flooded and non-flooded old-growth forest, secondary forest.

Use: Medicinal – anti-malarial, herpes virus and fungal infection remedy, many additional medicinal and ritual uses; hallucinogenic snuff in the northwest Amazon; highly prized international timber (Plotkin 1986, van Andel and Ruysschaert 2011).

Similar species: Distinguished from Annonaceae and Lecythidaceae families (many species) by bark slash with red sap, bark not peeling in strips, and horizontal branching pattern. The genus *Virola* is distinguished by stellate (star-shaped) hairs on twigs and leaves and fruits longer than wide.

Nymphaea ampla or *Nymphaea pulchella*
White waterlily
Pankukuwiwiri, watralelie (local names used for all *Nymphaea* spp.)
Nymphaceae

Above and below the waterline, *pankukuwiwiri* provides many benefits to local aquatic ecosystems (Figures 158, 159). The leaves shade and cool water bodies, increasing the percentage of dissolved oxygen. The plants provide important habitat for tadpoles, dragonflies, insect larvae, plankton and juvenile fish. Gastropods use the leaves as a 'foraging island', while waterbirds use the leaves as above-water platforms and nesting grounds. Decomposing *pankukuwiwiri* leaves and stems are colonized by several invertebrate groups, including the larvae of diverse fly species (Chironomidae, Diptera, Insecta) (Goncalves *et al.* 2003).

During the day, the white flowers of *pankukuwiwiri* open and are visited by scarab beetles (primarily Cyclocephala) (Prance 1980). The beetles are attracted by volatile oil scents emitted by the flowers at raised temperatures – a self-induced chemical process known as thermogenesis (Seymour and Matthews 2006). The beetles benefit from mating opportunities with other beetles that congregate within the floral chamber. In the evening, the flowers close and the beetle visitors are trapped overnight. To discourage self-pollination at this first stage, the female flower stigmas are receptive while the stamens remain inactive (protogyny). Once stigmas have been pollinated, the stamens begin releasing pollen to re-dust the beetles. Soon, the flowers open, releasing their beetle 'prisoners' to continue the cycle.

Figure 158. ***Nymphaea ampla*** **or** ***Nymphaea pulchella (pankukuwiwiri)*** **aquatic herb, flowers open during the day (image © Pieter & Siela Teunissen).**

Figure 159. Close-up of ***pankukuwiwiri*** **flower (image © Leen de Laender).**

Description: Aquatic plant with thick rhizomes and long-stemmed, round floating leaves, 15-45 cm in diameter; flowers white, solitary, with many yellow stamens; fruits round, black, up to 7 cm in diameter.

Distribution: Widespread in tropical and subtropical America – from central Mexico to South America and the Antilles; a dominant aquatic plant of coastal lagoons, slow-moving rivers, and canals in the Coastal Plain of Suriname.

Use: Medicine and ritual, often in combination with other plants (van Andel and Ruysschaert 2011). The name *pankukuwiwiri* ('pancake leaf' in English) refers to the leaf shape only – the leaves are inedible.

Similar species: Recent work on *Nymphaea ampla* suggests that two species should be distinguished – with *Nymphaea pulchella* in Suriname (Wiersema *et al.* 2008). The Amazon waterlily *(Nymphaea amazonum)* is also known as *pankukuwiwiri*. This lily is distinguished from *Nymphaea pulchella* by smaller leaves, smoother leaf margins, a circle of hairs on top of the leaf stalk, and night-time flowering.

Catasetum discolor
Flower-monk (English name used for many Catasetum spp.)
(no local name)
Orchidaceae

In most orchid species, flowers are bisexual and have fused male and female parts. In contrast, *Catasetum* spp. produce unisexual flowering stalks (inflorescences) of all male or all female flowers (Figure 160, 161, 162). Intermediate flowers are sometimes produced, but these are usually non-functional. Individual *Catasetum* plants may display flowering stalks of one or of both sexes, but each one is unisexual. Dodson (1962, 1978) found that the ratio of male-to-female inflorescences per plant changed with environmental conditions. Individuals were more likely to produce female or both female and male inflorescences under favorable conditions, such as full sunlight or nutrient-rich soil. Individuals in nutrient-poor or shady sites were more likely to produce only male inflorescences. Because male and female flowers look so different, populations with mostly female flowers have sometimes been mistaken as a different species from those with mostly male flowers.

Within truly bisexual flowers, pollen reception and dispersal events occur at a single location, although often at different times to avoid self-pollination. In *Catasetum* spp., pollinators must navigate between the opposite-sex inflorescences of different individuals to achieve cross-pollination. The showy male flowers of *Catasetum discolor* occur in greater abundance than female flowers and possess a trigger that fires a packet of pollen (pollinarium; plural: pollinaria) onto the bellies and legs of visiting euglossine bees (Romero and Nelson 1986). Once triggered, the male flowers soon wither. The hooded female flowers of *Catasetum discolor* are robust and longer-lasting than the male flowers. When a pollen-bearing euglossine bee enters a female flower, the pollen package becomes trapped and is yanked from the bee upon exit.

Figure 160. ***Catasetum discolor,*** **ground orchid with folded erect leaves; inflorescence of female flowers (image © Pieter & Siela Teunissen).**

Figure 161. ***Catasetum discolor*****, close-up of male flowers (image © Rui Oliviera-Santos).**

Figure 162. ***Catasetum discolor*****, close-up of robust female flower (image © Rui Oliviera-Santos).**

Description: Terrestrial orchids, forming clumps of fleshy, conical 'pseudobulbs' at base with narrow, folded leaves to 10-35 cm long; male inflorescences – stem bending, flowers 20-25, pink or greenish-red, corolla lip cup-shaped with showy, frilly margin; female inflorescences – stem erect, flowers few, large, hooded, mostly green, corolla lip deep, without showy, frilly margin.

Use: Ornamental.

Distribution: Widespread in tropical South America, from Brazil and Peru to the Guianas; in Suriname, occurs in savannas and Coastal Plain, in full sun or shade, on forest edges or in disturbed areas (Souza *et al.* 2012).

Similar species: *Catasetum macrocarpum* is similar, but with larger flowers. *Catasetum longifolium* is also similar but is an epiphyte that grows on *Mauritia flexuosa* (*morisi*) palms.

Passiflora coccinea
Passion flower vine (English name used for many *Passiflora* spp.)
Snekimarkusa, busimarkusa (local name used for many *Passiflora* spp.)
Passifloraceae

Passiflora coccinea is a very common, softly woody liana with conspicuous nectar glands on leaf margins, leaf stalks (petioles) and leafy bracts enclosing the flowers (Figures 163, 164, 165). These so-called 'extra-floral nectaries' are attractive and accessible to diverse ant species and other insects. This is a sharp contrast to the specialized plant-animal relationships that limit food and shelter benefits to one or a few insect species. Hossaer-McKey *et al.* (2001) found that different plant species bearing extra-floral nectaries, including *Passiflora coccinea*, were visited by different groups or 'suites' of ant species. In French Guiana, Leal *et al.* (2006) found that *Passiflora coccinea* extra-floral nectaries attracted 22 ant species, many more than other *Passiflora* species. These biologically diverse inhabitants were highly effective in repelling herbivores, egg-laying butterflies and nectar-thief butterflies. Remarkably, seed set was doubled in ant-protected versus unprotected fruits. Some ants were overly protective and repelled effective pollinators as well as nectar thieves. The primary pollinators of *Passiflora coccinea* in the Guianas are large bees, for example *Ptilotopus americanus*, and hummingbirds (Storti 2002).

Figure 163. *Passiflora coccinea (snekibusimarkusa)* liana, common in secondary forest (image © Pieter & Siela Teunissen).

Figure 164. Closed flowers of *snekibusimarkusa*, red bracts bearing nectar glands (image © Bruce Hoffman).

Figure 165. Open flower of *snekibusimarkusa* (image © Bruce Hoffman).

Description: Subwoody liana climbing via coiled tendrils, stems often trailing on the ground; leaves alternate, simple, 6-14 cm × 3-7 cm, three-veined at base, margins toothed; leaf stalk with two globe-shaped glands; flowers bright fluorescent red to red-orange, 7-9 cm wide, five-parted, enclosed by red, leafy bracts; sepals and petals long and narrow; flower with a central ring of white-pink filaments (corona) and a raised stalk bearing fused stamens and stigmas (gynophore) (Figure 165); fruits round to egg-shaped berries, to 6.5 x 5.5 cm, orange-yellow with green stripes; with tiny seeds embedded in pulp (Killip 1937).

Distribution: Widespread in tropical South America; common in disturbed areas, forest edges, treefall gaps, secondary forest, savannas.

Use: Sap from crushed stems used for conjunctivitis (Grenand *et al.* 2004); fruit pulp edible but not particularly tasty.

Similar species: *Passiflora quadriglandulosa* has three-lobed leaves and narrow (< 0.5 cm wide), fine-toothed leafy flower bracts. *Passiflora coccinea* has simple leaves and broad (ca. 3.5 cm wide), blunt-toothed leafy flower bracts (Tillett 2003).

Piper marginatum
cake bush, marigold pepper (English names used for many *Piper* spp.)
Uma-aneisi, aneisiwiwiri (local names used for many *Piper* spp.)
Piperaceae

Piper spp. in general have a similar vegetative appearance, which makes the genus easy to recognize (Figure 166). The assemblage of similar-looking *Piper* spp. individuals in a given area, however, can include a surprising number of unique species. One hypothesis concerning this trend is fine-scale adaptation to microhabitats and, thereby, avoidance of direct competition for the same resources (Daws *et al.* 2002). For example, *Piper marginatum* is a large-gap specialist that thrives in high temperatures and sunlight. If some *Piper marginatum* variants emerge with the ability to survive and reproduce in partial shade, this could result in a genetically distinct adjoining population.

As largely herbaceous plants, is it critical that *Piper* spp. successfully defend against being consumed by herbivores. In one study, *Piper marginatum* was found to possess 47 different carbon-based defensive compounds (Dyer *et al.* 1999). In addition, many *Piper* spp. have developed mutualisms with ants for herbivory defense. Ant colonies are provided housing in hollowed out petioles and fed with white structures known as 'food bodies'. Dyer *et al.* (1999) found that coccinellid beetle larvae, rather then ants, were housed and fed within *Piper marginatum* petioles.

Most Neotropical *Piper* spp. flowers are pollinated by non-specialist bees and flies, regardless of habitat. Seed dispersal, however, is a different story. For many forest-dwelling *Piper* spp., a few bat species (Carolliinae) are the only seed dispersers (Charles-Dominique 1991, Fleming 2004). For forest gap specialists like *Piper marginatum*, both birds and bats likely disperse seeds. Bats are considered better dispersers than birds because they defecate seeds during flight and spend more time flying in open areas.

Figure 166. *Piper marginatum (uma-aneisi)* herb, spike inflorescence composed of hundreds of tiny flowers (image © Chantal van den Bergh).

Description: Subwoody shrub, 0.5-3 m high; all plant parts smelling strongly of anise when crushed; leaves broad, 10-12 x 8-15 cm, base deeply heart-shaped, tip pointed, margin fringed; fruiting spikes upright, curled, white, 10-13 cm long; fruits tiny, green, oblong, with three stigmas.

Distribution: Widespread, southern Brazil to Mexico; common in open areas, secondary forest, forest gaps.

Use: Medicinal for multiple ailments, ritual (van Andel and Russchaert 2011); insecticide (Autran *et al.* 2009).

Similar species: *Piper* spp. are called *man-aneisi* (literally: man anise) in Suriname if they have narrow, pointed leaves; *uma-aneisi* (woman anise) such as *Piper marginatum*, when they have broader, rounded leaves.

Mourera fluviatilis
River weed (English name used for many Podostemaceae)
Danwiwiri, kumarunyanyan (local names used for many Podostemaceae)
Podostemaceae

Before the Podostemaceae plant family was known to science, Aublet (1775) described this species from a French Guiana specimen in the 18th century. Throughout the Interior of Suriname, the leathery, seaweed-like *danwiwiri* thrives in the dynamic conditions of river rapids and waterfalls (Figure 167). The dense mats or 'meadows' formed by this species play a profound ecological role by increasing structural and food web complexity within freshwater communities. A variety of aquatic animals obtain shelter and food amongst the *danwiwiri* leaves, including many insect larvae (Chironomidae, Pyralidae) and the fish that feed upon invertebrates or *danwiwiri* itself (Jäger-Zürn *et al.* 2013).

The life cycle of *Mourera fluviatilis* is directly tied to seasonal variation in water levels. During the rainy season, vegetative shoots and leaves remain deep underwater. In the dry season (August to October), the plants become exposed and are faced with drying and death. At this time, the plants produce masses of erect flowering spikes with showy pink flowers upon exposed rocks (Figures 168, 169). Sobral-Leite *et al.* (2011) found that *danwiwiri* flowers in French Guiana were cross-pollinated by several bee species (*Augochlora* spp., *Augochloropsis* spp. and *Trigona spinipes*). Wind plays a minor role in pollination, but a major role in seed dispersal. Mature capsules contain tiny, sticky seeds that are transported by wind when water is low and become attached to rocks. As waters rise with a new rainy season, the seeds are stimulated to germinate and new colonies are established upon the rocks (Rutishauser and Grubert 1999, Tippery *et al.* 2011).

Figure 167. ***Mourera fluviatilis* (*kumarunyanyan*) aquatic herb, typical habitat in rocky rapids (image © Peter & Siela Teunissen).**

Figure 168. Seaweed-like leaves and pink-flowered spikes of *kumarunyanyan* (image © Bruce Hoffman).

Figure 169. Close-up of *kumarunyanyan* flowers (image © Bruce Hoffman).

Description: Robust, perennial aquatic plant attached to rocks with claw-shaped hold-fasts, showing extreme morphological variation; leaves seaweed- or cabbage-like, spreading horizontally to 2 m long, surfaces rough and bumpy above, smooth below; inflorescence a flattened, erect spike with feathery edges, 15-70 cm long, bearing 20-90 flowers along the sides; flowers 2-4 cm long, with pink flower stalks and pink stamens; fruit capsules ovoid, to 1.5 x 0.4 cm, dehiscent, seeds tiny and numerous.

Distribution: The Guianas, Venezuela and northern Brazil; in Suriname, common, dominant species on rocks in fast-flowing waters of Interior rivers, waterfalls and some creeks.

Use: Leaves burned to obtain salt for subsistence; used in herbal baths by Maroon groups (van Andel and Ruysschaert 2011).

Similar species: This species is the largest and most common Podostemaceae species in Suriname and the only local member of the *Mourera* genus.

Triplaris weigeltiana (formerly *Triplaris surinamensis*)
(no English name)
Dreiten, mira-udu
Polygonaceae

Mira-udu is a slender tree resembling Eucalyptus tree species from a distance (Figure 170). It is a pioneer species of open, disturbed areas and stands are common throughout the Coastal Plain. The Sranantongo name, *dreiten,* indicates the seasonal cycle of the species, which produces beautiful, showy displays of flowers in Suriname's dry *(drei)* seasons (Figure 171). Late dry season is marked by the mass release and spinning flight of the three-winged fruits (Figure 172) (Parolin *et al.* 2013).

The name *mira-udu* reflects local recognition that this species is typically occupied by ant colonies – *mira* is the local (and Dutch) word for 'ant'. Ward (1999) documented at least five related ant species (of the *Pseudomyrmex viduus* group) that depend upon a *Triplaris* spp. for food and shelter. The ants live in chambers in the trunk and branches of *mira-udu* trees and very aggressively attack intruders.

Mira-udu is pollinated by generalist insects. Engel and Dingemans-Bakels (1980) found that *mira-udu* was an important source of pollen as food for several stingless bee species (*Melipona favosa, Melipona compressipes,* and *Trigona branneri*) in Suriname. Kerkvliet and Beerlink (1991) documented the importance of the tree in Suriname as a nectar source for bees and for commercial honey production.

Figure 170. *Triplaris weigeltiana (mira-udu)* tree in flower (image © Pieter & Siela Teunissen).

Figure 171. Female flowers of *mira-udu* (image © Pieter & Siela Teunissen).

Figure 172. Propeller-like fruits of *mira-udu* (image © Pieter & Siela Teunissen).

Description: Tree to 20 m high; with separate male and female plants (dioecious); trunk with small buttresses, bark gray, shallow-grooved, with thick round flakes; branchlets thick, inhabited by stinging ants; stipules fused into a hairy tube (ochrea) at nodes, leaving a ring-shaped scar upon falling; leaves alternate, simple, hanging, 20-35 x 5-9 cm; flowers in spikes, unisexual, white; female flowers forming a bowl with 3-4 cm long, papery wings, white or pink to rusty-red; fruits brown, three-winged.

Distribution: Widespread in tropical South America including the Guiana Shield; in Suriname, common in the Young Coastal Plain, swamps, rivers, open areas, and secondary forest.

Use: Herbal baths; medicine against dysentery; a porridge was made from the seeds in the 18th century (van Andel 2010).

Similar species: Highly distinctive – not easily confused with other species.

Rapatea paludosa
(no English name)
Tumikamisa (local name used for many *Rapatea* species)
Rapateaceae

Rapateaceae is a family of monocot herbs which originated and diversified on low fertility soils of the Guiana Shield highlands region (Givnish *et al.* 2000). A common family representative in Suriname is *Rapatea paludosa* (Figures 173, 174). As with many understory plants in the tropics, *tumikamisa* flowers are produced daily in low numbers throughout the year. This so-called 'steady state' strategy encourages the fidelity of pollinators by providing them with a dependable daily food source (Ackerman *et al.* 1982). By contrast, many tropical trees and lianas, such as *Doliocarpus dentatus* (*watratitei*), produce showy masses of flowers a few times per year. This so-called 'big bang' strategy aims to attract as many pollinators as possible and does not rely upon pollinator fidelity.

In French Guiana, Hentrich (2008) found that *tumikamisa* was pollinated primarily by females of a single bee species – likely *Euglossa chalybeata*. Pollinators visited *tumikamisa* flowers only in the early morning and flowers wilted by noon each day. Such a close plant-pollinator relationship is surprising, because *tumikamisa* plants present an open flower design. Pollinator fidelity usually involves closed flower designs that hide or protect floral rewards. As it turns out, *tumikamisa* is not overly generous, but limits access in an alternative way. As with some other plant species in this chapter, pollen is released only from small pores in the anthers and a specific buzzing frequency is required to 'shake' pollen out. *Tumikamisa* achieves pollinator fidelity by requiring a behavioral act that only certain bees are capable of performing (Buchmann 1983).

Figure 173. *Rapatea paludosa (tumikamisa)* herb, flower head enclosed by two, large leafy bracts (image © Pieter & Siela Teunissen).

Figure 174. Side-view of *tumikamisa* inflorescence stalk (image © Olivier Gaubert).

Description: Perennial herbs, 0.8-1.5 m tall, producing abundant mucilage – a thick, translucent, glue-like substance; leaves in fan-shaped arrangement; blades flat, 4-12 cm wide, with 15-25 cm long sheaths near base; flower heads globe-like or flattened, enclosed by two leafy, heart-shaped green bracts, on a 50-70 cm long stalk; flowers three-parted, petals yellow, delicate; fruit a brown, dehiscent capsule.

Distribution: Widespread, northern South America, Panama and Trinidad; in Suriname, Coastal Plain and Interior forests, mountains, common on white sands.

Use: (No known use)

Similar species: *Rapatea paludosa* is distinguished by an inflorescence stalk with opposing marginal 'wings' of different widths. Other *Rapatea* spp. have inflorescence stalks with wings of similar width or that lack wings altogether.

Paullinia pinnata
Sweet gum (English name used for many *Paullinia* spp.)
Feififinga (local name used for many *Paullinia* spp.)
Sapindaceae

More than twenty *Paullinia* species occur in Suriname. Their small, irregular flowers are pollinated by small insects, including stingless bees (Figure 175). Fruits open with three, spreading valves, which expose shiny seeds (Figures 176, 177). The conspicuous color contrast between the red fruit wall, black seed and attached white aril attracts birds, but monkeys also consume the fruits. Seeds of several *Paullinia* spp. are highly toxic due to alkaloids, including caffeine. The attached aril, however, is virtually alkaloid-free and has a sugar content close to 70 percent of its dry weight. This favors more effective seed dispersal by animals, which consume the aril presumably without damaging the toxic seed. The aril also strongly absorbs moisture, which preserves germination power by preventing seeds from drying out (Baumann *et al.* 1995, Weckerle and Rutishauser 2005).

Figure 175. *Paullinia pinnata (feififinga)*, a liana with pinnately compound leaves and tendrils borne on inflorescence stalks (image © Pieter & Siela Teunissen).

Figure 176. Flowers and fruits of *feififinga* (image © Pieter & Siela Teunissen).

Figure 177. *Feififinga* fruit, containing black seeds with white arils (image © Pieter & Siela Teunissen).

Description: Strongly branched woody liana with forked or coiled tendrils (Figure 175); older stems 3-5 angled in cross-section; scarce milky latex sometimes present; leaves alternate, pinnately compound, with 5-7 toothed leaflets on a winged leaf stalk; inflorescences in dense spikes, simple or compound; flowers bilateral symmetric, whitish-yellow, five-parted, petals free, ovary superior; fruits unevenly club-shaped, pinkish-red; containing one blackish seed with a white aril.

Distribution: Widespread in the Neotropics, also occurring in Africa; very common in Suriname, especially in Coastal Plain, swamps, riverbanks and disturbed sites.

Use: Binding and weaving material; edible aril; medicinal - treatment for diabetes, intestinal complaints, and malaria; fish and arrow poisons; several bioactive molecules documented (Annan *et al.* 2013).

Similar species: Sapindaceae climbers in the *Serjania* genus are similar but distinguished by winged fruits and typically have more leaflets than *Paullinia* spp.

Pouteria guianensis
(no English name)
Rode jamboka, jansnijder
Sapotaceae

A fruiting *rode jamboka* tree is a conspicuous sight in the forest, with hundreds of yellowish-orange, oval-to-round berries on the trunk and branches (Figures 178, 179, 180). The fruits have a tough, indehiscent fruit wall and large seeds, with delicious, sweet pulp attached tightly to the seeds. Such traits are particularly attractive to large monkeys, which break open the tough fruits with their strong teeth and swallow the seeds and pulp, eventually spreading the seeds near or far through defecation. In French Guiana, Regan *et al.* (1998) found the color vision pigments of the howler monkey (*Alouatta macconnelli*) perfectly adapted for detecting yellow-orange fruits amongst green foliage. This study provided some evidence for a co-evolutionary relationship between fruit eaters and fruit producers.

Estimating tree age is challenging in the tropics because tropical woods rarely have distinct annual growth rings. Laurance *et al.* (2004) circumvented this problem by estimating tree age from long-term growth rate data. *Pouteria guianensis* individuals were among the oldest species in the study at a mean 720 years. The wood of *rode jamboka* and many Sapotaceae is very dense and hard, which likely enhances resistance to decay and pests in the humid tropics. Biochemical defenses may play a role as well. Barbosa *et al.* (2007) found that an extract of *rode jamboka* wood and bark strongly repelled a termite species (*Nasutitermes* sp).

Figure 178. *Pouteria guianensis (rode jamboka)* tree trunk (image © Pieter & Siela Teunissen).

Figure 179. Mature fleshy fruit of *rode jamboka* (image © Pieter & Siela Teunissen).

Figure 180. Cross-section of *rode jamboka* fruit, seed with scar (hilum) along one side (image © Pieter & Siela Teunissen).

Description: Canopy tree 30 to 40 m tall with thin, steep buttresses; bark reddish-brown, soft and fibrous, with vertical grooves; slash with scant white latex; leaves simple, alternate, clustered at branch ends, with reddish-brown hairs; blades often wider towards the tip, 13-25 x 5-10 cm; leaf stalk margins recurved; inflorescences in leaf axils and below leaves; flowers tubular and four-lobed, 1-1.3 cm long, pale green; four stamens, ovary superior; fruit an orange-yellow berry, oval to round, 3-7 cm diameter; seeds 2-4, with dull brown scar (hilum) on one side (Pennington 2006).

Distribution: Central and northern Amazonia, including the Guianas and Trinidad; occasional to locally abundant in non-flooded old-growth forest and montane forests

Use: Medicinal – bark used against dysentery and diarrhea; fruit tasty, sweet; dense, rot-resistant wood used in docks and ships (Silva *et al.* 2009).

Similar species: The species-rich *Pouteria* genus includes many species similar to *rode jamboka*. *Pouteria melanopoda (blaka jamboka)* is a common species with orangish-brown bark and rectangular scales, slash with white latex, inner bark white with orange stripes, small flowers – less than 0.5 cm long, and fruits maturing red with white pulp.

Quassia amara
Bitterwood
Kwasibita
Simaroubaceae

K*wasibita* is an iconic medicinal and insecticidal plant in Suriname (Figures 181, 182). The genus was named by Carolus Linnaeus after an African-born healer and botanist, Graman Kwasi. Kwasi was brought to Suriname as an enslaved African in the 18th century and eventually gained freeman status. His most famous remedy was a *kwasibita*-based tea used to kill internal parasites (see also chapter on Use, Threats and Conservation).

The attractive red, tubular flowers of *kwasibita* are typical for a hummingbird-pollinated species. However, *kwasibita* flowers in nature often have holes at the corolla base where insects have chewed through the tissues to 'steal' nectar. Many ecologists view nectar robbing as deterimental to plants because nectar production requires a lot of energy and the robbers provide nothing in return. In Panama, Roubik *et al.* (1985) found that nectar robbing of *kwasibita* clearly diminished pollen transfer and seed production. By contrast, Maloof and Inouye (2000) concluded that nectar robbing is just as likely to benefit as harm the reproductive success of *kwasibita* and other plants. They found direct benefits – insects and birds defined as robbers may still pollinate plants – and indirect benefits, as robbers may change 'true' pollinator behavior in ways that ultimately increase reproductive success. An example of positive behavioral change is an increase in pollinator flight distance and switching between flowers after encountering too many 'robbed' flowers. This behavior could result in great cross-pollination between distant plants and, thus, greater reproductive success for a population.

Figure 181. *Quassia amara (kwasibita)* treelet, leaves pinnately compound, flowering stalks and flowers pinkish-red (image © Pieter & Siela Teunissen).

Figure 182. *Kwasibita* fruit (image © Pieter & Siela Teunissen).

Description: Shrub or small tree, 2-3 m tall; bark extremely bitter; leaves alternate, pinnately compound, 15-40 cm long, with 3-5 leaflets, main leaf stalk winged; inflorescences 10-25 cm long, rachis bright red, flowers tubular, 2.5-4.5 cm long, five petals, pinkish-red to bright red; five stamens, longer than petals; fruits 1-5 oblong drupes borne on a fleshy disc, maturing purplish-black, each 1-1.5 cm long.

Distribution: Widespread in the Neotropics, extending to Mexico, the West Indies and Argentina; widely cultivated, occurring in diverse forest types, including coastal and Interior regions of Suriname.

Use: Medicinal – for a wide range of internal ailments, including malaria, hypertension, diabetes, kidney problems, and flu; male aphrodisiac; insecticide against lice, fleas, and mosquito larvae (Brandão *et al.* 1992, van Andel and Ruysschaert 2011).

Similar species: The leaves of *kwasibita* can be confused with those of *Inga* species (*switibonki*, Fabaceae). The leaves of *switibonki* spp. are distinguished by cup-like extra-floral nectaries beween every pair of leaflets along the leaf axis. *Inga* species also do not have bitter tasting bark as in *kwasibita*.

Phenakospermum guyannense
South American traveler's palm
Bigipalulu
Strelitziaceae

Bigipalulu is a tall, showy, monocot herb, commonly seen in dense stands along roads in Suriname (Cron *et al.* 2012) (Figure 183). Vegetative growth is clonal and new areas are colonized in part through underground rhizomes. Flowering begins near the end of the rainy season (July-August) and individual plants may keep producing flowers and fruits for 2-4 months. The main stalk dies after setting seed, but new clonal stems are produced.

Kress and Stone (1993) presented a step-by-step description of *bigipalulu* bat (*Phyllostomus* spp.) pollination in French Guiana. Just before sunset, a few pointed, cream-colored flowers begin emerging from the large, boat-like bracts that protect inflorescences (Figure 184). These flowers produce abundant sugary nectar as hexose, the molecular form preferred by bats. The flower design forces bat visitors to push against a 'trigger' petal. Upon triggering, the flower opens explosively and the bat's body is hit by long, pollen-bearing stamens. Simultaneously, the receptive female part of the flower, the stigma, falls out of the flower and is left dangling far below on a long, cable-like style (see Figure 56 on flower terminology). This sequence of events reduced the chances of self-pollination, by maximizing the distance between male and female parts. Presumably, the hanging stigma is in a favorable position to be cross-pollinated as bats arrive bearing pollen from other *bigipalulu* populations. The dispersal of *bigipalulu* seeds, with bright orange, hair-like arils, has been observed for birds and capuchin monkeys (Figure 185). The arils have high oil content and are desired by many small birds.

Figure 183. *Phenakospermum guyannense* (*bigipalulu*), a large monocot herb (image © Pieter & Siela Teunissen).

Figure 184. Flowering stalk of *bigipalulu*, with leathery, boat-shaped bracts protecting flowers (image © Pieter & Siela Teunissen).

Figure 185. Open fruit capsules of *bigipalulu*, containing black seeds with orange arils (image © Pieter & Siela Teunissen).

Description: A tree-like herb to 12 m height; trunk of overlapping leaf bases and large banana-like leaves arranged in one plane (2D); inflorescences terminal on central leafy shoot, raised well above the leaves, with 3-10 boat-like bracts enclosing flowers or fruits (up to 25); flowers bisexual, cream-colored, with three free sepals and two fused petals, ovary inferior; fruit woody, three-parted capsule containing black seeds with hairy, orange arils (Kress and Stone 1993).

Distribution: Widespread in tropical South America east of the Andes; abundant in forests bordering savannas, marsh forest, forest margins and secondary forest.

Use: Mucilage from shoots used for wounds; leaves used for thatch roofs, seeds used in traditional crafts and jewelry.

Similar species: In cultivated bananas (*bana*), the basal stem/leaf sheaths are whorled (not aligned in one plane) and inflorescence stalks hang downwards (not erect and with boat-like bracts as in *bigipalulu*).

Cecropia sciadophylla
Pumpwood (English name used for *Cecropia* spp.)
Manbusipapaya
Urticaceae (Cecropiaceae)

Cecropia is a highly recognizable genus of Neotropical trees, with 6-7 species known in Suriname (Figure 186). *Manbusipapaya* (*Cecropia sciadophylla*) is the most common *Cecropia* species in the genus and is likely the most widespread and abundant tree species in the Neotropics as a whole (Berg and Franco 2005).

Cecropia spp., including *manbusipapaya*, play a critical role in tropical forest succession (Maury-Lechon 1991). In newly disturbed areas, they are typically the first 'pioneer' trees to become established as saplings. The shade and leaf litter of these sun-loving trees creates favorable microhabitat conditions for later succession plants such as *Goupia glabra* (*kopi*). The key to the success of *Cecropia* spp. is the massive production and wide dispersal of seeds that remain viable for many years. Prevost (1982) found that *Cecropia obtusa* and *Cecropia sciadophylla* accounted for 50 percent of the soil seed bank in French Guiana.

Cecropia spp. have intimate relationships with many animals in forest ecosystems. *Cecropia* spp. fruiting spikes are a highly desired food resource for rodents, bats, birds (76 species), monkeys, kinkajous (*netikeskesi*) and opossums (*busi-awari*) (Charles-Dominique *et al.* 1981, van Roosmalen 1985a). Nearly 80 percent of *Cecropia* spp. have a mutualistic relationships with *Azteca* spp. ants (Lobova *et al.* 2003). In exchange for protection, the ants are housed, and fed within specially grown 'trichilia' structures. *Cecropia sciadophylla* is one of the few *Cecropia* spp. that does not harbor ants. Lobova *et al.* (2003) observed bats in French Guiana dispersing fruits and seeds across different forest succession zones. Secondary forest fruits were transported into old-growth forest and old-growth forest fruits/seeds were transported into secondary forest.

Figure 186. ***Cecropia sciadophylla (manbusipapaya)*** **secondary-forest tree (image © Pieter & Siela Teunissen).**

Figure 187. Stilt roots of *manbusipapaya* (image © Pieter & Siela Teunissen).

Figure 188. Palmately compound leaves of *manbusipapaya*, leaflets stalked at base (image © Pieter & Siela Teunissen).

Figure 189. Male flower spikes of *manbusipapaya* (image © Pieter & Siela Teunissen).

Description: Tree, 20-35 m tall, trunk with leaf scars, hollow stems and stilt roots (Figure 187); bark gray, watery sap turning blackish, leaves alternate, palmately compound, blade 20-100 cm wide, leaflets silvery-white few-haired below, stalked (Figure 188); inflorescences clusters of spikes – male spikes 8-15 (Figure 189); female spikes 3-6, larger; fruits tiny, one-seeded (Berg and Franco 2005).

Distribution: Neotropics; common in secondary forest, disturbed sites, sometimes in non-flooded old-growth forest.

Use: Dried leaves used as 'bush tea' and for a wide range of ailments by Afro-Surinamese (van Andel and Ruysschaert 2011).

Similar species: *Cecropia sciadophylla* is distinguished from all other *Cecropia* spp. in Suriname by distinct leaf stalks (petiolules) on the leaflets.

Rinorea pubiflora (no English name) *Manaritiki* (local name used for all *Rinorea* spp.) Violaceae

Rinorea is a pantropical genus of understory trees and shrubs (Hekking 1988). *Rinorea pubiflora* (*Manaritiki*) is the most widespread, abundant and variable species of the genus *Rinorea* in the Neotropics (Figure 190). The small, hairy seeds of *Manaritiki* are nutritious and possess a fleshy structure at the base known as a 'caruncle'. Seed dispersal is accomplished primarily by ants and small birds that are attracted to the caruncle and seed. Because *Rinorea pubiflora* is very common in seasonally flooded forests, it is likely that the seeds are also dispersed by freshwater fish in the rainy season (Goulding 1980).

The underside of *Manaritiki* leaves typically possess domatia that provide shelter and food for tiny insects known as mites. In contrast with the large-cham bered petiole or stem domatia described in other species accounts, *Manaritiki* domatia are tiny, inconspicuous tufts of hair or pockets located in secondary vein axils (where secondary leaf veins merge with the central vein). The mites likely play a beneficial role for the host plant by feeding upon microscopic plant pathogens. Thrips and other tiny organisms may also take residence in domatia when mites are not present.

Figure 190. *Rinorea pubiflora (manaritiki)* treelet (image © Olivier Gaubert).

Figure 191. Three-sided fruits of *manaritiki* (image © Olivier Gaubert).

Description: Understory tree or treelet to 8 m tall; leaves opposite and simple, dark green and hairless above, lighter green below; flowers in small hanging inflorescences, yellow or white, five-parted; capsule oval to oblong, woody, 1.7-3 x 0.5-1 cm, green, hairless (Figure 191); seeds two to three per valve, with brown hairs.

Distribution: Widespread in the Amazon Basin to Costa Rica and the Guiana Shield region; often abundant in the understory of old-growth forests (non-flooded, seasonally-flooded and swamp), and sometimes in secondary forest.

Use: In Suriname, *Manaritiki* stems are used by local peoples to make a food-processing sieve or 'manari'. A mixing stick is sometimes fashioned by cutting out the stem at certain branching points with spoke-like lateral branchlets.

Similar species: *Rinorea pubiflora* is distinguished from other *Rinorea* spp. by densely reddish-brown hairs, domatia in leaf axils below, sepal veins, and hairy seeds.

Insects and their Relatives

Bart De Dijn, Hélène Hiwat

Insecta or insects are one the favorite subjects of macro photographers because they have that alien, otherworldly quality. The diversity of often spectacularly shaped and colored insect bodies is absolutely stunning, and nowhere more so than in humid tropical territories such as Suriname.

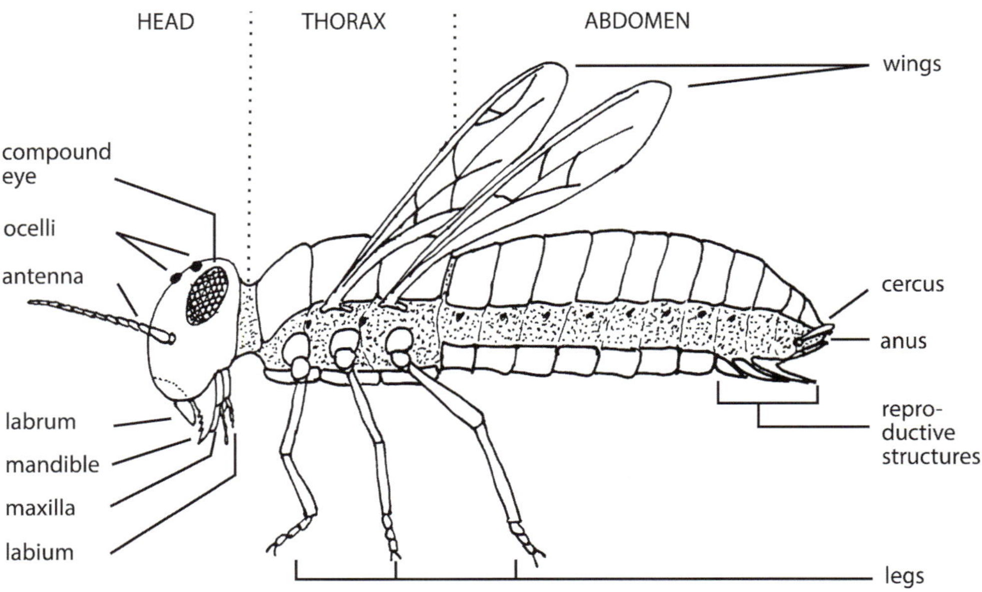

Figure 192. Schematic overview of insect body.

Insects have a segmented rump, three pairs of articulated legs, and a skin rich in chitin, a complex sugar molecule that allows the skin to function as a skeleton. The body has three main parts: the head, thorax (with the articulated legs), and abdomen (Figure 192). The head segments are fully merged and not obvious as such. The three thorax segments and in particular the many abdominal segments are often more obvious. Many insects have a narrow connection between the main body parts – very narrow in most Hymenoptera (wasps, bees and ants; Figure 193.a). Most adult insects are land animals with the ability to fly, allowing them to escape from danger, cover a large area to find food and mates, and disperse to colonize previously unoccupied habitat.

An overview of South American insects and other arthropods (see below) in English is presented by Hogue (1993), who also summarizes information on insect biology. Carrera (1980) covers similar ground specifically on Brazil. Nothing of this kind has been published on Suriname or the region, although a photo guide called 'Amazon Insects' (Castner 2000) merits mention, as well as a recent book on the butterflies of Suriname (Gernaat *et al.* 2012). General textbooks on insect biology include that by Johnson and Triplehorn (2004). Two important books on insects were published by Wilson (1971) and Eisner (2003); the first one focuses on social insects (such as ants and termites; see below), the second one on how insects use chemicals to defend themselves and communicate. The introduction below is mainly based on these sources, on older European textbooks (such as Snodgrass 1935, Wigglesworth 1950, Eidmann & Kühlhorn 1970, Jacobs & Renner 1974, Villiers 1979), and on personal observations in Suriname and elsewhere by the first author of the present chapter.

The makings of an insect

Eternal skeleton and three pairs of articulated legs. Chitin molecules provide strength to the insect skin by locking onto one another. Parts of the skin of the insect body are thick and have much interlocked chitin, thus forming tough plates, which collectively form an external skeleton (exoskeleton). Adult beetles (Coleoptera) and bugs (Hemiptera) (Figure 193.b,c) are examples of insects with a well-developed exoskeleton that provides not only support for the organs inside the body, but also some protection against predators. Between the skeletal plates there are soft-skin connections that allow some flexibility. Such flexibility is most needed in the abdomen, the hind part of the body. This part contains most of the digestive system, as well as the reproductive system. Its skin needs to be at least partly soft to allow it to expand when feeding, and in the female when eggs are maturing (in Figure 192, the dotted areas indicate soft skin on an expanded body). Most adult beetles and bugs cover the softer areas of their abdomen with their hardened front wings (elytra), which are no longer of much use in flying.

The three pairs of legs on the thorax, the middle part of the body, are articulated, as are many other appendages on an insect's body. They are linear arrangements of hard exoskeletal elements connected by small sections of softer skin at joints. Like in the body, muscles to make the elements move are attached on the inside.

Appendages on the abdomen. The abdomen may also have articulating appendages, meaning that they are connected to the body with a joint. These appendages typically occur in pairs, one on each side of the body. A pair of cerci (singular: cercus) may occur near the tip of the abdomen; they are typically small appendages with a sensory function. Immature stages, such as larvae (see below), may have additional articulating appendages on the abdomen; with immature aquatic insects these often function as external gills. Very different appendages can be found on the abdomen of terrestrial insect larvae: anything from welts to pseudo-legs that are not hard, or at least not articulating. They help with movement in burrows or on smooth surfaces; an example are the auxiliary abdominal legs of caterpillars (Figure 193.d).

Adult insects typically have well-developed male or female reproductive structures on the underside of the abdomen, near or at the tip (see Figure 192). These may be reduced and non-functional in the worker or soldier caste of social insects (see below). The male and female reproductive structures are often quite complex and differ a lot between different insect groups. Their essential purpose is

a
b
c
d
e
f
g
h

Figure 193. Diversity of insects and their behavior in Suriname (all images © Bart De Dijn):

a. Sphecid wasp couple mating at white-sand savanna (near Witsanti, Aug. 2016).

b. *Epicauta* sp. (Meloidae) feeding on leaves of *Solanum subinerme* (Troluco timber concession, near Tibitisoela, Oct. 2014).

c. *Anisoscelis foliacea* on leaf of *Passiflora* sp. (Paramaribo, Apr. 2017).

d. *Cocytius* sp. caterpillar feeding on soursop, *Annona muricata* (Paramaribo, Aug. 2014)

e. Unidentified ichneumonid parasitic wasp reared from pupae of *Megalopyge lanata* (Lepidoptera) (Paramaribo, Jun. 2015).

f. *Pterodictya* sp. (Homoptera) on tree trunk (Sipaliwini village, Aug. 2009).

g. Silverfish (Thysanura) on drawer in house (Paramaribo, Jan. 2017).

h. Elaterid beetle larva (on its back) found in stem of coconut palm, *Cocos nucifera* (Paramaribo, Mar. 2017).

i. Unidentified katydid (Tettigoniidae) laying eggs in twig using ovipositor (Republiek, Oct. 2006).

j. *Phoebis sennae marcellina* feeding on nectar of *Lantana camara* flowers (Paramaribo, Nov. 2016).

for copulation and egg-laying (female only). When insects are not sexually active, these structures, as well as the remnants of the often reduced last body segments, are often hidden by the extended skeletal plates of the preceding abdominal segment. On the other hand, an egg-laying structure (ovipositor) may be extremely well developed and not hidden at all, for example in female katydids (Figure 193.i) and many female parasitic wasps (Figure 193.e). In many other wasps, as well as in most ants and bees, the ovipositor has lost its original function, and acts as a stinging apparatus that is used for defensive purposes; many parasitic wasps use it to paralyze prey.

Head appendages and eyes. The head also has articulating appendages. Very simple ones are the paired mandibles, which in many insects are hard and massive structures with dents, used for biting and chewing. Apart from mandibles, insects have other mouthparts, such as the paired maxillae, which are much softer and consist of more, often softer elements. The maxillae in many insects serve to spoon food into the mouth, in conjunction with two other appendages (labrum and labium). The mouthparts are typically largely hidden and hard to distinguish (see Figure 192). In some insects they are merged into a single snout-like structure (proboscis), like the sucking 'tongue' of butterflies and moths (Lepidoptera) (Figure 193.j), and the stinging proboscis of mosquitoes, bugs, and cicadas, hoppers and plant lice (collectively known as Homoptera) (Figure 193.f).

Another distinguishing feature of insects is the single pair of articulated appendages on the front or top of the head, the antennae. These can be compared to our nose in terms of their olfactory function (sense of smell), but they are much more sensitive. Antennae also have a tactile function (sense of touch), and can serve to detect air movement (similar in function to hairs on the human body). The sensory function of the antennae is critical to most aspects of an insect's life, yet the type of sensors found on the antennae may also occur on other parts of the insect body. Many insects have poor vision, some none, but are adequately aware of their surroundings using their other senses.

Almost all adult insects have eyes, and most have two types: one pair of compound eyes, typically large and used for fine-grained color vision, and a trio of ocelli, visible as small hemispherical lenses on top of the head (similar to the eyes of spiders). Why seemingly less sophisticated ocelli are needed in addition to compound eyes is often not clear. But in certain bees, the ocelli seem to help with flight navigation and stabilization (see for example Zeil *et al.* 2014, Taylor *et al.* 2016).

Wings. The thorax in most adult insects has wings, in principle two pairs, although one pair is strongly reduced in Diptera (flies, mosquitoes and midges). An insect wing, at least one that actually serves for flying, is an extremely thin, strongly outward projecting, hardened skin fold, connected to the thorax with a complex joint. It has linear thicker parts called veins, which form a network or a branching pattern that divides the wing into 'cells'. Only in a few smaller insect groups – including lice, fleas, and silverfish (Figure 193.g) – adults never have wings.

Relatives. Three groups of small and inconspicuous six-legged animals that never have wings – Protura, Diplura, and Collembola or springtails – used to be considered insects, but molecular data (see Box 4 in the General Introduction / History chapter) indicates that they should rather be considered sister groups within the larger grouping of Hexapoda (literally: those with six-legs). Hexapoda in turn is considered a sister group of Crustacea (crabs, shrimps, crayfish, etc.), Chelicerata (spiders, mites, scorpions, etc.) and Myriapoda (centipedes, millipedes). The representatives of all these groups have articulated legs and chitin in the skin, but have more than three pairs of legs, and never wings. Chelicerata have four pairs of legs, Crustacea five, and Myriapoda more, often many more. The insects and all their relatives mentioned here are collectively referred to as Arthropoda or arthropods (literally: those with articulated legs).

Life cycle

Molting. What insects are maybe best known for is their metamorphosis, the change of form in the course of the life cycle. Let us first focus on something more fundamental, though: molting, meaning the shedding of old skin after the formation of a new one beneath. Insects, like other arthropods, must molt if they are to grow as individuals. Their skeleton-skin does not allow them to grow much, so it must be shed and replaced with a bigger one. The process of molting occurs repeatedly during an insect's lifetime: after leaving the egg, and as they grow, immature insects may molt up to 15 times (though much less in most species), until they reach the adult or mature stage, the stage capable of reproducing. New soft skin develops below the old hardened one, and by pumping and flexing the body, the old skin is ruptured. The insect wriggles out of its old skin, dilates its body, and allows the new, larger skin to harden. When you see an unusually pale insect, it is probably a freshly molted one, since the process of skin hardening typically goes hand in hand with the final tanning of the skin.

Metamorphosis. Strictly speaking, metamorphosis means that the body before and after molting is structurally and functionally different, not just in terms of reproductive structures and function. This is most dramatic when a larva becomes an adult insect, often via an intermediate 'resting' stage called a pupa or chrysalis. Maria Sibylla Merian (1705) beautifully illustrated this in the early 18th century based on observations of butterflies and moths which she reared in Suriname (see Box 1 in General Introduction / History chapter). Out of the egg comes what we call a larva, which in the case of Lepidoptera is better known as a caterpillar. Most caterpillars feed on green plant parts, such as leaves. Like other immature insects, they can be regarded as feeding machines: the larval stage of the life cycle is all about feeding and growing. The change into a pupa and then into a winged adult is a two-stage metamorphosis. The design of the resulting adult Lepidoptera is to mate, and lay eggs on a suitable larval host plant, meaning a plant on which the larva can feed. The structure of the mouth parts of the adults allows them to feed on liquids only, which they need mainly to hydrate, and get some additional energy and minerals. Like in other insects, the adult stage of Lepidoptera is not about growing at all, but about reproducing and dispersing, moving away from the location where the larval stage was spent, to colonize suitable places. Wings are very helpful for dispersal.

Life cycle with two main stages. The division of the life cycle into an immature growth stage and a mature, winged reproduction and dispersal stage is the norm in most insect taxa. While there are quite a few exceptions, this generally distinguishes insects from other arthropod groups. The typical insect life cycle is different from, for example, that of spiders and crabs. Adult insects don't really grow and molt, while adult spiders and crabs typically do. Most insect species have immatures that stay put, while adults disperse using their wings. Many spider species, on the other hand, have tiny immatures that produce silk threads on which they can be carried away by the breeze. Crab species typically produce tiny immatures which disperse by allowing themselves to be carried away by water currents.

The change from immature to adult is quite dramatic in many insects, but not in all. In some insects there is no metamorphosis, such as in silverfish (Thysanura). In others, the metamorphosis is not particularly dramatic, like in most Hemiptera and Homoptera, grasshoppers, katydids, crickets and mole crickets (Orthoptera), as well as cockroaches. Immatures of Hemiptera, Homoptera, Orthoptera and cockroaches are technically 'nymphs'; they have the same general build as the adults, but lack fully developed reproductive structures and wings (see Figure 194.h). Young nymphs have tiny wing buds, older ones have non-functional wings that can be about half as large as fully developed functional wings. Confusingly, many grasshopper and cockroach species lack wings as adults, or have rudimentary wings; in these species, only the full development of the reproductive structures on the abdomen gives the adults away. In other insects, like Hymenoptera, Lepidoptera, Coleoptera, and Diptera, the metamorphosis is dramatic indeed. Lepidoptera, most Coleoptera, and some Hymenoptera have immatures that do not look like adults at all, and are commonly referred to as caterpillars and grubs. Some Coleoptera, most Hymenoptera, and virtually all Diptera have larvae that are commonly called maggots or 'worms'; they have rudimentary legs or no legs at all. They may not even have a distinct head, and in that case it's hard to even recognize them as insects. The technical term for these caterpillars, grubs and maggots is 'larvae' (Figure 193.h). Many aquatic insects – they are called aquatic because the immatures live in the water, not necessarily the adults – have immatures that look particularly odd because they have external gills.

Species richness

Global arthropod dominance. The great biodiversity communicator Ed Wilson pointed out that arthropods, in particular insects, dominate our planet in terms of the number of species, and that the tropics, especially the tropical rainforest, is where this is most apparent (Wilson 1985, 1988, 1992). Wilson based this largely on Terry Erwin's pioneering work. Erwin knocked down insects from the crowns of Neotropical rainforest trees using insecticides. He assessed how many beetle species were unique to or shared between tree species (Erwin 1982, 1983, 1997). He also surveyed how many species were undescribed, meaning not yet formally described by scientists. Based on the proportion of beetle versus other arthropod species known to science, and data on the number of tree species in different regions, he extrapolated that there are likely millions of arthropod species in the Neotropics, and tens of millions globally, most of them undescribed. A similar extrapolation for Suriname yields an estimate of several hundred thousand species (De Dijn, unpublished). However, this may be a gross overestimate (see below), and somewhat exasperating to those who may have the burning ambition to inventory the entire arthropod fauna of a tropical country.

During the last decades, global species richness data has been compiled on a wide range of organisms, including marine and freshwater ones (Bouchet 2006, Balian *et al.* 2008). Researchers have applied a variety of methods to estimate global numbers of species (reviewed by Stork 1997, Costello *et al.* 2013, Stork *et al.* 2015), including in relation to micro-organisms (recently: Locey & Lennon 2016). Based on this, a reappraisal is in order: the global number of arthropod species likely does not exceed 10 million. Insects dominate the land in terms of numbers of species that are visible to the naked eye. In the sea, however, the dominant taxa are mollusks (mainly snails), which are not arthropods, and crustaceans, which are; fishes occupy a remote third place. These largely aquatic groups are also important in freshwater, yet insects dominate as much in freshwater as on land.

Ed Wilson and Terry Erwin, mentioned above, are entomologists or insect specialists, although Wilson may best be referred to more specifically as a myrmecologist, an ant (Formicidae) specialist, and Erwin as a carabidologist, a specialist in carabid beetles (Carabidae). An insect family often contains hundreds or thousands of species, or even more. There are so many insect species that a single person can only really gain adequate knowledge during his or her lifetime about a single family, or part of it.

Broad inventories. Entomologist and ecologist Dan Janzen came up with the concept of an 'all taxa biodiversity inventory' (ATBI), a broad, complete inventory of all biological species in an area, including their ecology (Janzen 2011). A true ATBI requires the involvement of at least hundreds of specialists over a long time, probably many decades. Janzen wanted to apply an ATBI to a conservation area in Costa Rica, but that did not materialize; the first ATBI is now being implemented in the USA, in the Smoky Mountains conservation area (see **www.dlia.org/about-atbi**).

A taxonomically and temporally more focused inventory was undertaken by entomologist Yves Basset and collaborators in lowland rainforest at San Lorenzo, Panama (Basset *et al.* 2012). They surveyed 6144 arthropod species in the course of a single year in a set of small plots scattered throughout the 6000 ha San Lorenzo forest area. They did not actually survey all arthropod taxa, but a subset that was presumed taxonomically and ecologically representative. By entering this data in various models, they arrived at an estimate of about 25,000 species of arthropods for the entire San Lorenzo forest area; each hectare of forest was estimated to hold more than 60 percent of the species on average. However, Harte and Kitzes (2015), using the same San Lorenzo data, but a different model, estimated that a single hectare would on average just have 22 percent of the species. By scaling up further, and projecting the Panama results onto Amazonia, they speculated that the Amazon region may have some 305,000 arthropod species, and that a region such as the Guiana Shield would have some 50,000 species. Still high numbers, but modest compared to those based on Erwin's extrapolation method (see previous section). So, it may be possible after all to nventory all of Suriname's insects and other arthropods with an army of specialists. We would like to hereby invite this yet-to-assemble army to spend one or more lifetimes in the country: welcome, *wan swit kon na ini*!

Ecological guilds and interactions

What guilds? Richard Root, the originator of the concept, defines a guild as 'a group of species that exploit the same class of environmental resources in a similar way' (Root 1967). The taxonomic position of the species is not relevant in this definition. An example would be the guild of floral nectar feeders, which contains certain birds, bats, bees and a wide range of other insects. Assigning species to guilds is often used as a framework to understand the ecological diversity within a taxon, and to identify which species are most likely to be competitors, namely those belonging to the same guild. Species of predatory bugs, for example, will likely compete with one another, but not with herbivorous, sap-sucking bugs. Animal guilds are mainly defined according to the type of food the species consume.

The guild concept was applied by Root (1973) to arthropods found on collard greens. It was later applied to arthropods occurring on trees in the temperate climate zone by Moran and Southwood (1982) and in the tropics by Stork (1987). The principles developed during these early studies are still applied in modern studies such as the one by Basset *et al.* (2012) mentioned earlier. This study distinguished the following main arthropod guilds:

- herbivorous chewers: with chewing mouthparts to consume living plant parts, such as roots, leaves, pollen and fruits; examples are grasshoppers, leaf beetles, butterflies and moths (the larvae do the chewing), and bees (the larvae chew on pollen);
- herbivorous sap-suckers: with a needle-like proboscis to extract sap from living plants; examples are cicadas and many mite species;
- fungal feeders, feeding on living fungi; many beetles feed on fungi;

a
b
c
d
e
f
g
h

Figure 194. Mutualistic and social behavior of Suriname arthropods (all images © Bart De Dijn):

a. Unidentified syrphid fly on *Mimosa pudica* inflorescence which offers abundant pollen as only reward for potentially pollinating visitors (Paramaribo, Jun. 2016).

b. Female *Xylocopa* sp. carpenter bee probing for nectar inside the base of the flower of *Passiflora* sp., and while doing so being dusted on the back with pollen (Paramaribo, Jan. 2017).

c. *Cyclocephala amazona* beetles entering inflorescence of soursop, *Annona muricata*, which offers feeding tissue as reward for its visitors (Paramaribo, Mar. 2015).

d. Female *Centris* sp. bee foraging for oil produced by glands at base of flowers of *Stigmaphyllon* sp., and while doing so being dusted with pollen on underside of body (Paramaribo, Feb. 2017).

e. Tiny worker *Plebeia* cf. *minima* stingless bee collecting floral resin from inflorescence of Dalechampia sp., likely not being a pollinator because of the size mismatch (Paramaribo, Dec. 2016).

f. Male *Eulaema meriana* (large euglossine bee) and *Eulaema cingulata* (smaller ones) extracting aromatic chemicals with front legs from interior of flowers of orchid *Catasetum discolor* (Paramaribo, Apr. 2007).

g. *Aristolochia* sp. vine with tiny fly visitor (likely a chloropid) on margin of opening of its inflorescence, which is assumed to provide no reward to its visitors, but does lure them in with an attractive smell (Paramaribo, Dec. 2016).

h. Unidentified adult female membracids (Hemiptera; formely Homoptera) guarding their eggs (below their bodies) on twig of a mango tree, *Mangifera indica*, alongside tiny nymphs, as well as a *Trigona* sp. stingless bee worker (only its black head is showing) that protects the membracids in exchange for a sugary secretion it gets from them (Paramaribo, Apr. 2017).

i. *Anelosimus* sp. social spiders in their shared web, attacking a membracid prey (Patamacca, Aug. 2008).

j. Communal nest of *Eulaema nigrita* in a nest box with several females building cells using a shared stockpile of mud and mammal excrement, and individually provisioning cells with pollen and nectar (Paramaribo, Jun. 2015).

- scavengers, feeding on decaying matter of plant or animal origin; this includes cockroaches (Blattaria), termites (Isoptera, though molecular data suggests these should be included in Blattaria), crickets, and many beetle and fly species;
- carnivorous predators, feeding on animal prey, mostly other arthropods, by immediately killing and consuming it; this includes many mite, wasp and beetle species;
- carnivorous parasitoids, as young immatures feeding on animal hosts (mostly other arthropods) without killing them, but as older immatures ultimately killing the host; this includes large numbers of parasitic wasp species, as well as many fly and some beetle species; and
- the ant guild, which includes all ant species, many of which are difficult to assign to the other guilds because many will not just kill and eat prey, but will also ingest plant sap and nectar, or feed on detritus (decaying organic material); some even actively grow fungi on parts of plants harvested for this purpose (see Agosti *et al.* 2000).

Mutualistic interactions. Guilds give us insight into interactions amongst arthropods and between arthropods and other organisms. The concept mainly highlights who does the eating and who is being eaten, but does not really convey that some interactions are mutually beneficial (mutualistic). Well-known mutualistic interactions involving insects are pollination (see below) and seed dispersal. Many ant species disperse seeds that have special feeding tissue attached to them (see review by Warren & Giladi 2014). Dung beetles, who make dung balls that they roll away from the spot where they found the dung, disperse seeds that are contained in the dung (see for example Vulinec 2000, Chapman *et al.* 2003). Several insect groups cultivate fungi (review in Mueller & Gerardo 2002), but the best-known are leaf-cutter ants, who harvest plant leaves to serve as growing medium for fungi which they cultivate and eat (see species account of *Atta cephalotes*). These ants cultivate species of fungi that cannot grow or even exist without them.

Insect flower visitation and pollination in the Guiana Shield has been reviewed by De Dijn (2005), who stressed that not all flower visitors actually pollinate. Pollination by insects means that they transfer pollen from the male parts of the flower to the female parts. A basic requirement for a pollinator of a particular plant species is that its body size and behavior fit the size and structure of the flower. Most tropical plants require cross-pollination, which means that fertile seeds will only develop after the transfer of pollen from one plant individual to another one that is not genetically identical (see Plants chapter). Flying insects are useful tools to effectuate such a transfer. To entice the insects to provide this service, the plants provide them with 'rewards'. These are mostly food rewards, such as pollen (produced in excess), nectar, feeding tissue and oils (Figure 194.a,b,c,d). Floral resin, used by bees to build nests, can also be a reward (Figure 194.e), as well as aromatic chemicals (Figure 194.f), used by male orchid bees for mating purposes (review in Roubik & Hanson 2004). Pollination can also be the result of deception. Plants of the genus *Aristolochia* produce inflorescences – complex structures with many flowers, often combined with leaf-like elements to function as a single unit that is visited by pollinators – which produce a scent that lures in small flies (Figure 194.g) (see Wolda & Sabrosky 1986 for a Neotropical case study). There is no obvious reward for the visitors, which appear to have been tricked (see Gonzalez & Pabon-Mora 2015 for some recent insights). The different floral rewards also attract opportunists who try to take the rewards without actually pollinating. Many flowers, especially small, shallow flowers, are visited by a mix of pollinating and non-pollinating visitors, mostly small-bodied insects. Large, tubular flowers, on the other hand, are typically built in such a way that easy access to the floral nectar reward is only possible for certain types of specialized insects, often large-bodied bees with a long proboscis or tongue. Many of these large flowers are nevertheless also visited by opportunists ('robbers'), such as large bees with a shorter tongue who pierce the flower from the outside to get to the nectar. To prevent this from happening, many large flowers have a protective set of tough, leaf-like structures that tightly envelop their base.

Social insects

Basic levels of sociality. Caring for eggs and immatures is insect social behavior at its most basic level. It occurs in many insect groups, including the treehoppers (Membracidae; see species account of *Membracis foliata*). Adult female treehoppers with their extravagantly developed thorax are often seen on young twigs and leaves, guarding their immature progeny as they suck plant sap (Figure 194.h). Parental care is also a well-known phenomenon with other arthropods, such as spiders. It may explain how communal nesting came about in spiders, such as *Anelosimus* spp. (Figure 194.i). Along roadsides in Suriname, their veil-like webs with a surface of several square meters can be found, draped on the vegetation. Hundreds of tiny reddish *Anelosimus* spiders build and live in the same web, and collectively attack entangled prey; they can be considered a colony. Communal, colonial nesting is also a basic form of social behavior in many nest-building wasp and bee species, such as the common orchid bee *Eulaema nigrita* (Figure 194.j). The individuals in such colonies are an adult female, her immature progeny, and her adult female progeny; adult males may be present, but typically leave the communal nest soon after reaching maturity.

Higher levels of sociality. Division of labor between the females in the colony is considered socially more advanced, and occurs in many nest-building bee and wasp species. Several species of sweat bees (Halictidae) – metallic green, reddish or blackish bees that nest in the soil – form colonies, which, when fully developed, include one or more females that lay fertilized eggs, while others forage. The highest level of insect sociality is one in which the division of labor is associated with differences in size and shape between the individuals living together (not differences between the sexes). In such cases we speak of 'highly social' insects with 'castes'. All termites and ants are highly social, as well as all honeybees (Apini) and stingless bees (Meliponini) (Wilson 1971; see also species accounts of *Nasutitermes ephratae*, *Atta cephalotes* and *Melipona favosa*). The castes in a termite colony include one or more queen castes (fertile, egg-laying females), a worker caste (infertile females with mainly non-defensive tasks) and a soldier caste (infertile males with mainly defensive tasks). Many ant species, such as harvester ants (*Atta* spp.), have workers that strongly vary in size.

These highly social insects have another important attribute: they forage socially, meaning that workers follow each other to food sources, or lay scent tracks to food sources. Honeybee workers even communicate flight distance and direction to a food source (see review in Roubik 1989). Colonies have socially foraging individuals that are all very close relatives, yet few that reproduce; this means that in many ways the colony is the biological unit, not the individual.

Stipomorpha spp.
(no specific English name)
(no specific local name)
Diptera, Syrphidae (hoverflies, flower flies), Microdontinae

Menno Reemer

Hoverflies (Syrphidae) occur on all continents, except Antarctica. Worldwide, over 6000 species are described, some 1800 of which occur in the Neotropics (Thompson 1999, Thompson et al. 2010); 183 species are currently known from Suriname (Reemer 2010, 2013, 2014, 2016). They can be tiny or large (3-30 mm), slender or stout, bald or furry, plain or colorful, and anything in between (Figure 195). Many share the general body shape and color pattern of stinging insects, such as wasps or bees, and thus engage in mimicry (they are visual mimics). They are true stunt pilots, capable of hovering in mid-air, flying backward, and accelerating from 0 to 10 meters per second within milliseconds. As adults, most species are flower-visitors, feeding on nectar and pollen. The lifestyles of the larvae are diverse. Some are predators, feeding on aphids (plant lice) or other insects, others feed on plant tissue (including leaves, roots or pollen). Yet others feeds on micro-organisms in decaying matter, like rotting wood, mud, or debris in water-filled bromeliads.

A peculiar group are the hoverflies of the genus *Stipomorpha* (Figure 195.a) (Reemer 2013). These will often be mistaken for small stingless bees (Meliponini), having long, hairy hind legs, and flying with legs dangling and abdomen upright. But hoverflies have only two wings (bees four), and don't bite, having no mandibles. So why look like a stingless bee? Well, although these bees don't sting, they do bite: stingless bees are social insects, and many species collectively attack intruders that venture near their colonies. Also, they taste very bad due to chemicals excreted by mandibular glands. Predators tend to remember what unpalatable food looks like, and what their attackers looked like; *Stipomorpha* spp. no doubt profit from the resemblance. The larvae of hoverflies of the subfamily Microdontinae, to which *Stipomorpha* spp. belong, engage in so-called chemical mimicry. The larvae live inside the nests of ants, where they feed on ant eggs, larvae and pupae. They get away with doing this without alarming the worker ants, by sharing their smell (chemicals emanate from their bodies that resemble those of their ant hosts).

Figure 195. Hoverfly diversity in our region (all images © Stephen A. Marshall):

a. *Stipomorpha tenuicauda*, a mimic of a small stingless bee (Meliponini); the fly has just emerged from an ant nest attached to a leaf.

b. the common *Palpada vinetorum*; larvae get food from decaying organic matter.

c. *Ornidia obesa*, a mimic of small metallic-green orchid bees (Euglossini).

d. *Salpingogaster nigra*, resembling a wasp; larvae are predators of spittle bugs (Cercopidae).

Description: *Stipomorpha* spp. have the first abdominal segments flared and flattened, wider than the thorax; the rest of the abdomen is not flattened and tube-shaped. The hind legs are widened towards the end and strongly hairy.

Distribution: *Stipomorpha* spp. occur in South America only; of the 25 described species, 9 are known from Suriname (2 from nowhere else).

Use: Hoverflies are important pollinators; predatory hoverfly larvae are bred in parts of the world, and deployed to destroy aphids that are crop pests.

Similar species: *Stipomorpha* spp. differ in color pattern, with the body either entirely black, largely orange, or in between (black-orange banded). They seem to mimic small stingless bee species with similar color patterns. Non-specialists can distinguish between *Stipomorpha* species by their color pattern.

Bactrocera carambolae
Carambola Fruit Fly
Carambola fruitvlieg
Diptera, Tephritidae (tephritid fruit flies)

Alies van Sauers-Muller

Bactrocera carambolae, the Carambola Fruit Fly (Figure 196.a), is one of many fruit fly species in the Guianas, but the only one of the genus *Bactrocera*, which is native to Asia (Allwood *et al.* 1999). It was undoubtedly brought to Suriname via imported fruit; economically, it is the most damaging fruit fly in the country (van Sauers-Muller 2008). Some of the more important host plants whose fruits are infested by the Carambola Fruit Fly are carambola, Curacao apple, mango (some varieties), sapodilla, West Indian cherry and guava. The native tephritid fruit flies in Suriname belong to the genus *Anastrepha* (Figure 196.b); they attack cultivated and forest fruits.

Two days after the Carambola Fruit Fly female deposits eggs in developing fruit, the larvae emerge and start feeding on the fruit pulp. When the fruit matures and drops to the ground, the larvae crawl out and pupate in the ground (Malavasi *et al.* 2013). After 8 to 12 days, an adult fly emerges from the pupa and starts looking for food, such as nectar, bird droppings, fruit sap on damaged fruits, and plant exudates (liquids produced by plants, such as at cuts in the bark). The flies reach full maturity in three to four weeks, after which they mate and females lay eggs. Carambola Fruit Flies follow strict mating rules (McInnis *et al.* 1999). Females choose males for mating based on how they smell (males smelling of aromatic methyl-eugenol are often preferred). Late in the afternoon (1830-1900 h), males sit in trees in a group, called a lek, each on its own leaflet; they wave their wings to spread their smell. Females search out the leks to choose their preferred male.

Native *Anastrepha* spp. have at least nine species of parasitoids that attack their larvae; only one of these has been found to attack the Carambola Fruit Fly to some extent. A parasite that severely attacks the Carambola Fruit Fly is an egg parasite found in the fly's area of origin in Asia (Chinajariyawong *et al.* 2000); this parasite is a candidate for introduction in the Guianas.

Figure 196. Tephritid fruit flies in Suriname:

a. *Bactrocera carambolae* reared from infested fruit in a cage in Paramaribo, with females and males on fruit of host plant carambola (*Averrhoa carambola*), and males on flowers of the tree *Couroupita guianensis* (produces aromatic chemicals attractive to the males) (image © Alies van Sauers-Muller).

b. *Anastrepha striata* fruit fly on immature fruit of host plant guava (Psidium guajava) at Uitkijk, Saramacca District (image © Alies van Sauers-Muller).

Description: Tephritid fruit flies are usually the size of house flies, but with much brighter colors (yellow, orange, brown); females with visible ovipositor (egg-laying apparatus), the needle-like tip of the abdomen. The Carambola Fruit Fly has a transparent wing with a dark front margin and a short dark streak diagonally across the hind part of the wing. It has a largely dark thorax, with a yellow border stripe on both the left and right margin when seen from above; the scutellum, a skeletal plate at the hind margin, is yellow too.

Distribution: In the Americas, the Carambola Fruit Fly is an invasive species found in Suriname, French Guiana, Guyana, and Amapa state (Brazil).

Similar species: *Anastrepha* spp. differ in having wings with extensive C-, S- or V-shaped bands, usually yellow-orange to brown colored; the body has similar colors and a very distinct, long ovipositor. Most species are a bit larger than the Carambola Fruit Fly.

Anopheles darlingi (no specific English name) *Maskita* (local name for all mosquitoes) Diptera, Culicidae (mosquitoes)

Hélène Hiwat

Anopheles darlingi (Figure 197) is native to the Amazon region, where it is one of the most important transmitters of human malaria because of its high biting rate. It is an ecologically flexible mosquito that adapts to changing environments (Hiwat & Bretas 2011). It transmits the single-celled parasites of the genus *Plasmodium* that cause human malaria; these human parasites are not native to the New World, but are thought to have been introduced from Africa during the transatlantic slave trade (Yalcindag *et al.* 2012).

Anopheles darlingi favors clear-water streams, rivers, inundated forests and swamps as breeding sites, but seem to have adapted to breeding in the increasingly abundant gold mining pits with still, hot and muddy water, where the aquatic larvae find their preferred shaded and stable micro-habitats between plant roots and debris (Rozendaal 1990, Achee *et al.* 2006, Moreno *et al.* 2007). Larvae feed just below the water surface, which is also where the pupae can be found. Once emerged from the pupae, female mosquitoes will mate and go look for their first blood meal, which enables them to produce eggs. *Anopheles darlingi* prefers human blood, but can also feed on other animals in the absence of humans. Biting behavior, including peak biting times, varies between localities and seasons (Hiwat *et al.* 2009).

In the past decade, the use of insecticide-treated bed nets (ITN) has become one of the pillars of malaria prevention worldwide. In Suriname, the mass distribution of ITNs to people in the Interior in 2006 negatively affected *Anopheles darlingi* populations. From 2006 onwards, following a number of interventions, including active malaria case detection surveys and treatment, malaria incidence decreased significantly in Suriname, reaching near-elimination levels in 2015 (Hiwat *et al.* 2012; unpublished data Suriname Ministry of Health Malaria Program).

Figure 197. *Anopheles darlingi* (images © Medical Entomology Unit, Institut Pasteur de la Guyane (IPG), Cayenne, French Guiana): a. adult female.

b. larva.

Description: Slender mosquito with long legs and palps (appendages on the mouthparts); tip of the hind legs entirely white. *Anopheles* spp. assume a unique position when resting or feeding, with the mouthparts and the rest of the body aligned (as in Figure 197.a). The aquatic larvae have short breathing tubes on the abdomen, which requires them to float parallel to the water surface to breathe (197.b).

Distribution: From Mexico to northern South America (Sinka *et al.* 2010); widespread in the Interior of Suriname.

Use: No known use, but transmits the human malaria parasite.

Similar species: *Anopheles nuneztovari* is very similar, but differs from *Anopheles darlingi* in wing spot pattern and leg coloration; while *Anopheles darlingi* bites the entire night with one or two peak biting periods, *Anopheles nuneztovari* has a distinct preference for biting at dusk.

Eacles guianensis (no English name) (*netikapelka*, local name used for all moths)) Lepidoptera, Saturniidae, Ceratocampinae

Hajo Gernaat

Five species of the genus *Eacles* are known from Suriname; three more may occur there, based on their regional occurrence. All are large moths, colored yellow with purple and brown, small milky-white or black eye-spots on both wings. Females are considerably larger than males (Lemaire 1988). *Eacles* spp. belong to the family Saturniidae. Adults (moths) in this family have no functional mouthparts and are not able to feed. Their metabolic stores (carbohydrates, proteins, lipids) have been accumulated in the caterpillar stage. The moths live only a few days, just long enough to reproduce (Janzen 1984, Scoble 1995). In Suriname, *Eacles guianensis* (Figure 198) is active at night and comes to light from about 1830 h to 0100 h.

When the heavy-bodied female emerges from the pupa at night, she is already stuffed with mature eggs. She exposes a gland at the tip of her abdomen that secretes a pheromone (a chemical to communicate with other individuals) to signal availability to mate. Males usually leave their pupae before the females, and disperse; this to avoid inbreeding. Their antennae are feather-like and extremely sensitive to female pheromones carried by the nightly air currents (Lemaire 1988). The male flies into the night, 'upstream' in the direction of pheromone source, the female (Howse & Wolfe 2011). After mating, the female lays fertilized eggs, usually in large clusters on foliage nearby, as she cannot afford to spend much energy and time to search for particular host plants (plants on which the larvae or caterpillars feed). Not coincidentally, Saturniidae use a wide range of host plants (Janzen 1984). Closely related species often live in the same general area, and use similar or even identical pheromones: so how to avoid a mix-up? One explanation is that they release a cocktail of pheromones, with the exact composition (proportion of different chemicals) being species-specific. Additionally, different species may release pheromones only briefly, during different hours of the night (Howse & Wolfe 2011).

Figure 198. *Eacles guianensis* at Brownsberg mountain (image © Joke van den Heuvel).

Description: Large, bright yellow moths; forewings, upper side of thorax and abdomen largely purplish black; on hind wing, an eyespot is broadly connected to a relatively dark irregular transverse purplish / brownish band.

Distribution: Ecuador, Venezuela, Guianas and Amazonian Brazil. In Suriname known from primary forest at Brownsberg mountain (485 m elevation) and Raleighvallen (50 m elevation; middle Coppename River).

Use: (No known use).

Similar species: *Eacles* spp. are unique in Suriname for their large size, absence of mouth parts, and yellow with purple color. Other regional *Eacles* spp. have a more subdued yellow color and/or the transverse band on the hind wings more regular and merely skirting the eye-spot or passing below it.

Hyposcada zarepha zarepha (no specific English name) *Kapelka* (local name used for all butterflies) Lepidoptera, Nymphalidae, Ithomiinae

Hajo Gernaat

Hyposcada zarepha zarepha is an uncommon ithomiine butterfly species of the dense understory of primary forest on bauxite plateaus in the Guiana Shield. It has a weak, fluttering flight, rarely more than 1.5 m above the ground, and feeds primarily on nectar. Its wings are largely transparent (Figure 199; note: transparent wing parts appear whitish here due to flash photography), a useful camouflage in its dark habitat (Neild 2008). When disturbed, this species makes no attempt to escape, but flies leisurely in full view. It is chemically protected; the body contains pyrrolizidine alkaloids (PA), chemicals that are distasteful and potentially toxic to predators. The latter learn to associate the butterfly's appearance with an unpleasant experience to avoid (Brower 1984). The butterflies' protection lies, therefore, in being noticed (DeVries 1987). Most distasteful butterfly species acquire their protective chemicals when in the caterpillar stage, often from ingested plant tissue. Ithomiinae, however, ingest PA as adult butterflies by pharmacophagy (Hartmann 2009), meaning that they suck PA-containing flower nectar of some Asteraceae (like *Eupatorium* spp.) or Heliotropicaceae (like *Heliotropum* spp.), or suck juices from dead or withered Boraginaceae (like *Cordia* spp.). They may even ingest PA from the rotting carcasses of other insects, such as other ithomiine butterflies (see Figure 199).

Species that resemble (mimic) a common distasteful species gain protection. By means of natural selection, several similar-looking species are thought to have evolved; such a group of species is called a mimicry ring (Mallet & Gilbert 1995). In Suriname, the transparent-wing mimicry ring consists of at least 23 species in different butterfly and moth families: Ithomiinae (species in genera *Napeogenes, Hyposcada, Oleria, Episcada, Hypoleria, Mcclungia*), Pieridae (*Dismorphia*), Riodinidae (*Ithomiola, Stalachtis*), Notodontidae (*Dioptis*) and Arctiinae (*Agyrta, Dysschema, Hyalurga, Calodesma, Isostola*) (Gernaat *et al.* 2012).

Figure 199. *Hyposcada zarepha zarepha* extracting PA chemicals (see text) from the dead body of another ithomiine species, *Oleria aegle aegle* at Brownsberg mountain (image © Joke van den Heuvel).

Description: Wings with large transparent patches; color pattern as in Figure 199 (note absence of white spots at the margin of the hind wing underside).

Distribution: Venezuela and Guianas. In Suriname, known from Brownsberg (485 m elevation) and Nassau mountains (550 m elevation).

Use: (No known use).

Similar species: Similar to 22 other butterfly and moth species with transparent wings. The genus *Hyposcada* is distinguished by the unique pattern formed by the wing veins. Other *Hyposcada* spp. have white spots at the margin of the hind wing underside.

Acrocinus longimanus
Harlequin Beetle
Sege (local name used for large beetles and similar large insects)
Coleoptera, Cerambycidae

Hélène Hiwat

The Harlequin Beetle, *Acrocinus longimanus* (Figure 200), is a cerambycid or longhorn beetle (named after the very long antennae in most species). Harlequin beetles have not only long antennae, but also very long forelegs, especially the males. The males use their extra-long legs to fight during mating competition; they can hook their rivals and catapult them out of the way. Mating success is positively correlated with the size of the males (Zeh *et al.* 1992).

For a long time the bright colors of the adult beetles were thought to make the beetles extra vulnerable to predators. However, in the early 1900s, the Brazilian entomologist Dr. Gregório Bondar discovered that in their preferred habitat, which is recently fallen *Ficus* spp. trees and trees of the Apocynaceae family, the beetles are perfectly camouflaged due to the colorful *Fomes* spp. fungi growing on the tree trunks (Bondar, 1926). *Fomes* spp. are bracket fungi (Basidiomycota) that develop only after the sap of the fallen tree starts fermenting. They are an important food source for the beetle larvae. Female Harlequin Beetles seek out old trees infested with these bracket fungi. They make incisions in the tree bark in which eggs are deposited. Larvae bore into the wood for 7-8 months, then pupate and finally emerge as adult beetles.

Harlequin Beetles serve as a mode of transportation for a number of smaller invertebrates, including pseudoscorpions (Aguiar & Bürnheim 1998). Tiny *Cordylochernes scorpioides* (see species account) will crawl onto the beetle's back, under the wings, holding on with their pinchers and with silk threads when the beetle is flying. While their ride moves from one tree to the next, the male pseudoscorpions take the opportunity to check out their female co-passengers for mating, kicking off the unwanted male competition whenever possible (Zeh & Zeh 1992).

Figure 200. Male *Acrocinus longimanus* attracted to light at Brownsberg mountain (image © Auke Hielkema, Mar. 2016).

Description: Adults are large (up to 75 mm length, but sometimes only half this size) with very long front legs, especially in males; elytra (hard front wings that cover the hind wings and much of the body) with a unique pattern of black and bright red and yellow markings.

Distribution: From the south of Mexico to Brazil, and in some of the Caribbean islands; widespread in Suriname.

Use: Used in the pet trade; dried and pinned they are a collector's item.

Similar species: One other very large longhorn beetle in our region has similar markings on the elytra, *Macrodontia cervicornis*. Instead of long forelegs, it has enormous mandibles. Most other cerambycids are much smaller, or have no or very different markings on the elytra.

Rhynchophorus palmarum
South American Palm Weevil
Kabisi-sege (adult), *kumbuworon* or *grugru* (larva)
Coleoptera, Curculionidea

Hélène Hiwat

The South American Palm Weevil, *Rhynchophorus palmarum*, is an armored giant among the beetles (Coleoptera) of the family Curculionidae (weevils). It is best known as a pest of palms and sugarcane (van Dinther 1960, Maas 1970). Adults (Figure 201.a) are attracted to the chemicals released by wounded or freshly cut palms. A male can rally other males and females to a food source by releasing a pheromone (chemical that affects behavior of other individuals) (Khan 1993). Females lay eggs at the base of the leaf-sheaths or shoots of palms, or in wounds or burrows made by other insects in the trunk. The larvae (Figure 201.b) reduce much of the trunk tissues to a jumbled mass of fibers; after several molts, they make pupal cells from these fibers. The adult weevils are long-lived (about three months).

As early as 1705, Maria Sybilla Merian reported that the indigenous people of Suriname readily consume the larvae of this beetle, calling them *grugru*; today the name *kumbuworon* is often used in Suriname. The larvae are prized food items for many forest-inhabiting people, and large palms may be cut for the express purpose of 'breeding' the weevil's larvae. Final instar larvae (just before they pupate) can be 14 grams apiece. Their nutritional value is high and can be an essential addition to a low-protein diet. The taste has been compared to coconut-cured bacon (Howard et al. 2001).

The South American Palm Weevil often carries a number of nematodes (tiny wormlike animals, but not related to earthworms) in its guts, such as *Bursaphelencus cocophilus*, the cause of red ring disease. Plants infected with the nematode suffer from permanent wilting. The geographical distribution (range) of the disease closely overlaps with that of the weevil. Both weevil and nematode propagate in coconut palms and a number of wild palms that are abundant in our region, such as *Mauritia flexuosa* (*morisi*) and *Attalea maripa* (*maripa*) (Magalhães *et al.* 2008).

Figure 201. ***Rhynchophorus palmarum*:**
a. adult in pupal cell made in stem of *Astrocaryum sciophilum* palm at Bakhuis Mountains (image © Bart De Dijn, Mar. 2007).
b. larva, ready to be used as food (image © Reinaldo Aguilar; Creative Commons License; source: www.flickr.com/photos/plantaspinunsulaosa/6964371063\).

Description: Large (up to 50 mm length), robust black beetle; longitudinal grooves on the elytra (hard coverings of the functional wings and hind part of the body); head with a long-thin 'snout' on which the antennae are located, as well as small mandibles (typical of weevils; well developed in this species).

Distribution: Central America to northern South America, expanding its range northward (Esparza-Diaz *et al.* 2013); common throughout Suriname.

Use: Larvae are a popular food in the Interior of Suriname.

Similar species: The similar but mostly reddish *Rhynchophorus ferrugineus* (Red Palm Weevil) is an invasive Asian species that may reach the South American mainland soon (Fiaboe *et al.* 2012). Other weevils in our region are either less robust or (much) smaller; other beetles lack the pronounced snout.

Coprophanaeus lancifer

Hélène Hiwat

Scarab dung beetle (refers to all scarabeids that feed on dung)
Sege (local name used for any large beetle and similar large insect)
Coleoptera, Scarabaeidae, Scarabeinae, Phanaeini

Beetles of the sub-family Scarabeinae are commonly referred to as dung beetles, but not all feed on dung (see below). The metallic blue *Coprophanaeus lancifer* (Figure 202) is a giant among Neotropical dung beetles, and looks like a futuristic warrior from a video game. Both males and females have a massive horn on the head, and a broad, saddle-shaped pronotum (upper side of the thorax) (Edmonds 2010). They are territorial in the sense that they use these features to fight and defend a food item against same-sex competitors. The metallic coloration helps them discern potential mates on sight. This species is mainly active during dawn and dusk. At this time, the dim light optimizes the contrast of their coloration against the forest background, and helps them find a mate (Théry *et al.* 2008).

Scarabeinae are ecologically varied (Hanski & Cambefort 1991). While most species feed on dung, others hardly if ever do so, instead feeding on carrion, rotting fruits, etc. So-called 'rollers', recognizable by their long, bent hind legs, fashion dung portions into balls, which they roll away to bury far from the rest of the dung source, to avoid theft. 'Tunnelers', seemingly less concerned about theft, bury their share of the dung in the soil below the dung source. 'Dwellers' simply breed in the dung or directly below it, not burying it at all.

Coprophanaeus lancifer tunnels, but is attracted to carrion of vertebrate animals more than to their dung. Males and females cooperate to bury a piece of carcass below the soil surface. Their great strength allows them to bury pieces many times their own weight (Uruhary-Rodrigues *et al.* 2008). They tear flesh from the carcass to make brood masses (rotting flesh fashioned into balls or lumps), which are buried and in each of which one egg is laid. The larvae complete their development in the brood masses.

Figure 202. ***Coprophanaeus lancifer*** **(image © Trond Larsen; Creative Commons License; source: scarabaeinae.myspecies.info/taxonomy/term/18701).**

Description: Large (up to 50 mm length, but quite variable), robust beetle with a massive backward-curved horn on the head; two smaller but sharp spines on the front of the head, which probably help with the tearing of flesh; strong saw-toothed legs adapted to digging; body black but with overpowering blue and purple metallic hues.

Distribution: Amazonia and Guiana Shield; occurs in forests throughout Suriname, but apparently more abundant in the Interior.

Use: Dung beetles are used as ecological indicators (Herzog *et al.* 2013), and in forensics (Mayer & Vasconcelos 2013, Uruhary-Rodrigues *et al.* 2008).

Similar species: *Coprophanaeus ensifer* looks similar but has green-metallic hues. Other Scarabeinae in our region may have blue hues, but are smaller; Scarabeidae that belong to different sub-families, as well as other beetles, may be large, but lack the blue hues in combination with the single, large horn (they may have a shorter horn or multiple horns).

Corydalus nubilus

Simon Clavier, Atilano Contreras-Ramos, Nicolas Dedieu

Adult *Corydalus* spp. are known as dobsonflies, larvae as hellgrammites (no specific local name)

Megaloptera, Corydalidae

Figure 203. *Corydalus nubilus*:

a. larva in a small clear water stream (Crique Gregoire) in French Guiana, searching for prey (image © Nicolas Dedieu)

b. adult male (dried museum specimen; image © Atilano Contreras-Ramos).

c. adult female (dried museum specimens; image © Atilano Contreras-Ramos).

Megaloptera is a small, 'primitive' group of insects with some 350 species worldwide. It is currently regarded as separate from Neuroptera (lacewings, antlions, etc.) and consisting of two families: Sialidae (alderflies) and Corydalidae (dobsonflies and fishflies). From the Neotropics, 66 species of Corydalidae are known (Contreras-Ramos 2007, Ardila-Camacho 2014). The genus *Corydalus* is the most diverse with 23 species reported from the Guiana Shield. Many species are only known from Venezuela (8 spp.) or Brazil (3 spp.) suggesting restricted range endemism (Contreras-Ramos & Dunk 2010). The Suriname fauna remains poorly known,with records of only three species of *Corydalus* and one species of *Chloronia* (Geijskes 1984).

Adult males of *Corydalus nubilus* are fierce-looking (Figure 203.b) with large menacing mandibles to attract females (Figure 203.c) and fight off competing males. They are nocturnal and secretive, and live for just a few days, staying in streamside vegetation near the larval habitats. The larvae (Figure 203.a) are spectacular aquatic insects: large and voracious predators, with strong mandibles for grasping prey, and with many paired appendages and gill tufts on the abdomen. Larvae occur in a wide range of unpolluted streams, except those with a slow current and low levels of dissolved oxygen. They take shelter under stones or pieces of wood in forest creeks, or in rivers among leaves of Podostemaceae at rapids. They feed on invertebrates but may also scavenge. Residing at the top of aquatic invertebrate food chains, they are important links in aquatic food webs. Corydalidae larvae are useful bio-indicators of water quality, such as level of oxygenation, and appear to be more sensitive than larvae of sister-family Sialidae (Clavier *et al*. 2010).

Description: *Corydalus* spp. larvae are up to 8 cm long, with a dark-brown patterned head and thorax. They have many paired thread-like appendages and gill tufts on the abdomen, and a pair of small leg-like appendages (prolegs) at the abdomen tip. Adult *Corydalus* spp. have net-veined wings and are typically reddish to grayish brown. Some species, including *Corydalus nubilus*, have sexual dimorphism, males having larger mandibles than females.

Distribution: In unpolluted freshwater streams of the Guiana Shield, including the Interior of Suriname (not in the Coastal Zone).

Use: Corydalidae larvae are used as bio-indicators.

Similar species: Adult *Corydalus* spp. can easily be distinguished from those of related *Chloronia* spp., which are yellow and have smaller mandibles. *Corydalus* spp. larvae differ from other Neotropical genera of Corydalidae by having head and thorax patterned. Sialidae larvae are also similar, but have no prolegs; they instead have a long thread-like appendage at the abdomen tip.

Atta cephalotes

Hélène Hiwat

Leafcutter ant (general name for *Atta* spp. and related species)
Mira (local name for ants in general)
Hymenoptera, Formicidae, Myrmicinae

In the Interior of Suriname, leafcutter ants (*Atta* spp.) are best known as agricultural pests, especially where swidden agriculture takes place. *Atta* spp. damage crops such as citrus and cassava (Van Dinther 1960, Cherrret 1968). They cut leaf fragments from cultivated and wild plants, chew them up, and use the resulting pulp to raise specific fungi (Agaricales: Lepiotaceae) in underground nest chambers. They have a symbiotic relationship with these fungi, which they cultivate in 'fungus gardens' and feed on; the ants protect the fungi against natural enemies, especially micro-organisms (Cherrett & Quinlan 1979, Guedes *et al.* 2012, Pagnocca *et al.* 2012, Garret *et al.* 2016).

The life cycle and social lifestyle of *Atta* spp. is described by Wilson (1984) and Holldobler and Wilson (2011). A virgin queen leaves the maternal colony in the rainy season to mate, taking along a small bit of symbiotic fungus. The mated queen starts a new nest and ultimately produces millions of ants that form a new colony. Most of the ants in the colony are 'workers' that do not reproduce. Small 'minor' worker ants cut and transport leaves (Figure 204.a), as well as tend to the queen, the brood and the fungus gardens; large 'major' worker ants (soldiers; Figure 204.b) are fierce protectors against predators and other threats. *Atta* spp. suffer heavily from attacks by certain parasitoid phorid fly species (Phoridae) (Folgarait 2013, Uribe *et al.* 2016). A female phorid fly lays eggs in the body of living ants; their larvae develop inside the ants. A total of 39 species of phorids from 9 genera are known to parasitize *Atta* spp. Different phorid species target different parts of the ant body at different times and places (day or night; close to or away from the nest). The ants show evasive or defensive behavior (as in Figure 204.a) toward these parasitoids which helps them survive, but has a negative impact on their foraging activity.

Figure 204. ***Atta cephalotes*** **in a forest fragment at Paramaribo (images © Bart De Dijn, Jan. 2017):**

a. minor worker ants transporting leaf fragments, with tiny minor on one of the leaves, guarding the larger ant against phorid parasitoid attacks.

b. minor and major workers entering and leaving one of the nest entrances.

Description: *Atta cephalotes* is rusty brown; workers with two petiole (waist) segments and three pairs of spines on the body. Major workers have small eyes, the upper part of the head swollen (resembling two merged globes), and long woolly hair on the front side of the head. The queen is 22 mm in length, minor workers up to 7 mm, major workers (soldiers) up to 14 mm.

Distribution: Southern Mexico till Atlantic forest zone of southeast Brazil, as well as some Caribbean islands (Solomon 2007); common throughout Suriname in forested areas and small manmade clearings for agriculture.

Use: Males are used as a snack, as well as fish bait by indigenous people in the Interior of Suriname (Bart De Dijn, personal communication).

Similar species: *Atta cephalotes* can be distinguished from the very similar *Atta sexdens* by the woolly hairs on the heads of major workers. *Atta cephalotes* is considered a wet forest species; *Atta sexdens* seems more tolerant to disturbance and dryer conditions, and can survive in open grassland (Solomon 2007).

Melipona favosa
(no specific English name)
Parwa-oni (*kawfre-oni* refers to any *Melipona* spp.)
Hymenoptera, Apidae, Meliponinae (stingless bees)

Bart De Dijn

Melipona spp. are Neotropical bees known locally as *kawfre-oni* (meaning: horsefly bee); *Melipona favosa* is known as *parwa-oni*, as it is abundant in coastal mangrove forest (*parwa*). *Melipona* spp. are social bees that produce honey, like the common honeybee (*Apis mellifera*; *brasyon-oni*) (Roubik 1989). Colonies inhabit nests located in pre-existing hollows in living tree boles or branches where they are protected against mammal predators (Wille & Michener 1973; Roubik 1979, 2006). *Melipona* spp. have no sting, but deter enemies by biting, and – based on how they smell – tasting very bad.

The reproductive biology of *Melipona favosa* (Figure 205) is well known (see Sommeijer 1984, 1985; Sommeijer *et al.* 1999, 2003, 2004). Brood cells under construction in the nest are inspected by the queen, attended by workers engorged with honey. At the queen's 'OK' signal, workers discharge honey into the cell, and at least one adds an unfertilized egg. The worker egg is eaten by the queen, who then lays a fertilized egg, which will develop into a worker (in three-quarters of cases) or a daughter queen. Males develop from unfertilized eggs, mostly eggs laid by workers. Virgin queens leave the maternal nest to get inseminated by male bees which often congregate near an existing nest. Fertilized queens may move into an 'old' nest to try and supersede the long-resident queen, or into a 'new' nest that has no fertilized queen. A new nest is built by the workforce of an existing colony, within the flight range of that colony's nest. *Melipona* workers trigger the development of queens by adding geraniol, a compound produced by glands in their head, to the honey in a cell (Jarau *et al.* 2010).

Figure 205. Worker of *Melipona favosa* (images © Bart De Dijn, Nov. 2015):

a. returning with pollen load to nest in base of mimosoid tree in Paramaribo.

b. foraging for pollen on flowers of ornamental palm *Adonidia merrillii* in Paramaribo.

Description: Body 7-9 mm long, lacking metallic hues; black with red hairs on thorax, and broad yellow markings (bands) on abdomen. As in all *Melipona* spp., wings short (not extending beyond abdomen), thorax densely hairy, and hind legs of workers wide and sideways flattened towards the end. The flattened hind leg part is the corbiculum (plural: corbicula), a (virtually) hairless central depression with long hairs at the margins; it serves to carry pollen.

Distribution: Colombia till coastal French Guiana, also Trinidad and Tobago (Camargo & Pedro 2013); in Suriname, found only in coastal forests, and neighboring urban and rural areas.

Use: This bee is traditionally kept in small hive boxes in Coronie (along the west coast of Suriname) to harvest honey; the honey is used medicinally.

Similar species: Other *Melipona* spp. do not have both bright red thoracic hairs and broad, yellow abdominal bands. Stingless bees other than *Melipona* spp. are smaller and/or have wings that extend well beyond the abdomen; workers or females of many bee species other than stingless bees also have corbicula, but have a sting, such as some orchid bees (Euglossinae) that resemble *Melipona* spp. (their face has a green-gold metallic hue).

Eufriesea surinamensis (no specific name in English) (*segewaswasi refers to any large bee)* Hymenoptera, Apidae, Euglossinae (orchid bees)

Bart De Dijn

The Neotropical Euglossinae or orchid bees are the sole pollinators of some 10 percent of the regional orchid flora. The male bees produce an oily material to collect floral 'scents' – aromatic chemicals produced by plants in their flowers. When visiting flowers, the males apply the oily material to flower tissue with their front legs, and transfer it – loaded with scents – to slits in their very large hind legs. Each male patrols a small territory, centered on a tree stem. It frequently lands on the stem and extracts scented material from its hind legs and rubs it on its wings; it then alights to hover nearby and presumably disperses scents. The function(s) of the scents is not yet clear: they may help bees to recognize mates of their own species, they may be sexual lures. The biology of these bees has been reviewed by Dressler (1982), Ramirez *et al.* (2002) and Roubik and Hanson (2004); recent work on their biology includes Eltz *et al.* (2005, 2007) and Zimmerman *et al.* (2009).

Female *Eufriesea surinamensis* collect pollen and nectar to provision brood cells built from plant resin and bark chips (Figure 206). 'Nests' – clusters of brood cells – are built at semi-exposed locations, commonly in gaps in the floors and walls of houses (as described in Dodson 1966). *Eufriesea* spp. do not form social colonies, but aggregate, as daughters re-use the maternal nesting site (Myers & Loveless 1976). The nesting aggregations attract parasites, such as the large green *Exaerete* spp. (Euglossinae) (Bennett 1972). Most of the about 20 *Eufriesea* spp. known from Suriname are large, long-tongued bees with striking black, yellow and/or red hair bands. Each species has one of three distinct banding patterns, and *Eulaema* spp. (Euglossinae) share these patterns; species with similar patterns are considered mimics (Dressler 1979). By looking alike, these bees should enjoy added protection. The female bees have a nasty sting, and predators will tend to avoid anything resembling what stung them before. Male bees have no sting, and undoubtedly benefit from resembling the fierce females.

Figure 206. Female of *Eufriesea surinamensis* building brood cells with resin and wood chips at nest (aggregation) in Paramaribo (image © Bart De Dijn, Mar. 2007).

Description: Body large (16-20 mm), and tongue long (14-15 mm when folded under body). Hairs from head till first segment of abdomen black, on rest of abdomen hairs ochre-yellow. Below the dense hair cover, the abdomen is black with a green-bronze metallic hue above (except on less hairy first abdominal segment); males with green metallic hue on face.

Distribution: Mexico till Peru and Sao Paulo state (Brazil), also Trinidad (Moure *et al.* 2012); widespread in Suriname common in inhabited areas.

Use: No known use, but important pollinators (in particular of certain orchids).

Similar species: Many other *Eufriesea* spp. look similar, but differ in tongue length and/or color pattern details (hair color, hue of the first two abdominal segments); the similar, more common *Eulaema cingulata* has yellow hairs on the first abdominal segment, and a band of black ones on the second.

Rhyncholepta grandicallosa

Jocelia Grazia, Cristiano Schwertner

Stink bug (English name used for all Pentatomidae)
Fyofyo (local name used for all Pentatomidae)
Heteroptera, Pentatomidae, Pentatominae

The family Pentatomidae (stink bugs) occurs worldwide, with near 5000 species in about 900 genera (Grazia & Schwertner 2017). *Rhyncholepta grandicallosa* (Figure 207) belongs to the subfamily Pentatominae (Becker & Grazia 1971), which has about 600 species in the Neotropics (Rolston & McDonald 1979, Rolston *et al.* 1980, Rolston & McDonald 1981, 1984; Grazia *et al.* 2015). There is no published data on the biology of *Rhyncholepta grandicallosa*, but its biology may be inferred from what we know about stink bugs in general.

Phytophagy (feeding on plants) is the ancestral condition in the family (Schuh & Slater 1995); stink bugs feed on different parts of plants, but mainly on developing fruits and seeds (Schuh & Slater 1995, Panizzi 1997, Schaefer 2009). Most stink bug species feed on a wide range of plants, but some have a preference for certain plant families (Panizzi 1997, 2000; Yonke 1990). The mating behavior of several species has been studied, and involves attraction by means of pheromones (chemicals released in the air) and by sound; it also involves courtship that includes physical interaction (Borges *et al.* 1987, Blassioli-Moraes *et al.* 2013). Female stink bugs lay eggs on the underside of leaves in clusters ranging from 12 to 100 eggs. The immatures are nymphs; they resemble the adults but do not have fully formed wings. After the fourth molting event, the stink bug nymph becomes an adult with fully formed wings. The development time from egg to adult ranges from 45 to 90 days; it is strongly influenced by abiotic conditions (temperature and humidity) and food quality. There may be up to seven generations per year. Parental care is given by adults of several species (Eberhard 1975), but has not been observed with *Rhyncholepta grandicallosa*.

Figure 207. *Rhyncholepta grandicallosa* at Rio Negro, Amazonas state (Brazil) (image © Cristiano Feldens Schwertner).

Description: Body length 11-12 mm. Upper side of body mostly red-brown with rust-red dots (densely packed, irregularly distributed), but parts are contrasting shiny green-yellowish (such as parts of the scutellum, the large triangular plate at the middle of the thorax, partly covering the wings). Head triangular, eyes prominent, pronotum (front part of the thorax) with a pair of large spines.

Distribution: Imperfectly known, but roughly from Panama to central Amazonia and the Guianas, including Suriname. In the Amazon region collected using light traps up to 25m high in the forest canopy.

Use: (No known use).

Similar species: Body shape and especially color pattern of this species is fairly unique; closely related *Rhyncholepta meinanderi*, recorded from northern South America, although never collected in Suriname, looks deceivingly similar though (see Becker & Grazia 1971).

Panstrongylus geniculatus
Kissing bug (English name used for all Triatominae)
(no specific local name)
Hemiptera, Reduviidae, Triatominae

Hélène Hiwat

Kissing bugs (Triatominae) suck the blood of mammals, and in the process may transmit diseases such as Chagas, a lesser-known insect-transmitted disease in our region that affects humans. Chagas disease is caused by *Trypanosoma cruzi*, a single-celled parasite. Infected people may develop life-threatening heart or intestinal problems. Kissing bugs belong to the family of assassin bugs (Reduviidae), a diverse group of mostly insect-eating predators, except for the sub-family Triatominae, which suck the blood of larger vertebrate animals. Triatominae nymphs as well as adults feed on blood (Schofield & Dolling 1993). During daytime, the bugs remain in palm trees and burrows of the animals they target (which typically rest there during the day); at night they walk or fly around (Miles *et al*. 1981).

Panstrongylus geniculatus (Figure 208) is a commonly encountered triatomine species in Suriname and French Guiana (Berenger *et al.* 2009, Hiwat 2014, Peneau *et al.* 2016). It occurs in forested habitats, and spends much time in the burrows of its blood hosts, such as armadillos of the genus *Dasypus* (Miles *et al.* 1981). It regularly comes in contact with people (Carrasco *et al.* 2005, Hiwat 2014), and may be a wild species that is in the process of adapting to areas inhabited by people. This may in part be due to the increasing destruction of its natural forest habitat, but may also be due to the availability of alternative blood hosts (people, domesticated animals). This may lead to increased disease transmission to people, as was for instance recorded in a small riverine settlement in Brazil, where pigsties adjoining houses were heavily infested with *Panstrongylus geniculatus* (Valente *et al.* 1998). In Suriname, the first acute case of Chagas disease ever reported was thought to be due to transmission by *Panstrongylus geniculatus*, since the bugs were found in the patient's yard, near his dogs (Oostburg *et al.* 2003). The proportion of kissing bugs that carry *Trypanosoma cruzi* can be very high, even in urban areas (Carrasco *et al.* 2005, Peneau *et al.* 2016).

Figure 208. Adult *Panstrongylus geniculatus* attracted at night to a light trap at Colakreek (Vierkinderen plantation) (image © Auke Hielkema, Mar. 2014).

Description: Body about 20 mm long; mouthparts stinging-sucking, abdomen wide. Overall light brown with a pattern of darker spots and stripes, wings brown, pronotum (upper, first part of thorax) with a clover-shaped spot; the abdominal segments have a distinct black pattern on the margin.

Distribution: From southern Mexico to northern Argentina; presumed widespread in Suriname (certainly in the Coastal Zone).

Use: No known use, but transmits *Trypanosoma cruzi*, the parasite causing Chagas disease.

Similar species: A closely related, less commonly encountered species in our region is *Panstrongylus rufotuberculatus*, which has more black spots on the margin of the abdominal segments, and wings with a greenish hue.

Membracis foliata
Treehopper (name used for all Membracidae)
(no specific local name)
Hemiptera, Membracidae

Hélène Hiwat

Treehoppers (Membracidae) are distant relatives of aphids (Aphidoidea, another group within the Hemiptera). Like aphids, membracids have stinging-piercing mouthparts to suck plant sap. Their hallmark is the often bizarrely shaped pronotum (upper, front part of thorax). As a by-product of ingesting large quantities of plant sap, treehopper nymphs (wingless immatures) secrete substantial amounts of a sugary substance called honeydew. Aphids are also prolific producers of honeydew. The sugary secretion attracts a variety of ants, bees and wasps that opportunistically feed on it. Ants seem to protect the treehoppers from predators, to maintain the honeydew supply (Wetterer *et al*. 2000). Plants may actually benefit indirectly from this, since the ants may harass other insects that are more damaging to the plants than the treehoppers (Styrsky & Eubanks 2007).

Membracis foliata (Figure 209) was originally described by Linnaeus in 1758; ever since there has been much confusion as to its identity (Sakakibara & Evangelista 2010). It is a common treehopper in the Guianas, well known as a pest of *Annona muricata* (soursop; *sunsaka*) (Van Dinther 1960, Hamada *et al.* 1998), a Neotropical fruit tree that produces large, juicy fruits. The females deposit oblong-shaped eggs in small batches inside the twigs and branches of soursop. At the exterior of the twig or branch, the place of oviposition is protected by an oval patch of a white gluey substance. The nymphs and adults suck sap from the twigs, branches and fruits of the soursop trees. They often congregate and show rudimentary social behavior, such as brood care and parent-off-spring communication (Lin 2006). If young soursop fruits are infested they may remain small and deformed, which is why *Membracis foliata* is considered a pest.

Figure 209. Adult *Membracis foliata* at the Paramaribo botanical garden (Cultuurtuin) (image © Auke Hielkema, Mar. 2013).

Description: Body 9-12 mm long, and about as tall; pronotum (first, upper part of thorax) tall and leaf-shaped, blackish, with an oblique white band near the front margin (but not touching it), and a U-shaped spot at the center. Nymph caramel-brown with black spots, and a pair of fleshy spines on the upper side of each abdominal segment; nymphs are often largely covered in a white, waxy substance.

Distribution: Mexico till Argentina and southern Brazil; widespread in Suriname, common in yards in rural and urban areas, such as Paramaribo.

Use: Not used, but a pest of soursop.

Similar species: A closely related, similar species is *Membracis foliatafasciata,* which has a white band with an outline that overlaps with the front margin of the pronotum. Other treehoppers have a pronotum with a very different shape and/or different color pattern.

Nasutitermes ephratae

David Sillam-Dussès, Virginie Roy

Nasute termite (English name used for all Nasutermitinae)
Uduloso (local name used for all Isoptera or termites)
Isoptera, Termitidae, Nasutitermitinae

Nasute termites (Nasutitermitinae), like other termites (Isoptera), establish complex social colonies. Mature individuals that are part of the colony are either fertile females (queens), males (kings), or infertile individuals that mostly defend the colony (soldiers) or engage in all other work to maintain the colony (workers). Many nasute termite species build exposed carton nests which house the colony, such as *Nasutitermes ephratae* (Figure 210), one of the most common Neotropical species. This species' colonies are typically headed by a single queen and king, but occasionally there are additional queens (Becker 1961, Mathews 1977, Thorne 1982, Thorne 1985). Queens and king are protected in a central portion of the nest ('royal chamber'). This species has been recorded from primary forests, secondary forests and disturbed sites. It seems to thrive in open forests with low humidity and high abundance of dry wood, where density was estimated at 25 nests per hectare (Bandeira et al. 2003). It feeds on dead branches and twigs in the forest litter (Mathews 1977).

Nasutitermes ephratae builds nests attached to small trees and saplings at a height of 1.5 to 3 m above ground level. The arboreal nests are spherical or ellipsoidal, with a diameter of 25-60 cm (Mathews 1977, Thorne 1980). The exterior of the nest is light brown and smooth, with a leathery appearance. The thin surface layer of the nest is easily removed in large sections; it covers an inner structure of more or less dense carton with lots of galleries and chambers (Thorne 1980). *Nasutitermes ephratae* is also capable of building mounds (structures built on and partially in the soil); this has been observed in savannas in Venezuela (Thorne 1980, López-Hernández 2001). Throughout its geographical range, it is considered as a minor pest, damaging manmade wooden structures and occasionally fruit trees (Constantino 2002).

Figure 210. Workers of *Nasutitermes ephratae* foraging under the protection of soldiers (image © David Sillam-Dussès).

Description: Soldier: head capsule (bulk of the hard external skeleton of the head) chestnut brown, sometimes very dark with a paler nasus ('nose': sharp, protruding part of the head capsule); nasus long and slender, only slightly conical seen from above; basal third of nasus broadened abruptly below; top of head capsule with raised lump in the middle.

Distribution: From Mexico till Peru and Brazil, as well as much of the Caribbean; occurs throughout Suriname.

Use: Brood and immatures (nymphs) are used as food by indigenous people in parts of Venezuela (Araujo & Beserra 2007).

Similar species: Soldiers of *Nasutitermes corniger* can be distinguished from those of *Nasutitermes ephratae* by the presence of short hairs above on the abdomen (Hartke & Rosengaus 2011). The exterior surface of *Nasutitermes corniger* nests is dark brown and coarse with small bumps, while that of *Nasutitermes ephratae* nests is lighter brown and distinctly smoother.

Eurycotis sp.
(no specific name in English)
Kakalaka (local name use for all cockroaches)
Blattodea, Blattidae, Polyzosteriinae

Dominic Evangelista

There are as many as 59 species of *Eurycotis* cockroaches in the Neotropics, with a large number in the Caribbean (Beccaloni 2014). Based on published sources, *Eurycotis blattoides* is the only known Guianan species of the genus *Eurycotis*, of the sub-family Polyzosteriinae (Evangelista *et al.* 2015). However, most species of the genus *Eurycotis* that occur in this region have probably yet to be recorded. These cockroaches fire a defensive chemical from their abdomen that is irritating to the touch and has an unpleasant, acrid smell (Gutierrez 1999; Turnbull & Fashing 2002). The secretion smells similar to that of a stink bug (Pentatomidae; *fyofyo*) and causes a distinct burning sensation to human skin. It is no doubt useful to deter predators. All *Eurycotis* are entirely terrestrial and cannot fly. The adults have inconspicuous vestigial wing pads as opposed to large functioning wings.

As a result, the adults can be confused with immatures (nymphs). The individual in Figure 211 is an unidentified species of *Eurycotis* found in Guyana. It looks very similar to *Eurycotis blattoides* but the original description of this species makes identification difficult. The species pictured here has been observed dropping its wings when grabbed (Evangelista, personal observation). This clever behavior probably serves to escape from predators, and has so far not been noted in any cockroach. The unidentified *Eurycotis* sp. is found in leaf litter, under logs, and hiding in small holes in wood or in bromeliads. *Eurycotis* spp. are otherwise mysterious and almost nothing else is known about them, including about their diet or ecological relationships.

Figure 211. An unidentified *Eurycotis* sp. from Madewini, Guyana (images © Dominic Evangelista):

a. adult escaping capture on the white sands of the coastal rainforest.

b. dried museum specimen.

Description: Oily black appearance overall, but more yellow or red where the external skeleton is thin. Its body is fairly cylindrical compared to the typical cockroach, which is flattened. The adults have only short vestigial wing pads.

Distribution: *Eurycotis* sp.: Guyana; *Eurycotis blattoides:* Guyana, Suriname, French Guiana.

Use: (No known use).

Similar species: The coloration of *Eurycotis blattoides* is distinct from other species of the *Eurycotis* genus (Gutierrez 2013). The unidentified species described above resembles it though (Figure 211), and superficially resembles cockroach species of the genus *Lamproblatta* and juvenile *Nyctibora* (which also have robust bodies like *Eurycotis* spp.). *Lamproblatta* spp. look very similar but lack any trace of wings (for images see Evangelista *et al.* 2016); *Nyctibora* spp. are uniquely covered in a carpet of short silky hairs.

Choeradodis rhomboidea

Julio Rivera, Antonio Agudelo

Green shield mantis, hooded mantis (name used for *Choeradodis* spp.)
(no specific local name)
Mantodea, Mantidae, Choeradodinae

Choeradodis rhomboidea (Figure 212) is one of Suriname's more distinctive predatory praying mantises (Mantodea). The Neotropical genus *Choeradodis* has only four other known species (Roy 2004). *Choeradodis* spp. resort to phyto-mimicry (looking like plant parts, such as leaves) to avoiding detection by the small invertebrates they prey upon, as well as their natural enemies, such as birds. A *Choeradodis* species was one of the first Neotropical praying mantises portrayed by European naturalists (see plate 27 in Merian 1705).

This species can be found sitting on leaves in the forest, pressing its body down to conceal its head and formidable forelegs, which are built to lash out and grasp prey. The large, shield-like prothorax (front part of thorax) helps with this concealment. In males, it has a rhomboidal shape (Figure 212.a), pentagonal in females (Figure 212.b); in hatchlings, it is much narrower, becoming progressively larger after each successive molt. This species hosts so-called epizoic liverworts, lichens and fungi, meaning that they naturally grow on the animal's body, in this case on the prothorax. This is a clue as to the habitat preferences of this elusive species, since such epizoic organisms typically grow on the surface of leaves in more or less exposed areas within the rainforest, such as treefall gaps or the inner canopy (Lucking *et al.* 2010). The prothorax may have brown marks, making it appear like a partially dried leaf. When on the move, this mantis rocks its body back and forth, imitating a leaf shaking in the wind. All this makes the species extremely hard to detect. *Choeradodis* spp. resort to multiple lines of defense when detected by a predator (Robinson 1969). For instance, the insect can dodge, drop or run, but it may also go for confrontation, engaging in so-called deimatic display: raising its head, prothorax and forelegs and facing the predator (Figure 212.b). To appear more menacing, the insect rubs the finely toothed femur of the front leg against the margins of the enlarged prothorax, producing a rasping sound, while flashing contrastingly colored markings on the inside of the legs. This is assumed to startle a predator.

Figure 212. *Choeradodis rhomboidea*

a. male from the Amazon in search of prey (image © Antonio Agudelo).

b. female from the Amazon in deimatic display (image © Pedro Ivo).

Description: Adult is approximately 70 mm long, with a greatly enlarged prothorax, leaf-like forewings, and a vivid green overall color; the inner side of the front leg has a large black mark.

Distribution: Forested areas across Suriname, Venezuela, Guyana, French Guiana and northern Brazil (Para and Amazonas states).

Use: (No known use).

Similar Species: Similar to *Choeradodis strumaria*, also found in Suriname, which lacks the large black mark on the inner side of the foreleg.

Tropidacris collaris
Giant Violet-winged Grasshopper
Spinka (local name used for all grasshoppers)
Orthoptera, Acridoidea, Romaleidae

Hélène Hiwat

The Romaleidae are a Neotropical family of grasshoppers (Acridoidea) with 467 species in 109 genera (Poderoso *et al.* 2013). They are herbivorous; most species eat grasses. They damage crops such as sugarcane and corn, beans and cassava (yucca). In grassy, vegetated areas where much light penetrates to soil level, Romaleids may occur in large numbers (Almeidae & Câmara 2008). Adults communicate with sound and sight, as well as scent and touch. Romaleidae males, which are slightly smaller than females, make loud sounds by vibrating the wings to attract females. Females lay egg packages in the ground that each contain 50-100 eggs. After about 30 days, the eggs hatch. The immatures (nymphs) lack functional wings; those from the same egg batch tend to stay together. They molt every four to five days. After the fifth and last molt, they become adult locusts, winged, and very mobile and sexually active.

Tropidacris spp. have a specialized digestive system that allows them to eat the driest of plants (Carbonell 1986). *Tropidacris collaris* (Figure 213) causes crop damage throughout much of South America (Chagas 1995, Pelizza 2012, Afonso 2014). Exotic plant species, such as most crops, are often naïve in an evolutionary sense, meaning that they have not co-existed with this voracious native herbivore long enough to develop effective defenses (Poderoso *et al.* 2013, Afonso 2014). Entomophagous fungi may play a prominent role as a biological control agent against this grasshopper in the near future (Pelizza *et al.* 2012). Such fungi produce spores that infect insects; the fungus develops inside the insect, and influences its behavior, causing it to move up high in the vegetation, grasp a leaf, twig or stem, and die there. Once dead, the insect is digested by fungal enzymes; fungal spore-bodies (tiny mushrooms) grow out of the dead insect. The dead insect, high up, is well positioned for the release and spread of fungus spores (Shang *et al.* 2015).

Figure 213. *Topidacric collaris*:

a. adult at Tafelberg, in defensive posture (raised legs) after being disturbed (image © Bart De Dijn, Dec. 2005).

b. nymphs at Colakreek (Vierkinderen plantation) (image © Auke Hielkema, Jan. 2013).

c. adult with wings opened to show hind wing color (specimen from Colakreek; image © Auke Hielkema, Sep. 2016).

Description: This is a large species: body length up to 10 cm, wingspan 18 cm. Forewings with leaf-like venation pattern (a form of camouflage); hind wings (hidden between front wings when the grasshopper is not flying) with a striking green or blue color pattern. Nymphs are aposematic (show bright warning colors to indicate toxicity) and are gregarious (they stay close together), which further enhances the warning message (Joron 2009).

Distribution: From Colombia to central Argentina; in Suriname they are widespread, often found in savannas and slash-and-burn fields.

Use: Sold live as pets, and dead as collector's items.

Similar species: *Tropidacris cristata* looks similar, but has striking orange-colored hindwings; it is also common throughout much of South America. Most other romaleids are smaller and have different color patterns.

Scapteriscus didactylus
West-Indian Mole Cricket, changa
Kotkoti (local name used for all Gryllotalpidae)
Orthoptera, Grylloidea, Gryllotalpidae

Hélène Hiwat

Mole crickets (family Gryllotapidae) are winged insects with massive shovel-shaped fore limbs and hind limbs adapted for jumping. Their lives are mostly spent underground, digging tunnels. The females dig 'chambers' in which they lay their eggs. Hatchlings eat their egg shells and sometimes also their siblings (Nickle 2003). Nymphs (immatures) resemble adults, but do not have fully developed wings. After 8 to 10 molts, nymphs transform into adults. The males of several species engage in a musical courtship at night. The songs are made with ribbed, so-called stridulation structures on the fore wings. Different species have slightly different stridulation structures and as a consequence produce different songs. The enlarged entrance of the male burrow increases the volume of the call, attracting females (Walker & Moore 2013). Mating occurs in the male burrow, which the female may take over after mating.

Mole crickets are notorious for eating plant roots and seedlings, and may destroy entire lawns and fields of crops (Van Dinther 1960, Frank *et al.* 1995, Rentz 2007). *Scapteriscus* spp. have been called 'the most damaging crickets of the New World' (Walker & Moore 2013). They are not strict vegetarians though; some species are mostly carnivorous. The Southern Mole Cricket, *Scapteriscus borellii*, for instance, feeds largely on insects and other soil-inhabiting animals. Biological pest control of mole crickets using parasitic nematodes, wasp and flies has been applied with variable success (Frank *et al.* 1995, Parkman *et al.* 1996). In French Guiana and Suriname, *Scapteriscus didactylus* (Figure 214) is a significant predator of the eggs of marine turtles, including the critically endangered Leatherback Turtle (*Dermochelys coriacea*) (Maros *et al.* 2003, 2005; Hilterman & Goverse 2007). Only the largest immature stage of this mole cricket preys on the eggs, leaving behind characteristic round holes in the shells. Other species of mole crickets (*Scapteriscus borellii* and *Neocurtilla hexadactyla*) do not seem to attack turtle eggs in our region (Auguste *et al.* 2003).

Figure 214. *Scapteriscus didactylus* on sandy soil in Paramaribo (image © Hélène Hiwat).

Description: Cricket-like, elongate body (males 19-34 mm long, females 24-36 mm). Massive shovel-like structures on front legs with several blades, used for digging. Species identification is based on detailed measurements of body parts, and analysis of mating sounds (Nickle 2003).

Distribution: Native to South America, introduced to the Caribbean islands and Australia (Maros *et al.* 2006, Rentz 2007); common throughout Suriname.

Use: No known use, but major agricultural pest.

Similar species: True crickets (Gryllidae) are less elongate and lack the massive shovel-like forelegs. *Scapteriscus borellii* resembles *Scapteriscus didactylus*, but the former has more widely spaced blades on the forelegs (Maros *et al.* 2005).

Mecistogaster lucretia
Bumpwinged Helicopter Damselfly
Grasbarki (local name used for all Odonata)
Odonata, Zygoptera, Coenagrionidae, Pseudostigmatinae

Marcel Wasscher, Johan van't Bosch

In flight, *Mecistogaster lucretia*, the Bumpwinged Helicopter Damselfly (Figure 215), slowly moves through the forest understory like a sprite. With a total body length of 15 cm, it is the longest dragonfly species in the world, and has a wing span (Allaby 2009) of up to 12.5 cm (20 cm in some other Pseudostigmatinae) (Heckman 2008, Wilson 2009). About 300 million years ago, their now extinct relatives (also Odonata; dragonflies and damselflies) had a 60 cm wing span, at a time when the atmosphere was richer in oxygen, which must have allowed large insects to better oxygenate deep tissues and be more active.

Mecistogaster lucretia lives in mature forests, and is the commonest species of giant damselfly (Pseudostigmatinae) in Suriname. Pseudostigmatinae are active away from any obvious surface water bodies, which may come as a surprise since all Odonata have aquatic larvae. They breed in very small water-filled cavities in the forest, such as those in living or dead tree stems (Fincke 1984). Such cavities are more common in old than in young trees. A giant damselfly male can be seen defending a water-filled cavity against other males. Females have been seen throwing their eggs with their long abdomens into the far, dark corners of water-filled cavities. Giant damselfly nymphs feed mainly on mosquito larvae, though cannibalism has been observed. When a cavity is very small, only one nymph may be able to survive. The maximum known lifespan of an adult giant damselfly is 165 days; the abdomen of long-lived individuals in the humid rainforest may be covered with algae.

The diet and feeding behavior of adult giant damsels is different from that of other Odonata. They feed on small web-building spiders. A giant damsel will hover slowly in front of a web, and cleverly pick the owner out of it, without getting entangled. Based on data from specimens in the Naturalis insect collection (at Leiden, The Netherlands), *Mecistogaster lucretia* is most abundant in Suriname from July to November (but this may be because this is when most collecting expeditions have taken place).

Figure 215. Male *Mecistogaster lucretia* captured near Uitkijk, with the collector's hand as a reference for size (image © Klaas-Douwe Dijkstra).

Description: Very large, but also very slender damselfly; distinctive black bump on the hind wing.

Distribution: Northern half of South-America, till northern Argentina; common throughout Suriname in mature rainforest.

Use: (No known use).

Similar species: Of the Pseudostigmatinae (see Garrison 2010) six species are known from Suriname; these resemble *Mecistogaster lucretia*, but lack the black bump on the hind wing. Other Odonata have shorter bodies and wings.

Perithemis lais
Fine-banded Amberwing
Grasbarki (local name used for all Odonata)
Odonata, Libellulidae

Marcel Wasscher, Johan van't Bosch

Perithemis lais (Figure 216) is a brightly colored, restless dragonfly that mimics a wasp (von Ellenrieder & Muzón 1999, Heckman 2006). It typically flies low over the water, at the edge of creeks and rivers with a slow current, less so around trenches and pools. Males are restless even when they sit, first moving their wings up and down, then the abdomen. Males scanning for a mate outnumber the females. When they copulate, they first form a tandem, the male clasping the female by the head with the tip of the abdomen. The tandem becomes a mating wheel when the female bends the tip of her abdomen forward to connect the genitals to those of the male at the base of his abdomen. After disengaging, the male guards the female while she dips her abdomen in the water to drop her eggs, such as in the water that forms tiny pools on the leaves of water lilies (Nymphaeaceae).

The adults are active throughout the year in Suriname, but appear to be most abundant in the long dry season (August-November, based on data from specimens in the Naturalis insect collection; see also previous account). Most *Perithemis* species, like most dragonflies, have their DNA organized in 13 chromosome bundles (Ferreira *et al.* 1979). *Perithemis lais*, however, has only nine chromosome bundles, as several have fused together. A low number of chromosomes means that this species will be less able to adapt to environmental changes, such as those predicted when the climate changes. A low chromosome number is not uncommon in tropical Odonata (dragonflies and damselflies), making them more vulnerable to such changes than species from temperate climate regions which have higher chromosome numbers. So although *Perithemis lais* is at the moment a rather common species in Suriname, this may change when the climate changes.

Figure 216. Male *Perithemis lais* at Montjoly, French Guyana, near the shoreline (image © Denis Gaschignard).

Description: Small dragonfly (17-20 mm body length). Males have wings with a strong golden hue, and a few brown spots and lines; the abdomen is orange, and the thorax is yellow, with two brown stripes on the sides. Females look similar, but more drab, also lacking the golden hue in the wings.

Distribution: The northern half of South America: from Venezuela to northern Argentina; widespread in Suriname.

Use: (No known use).

Similar species: Of the amberwings (genus *Perithemis*; see Garrison *et al.* 2006) four other species occur in Suriname, but *Perithemis lais* is the smallest and commonest. The other species lack the yellow thorax combined with the two brown lines. The males of the other species also have gold colored wings, but either lack the brown wing spots, or these spots are more obvious.

Callinectes boucorti

Hélène Hiwat

Swimming crab, pilar soft crab (English name for *Callinectes* spp.)
Srika (local name for *Callinectes* spp.)
Crustacea, Decapoda, Portunidae

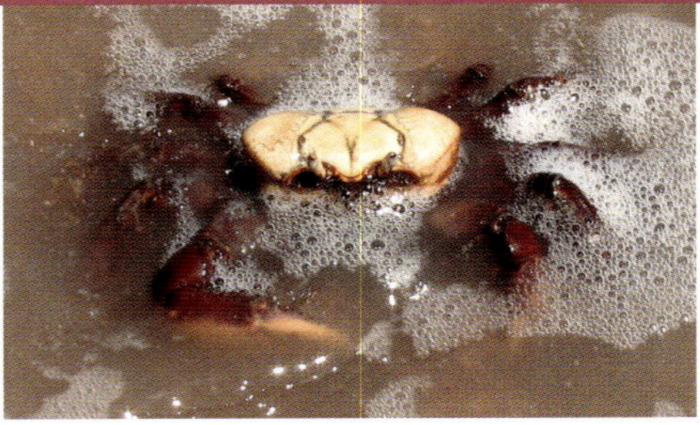

Figure 217. ***Srika*** **(*Callinectes* spp.):**

a. likely *Callinectes boucorti*, on shell beach at Braamspunt (image © Bart De Dijn, Aug. 2014).

b. likely *Callinectes ornatus*, on shell beach at Galibi (image © Vanessa Kadosoe, Oct. 2013).

True crabs (Decapoda, Brachyura) have a broad, flattened cephalothorax or carapace (merged head and thorax), and a small abdomen which is folded against the underside of the cephalothorax. They have five pairs of legs, the first of which are chelipeds, which means that they end in pincers; chelipeds are often massive compared to the other legs. In the female, during the spawning season, the eggs are attached to the abdominal appendages (egg-carrying females are called berried crabs). Swimming crabs (Portunidae) have hind legs adapted to paddling. With these paddles they swim about in estuarine and marine ecosystems to feed on other crustaceans, fish, mollusks, algae and vascular plants. They are a key food source for coastal birds and fishes (Nevis *et al.* 2009).

Besides being in the mangroves, *Callinectes boucorti* or *srika* (Figure 217.a) can be found in coastal lagoons, for instance at Bigi Pan (northwest Suriname) where it is one of the dominant crab species (Mol 2012b). It is an estuarine crab that digs burrows in the sediment, and tolerates waters with a low salinity. The distribution and abundance of swimming crabs (*Callinectes* spp.) can vary depending on the season, because of the seasonal changes in salinity and pH of the water in the Estuarine Zone (Nevis *et al.* 2009). *Callinectes ornatus* (Figure 217.b), which has a lower tolerance for low salinity, is found further away from the coast than *Callinectes boucorti*, and may sometimes end up as by-catch in the offshore shrimp trawler fisheries (Willems *et al.* 2016a). *Callinectes boucorti* are used for human consumption and are apparently a very healthy food with low levels of fat and cholesterol (Lira *et al.* 2007).

Description: *Callinectes* spp. have a carapace that has 5 to 9 teeth on the front margin, and a well-developed lateral spine at its greatest width; the last pair of legs is wide and flattened, adapted for swimming. All species in the genus have the 'hands' of the chelipeds modified into a major chela (large pincer) on one side of the body, and a minor chela (small pincer) on the other side. Males have a T-shaped abdomen.

Distribution: Atlantic intertidal zone (between high and low tide) till sea up to 20 meters depth, from Florida, Mexico and the West Indies till Santa Catarina (Brazil); widespread in the Estuarine Zone of Suriname.

Use: Popular food item in Suriname (cooked and curried.

Similar species: The somewhat smaller *Callinectes ornatus* is common in the deeper coastal waters of the Guianas (Willems *et al.* 2016a). Species of the genus *Callinectes* are very similar, but can be differentiated based on the length and shape of the (well hidden) male copulation organs (Williams 1974).

Braga patagonica
Fish louse (general name for all isopod fish parasites)
(no local name)
Crustacea, Isopoda, Cymothoidae

Hélène Hiwat

Isopods are highly diverse and have a worldwide distribution, both on land and in the water. They have two pairs of antennae, seven pairs of jointed limbs on the thorax, and five pairs of branching appendages on the abdomen that are used in respiration. Cymothoidae, isopod fish lice, are parasites of marine and freshwater fish, and have 40 recognized genera worldwide (Smit *et al.* 2014). Fish are not totally defenseless against these lice. Many marine host fish, for instance, have a mutually beneficial (mutualistic) relationship with cleaner shrimps, such as *Periclimenes pedersoni*, that remove immature lice and feed on them (Bunckley-Williams & Williams 1998). South America has the greatest variety of freshwater cymothoids, with at least 27 recorded species (Thatcher 2006).

Braga patagonica (Figure 218) occurs in both freshwater and marine fish (Tavares-Dias *et al.* 2014, Chagas *et al.* 2015). It has a preference for entering the body of the host fish via the gills and settling in its mouth cavity, replacing its tongue after causing it to wither. The species attacks a limited set of hosts, contrary to its fellow *Braga* species (Tavares-Dias *et al.* 2014). In Suriname it has been recorded from *anyumara* (*Hoplias aimara*), *dagufisi* (*Acestrorhynchus falcatus*), and some freshwater shrimps. According to Tilles (1979), the specimens of *Braga fluviatilis* and *Braga patagonica* collected in the fresh waters of Suriname in 1910 and 1964 respectively, and stored at the Leiden Museum of Natural History in the Netherlands, were the first records for these species from the Northern Hemisphere. *Braga patagonica* is a protandrous hermaphrodite; this means that each juvenile first develops into a male, but in the absence of females, eventually becomes a female and attaches itself permanently to the host. Females produce and disperse chemicals (pheromones) which prevent nearby males from changing sex.

Figure 218. ***Braga*** **cf.** ***patagonica*** **(male) found as parasite on the shrimp** ***Macrobrachium brasiliense*** **in the Mindrineti Creek (image © Vanessa Kadosoe):**
a. upper side of the body, b. underside.

Description: Adults about 14 mm (males) to 18 mm (females) long (Tavares-Dias *et al.* 2014). Head triangular, with small eyes, and with side lobes that have hairy bristles. Last segment of the body (pleotelson) wider than long, with a pair of flattened branched appendages (inner branch = endopod, outer branch = exopod). Females brood their young in a pouch under the thorax.

Distribution: Coastal waters and freshwaters of the Amazon region, including Suriname, and from Para State (Brazil) to Patagonia (Argentina).

Use: No use, but detrimental to fish and fish stocks.

Similar species: In the closely related *Braga fluviatilis* the above-mentioned endopod and exopod are the same length; in *Braga patagonica* the exopod is longer than the endopod (de Castro & de Loyola e Silva 1985). Also, with *Braga fluviatilis* the last segment of the first pair of legs on the thorax is longer than in the subsequent pairs; this is not the case with *Braga patagonica*.

Scolopendra gigantea and other *Scolopendra* spp.
Amazonian Giant Centipede
(no specific local name)
Myriapoda, Chilopoda, Scolopendromorpha

Hélène Hiwat

Centipedes (Chilopoda), giant centipedes of the genus *Scolopendra* (unidentified species in Figure 219) in particular, are fierce predators, able to eat prey much larger than themselves. Prey includes large insects and spiders, amphibians, reptiles, rodents and even bats (Molinari *et al.* 2005). To subdue the prey, centipedes use venom, available in a pair of venomous fangs called forcipules, which are considered a modified first pair of legs. They can also bite in defense; humans that are bitten by large centipedes experience excruciating pain, but do not die. Giant centipedes have peptides (short strings of amino acids) in their bodies which have potential to be used in the development of antibiotic, anti-inflammatory and anti-cancer treatments, and may be used as natural insecticides (Hakim *et al.* 2015).

Globally, Scolopendromorpha includes a total of 600 species in 33 genera (Adis & Harvey 2000), including the Amazonian Giant Centipede, *Scolopendra gigantea*, which can become 30 cm long and 10 years old. It has a set of modified, enlarged hind legs, which it uses as hooks to suspend itself, for example from cave ceilings when hunting bats. The hind legs are raised and splayed to reveal spines or spurs as a warning to enemies; they can be used to stab in defense, especially when the head cannot be used to carry out any defensive action. Last but not least, the hind legs are also used for grasping during the mating ritual (Kronmüller & Lewis 2015).

Figure 219. *Scolopendra* sp. at Tafelberg, hiding in a shirt hanging to dry in a field camp (image © Bart De Dijn).

Description: *Scolopendra* spp. are distinct due to their large size (mostly larger than 10 cm). *Scolopendra gigantea,* a dark orange-brown colored centipede with orange legs, can grow to over 30 cm long; the base of most of its 21 pairs of legs has one or more spurs.

Distribution: *Scolopendra gigantea:* forested areas in northern South America and some Caribbean islands (Shelley & Kiser 2000).

Use: Used in the pet trade.

Similar species: Other centipedes in Suriname are smaller or at least less robust than *Scolopendra gigantea*. *Scolopendra morsitans* (6-13 cm length) is a fairly small, red-headed centipede; it's body is light brown with dark bands; this is an introduced species in our region (Shelley *et al.* 2005).

Avicularia avicularia
Pinktoe Tarantula
Bus'anansi (local name for tarantulas)
Aranea, Theraphosidae

Hélène Hiwat

Tarantulas (Theraphosidae) are big, scary-looking spiders (Aranea), yet their bite is of about the same caliber as a bee sting (Barceloux 2008). They are mostly nocturnal predators; insects are their main prey, alongside frogs and mice. The name *Avicularia* means 'bird-eating', and is due to the first depiction of a tarantula, eating a hummingbird, by Maria Sybilla Merian (1705) (Marshall 2001, Campos e Silva & De Almeida Meirelles 2015). *Avicularia avicularia* (Figure 220) is a night-active 'sit and wait' predator (Cloudsley-Thompson & Constantinou 1985); during the daytime it takes shelter in its dense, tunnel-shaped web, spun between large leaves, or against a tree stem or the wooden beams of a roof. Males mature in two years, after which they die in about three months; females take a year longer to mature, continue molting while mature, and may live up to seven years (Stradling 1978).

A recent review concludes that there are 14 species of *Avicularia*; the species are reliably differentiated based on the shape of an obscure structure called the tegulum (Fukushima 2011). The tegulum is a tiny part of the male sexual organ, located inside the pedipalps, which are leg-like appendages located near the mandibles (in males these are larger than in females, typically club-shaped). The tarantula's mating ritual begins when the male spins a small web and deposits sperm drops on it, picking up these drops with his pedipalps. He lures a female out of her shelter with rhythmic foot tapping and seduces her with dancing moves. Once she accepts him, he copulates by depositing sperm, using the pedipalps, in slits in the underside of the female's abdomen (Feretti *et al.* 2013). Females lay fertilized egg batches in a dense web (cocoon) and guard them for six to nine weeks, after which several hundred juvenile tarantulas hatch. Tarantulas have few natural enemies beyond the formidable tarantula hawk wasps (Pompilidae: genus *Pepsis* and *Hemipepsis*) (Petrunkevitch 1926, Rego *et al.* 2004). These exceptionally large, parasitic wasps will paralyze a tarantula with their sting, drag it into a burrow and lay an egg on the spider's body. When the egg hatches, a wasp larva emerges. The larva enters the body of the immobile but still living spider to feed on its tissue; it avoids vital organs as long as possible, to prevent the premature death of the spider and the spoilage of its tissues.

Figure 220. *Avicularia avicularia* leaving its shelter for a night's hunting in Colombia (image © Dick Culbert, Jan. 2014; www.flickr.com; Creative Commons License).

Description: Large (males up to 9 cm, females 13 cm), entirely black to grey, except the contrastingly colored leg tips, which are light pink to violet or red.

Distribution: Northern South America and the Caribbean islands (Fukushima 2011); widespread and common throughout Suriname.

Use: Popular in the pet trade; docile and easy to handle.

Similar species: *Avicularia avicularia* may be confused with *Avicularia metallica*; adults of the latter are larger and have white hairs intermixed with the black hairs. Other tarantulas in Suriname are not predominantly black.

Latrodectus curacaviensis

Hélène Hiwat

Widow spider (English name used for all *Latrodectus* spp.)
(no specific local name)
Aranea, Theridiidae

Latrodectus spp. or widow spiders confuse biologists, as evident from the enormous variability in the number of recognized species during the past century (Levi 1959, McCrone & Levi 1964). They are relatively small spiders: the female is 12-16 mm long, much more robust than the smaller male. These spiders are named after the phenomenon of the female eating the male after mating. Such sexual cannibalism does indeed occur in *Latrodectus* spp., but is not the rule. Widow spiders are active during the day. Many species live in inhabited areas, but seem to prefer outdoor environments. The females make three-dimensional webs in corners of gardens, sheds and vegetation near houses; they will hang upside-down in their webs. Their diet consists largely of beetles. Evidence exists that female widow spiders exhibit social behavior, such as peer tolerance, and maybe brood care (Bertani *et al.* 2007). The males do not make a web, and can be regarded as little more than sperm donors, and the occasional light snack for females.

While the males are not considered harmful to humans, the bite of the female spiders is considered dangerous. Females inject venom via the fangs that contains latrotoxin (Grishin 1998). This can cause latrodectism: a very painful condition, which may be accompanied by sweating, muscle spasms, local paralysis, temporary mental disorders (fear, hallucinations), rash and local swelling of the body. The condition may last a couple of days, but is seldom fatal. In 1998, three cases of latrodectism were reported from Cassipora, an Indigenous village in Suriname (Oostburg & Hiwat 1998). The condition was due to the bites of *Latrodectus curacaviensis* (Figure 221). The spiders were found in their webs in low shrubs surrounding the houses, and on wooden crosses at the cemetery. Clearing the shrubs and educating the people solved the problem; no new cases of latrodectism have been reported since, although the spiders are still present in the general area (Hiwat & De Dijn, personal observation).

Figure 221. Large female and two small males of *Latrodectus curacaviensis* in their web on a wooden cross at the Redidoti village cemetery (image © Bart De Dijn, Jan. 2017).

Description: Small (up to 16 mm); females with bright orange-red and black abdomen; the red markings on the underside often shaped like an hourglass.

Distribution: Lesser Antilles and South America; in Suriname, the only records are from the villages Cassipora and Pierrekondre, but the species is likely more widespread (in similar habitats: inhabited shrub savanna).

Use: (No known use).

Similar species: The similar cosmopolitan species *Latrodectus geometricus* is known from Brazil, not (yet) Suriname. *Latrodectus* spp. are hard to differentiate; identifications often require the study of reproductive organs, but molecular (genetic) data can now also be used (Garb *et al.* 2004).

Amblyomma cajennense
Cayenne Tick
Kupari (local name used for any tick)
Arachnida, Ixodida, Ixodidae

Hélène Hiwat

A nuisance in areas with natural vegetation are ticks. Ticks (Ixodida) include those of the family Ixodidae, with 28 species recorded as blood-feeding external parasites of humans in South America (Gugliemone *et al.* 2006). The Cayenne Tick, *Amblyomma cajennens* (Figure 222), is one of them; it is mainly found in grasslands with an abundance of potential hosts, such as man and domestic and wild animals (Rojas *et al.* 1999). Immatures emerge after heavy rainfall, and do not tolerate high temperatures. During dry periods, they may survive by going into diapause (inactive state with lowered metabolic rate) (Labruna *et al.* 2003, Beck *et al.* 2011).

Male and female ticks find each other on the blood hosts. The males produce volatile chemicals (pheromones) to attract females to hosts. Females are much larger than the males. When feeding, fertilized females may swell to several times their original size ('engorged' and resembling a white pea). Hundreds to tens of thousands of eggs are laid by each female about a week after fertilization, typically in areas with dense vegetation. After this the female dies, and about two months later, immatures hatch from the eggs. Nymphs seem to have a lower host specificity than adults, and have for instance been found on a variety of birds (Rojas *et al.* 1999). The complete life cycle takes about six months (Labruna *et al.* 2003). Carbon dioxide, heat and movement from a passing host serve as stimuli for 'questing' behavior. Questing ticks crawl up the stems of grass or perch on the edges of leaf litter with front legs extended. The ticks will then grab a potential host that brushes them.

Surprisingly, Cayenne Ticks may not be all bad for people. Researchers from the Instituto Butantan in São Paolo (Brazil) found that their saliva contains compounds which can slow down the growth of cancer cells, while not affecting healthy cells. These compounds may ultimately be used in the development of new cancer treatments (Simons *et al.* 2011).

Figure 222. Female *Amblyomma cajennense* looking for a host near the Mazaroni waterfall at Brownsberg mountain (image © Auke Hielkema, Aug. 2015).

Description: Cayenne Ticks are 'hard ticks', with a tough body and an easily visible head. Body flat, oval-shaped, small (about 7 mm), but much larger and swollen in engorged females; four pairs of wiry legs and small mouthparts. Body brown, with large black region on abdomen of adult females.

Distribution: Southern USA to northern Argentina, and the larger Caribbean islands; widespread in Suriname.

Use: Parasites of vertebrates, including man; potential for use in cancer treatment.

Similar species: *Amblyomma ovale* and *Amblyomma oblongoguttatum* are similar species from our region. Taxonomists argue about the status of the various species of the genus *Amblyomma*; studies using molecular data may help solve the taxonomic issues (such as Beati *et al.* 2013).

Tityus carolineae

Thick-tailed scorpion (English name used for all Buthidae)

Kruktutere (local name used for all scorpions)

Arachnida, Scorpiones, Buthidae

František Kovarík

Scorpions are often feared for their painful and possibly venomous stings, but most of the species are relatively harmless. Only about 25 of the around 2000 species known worldwide have venom capable of killing a human being; none if these occur in Suriname (Kovarik 2009, Kovarik & Affilastro 2013). Scorpions are predatory animals, easily recognized because they have four pairs of legs, a pair of grasping pedipalps (enlarged, grasping appendages at the front of the head), and a narrow, segmented tail ending with a venomous stinger. The tail is often positioned in a characteristic forward curve over the back. The pedipalps are used for prey capture, defense, mating and sensing the environment. Some species additionally use their pedipalps for burrowing or climbing. Scorpions range in size from 9 mm (*Microtityus minimus*) to 20 cm (such as *Hadogenes troglodytes*). They are adapted to a wide range of environments, and can be found on all continents except Antarctica.

Tityus is by far the most diverse scorpion genus of the world, with some 230 species currently recognized. *Tityus carolineae* (Figure 223) belongs to the '*Tityus asthenes*' group of species, part of the '*Tityus metuendus*' species complex. *Tityus carolineae* is the largest species of this complex; males are 82–100 mm, and females 70–84 mm total body length (Kovarik *et al.* 2013). Species of the *Tityus asthenes* group require a high humidity, but can be kept in captivity without any problems. Individuals tolerate each other and cannibalism is relatively rare.

Figure 223. *Tityus carolineae* (images © František Kovarík):

a. male.

b. male, dried museum specimen (type specimen).

c. female, dried museum specimen (type specimen).

Description: The body of *Tityus carolineae* is uniformly dark reddish brown to blackish; the pedipalps are reddish, with blackish 'fingers'. The pedipalp chela (pincher) is large and robust in males, small and oval in females; in both sexes the fingers of the chela are very long. The tail (metasoma) is moderately elongated, more elongated in the males than in the females.

Distribution: *Tityus* spp. occur mainly in South America, but species are also known from North America and the Caribbean; *Tityus carolineae* (Figure 223) is known only from Suriname and Guyana.

Use: (No known use).

Similar species: *Tityus asthenes*, after which the taxonomical group to which *Tityus carolineae* belongs was named, looks very similar; it is common in western Amazonia. The distinguishing feature for *Tityus carolineae* is the number of teeth on each pecten (males 18-20, females 19-20); the pecten is a paired, comb-shaped appendage on the underside of the body, unique to scorpions.

Cordylochernes scorpioides
Harlequin Beetle-riding Pseudoscorpion
(no specific local name)
Arachnida, Pseudoscorpiones, Chernetidae

Volker Mahnert

The Arachnid order of Pseudoscorpiones (pseudoscorpions) has 3500 species in 26 families (Harvey 2013a), including the Chernetidae, to which *Cordylochernes scorpioides* belongs (Figure 224). They resemble actual scorpions (Scorpiones), but lack the stinging tail. Only five species have been recorded from Suriname (Harvey 2013b). The knowledge of the pseudoscorpion fauna of the Guianas is very inadequate. Even the species number recorded from Brazil (166) (Harvey 2013b) is considered an underestimate. Pseudoscorpions are predators, and mostly small (1 to 5 mm length); they feed on small arthropods such as insects and mites. Most species are gregarious; some form colonies and exhibit social behaviour (Zeh & Zeh 1990). In the seasonally flooded Amazonian forests, during the dry season pseudoscorpions live in the soil and leaf litter, where they breed; during the rainy season, they migrate to tree trunks and crowns to complete their developmental cycle (Adis 1993, de Morais *et al.* 1997). Many species are associated with other animals (first summarized in Beier 1948). Some colonize the nests of social insects, such as stingless bees (Gonzalez & Mahnert 2008), others the nests of birds (Turienzo *et al.* 2010) or mammals. Their small body size allows many species to hitch rides on other animals (phoretic behavior) (Aguiar & Bührnheim 1998); this was first recorded by Poda (1761). *Cordylochernes scorpioides* has been found under tree bark, and is frequently phoretic, mainly on the large longhorn beetle *Acrocinus longimanus*.

The developmental cycle comprises the embryonic stage (embryos are carried by the female in a brood sac), three post-embryonic immature stages (nymphal stages), and the mature (adult) stage. The adult male produces spermatophores (packages of sperm), which it deposits; these are taken up by the adult females which are guided to them by chemical cues. An intricate mating behavior has developed in some families (such as Chernetidae, Cheliferidae): the male performs an introductory mating dance, after which he guides the female to the spermatophore (Weygoldt 1969). The development of immature stages takes place in silk constructions produced by the females and nymphs.

Figure 224. *Cordylochernes scorpioides* in Panama (images © Jeanne Zeh):
a. female with very young nymphs.
b. male.

Description: Body length 3-4 mm; color intense reddish brown. Carapace (front part of body) with a deep transverse furrow, and one indistinct eye-spot on each side near the front border; pedipalps (appendages with pinchers) robust.

Distribution: Southern Mexico through Central America and Amazonia down to northern Argentina; presumed widespread in Suriname.

Use: (No known use).

Similar species: Similar to species of the genus *Lustrochernes* (Chernetidae), but those are smaller, with more slender pedipalps and stouter walking legs. Most of the pseudoscorpions living under elytra of the Harlequin Beetle *Acrocinus longimanus* (see species account) belong to the featured species.

Freshwater Fishes

Jan Mol, Kenneth Wan Tong You

If you go snorkeling in a shallow reach of a clear-water river in the Interior at the end of the dry season, when the water level is low, you can get a first impression of the colors and diversity of freshwater fishes in Suriname (Figure 225). However, to understand the true extent of freshwater fish diversity in Suriname you need to sample other river systems and other habitats – because many Surinamese freshwater fishes are narrowly restricted in their distribution. Some are found in just a single river system or even creek – in other words, they are endemic to that stream (Mol 2012b).

Fishes are the oldest group of vertebrate animals, dating back more than 500 million years (Shu *et al.* 1999), and the most successful vertebrates in aquatic habitats (Box 8). Most freshwater fishes of Suriname are poorly known, not only to the man in the street, but also to science. Most species (65%) are not more than 15 cm in length and 90 percent live in the Interior of the country, most of which is remote and not easily accessible. However, some large fishes show up in books or on film, like the piranha (*piren*), Electric Eel (*stroomfisi*), wolf fish (*anyumara*) and Atlantic Tarpon (*trapun*), while several of the small, often colorful species are well known to aquarium hobbyists all over the world.

In Suriname, freshwater fishes are important for food, sport fishing, and as pets. In the remote Interior of Suriname, fish makes up the main portion of daily protein intake for tribal people, and even in densely populated coastal areas swamp fishes are regularly caught and consumed. If there is a national fish, it is the *hei-ede kwikwi (Hoplosternum littorale)*, a type of catfish targeted in local sport fishing tournaments. Tourist sport fishers mainly target large river fishes such as Atlantic Tarpon or *trapun* (*Megalops atlanticus*), goliath catfish or *lalaw* (*Brachyplatystoma filamentosum*), Redtail Catfish or *mototyar* (*Phractocephalus hemioliopterus*), peacock bass or *tukunari* (*Cichla* spp.) and wolf fish or *anyumara* (*Hoplias aimara*) (see species accounts). In the 1970s, aquariums became a popular hobby in Suriname, and the country exported many of its small but beautiful fishes, such as hatchetfishes, pencilfishes, tetras and catfishes of the genus *Corydoras* (see species accounts). But surprisingly, this small country was at the same time also a major importer of Amazonian aquarium fishes (Junk 1984). This resulted in the introduction of several exotic species, among them the Cardinal Tetra (*Paracheirodon axelrodi*).

Fishes are important parts of aquatic food webs and healthy fish populations indicate healthy aquatic habitats that are safe for fishing, swimming and drinking (Holmlund & Hammer 1999). In Suriname and elsewhere, the presence, abundance and health of freshwater fishes and characteristics of fish communities are used as indicators of ecosystem integrity in environmental impact assessment and monitoring studies.

Diversity

Regional context. Fishes include a tremendous diversity of species and groups with different ancestries; this is reflected in their classification system (Box 9). Suriname has some 500 fresh and

Figure 225. Underwater impressions of fishes of the Upper Suriname River (image © Willem Kolvoort):

a. a large school of *Poptella brevispina*,

b. a solitary *Hypostomus plecostomus*,

c. a mixed school of the anostomids *Leporinus fasciatus*, *Leporinus friderici*, *Leporinus maculatus* and *Leporinus granti*, and

d. a pair of *Guianacara owroewefi* cichlids.

brackish-water fish species (Mol 2012b, Figure 229). These are part of the fish fauna of the Guiana Shield (Abell *et al.* 2008), which in turn is part of the Neotropical fish fauna – the richest freshwater fish fauna on Earth, with an estimated 5000-8000 species (Lundberg *et al.* 2000, Carvalho *et al.* 2007). The rivers of Suriname are currently not connected to the Amazon River in the way the Casiquiare Canal connects the Orinoco River or the seasonally flooded Rupununi Savanna connects the Essequibo River, but such connections may have existed in the past, for example via the headwaters of the Marowijne River (Cardoso & Montoya-Burgos 2009, Fisch-Muller *et al.* in press). This would explain the presence of some small-sized fishes in the headwaters of both the Marowijne River and nearby tributaries of the lower Amazon, such as *Parotocinclus halbothi* and *Stenolicmus ix* (see species accounts).

Suriname fish fauna. Approximately 400 Surinamese fish species are restricted to fresh waters; 90 of these (22.5%) occur solely in Surinamese river systems including the Corantijn and Marowijne watersheds, which in part extend to Guyana and French Guiana, respectively. As such, freshwater

BOX 8 Living in water

Organisms that live in aquatic and terrestrial habitats face distinct challenges and, not surprisingly, the life aquatic is very different from life on land. The density of water is 830 times higher than the density of air, and aquatic organisms can achieve neutral buoyancy by storing low-density materials (such as air in the swim-bladder of fishes). In contrast land organisms need to support their body against collapse by gravity (compare woody trees with algae, or the heavy bones of a cow with the light-weight fish skeleton). Moving through a dense and viscous medium like water, pushing away the water, costs a lot of energy compared to movement on land; fishes reduce the resistance force of water by streamlining their body (Figure 226).

Water has a high heat capacity – it absorbs lots of heat – and easily conducts heat, making it difficult for aquatic organisms to maintain a steep temperature difference between their bodies and their surroundings. This is why fishes are referred to as ectotherms ('cold-blooded'): their body temperature tends to match that of the environment. On land, temperatures are often more extreme than in the water; in Suriname, Geijskes (1942) measured a 10 °C amplitude in air temperature against only 2 °C in streams. The low heat capacity and heat conductivity of air explain why many land animals are endotherms ('warm-blooded'): the body temperature is regulated by metabolic means, as in birds and mammals.

Water holds no more than one-twentieth of the oxygen that air can hold, and the diffusion rate of oxygen in water is 33,000 times lower than in air (Verberk *et al.* 2011). Biological and chemical processes can further lower the oxygen content of water. Fishes thus face the challenge of capturing sufficient oxygen from their environment. On the other hand, prevention of water loss (desiccation) is important on land, but not in fresh water. This said, fishes devote 25 to 50 percent of their total energy output to maintaining the right amount of water in their bodies (Bone & Moore 2008).

Light is scattered and absorbed easily in water, and does not penetrate to great depths. This has consequences for vision in aquatic animals, and, with aquatic plants, for photosynthesis, which is restricted to the zone immediately under the water surface (photic zone). The way aquatic animals sense their environment for navigation, communication and finding food is quite different from the way terrestrial animals do. In the aquatic environment, sound and dissolved chemicals can be detected over longer ranges than light. To detect vibrations and movement in the surrounding water, fishes have an elaborate lateral line system (Figure 226); it is a perforated tube with built-in motion sensors, located lengthwise on either side of the body below the skin surface, often continuing onto the head and branching. It is very sensitive; some fishes can use it to track the wake of their prey even ten seconds after the prey has passed (Pohlmann *et al.* 2004). Many fish species have a modified lateral line system with which they can register electromagnetic fields. The adipose fin (Figure 226), present in 20 percent of ray-finned fishes, has been shown to function as a water flow sensor (Aiello *et al.* 2016).

Suction feeding by shaping (often protruding) the mouth, expanding the oral cavity and creating a negative pressure – for the purpose of sucking in the prey with the water – is the norm in bony fishes, but is only effective in an aquatic environment.

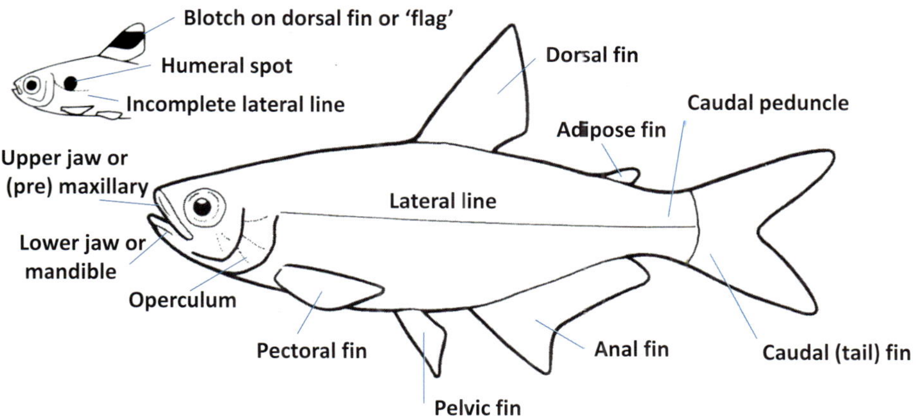

Figure 226. External anatomy of a bony fish as exemplified by a characin. Note streamlined body, two paired fins (pectoral and pelvic) and three unpaired fins (dorsal, caudal and anal). The unpaired adipose fin is often, but not always present in characiform fishes and catfishes. Characteristic spots or blotches may be present posterior of the head (humeral spot), on the caudal peduncle or on the dorsal fin (sometimes referred to as a 'flag'). Small characins may have an incomplete lateral line. Catfishes often have barbels (wiry appendages) on the upper and lower jaw and strong spines leading the dorsal and pectoral fins. Cichlids have numerous fin spines in the anterior part of both the dorsal and anal fins and a divided lateral line (with the anterior part dorsally placed and the posterior part mid-lateral). (Drawings modified after Nelson et al. 2006 and Planquette et al. 1996).

species better characterize the fish fauna of Suriname than marine or brackish-water species, most of which have a wide distribution off northern South America and beyond (none is endemic to Suriname). In recent years, one or two new freshwater fish species have been recorded in Suriname each year. Several of these species were already known from neighboring countries, and were collected in Suriname during inventories of remote areas. However, other species are new to science (Box 9), and potentially endemic to Suriname.

All Surinamese freshwater fishes are grouped in the subclass Actinopterygii or ray-finned bony fishes, with the exception of a number of species in the class Chondrichthyes or cartilaginous fishes, namely the river stingrays (Potamotrygonidae; *libaspari*) and the critically endangered Largetooth Sawfish *(Pristis pristis)*, sporadically caught in the estuaries of large rivers (Table 5, Figure 229).

The most species-rich groups of freshwater fishes in Suriname are three orders of otophysan fishes, which together comprise three-quarters of the species: characins (order Characiformes; 150 spp., Figure 229b), catfishes (Siluriformes; 170 spp., Figure 229a) and Neotropical knifefishes (Gymnotiformes; 21 spp., Figure 229a) (see also Figure 230).

BOX 9 Names and classification of fishes

Biologists classify organisms into species defined as populations or groups of populations having common attributes and the ability to interbreed and produce live, fertile offspring. Different species are reproductively isolated from one another. In the hierarchical classification system of Linnaeus (1758), related species that evolved from a common ancestor are organized into larger and larger groups or taxa such as genus, family, order and class. Currently 536 families of fishes with extant species are recognized worldwide (Nelson *et al.* 2016); of these 64 are represented in the fresh and brackish waters of Suriname (Table 5). At the highest level, living fishes are grouped in four classes: the jawless hagfishes (Myxini, 78 species) and lampreys (Petromyzontida, 46 species), the cartilaginous sharks and rays (Chondrichthyes, 1200 species) and the bony fishes (Osteichthyes, 2 coelacanths, 6 lungfishes and more than 30,500 ray-finned fishes) (Nelson *et al.* 2016).

Fish species often, but not always, have local names, and widespread species may have many local names in different languages. However, in the classification system of Linnaeus all species have a single valid Latin name that is composed of two parts, genus followed by species. The Redeye or Black Piranhas, for example, are called *Serrasalmus rhombeus*, the species *rhombeus* of the genus *Serrasalmus*. Valid names refer to the first description of a species starting with the 10th edition of Linnaeus (1758); 19 Surinamese fish species were described by Linnaeus himself (Figure 227; Mol 2012b).

Figure 227. Dried skin of *Callichthys callichthys* (*platkop kwikwi*) from the Gronovius collection, now in the collection of the Natural History Museum, London (BMNH1853.11.12.194; image © Mark Allen, All Catfish Species Inventory). This specimen was probably collected in Suriname in the first half of the 18th century and used by Linnaeus to describe the species.

When a 'new' fish species is discovered it has to be formally described in a scientific paper. A new species (see for example Covain *et al.* 2012; Figure 228) is not a species that has recently evolved, but a species that has not yet been formally described. Species new to science may have been living on Earth for millions of years, and may be well known to local people. The formal species description not only includes a detailed description of the 'new' species, but also a diagnosis explaining how it differs from related species which have been described previously. The species description often contains photographs or other illustrations of the type material and states in which museums it has been deposited. Modern species descriptions also include DNA sequences to identify the species, for example the so-called barcode gene (Hebert *et al.* 2003, Pereira *et al.* 2013). The type specimen or type is a particular specimen or series of specimens of an organism to which the scientific name of that organism is formally attached; the type locality is the place where a type was collected. The publication in which the species is described gives the new species a formal scientific name. Once the manuscript has been accepted for publication, the new species name is officially created; the name can generally not be changed except in the case of error. New species (including fishes) are constantly discovered and described.

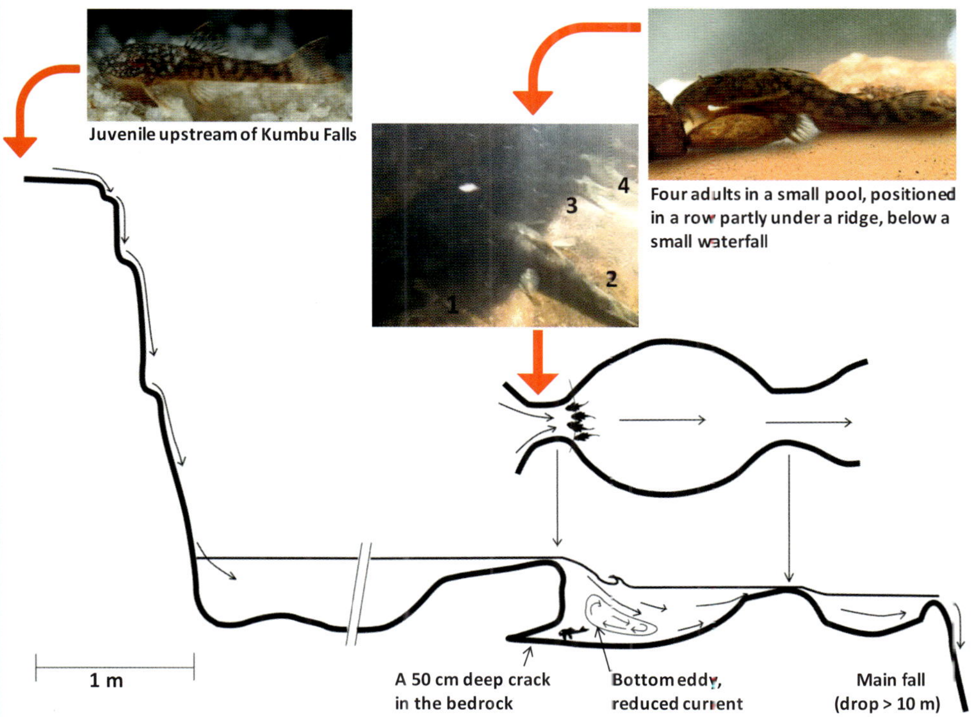

Figure 228. A new *Guyanancistrus* catfish (6 cm) with light-yellowish spots was discovered in 2011 at Kumbu Falls in Brownsberg Nature Park. Analysis of its DNA in 2014 confirmed that it is a valid species. In 2017, its formal description has been submitted to a scientific journal (Fisch-Muller *et al.* in press). The restricted distribution, behavior and ecology of this new species provoke interesting scientific questions.

Figure 229. Freshwater fishes of Suriname:

a. Catfishes, knifefishes, killifishes, cichlids and miscellaneous groups; A) *Potamotrygon marinae* (length 41 cm, Potamotrygonidae); B) *Synbranchus marmoratus* (100 cm, Synbranchidae); C*) *Anablepsoides fransvermeuleni* (7 cm, Rivulidae); D) *Micropoecilia picta* (4 cm, Poeciliidae); E) *Potamorrhaphis guianensis* (29 cm, Belonidae); F) *Electrophorus electricus* (250 cm, Gymnotidae); G*) *Sternarchorhynchus galibi* (20 cm, Apteronotidae); H) *Cichla ocellaris* (50 cm, Cichlidae); I) *Crenicichla multispinosa* (24 cm, Cichlidae); J*) *Geophagus harreri* (22 cm, Cichlidae); K) *Polycentrus schomburgkii* (7 cm, Polycentridae); L) *Ochmacanthus* sp (5 cm, Trichomycteridae); M) *Callichthys callichthys* (25 cm, Callichthyidae); N*) *Pseudancistrus kwinti* (13 cm, Loricariidae); O) *Doras micropoeus* (35 cm, Doradidae); P) *Pseudoplatystoma fasciatum* (95 cm, Pimelodidae); Q) *Ageneiosus inermis* (50 cm, Auchenipteridae).

An asterisk (*) denotes species that are endemic to Suriname; images A, B, E, G, H, I, J, K, L, M, O, P, Q © Mark Sabai-Pérez; C © Frans Vermeulen; D © Kenneth Wan Tong You; F © Sandra Raredon; N © Phillip Willink.

b. Diversity in characiform fishes. A) *Crenuchus spilurus* (length 6 cm, Crenuchidae); B) *Hoplias aimara* (110 cm, Erythrinidae); C) *Parodon guyanensis* (12 cm, Parodontidae); D*) *Cynodon meionactis* (30 cm, Cynodontidae); E) *Myloplus rubripinnis* (35 cm, Serrasalmidae); F) *Pristobrycon striolatus* (30 cm, Serrasalmidae); G) *Bivibranchia bimaculata* (15 cm, Hemiodidae); H) *Leporinus fasciatus* (35 cm, Anostomidae); I) *Curimatopsis crypticus* (6 cm, male, Curimatidae); J*) *Semaprochilodus varii* (33 cm, Prochilodontidae); K) *Copella arnoldi* (4 cm, Lebiasinidae); L) *Acestrorhynchus falcatus* (33 cm, Acestrorhynchidae); M) *Brycon pesu* (14 cm, Bryconidae); N) *Chalceus macrolepidotus* (26 cm, Chalceidae); O) *Hemigrammus bellottii* (3 cm, Characidae); P) *Roeboides affinis* (13 cm, Characidae, Characinae); Q) *Carnegiella strigata* (4 cm, Gasteropelecidae).

An asterisk (*) denotes species that are endemic to Suriname; images A, M © Phillip Willink; B, C, D, E, F, G, H, I, J, K, L, O, Q © Mark Sabaj-Pérez; N © Ross Smith; P © Anil Gangadin.

Other major groups in the country belong to the recently described Ovalentaria (Wainwright *et al.* 2012; 13% of the country's species), including 30 cichlids (family Cichlidae), 22 killifishes and livebearers (Cyprinodontiformes), 4 needlefishes and halfbeaks (Beloniformes), 3 mullets, and 1 polycentrid Leaf Fish (Figure 229a, Table 5).

Finally, more than 50 species occur in Suriname that are fresh- or brackish-water representatives of predominantly marine groups, such as river stingrays (*libaspari*; Figure 229a), anchovies, gobies, eleotrid sleepers, achirine flatfishes (*boki*), snooks (*snuku*), sciaenids (*kubi*), puffers (*bosrokoman*), a toadfish (*lompu*), a swamp eel (*snekfisi*; Figure 229a) and a freshwater pipefish. These make up the remaining 12 percent of the country's fish fauna. Many Surinamese freshwater fish species are small and very similar in appearance, and thus hard to identify, such as the 'tetras' in the family Characidae (see species account of *Hyphessobrycon simulatus*).

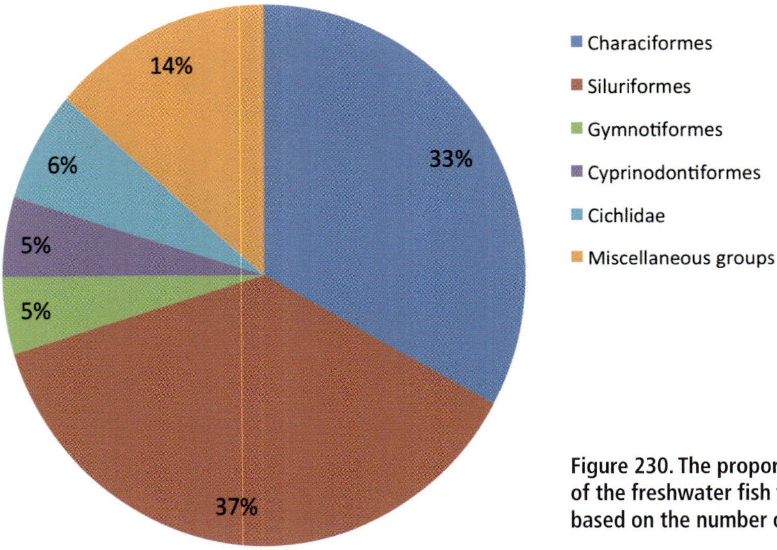

Figure 230. The proportional composition of the freshwater fish fauna of Suriname based on the number of species.

Life histories and reproduction

Size. The largest freshwater fishes of Suriname, such as the goliath catfish (*lalaw*) and Redtail Catfish (*mototyar*), live in large rivers; large female Atlantic Tarpon (*trapun*) spawn at sea (Table 6). However, even small rainforest creeks may be inhabited by surprisingly large fishes, such as two-meter-long electric eels or meter-long wolf fish. Miniature fishes often live in small-sized habitats such as headwater creeks, for example *Parotocinclus halbothi*, *Stenolicmus ix* and *Hyphessobrycon heterorhabdus*, black-water creeks, such as *Nannostomus marginatus* and *Micropoecilia bifurca*, small creeks in *terra firme* forest, such as *Hemigrammus geisleri* and *Microcharacidium eleotrioides*, or mountain brooklets, such as *Harttiella crassicauda*.The large species such as *lalaw* and *trapun* show enormous variation in size from tiny larvae (fish larvae are the smallest vertebrates on Earth) to giant adults (Table 6). The size of the fish impacts the water flow around its body: as they grow, fish experience a gradual change from viscous ('sticky') flow as larva to inertial, even turbulent flow as adults (Müller *et al.* 2000, Maugiet *et al.* 2010).

Order: (sub)family	English name	# spp. in SU	Remarks
Pristiformes:			
Pristidae	sawfishes	1	occasionally in estuaries, rare
Myliobatiformes:			
Potamotrygonidae*	river stingrays	3	FW
Elopiformes:			
Elopidae	tenpounders (ladyfishes)	1	M and BW
Megalopidae*	tarpons	1	M, juveniles in FW
Clupeiformes:			
Pristigasteridae	longfin herrings	3	M, E, lower freshwater reaches
Engraulidae	anchovies	12	M, E, lower freshwater reaches; 1 species FW
Clupeidae	herrings and sardines	3	M, E, lower freshwater reaches
Characiformes:	*characins*		*all species FW*
Crenuchidae	South American darters	6	
Erythrinidae*	trahiras (wolf fishes)	5	
Parodontidae*	parodontids	1	
Cynodontidae	cynodontids	2	
Serrasalmidae*	pacus and piranhas	13	
Hemiodontidae	hemiodontids	7	
Anostomidae*	toothed headstanders	17	
Chilodontidae	spotted headstanders	4	
Curimatidae	toothless characiforms	8	
Prochilodontidae*	flannel-mouth characiforms	2	
Lebiasinidae*	pencilfishes	7	
Acestrorhynchidae	acestrorhynchids	2	
Chalceidae	characins, tetras	1	
Iguanodectidae	characins, tetras	6	including *Bryconops*
Triportheidae	characins, tetras	1	
Bryconidae*	characins, tetras	2	
Characidae (a-g)	characids	64	
a.genera not assigned to subfamilies below*	characins, tetras	44	
b.Stethaprioninae	silver dollar tetras	3	
c.Characinae*	characins, tetras	8	
d.Tetragonopterinae	characins, tetras	2	
e.Aphyocharacinae	characins, tetras	1	
f.Cheirodontinae	characins, tetras	1	
g.Stevardiinae	characins, tetras	5	
Gasteropelecidae*	freshwater hatchetfishes	2	
Siluriformes:	catfishes		all species FW, except Ariidae
Cetopsidae*	whale catfishes	4	
Trichomycteridae*	pencil or parasitic catfishes	7	
Callichthyidae*	armored catfishes	23	
Loricariidae (a-c)	suckermouth armored catfishes	58	
a.Hypoptopomatinae*		6	
b.Loricariinae*		17	
c.Hypostominae*		35	including 'Ancestrinae'
Ariidae	sea catfishes	13	M or BW
Aspredinidae*	banjo catfishes	8	4 *Bunocephalus* spp FW, other E
Doradidae*	thorny catfishes	7	
Auchenipteridae*	driftwood catfishes	13	
Heptapteridae	heptapterids	19	
Pimelodidae*	long-whiskered catfishes	12	
Pseudopimelodidae	bumblebee catfishes	6	
Gymnotiformes:	Neotropical knifefishes		all species FW
Gymnotidae*	nakedback knifefishes	4	includes electric eel *Electrophorus*

Order: (sub)family	English name	# spp. in SU	Remarks
Rhamphichthyidae*	sand knifefishes	2	
Hypopomidae	bluntnose knifefishes	7	
Sternopygidae	glass knifefishes	5	
Apteronotidae*	ghost knifefishes	3	
Batrachoidiformes:			
Batrachoididae	toadfishes	1	M, E
Gobiiformes:			
Eleotridae	sleepers	5	
Gobiidae	gobies	8	
Mugiliformes:			
Mugilidae	mullets	3	E, BW
Cichliformes:			
Cichlidae*	cichlids	30	FW
family not placed in order:			
Polycentridae	Afro-American leaffishes	1	FW
Beloniformes:	needlefishes		
Hemiramphidae	halfbeaks	1	
Belonidae	needlefishes	3	
Cyprinodontiformes:	killifishes and livebearers		
Rivulidae*	New World rivulines	14	FW, 1 species in mangroves
Anablepidae*	four-eyed fishes	2	mainly E
Poeciliidae	livebearers	6	mainly Coastal Plain
Synbranchiformes:			
Synbranchidae	swamp eels	1	FW
Pleuronectiformes:			
Achiridae	American soles	4	
Syngnathiformes:			
Syngnathidae*	pipefishes	1	mainly M, 1 species FW
Perciformes:			
Centropomidae	snooks	3	M and BW
Acanthuriformes:			
Sciaenidae	drums (croakers)	5	mainly M, 5 species FW
Tetraodontiformes:			
Tetraodontidae	puffers	2	mainly M, 2 species E

Table 5. Fish families with freshwater representatives in Suriname (SU).

Classification largely follows Nelson *et al.* (2016); the order-level classification of the leaffish family Polycentridae is uncertain (Collins *et al.* 2015). Families and subfamilies marked with an asterisk (*) are represented with species accounts in this chapter. FW = fresh water, BW = brackish water, E = estuarine, and M = marine.

Age. Like growth rings in trees, growth marks in fish scales and otoliths (literally: 'ear stones') can identify life history events such as age at metamorphosis, age at maturity and lifespan (Panfili *et al.* 2002). In Suriname, the growth of young *hei-ede kwikwi* (*Hoplosternum littorale*; Ponton *et al.* 2001) and *katrina kwikwi* (*Megalechis thoracata*; Mol & Ponton 2003) has been studied by reading daily growth increments in otoliths. The lifespan of *hei-ede kwikwi* was estimated to be four years (Meunier *et al.* 2002). In the Marowijne River, the large pacu *Tometes lebaili* was estimated to live 8-9 years and reach sexual maturity at 3-4 years (Lord *et al.* 2007). The age and growth of other freshwater fishes that occur in Suriname have been studied in other countries (Table 7).

Scientific name	Local name	Habitat	Length (cm)	Wet mass (g)
Brachyplatystoma filamentosum, **adult**	*lalaw*	Large rivers	300	200,000
same species, **larva**		Large rivers	0.4-1	
Megalops atlanticus, **adult female**	*trapun*	Ocean, estuaries	250	120,000
same species, larva		Estuaries	2-3	
Electrophorus electricus	*stroomfisi*	Rivers and forest creeks	250	20,000-
Phractocephalus hemioliopterus	*mototyar*	Rivers	130	50,000
Hoplias aimara	*anyumara*	Rivers and forest creeks	100	40,000
Cichla ocellaris	*tukunari*	Rivers	50	7,000
Pseudophallus aff. *brasiliensis*		Rivers	7.1	0.3
Harttiella crassicauda		Mountain brooklet	4.7	0.8
Parotocinclus halbothi		Headwater forest creeks	1.9	<1
Stenolicmus ix		Headwater forest creeks	2.0	<1
Hyphessobrycon heterorhabdus		Headwater forest creeks	3.0	<1
Hemigrammus geisleri		Clear water forest creeks	1.9	<1
Microcharacidium eleotrioides		Clear water forest creeks	2.1	<1
Nannostomus marginatus		Black and clear water creeks	2.6	<1
Micropoecilia bifurca		Black and clear water creeks	1.9	<1

Table 6. Variation in size among freshwater fishes of Suriname (Mol 2012b, www.fishbase.com; length of larvae of goliath catfish (Garcia-Davila *et al.* 2015) and tarpon (Smith 1979).

Fecundity. With the exception of auchenipterid catfishes (Mazzoldi *et al.* 2007), river stingrays and livebearers (Wourms 1981), all Surinamese freshwater fishes have external fertilization outside the mother's body. In the armored catfishes *seseiguse* (*Corydoras aeneus*; Kohda *et al.* 1995) and *hei-ede kwikwi* (Mazzoldi *et al.* 2007) spawning involves "sperm drinking" by the female. Fecundity, the number of eggs ripened by a female fish during a spawning season or spawning event, varies widely. Atlantic Tarpon spawn millions of eggs (Crabtree *et al.* 1997), and prochilodontid fishes

Scientific name	Local name	Habitat	Longevity (years)	Country of study
Megalops atlanticus	*trapun*	Ocean, estuaries	>50	Florida, USA
Leporinus friderici	*waraku*	Rivers, creeks in terra firme forest	4	French Guiana
Schizodon fasciatus	*nyamsifisi*	Rivers, creeks in terra firme forest	7	Central Amazonia, Brazil
Myloplus rhomboidalis	*kumalu*	Rivers	5	French Guiana
Tometes lebaili		Rivers	8-9	French Guiana *and Suriname*
Hoplosternum littorale	*hei-ede kwikwi*	Coastal freshwater swamps	4	French Guiana
Sciades couma	*kumakuma*	Estuaries	4.5	French Guiana
Sciades parkeri	*yaribaka*	Estuaries	4	French Guiana
Sciades proops	*kupila*	Estuaries	3.5	French Guiana
Pseudoplatystoma fasciatum	*spigrikati*	Rivers	10	Bolivia
Plagioscion squamosissimus	*kubi*	Rivers	10	Bolivia

Table 7. Longevity or lifespan of some freshwater fishes of Suriname (Panfili *et al.* 2002, Dei Tos *et al.* 2010).

(*kwimata*) more than 100,000 (Roberts 1973). *Harttiella crassicauda* spawn just a few dozen eggs (Mol *et al.* 2007b), and fishes that give birth to free-swimming young such as the freshwater pipefish (*Pseudophallus* aff. *brasiliensis;* Mol 2012a), river stingrays (Lasso *et al.* 2014), guppies and four-eyed fishes also have a low fecundity (Breder & Rosen 1966; see species accounts). In general, fecundity varies among species inversely with the amount of parental care given to the individual progeny. Live-bearing fishes thus have lower fecundity than egg-laying fishes, and nest builders and mouth brooders have a lower fecundity than fishes that do not guard their spawn (Balon 1975). Within species fecundity is positively related to size.

Spawning and courtship. In many species several batches of eggs are produced during the spawning season at intervals of a few days to weeks, as for instance in the *kwikwi* species *Callichthys callichthys*, *Hoplosternum littorale* and *Megalechis thoracata* (Mol 1993). Most Surinamese freshwater fishes spawn at the start of or during the rainy season in inundated floodplains and swamps, thus providing plenty of space and food for their offspring (Lowe-McConnell 1964). However, cichlids and livebearers often spawn or give birth throughout the year and, to a lesser extent, this may also be the case for small fishes in rainforest creeks (Schwassmann 1992). Spawning migrations over some distance are known in Atlantic Tarpon, which spawn at sea, marine anchovies (Engraulidae) spawning in lower freshwater reaches of rivers, and goliath catfishes (*Brachyplatystoma* spp.), *botromanki* (*Pseudauchenipterus nodosus*) and *kwimata* (*Prochilodus rubrotaeniatus*) that spawn upriver (Carolsfeld *et al.* 2003, Vari 1982; Jan Mol, personal observations).

Male fishes are often not easily distinguished from females, but some species exhibit sexual dimorphism (observable differences between the sexes), sometimes only in the breeding season. Male river stingrays of the genus *Potamotrygon* have claspers, while male livebearers and auchenipterid catfishes have a gonopodium or modified anal fin, and sometimes also differences in color pattern. In *kwikwi* species, males have an enlarged pectoral spine, and *plar'plari* (*Ageneiosus inermis*) males have an enlarged spine with hooks in the dorsal fin and ossified, fringed barbels (wiry appendages) on the upper jaw (Figure 253c). *Ancistrus* spp. males carry branched tentacles on the snout (Sabaj *et al.* 1999) and *Pseudancistrus* spp. males carry long bristles on the side of the head (Figure 229a). *Curimatopsis crypticus* sexes differ in color pattern (Mol 2012b, p. 114). In cichlids, sexes may differ morphologically, such as *Cichla ocellaris* whose males have a fatty hump on the nape, or in color pattern, as with *Nannacara anomala*. The gender of individual electric knifefishes (Gymnotiformes) can be distinguished through characteristics of their electric organ discharge (Hopkins *et al.* 1990), with males often producing stronger discharges and females displaying higher discharge frequencies. Courtship of Surinamese freshwater fishes is best known in cichlids and guppies, often from aquarium observations (Keenleyside 1979, Reebs, 2001), but for some cichlids also from observation in the field (Lowe-McConnell 1969, Keenleyside & Bietz 1981).

Parental care. Many Surinamese fishes provide some parental care to their eggs and/or newborn. Cichlids are famous for their territoriality and parental care by both sexes (Figure 260). Suckermouth armored catfishes (Loricariidae) are another group with marked territorial habits, which guard their eggs and newborn in cavities in submersed woody debris (*Ancistrus* spp.; see Sabaj *et al.* 1999) or in nest burrows dug in the river bank (*Hypostomus* spp.). Female Splashing Tetras *(Copella arnoldi)* deposit their eggs above the water surface – out of reach of aquatic egg predators – on overhanging vegetation, and the male then keeps the developing eggs moist by splashing water droplets on them (Krekorian 1976). *Kwikwi* species also position their eggs above the oxygen-poor swamp water in floating bubble nests; these nests are guarded by the male (Mol 1993) (see species account). *Pataka* (*Hoplias malabaricus*) guard and defend their eggs in a simple nest pit in newly flooded swamp (Prado *et al.* 2006, Jan Mol personal observation). Most cichlids guard both eggs and young (see species accounts *Cichla ocellaris, Krobia guianensis*), while *songeh* (*Geophagus* spp.) and *Satanoperca* spp. brood the eggs in their mouth (Lowe-McConnell 1969). Ariid catfishes, usually the male, also brood their large eggs in the mouth (see Figure 32b in Ecosystems chapter) (Burgess 1989). Male pipefish of the genus *Pseudophallus* brood eggs in a pouch (Mol 2012a) and may provide nutrients to the developing embryos. Male *basyafisi* (*Loricariichthys maculatus*) carry the eggs with them on the enlarged lower lip (Burgess 1989) and females of the banjo catfishes of the genera *Aspredinichthys* and *Aspredo* (*banyaman* or *trompetfisi*) carry the eggs attached to the ventral side of the body (Burgess 1989). Female livebearing poeciliids (guppies), anablepids (*kutai*) and *Potamotrygon* spp. river stingrays (*libaspari*) even provide nutrients to developing embryos via placenta-like structures (see species accounts *Anableps anableps, Potamotrygon boesemani*) (Wourms 1981).

Fishes living together in communities

Dilemma. Fish species co-exist in what ecologists call communities. Compared to temperate zone communities, tropical fish communities, like those in Suriname, have large numbers of species which interact in complex ways. Medium-sized rivers of Suriname have approximately 100 fish species living together in the same river system (see for example Mol *et al.* 2006 for Upper Coppename River, Mol & Wan Tong You 2013 for Paloemeu River, and Willink *et al.* 2011 for the Sipaliwini and Koetari rivers). In the Interior, more than 60 fish species can live together in the same forest creek (see for example Mol & Ouboter 2004) or rapid complex (Mol *et al.* 2006). How do all these fishes share the various resources? What prevents one competitor from eliminating another? This is largely explained by differences in habitat, diet and behavior between fish species (see next). To a considerable extent, they avoid being at the same type of location within a stream, and/or eating the same kind of food, and/or obtaining it in the same way. Thus they essentially avoid or at least limit competition, and do not eliminate one another.

Different activity periods. Surinamese freshwater fishes are active at different times of the daily cycle. Characiformes and cichlids, which orient themselves mostly visually, dominate the diurnal (daytime) fish fauna (Lowe-McConnell 1964, Sabino & Zuanon 1998, Arrington & Winemiller 2003). They employ vision in the formation of shoals and schools, during courtship, and for foraging and defensive tactics. At night, the fish fauna is composed mainly of catfishes and electric knifefishes that use chemical, tactile, sound and electric stimuli for orientation. These nocturnal species hide in crevices or sheltered places during the day. Other fishes, such as *kwikwi*, are most active in the twilight hours of early morning and late afternoon (Boujard *et al.* 1990b; Jan Mol personal observations).

Different habitats. Fishes often prefer different habitats or microhabitats, meaning different ecosystems or environments or different parts of these (Romero & Casatti 2012, Sabino & Zuanon 1998). Within a species, habitat use may also differ with life stage or season (see below). Freshwater habitats can be categorized as either lentic or lotic. Lentic refers to standing water, as in swamps, flooded forest, lakes and reservoirs, while lotic refers to running water, as in rivers and creeks. The stream habitat can be further divided into bottom (home to catfishes), surface (needlefishes, hatchetfishes, and *Triportheus* spp.) and mid-water (many characins). Important aquatic habitats in Suriname are estuaries, mangroves, freshwater swamps, lower freshwater reaches of rivers with associated floodplain forest, black-water streams, clear-water rivers of the Interior with rapid complexes, small creeks in *terra firme* forest, headwaters and mountain streams (see Ecosystems chapter). Within these large-scale habitats there are many micro-habitats such as sand banks (Arrington & Winemiller 2003, Zuanon *et al.* 2006), floating vegetation (Henderson & Hamilton 1995, Schiesari *et al.* 2003), floating litter (Carvalho *et al.* 2013), woody debris (Lowe-McConnell 1964, Taylor *et al.* 2008), litter of dead leaves (Henderson & Walker 1990), stands of large aquatic plants (macrophytes), submerged root masses, and more. It is clear that the natural habitat of our beloved aquarium fishes (Figure 225) (Lucanus 2009, Mikolji 2013) is often quite different from that presented in pet shops or public aquarium exhibits.

Mobility and migration. Lowe-McConnell (1987, p. 130) notes that 'the key to Amazonian fish community structure lies in the mobility of the fishes', as 'many species make separate movements in relation to oxygen levels and trophic, breeding, and dispersal migrations'. Thus communities are dynamic, with the species composition and relative abundances of different species continually changing. Many fish species need large areas to complete their life cycles, whereas others may do so in a few square meters of a forest creek.

Fishes migrate between different habitats for spawning or feeding, and often live in different habitats between their early stages and adulthood. For example, juvenile goliath catfishes (*Brachyplatystoma* spp.) live in estuaries, but as adults they travel large distances upriver to spawn (Barthem *et al.* 1991) and, in Suriname, large specimens of *Brachyplatystoma filamentosum* and *Brachyplatystoma rousseauxii* have been collected in rapids of the Suriname and Corantijn rivers (see species account). Atlantic Tarpon spawn at sea, but their so-called leptocephalus larvae are found in the estuaries, while (older) juveniles live in lower freshwater reaches of rivers and freshwater swamps (see species account). Several marine anchovies spawn in lower freshwater reaches; a known spawning site is a mid-river bank with aquatic macrophytes (water plants) 105 km up the Suriname River near Overbridge (Mol 2012b, p.52). *Botromanki* catfish (*Pseudauchenipterus nodosus*) live and feed in the estuaries, but spawn upriver in freshwater reaches (Jan Mol personal observation). *Prochilodus rubrotaeniatus*, and possibly *Semaprochilodus varii*, migrate from lower freshwater reaches upriver to spawn. These longitudinal migrations transfer nutrients between marine and freshwater habitats, or lower and upstream reaches, similar to the well-known spawning migrations of salmon in cold-temperate regions (Schindler *et al.* 2002). Shorter lateral movements from the river channel into the flooded forest and swamps are also known in Surinamese freshwater fishes, both for feeding and reproduction, as for example with *kwikwi* species and *pataka*, and with *Leporinus nijsseni* in floodplain forests of the upper Nickerie River.

On the other hand, minute sand-dwelling pencil catfishes such as *Stenolicmus ix*, dwarf cichlids (*Apistogramma* spp., *Nannacara anomala*) and small characin species associated with submerged leaf litter in forest streams may have very small home ranges restricted to shallow, marginal stream habitats. Carvalho *et al.* (2007) point out that the small home ranges and sedentary habits of small fish species may have an important role in creating and maintaining diversity in the Amazon fish fauna as exemplified by the many species in the genus *Corydoras* (known locally as *seseiguse*).

Different diets. The fishes of Suriname exploit a wide variety of food types, and ontogenetic (developmental) changes in diet are common. The larvae of freshwater fishes mainly feed on the same micro-crustaceans, rotifers and single-celled protozoans; later as juveniles they also include aquatic insects in their diet (see for example Mol 1995 for three *kwikwi* species in Suriname; see also Winemiller 1989b, Ponton & Mérigoux 2001). This implies that these young fishes compete with each other for food, while being subject to high predation pressure (Mol 1996). This may keep populations of early life stages below densities at which resource competition becomes important. Expanded aquatic habitat in floodplains and swamps in the rainy season, the main reproductive season (see Ecosystems chapter), and adverse abiotic (physical and chemical) conditions may also keep the population densities of competitors low.

Food availability is subject to strong seasonal changes (Junk *et al.* 1989), resulting in a predominance of species with generalist or opportunist feeding habits, eating whatever they encounter and can swallow (Carvalho *et al.* 2007). For fishes in their early life stages and those that do not feed on fish, food abundance peaks in the rainy season. By contrast, fish eaters consume most prey in the dry season when the waters recede and the prey is more concentrated. In lean times, often the dry season, diet overlap among species can increase, but many fishes stop feeding altogether (as shown by their empty stomachs; Arrington *et al.* 2002) and survive on fat deposits accumulated during better times (Junk 1985). Adults may prefer more specific food items, thus diminishing competition for food between species.

Many adult freshwater fishes are piscivores, feeding on fish (Winemiller 1989b): erythrinids (*anyumara*, *pataka*, *walapa*), piranhas (but see Sazima & Machado 1990), *Acestrorhynchus* spp. (*dagufisi*), the Electric Eel, Leaf Fish *(Polycentrus schomburgkii)*, pike cichlids or *datrafisi* (*Crenicichla* spp.), peacock bass or *tukunari* (*Cichla* spp.), and most medium- to large-sized catfishes (see Barbarino Duque & Winemiller 2003). Adult *kwasimama* catfish (*Hypophthalmus marginatus*) are mid-water plankton feeders with long sieve-like gill rakers (Carvalho 1980). Prochilodontids, curimatids and hemiodontids with reduced dentition (Géry 1977) and often long coiled intestines feed on detritus (Bowen 1983; Forsberg *et al.* 1993). Suckermouth armored catfishes (Loricariidae) mainly feed on algae (periphyton) and detritus (Delariva & Agostinho 2001), but *Hypostomus taphorni* and species of the *Panaque* group (*Panaqolus* spp., *Peckoltia* spp.) feed on decaying wood (Schaefer & Stewart 1993, German 2008). Examples of Surinamese freshwater fishes that feed on leaves of aquatic macrophytes or trees include *Myloplus ternetzi* (Boujard *et al.* 1990a) and *nyamsifisi* (*Schizodon fasciatus*) (Carvalho *et al.* 2007). Serrasalmine species in the genera *Acnodon*, *Metynnis*, *Myleus*, *Myloplus* and *Tometes* (*kumalu* and *pakusi*) feed on Podostemacae (*kumarunyanyan*) leaves, fruits and seeds (Boujard *et al.* 1990a). *Moroko* (*Brycon falcatus*), the freshwater sardine or *sardin Triportheus brachipomus* and the large doradid catfish *Pterodoras granulosus* include seeds in their diet without destroying them; the seeds that pass through the digestive tract remain viable, and germinate under adequate conditions (this is seed dispersal; Goulding 1980). The characins *Roeboides affinis* and *Roeboexodon guyanensis* feed on fish scales using external teeth on the snout (Sazima 1983; Figure 229b; see species account), while juvenile piranhas *Serrasalmus* feed on fins (Sazima & Machado 1990). Both scales and fish fins can regenerate after a relatively short time, and may thus be considered renewable food sources. Many fishes have an opportunistic, omnivorous diet that includes aquatic invertebrates and, in lean times, living plant matter and detritus; these include *kwikwi* species, many characins, gymnotiforms, cichlids and smaller catfishes (Knöppel 1970, Gonzalez & Vispo 2003). Many fishes, including piranhas, catfishes and characins, also feed opportunistically on carcasses (scavenging) (Sazima & Machado 1990). Feeding on external parasites of large predatory fishes ('cleaning') was observed in piranhas (Sazima & Machado 1990), *Anostomus* spp. (Lucanus 2009) and juvenile *soké* (*Platydoras costatus*; Carvalho *et al.* 2003).

Predators and prey. Fish mortality by predation is intense. Most piscivorous fishes only eat fishes up to a size as large as the gape of their mouth, so size is a risk factor for most potential prey: smaller fish are more vulnerable to predation. Piscivorous fishes that rely on vision use tactics including stalking, ambushing, attacking under disguise and chasing to secure their prey. Pike cichlids *Crenicichla* spp. (*datrafisi*) stalk their prey with use of cover. Wolf fishes (*Hoplias* spp., known locally as *anyumara* or *pataka*) ambush their prey from within vegetation or woody debris. In habitats with little cover, predators approach their prey under disguise, such as scale-eating *Roeboides* and fin-eating juvenile piranhas *Serrasalmus* (Sazima 1983, Sazima and Machado 1990). Chase is a tactic employed by fast, streamlined fishes in open water such as *dagufisi* (*Acestrorhynchus* spp.; Figure 229b) and *Cynodon* spp. (Figure 229b). Nocturnal knifefishes generate weak electric discharges to locate their prey and electric eels (*Electrophorus* spp.) deliver strong electric discharges to stun their prey (see species account). Other non-fish predators in Suriname include aquatic as well as aerial and terrestrial vertebrates: the estuarine Guiana dolphin *Sotalia guianensis*, the otters *Pteronura brasiliensis* and *Lontra longicaudis*, the water opossum *Chironectes minimus*, the crab-eating raccoon *Procyon cancrivorus*, the fish-eating bat *Noctilio leporinus*, the caimans *Caiman crocodilus*, *Paleosuchus palpebrosus* and *Paleosuchus trigonatus*, some fish-eating water snakes (such as the anaconda *Eunectes murinus*), and a number of bird predators such as kingfishers, herons, the Ahinga (*Anhinga anhinga*) and the Neotropical Cormorant (*Phalacrocorax brasiliensis*).

Avoiding, hampering or confusing predators is one of the most efficient and low-cost defense strategies, which includes camouflage, transparency (in open water) and mimicry. Day-active suckermouth armored catfishes, for example, blend with the background by color or resemblance to immediate surroundings, or they have a disruptive pattern that obliterates the form of their body, and move inconspicuously while feeding (Carvalho *et al.* 2007). Other catfishes resemble sticks or dead leaves – which are abundant in streams – for example *Helogenes marmoratus* and *Bunocephalus* spp. (see species accounts). Small fishes of open water, such as *Tomeurus gracilis*, and some some fish larvae, such as the leptocephalus larvae of Atlantic Tarpon (Figure 232b), use translucency or transparency to make themselves invisible to predators. Mimicry is illustrated by the association of two or more species of small characins that are similar in color and form and which have no evident defenses such as venom or spines, for example mixed schools of X-ray Tetras and False X-ray Tetras (see species account *Hyphessobrycon simulatus*): by increasing their numbers, individuals of similar-looking fishes lessen their chances of being preyed upon. A common tactic to avoid visually guided predators is to remain out of sight by hiding in hollows in the bank or in submersed logs or leaf litter, or by sheltering in soft bottoms such as sand, as exemplified by

Mastiglanis asopos and Gymnorhamphichthys rondoni (see species account). Some fishes have 'invested' in defensive spines or venom, such as catfishes and stingrays, or protective armor, like armored catfishes; these features make them harmful prey to predators and/or more difficult to swallow.

A look into the future

Ever since the collection efforts of Boeseman, Nijssen and Mees in the 1960s and 1970s first made specimens of Surinamese freshwater fishes widely available, many talented ichthyologists have devoted considerable efforts to increase our understanding of the freshwater fish fauna of Suriname (Mol 2012b). Recent, ongoing sequencing of the DNA of Surinamese freshwater fishes in the comprehensive Guianas Freshwater Fishes Barcoding of Life project (Gui-BOL, Raphael Covain *et al.*, see for example **www.boldsystems.org/index.php/Public_RecordView?processid=GBOL015-12**; Ratnasingham & Hebert 2007), and studies of specific groups of fishes (as by Covain *et al.* 2012, Fisch-Muller *et al.* in press) have revealed the occurrence of many 'new' species in the rivers of Suriname (Box 9). Most of these are small and difficult to distinguish morphologically or even genetically from other, often closely related, species (Bickford *et al.* 2007; see for examples Covain *et al.* 2012, Benzaquem *et al.* 2015), but they also include a new species of electric eel (see species account) and several *Corydoras* spp.

In Amazonia, unique fish species have been discovered in two unexpected environments: the deep channel of large rivers (see for example Crampton 2008) and underground running waters in the subsoil of terrestrial environments (Ohara *et al.* 2016), two aquatic habitats that have not been studied in Suriname. Thus many fish species probably remain to be discovered in Suriname. For most of the known, formally described species, we still know next to nothing of their ecology (as evident from many of the species accounts included here). Large parts of western, central and south Suriname are still pristine, and the future of the fishes in these areas looks bright. But in the east of the country, where gold mining occurs and where the Brokopondo Reservoir is located, and in the northern part of the Coastal Plain where most people live, the habitats of fishes are being destroyed at an alarming rate (see Use, Threats and Conservation chapter). Even in pristine river systems, fish species with low fecundity, a restricted distribution, or both can be threatened by overfishing for aquariums, such as the beautiful river stingray *Potamotrygon boesemani*, which is endemic to the Corantijn River (see species account). Conservation efforts directed towards freshwater fishes and their habitats are badly needed (Helfman 2007, Carvalho *et al.* 2007, Reis 2013).

Potamotrygon boesemani
River stingray (English name used for all Potamotrygonidae)
***Libaspari* (local name used for all Potomotrygonidae)**
Chondrichthyes, Myliobatiformes, Potamotrygonidae

Potamotrygonidae is the only living family of cartilaginous fishes (sharks, rays, etc.) restricted to freshwater habitats. Adaptations to freshwater include blood with a low urea concentration and a rudimentary gland in the rectum that is normally used for salt excretion (Thorson *et al.* 1983a). Male river stingrays have claspers or modified pelvic fins (below, near the tail base) that are used to transfer sperm to the cloaca (single opening for the digestive, urinary and reproductive tracts) of the female during copulation. Developing embryos are nourished by uterine milk (Thorson *et al.* 1983b). Females give birth to between one and eight young in each gestation (Charvet-Almeida *et al.* 2005). River stingrays are much feared because of their venomous stings, located on the upper side of the tail, near the tip (Haddad *et al.* 2004). However, the stings are purely defensive, and river stingrays actually pose little threat if not stepped on or directly interfered with. The injury is painful and heals slowly due to the death of body tissue and secondary bacterial infections, but it is rarely lethal.

Potamotrygon boesemani (Figure 231) lives in shallow water, where it often lies camouflaged on the bottom, or hidden under the sand with only the eyes, spiracles (small holes behind the eyes that open to the mouth) and tail exposed (Mol 2012b, p. 71). Although not much is known about the ecology of this particular species, the biology of the family of freshwater stingrays is fairly well known (Lasso *et al.* 2014). For example, Shibuya *et al.* (2009) found that freshwater stingrays of the Rio Negro fed on aquatic insects, crabs, shrimps and fishes, but they also noted differences in the diet of different species that live together (food partitioning).

Figure 231. The colorful river stingray *Potamotrygon boesemani* (*libaspari*) is endemic to the Corantijn River (image © Ross Smith).

Description: Medium-sized stingray; disc-shaped body 43 cm wide, with many deep orange to red spots of irregular form, encircled by a relatively broad black band.

Distribution: Many river stingray species have distributions restricted to a single river system; *Potamotrygon boesemani* is only known from the Corantijn River where it occurs both upstream and downstream of the rapids that are furthest downstream.

Use: Of great interest to the aquarium trade (and very expensive!), but its restricted distribution, slow growth, late maturity and low fecundity make it very vulnerable to over-exploitation.

Similar species: *Potamotrygon marinae* of the Marowijne and Suriname rivers, with a 40-cm wide body and circular yellowish spots (each composed of smaller patches; see Figure 229a). *Potamotrygon* cf. *orbignyi*, a uniform brownish species with vague dark spots or a dark mesh pattern in the Marowijne, Suriname, Coppename, Nickerie and Corantijn rivers (Mol 2012b).

Megalops atlanticus
Atlantic Tarpon
Trapun
Actinopterygii, Elopiformes, Megalopidae

Trapun or Atlantic Tarpons are large fish whose juveniles live in fresh or brackish-water habitats such as estuaries, lagoons, coastal swamps or lower reaches of rivers. Adults spawn huge numbers of eggs in the sea (12 million eggs in a large female; Crabtree *et al.* 1997). The transparent, ribbon-like larvae move inshore to estuaries (Crabtee *et al.* 1992). Juvenile *trapun* are facultative air-breathers: they can use their modified swim bladder to extract oxygen from the air. This allows them to survive in swamp water that is poor in oxygen (Geiger *et al.* 2000). *Trapun* feeds on fish, shrimps and crabs. They are long-lived (at least 48 years; Crabtree *et al.* 1997).

In Suriname, juvenile *trapun* (Figure 232a) are found in brackish ponds (such as in the Bigi Pan area), freshwater swamps and lower freshwater reaches of rivers. The so-called leptocephalus larvae (Figure 232b), a thin and transparent type of larva of eels, tarpons and related elopomorph fishes (Smith 1979, Miller 2009), are observed in estuaries.

Figure 232. Atlantic Tarpon *Megalops atlanticus* (*trapun*): a) juvenile with posterior ray of the dorsal fin prolonged, b) transparent leptocephalus larvae (lower Commewijne River) (image a © Jan Mol; image b © Kenneth Wan Tong You).

Description: Large, up to 250 cm length and 120 kg, but males are smaller and rarely reach 50 kg; moderately elongated, silvery fish with a highly compressed body and very large scales. The last ray of the dorsal fin (on the back) is greatly prolonged. The mouth is large and obliquely upturned.

Distribution: This species occurs mainly in the periphery of the tropical Atlantic Ocean, along the coasts and some way up the main rivers that discharge into the sea. In Suriname, it is not found beyond the furthest downstream rapids.

Use: *Trapun* meat is used to make fish balls in Suriname. They are not considered particularly valuable food fish, but are very popular with sport fishers. Dried *trapun* is exported to the Netherlands. Because *trapun* cannot be filleted, it is cooked whole; eating this fish is not without risk since it has so many bones (Mahajan *et al.* 2016).

Similar species: The nearest relative of *trapun* in Suriname is the Tenpounder or Ladyfish *Elops saurus,* known in Suriname as *daguboi* or *landyan*. This is a much smaller species (< 50 cm), with much smaller scales, and a less compressed body. It lives in mangrove lagoons, estuaries and in the sea (Mol 2012b).

Parodon guyanensis
(no English name)
(no local name)
Actinopterygii, Characiformes, Parodontidae (parodontids)

Figure 233. ***Parodon guyanensis*** **in a rapid complex of the upper Suriname River (image © Willem Kolvoort).**

Parodon guyanensis (Figure 233) are specialized for life on the rocky river bottom in rapids with fast currents. The enlarged pectoral fins are inserted low on the body and press the fish down on the bottom in strong currents; they are also used to 'walk' on the bottom. In its early life stages the species is elongated, laterally compressed (flattened sideways), with a pointed snout, and lacks the black bar on the dorsal fin; it occurs on muddy substrate and feeds on tiny crustaceans and insect larvae (Ponton and Mérigoux 2001). In rapids, *Parodon guyanensis* can be seen grazing algae from rocks during the day. The biology of Parodontidae is largely unknown; the behavior and reproductive ecology have not been studied (but see Sazima 1980).

Description: This species is 12 cm long and has a spindle-shaped (fusiform) body with the inferior (downward opening) mouth of a bottom feeder; it also has a flat and broad breast, well-developed paired fins, a characteristic black bar (stripe) on the dorsal fin (on the back) and a V-shaped zigzag mark along either side of the body (following the lateral line system).

Distribution: Occurs in the rapids of the Orinoco River and other rivers of northern South America (Guianas); in Suriname in those of the Corantijn, Coppename, Saramacca, Suriname and Marowijne rivers.

Use: Traded as an ornamental fish.

Similar species: None. With their flat breast, broad paired fins, black bar on the dorsal fin and zigzag mark along the lateral line system, they cannot be confused with other fish species in Suriname.

Prochilodus rubrotaeniatus and *Semaprochilodus varii* (no English names) *Kwimata, kurimata* (local names used for both species) Actinopterygii, Characiformes, Prochilodontidae (flannel-mouth characiforms)

Figure 234. The two *kwimata* species of Suriname: a) *Prochilodus rubrotaeniatus* (specimen on alcohol) and b) *Semaprochilodus varii* (Marowijne River) (images a and b © Mark Sabaj-Pérez).

With weak teeth and long coiled intestines, *kwimata* (*Prochilodus rubrotaeniatus* and *Semaprochilodus varii*; Figure 234a,b) feed on the river bottom on detritus, algae and associated micro-organisms (Bowen 1983). They are well known for their spawning migrations in large rivers (Carolsfeld *et al.* 2003). Spawning schools of females and courting males emit a low-pitched drumming sound that can be heard in a boat. In the Rupununi Savanna in southern Guyana, *Prochilodus rubrotaeniatus* spawns at the start of the rainy season: the numerous eggs are deposited among grass roots in shallow water (about 30 cm deep) (Lowe-McConnell 1964). According to the Saramacca Maroon people of the middle Suriname River, this species spawns in flooded vegetation along the inner bends of the river.

By ingesting large quantities of fine bottom sediment the *kwimata* species modify the riverine ecosystem, in particular the algal and invertebrate communities: exclusion of *kwimata* resulted in increased sediment accrual on bottom substrate, increased invertebrate densities and diatoms, but a decrease in cyanobacteria (Flecker 1996). This disturbance of bottom substrate is a form of ecosystem engineering. With prochilodontids comprising the bulk of the fish biomass in many South American rivers, the *kwimata* are functionally dominant species in these rivers. A functionally dominant or 'keystone' species is a species with a disproportionately large effect on its environment. Such species play a critical role in maintaining the structure of an ecological community, affecting many other organisms in an ecosystem and helping to determine the types and numbers of various other species in the community they are a part of.

Description: Medium- to large-sized, robust fishes (35-43 cm long) with large scales and a somewhat downward oriented (subterminal) mouth; two series of numerous small teeth are attached to the fleshy lips. The dorsal fin is preceded by a forked (pre-dorsal) spine. *Semaprochilodus varii* has black and yellow stripes on the tail and anal fins, and the edge of the operculum (gill cover) is colored black.

Distribution: *Semaprochilodus varii* is endemic to the Marowijne River, but *Prochilodus rubrotaeniatus* is widespread in large rivers of Amazonia, the Guianas and Venezuela. In Suriname, the latter species no longer occurs in the Suriname River downstream of Brokopondo Reservoir (Mol *et al.* 2007a).

Use: Prochilodontids are important food fishes in all South American countries where they occur (Lowe-McConnell 1984).

Similar species: Several species in the family Curimatidae resemble prochilodontids in behavior (feeding on detritus and occurring in large schools) and in body form, but they are smaller (except *Curimata cyprinoides*; *makasriba*), and they lack both teeth and pre-dorsal spine.

Anostomus anostomus
Striped Headstander
(no local name)
Actinopterygii, Characiformes, Anostomidae (toothed headstanders)

Anostomus anostomus (Figure 235a) was described by Linnaeus in 1758, and the type locality (see Box 9) is probably Suriname. It feeds on algae and small invertebrates with its body in an inclined (oblique) position, hence the name headstander (Géry 1977). Species of the genus *Anostomus* are known to pick parasites off large predatory fish like *Hoplias aimara* (*anyumara*) (Lucanus 2009, p.149), just like marine cleaner wrasses. These inquisitive fish will even enter the mouth of snorkelers to clean their teeth.

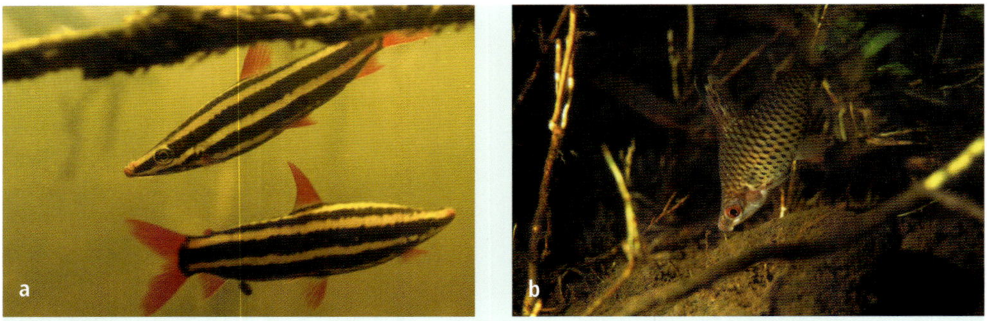

Figure 235. Headstanders in Suriname (images © Willem Kolvoort):
a. Striped Headstander *Anostomus anostomus* (Anostomidae) in the upper Suriname River, and
b. a spotted headstander *Chilodus zunevei* (Chilodontidae).

Description: Spindle-shaped (fusiform), small fish, up to 16 cm long, with a series of just 3-4 teeth in each half of both the upper and lower jaw, arranged as steps of a stair. The mouth is small, terminal, and upturned in such a way that the lower jaw is vertically aligned when the mouth is closed. The body has a beautiful coloration with bright red fins, and three longitudinal black bands on each side, separated from each other by narrow yellow stripes.

Distribution: Widespread in tropical South America, from Amazonia to Venezuela and the western and central part of the Guianas. In Suriname, it is present in the Suriname, Saramacca, Coppename and Corantijn rivers, both in the main river channel and in large tributaries. The species no longer occurs in the Brokopondo Reservoir (Suriname river basin) (Mol *et al.* 2007).

Use: Headstanders are well-known and popular aquarium fishes.

Similar species: Two additional striped headstander species (*Anostomus* spp.) with a slightly different color pattern occur in Suriname, as well as four species of spotted headstanders (family Chilodontidae; Figure 235b), also diverging in color pattern (see Mol 2012b).

Gasteropelecus sternicla
Silver or Common Hatchetfish
Banketman, bijlzalm (local names also used for the Marbled Hatchetfish)
Actinopterygii, Characiformes, Gasteropelecidae (freshwater hatchetfishes)

Freshwater hatchetfishes (Figure 236) swim near the water surface where they feed on terrestrial invertebrates that have fallen in the water (Netto-Ferreira *et al.* 2007). They have upturned mouths and locate their prey with their lateral line system, which detects vibrations in the water from prey struggling to escape from the water surface (Bleckmann 1993). Hatchetfishes are capable of jumping relatively far and high out of the water by use of their modified pectoral fins (see Figure 226) powered by strong, enlarged muscles; they however do not seem to move their pectoral fins when airborne ('flying') (Wiest 1995). In the Interior, entire schools of Silver Hatchetfish can be observed jumping out of the water, apparently to escape from an approaching predator.

Figure 236. The two freshwater hatchetfishes of Suriname (images © Mark Sabaj-Pérez):

a. Silver Hatchetfish *Gasteropelecus sternicla*, and

b. Marbled Hatchetfish *Carnegiella strigata*.

Description: Freshwater hatchetfishes or 'flying' characins are small surface fishes with a rounded, trenchant, hatchet-shaped belly, a superior (upturned) mouth and very large pectoral fins. *Gasteropelecus sternicla* (4 cm long) has a uniform silvery body with a thin longitudinal black band at each side of the body and a black stripe at the base of the anal fin; an adipose fin is present (Figure 236a).

Distribution: The Silver Hatchetfish is widespread in the Amazon and northern South America: Brazil, French Guiana, Guyana, Peru, Suriname and Venezuela. In Suriname, it occurs in both small black-water streams of the Zanderij Belt and large rivers and forest creeks of the Interior; it is present in all river systems of Suriname.

Use: A well-known and popular aquarium fish.

Similar species: The Marbled Hatchetfish (*Carnegiella strigata*) (Figure 236b) has a marbled body with broad bands extending from the keel to the back and lacks an adipose fin. It is most often found in small black-water streams, where it lives together with *Gasteropelecus sternicla*. The silvery *Triportheus brachipomus* (*sardin*) is not a hatchetfish, but a characoid of the family Triportheidae; it is much larger (30 cm) and elongated, lacking the trenchant, hatchet-shaped belly.

Hyphessobrycon simulatus
False X-ray Tetra
Reditere sriba (the local name *sriba* is used for many small characins)
Actinopterygii, Characiformes, Characidae (characids, tetras)

Surinamese species of the large and diverse family Characidae (Géry 1977, Oliveira *et al.* 2011) include small-sized tetras with an incomplete lateral line (linear series of perforated scales at the side of the body that allow water into the lateral line system) (Figure 226) in the genera *Hemigrammus* (Figure 229b), *Hyphessobrycon* (Figure 237a), *Pristella* (Figure 237b) and *Thayeria* (the Penguin or Hocky-stick Tetra) and somewhat larger fishes with a complete lateral line in *Astyanax*, *Jupiaba* and *Moenkhausia* (Mol 2012b). *Hyphessobrycon simulatus* is a small tetra (3 cm) of coastal freshwater swamps and black-water streams of the Zanderij Belt. Its characteristic pigmentation pattern includes a vertically elongated humeral spot and a black flag on the dorsal fin (Figure 237a). In Suriname, *Hyphessobrycon simulatus* and *Pristella maxillaris* (Figure 237b) mimick each other and often live together in mixed shoals (a case of mimicry); both species have a flag on the dorsal fin. *Hyphessobrycon simulatus* jerks the flag rapidly when alarmed, while *Pristella maxillaris* uses its black flag to lead other members of the same species and school to follow (Keenleyside 1955).

a

b

Figure 237. Two common and very similar *reditere sriba* tetras from Para River with a 'flag' on the dorsal fin, a humeral spot and a red tail fin; in Suriname, these two species often live together in mixed shoals (images © Mark Sabaj-Pérez):

a. the False X-ray Tetra *Hyphessobrycon simulatus*, and

b. the true X-ray Tetra *Pristella maxillaris*.

Description: Most Surinamese tetras can be distinguished by their pigmentation such as spots, stripes or bands and the location on the body of these markings. *Hyphessobrycon simulatus* has a black flag on the dorsal fin, a vertically elongated humeral spot (located behind the operculum or bony gill cover), a vertical black band through the eye, and a bright red caudal fin.

Distribution: The False X-ray Tetra is known from French Guiana and Suriname; in Suriname it occurs in the Marowijne, Commewijne, Suriname and Saramacca rivers.

Use: An aquarium fish, but not as popular as the true X-ray Tetra *(Pristella maxillaris)*.

Similar species: Other Surinamese tetras with a black flag on the dorsal fin are *Hemigrammus unilineatus* (first rays of anal fin colored black), *Hyphessobrycon copelandi* (rounded humeral spot), *Hyphessobrycon rosaceus* (lacking a humeral spot), *Pristella maxillaris* (round humeral spot, tail fin colored bright red, black spots on the pelvic and anal fins; Figure 237b), and *Moenkhausia hemigrammoides* (first rays of anal fin black, complete lateral line) (Mol 2012b).

Brycon falcatus
South American trout (English name used for all *Brycon* spp.)
Moroko
Actinopterygii, Characiformes, Bryconidae

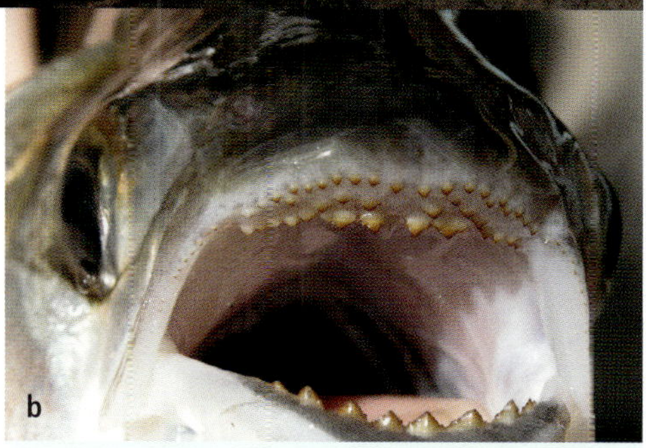

Figure 238. *Brycon falcatus* (*moroko*) of Coppename River (a) with detail of teeth (b) (images © Barry Chernoff).

The *moroko* (*Brycon falcatus*) (Figure 238) feeds on food items that have fallen into the water such as fruits, seeds, flowers and insects (Goulding 1980). It is able to jump out of the water to pluck low-hanging fruit directly from trees. In the low-water season *moroko* can easily be observed in shallow water near shore due to their characteristic V-shaped marking in the tail fin. *Brycon* spp. are sometimes called South American trout, but although they resemble true trouts – which also have an adipose fin – they are not closely related to them; true trout belong to the family Salmonidae, and occur in cold water in the north temperate climate zone.

Description: Medium-sized fish, up to 37 cm long (1 kg), with large mouth; there are three rows of teeth on the upper jaw (larger teeth in the inner row), and on the lower jaw there are two symphyseal teeth (located near where the jaws articulate) behind the main tooth row (Figure 238b). There is a V-shaped black band at the base of the caudal fin, a black band at the base of the anal fin, and a humeral spot with the size of the pupil of the eye; the lateral line (series of perforated scales) is situated low on the body.

Distribution: This species is widespread in rivers of northern South America and the Amazon and Orinoco rivers. In Suriname, it is known from most if not all rivers and large tributaries.

Use: *Brycon* spp. are important food fishes throughout South America (Lowe-McConnell 1984); they are cultivated in several South American countries, but not in Suriname. They are also popular with sport fishers.

Similar species: *Brycon pesu* is a similar but smaller (12 cm) species of the Marowijne River; it has a black adipose fin and a yellowish tail fin with small dots more or less arranged in two vertical rows (Figure 229b). Another related species is the entirely silvery *Triportheus brachipomus* (*sardin*), which lives near the water surface and can jump several meters out of the water using its wing-like pectoral fins.

Roeboexodon guyanensis
(no English name)
(no local name)
Actinopterygii, Characiformes, Characidae, Characinae

Figure 239. The scale-eater *Roeboexodon guyanensis*, Marowijne River (image © Mark Sabaj-Pérez).

Roeboexodon guyanensis (Figure 239) is a relatively rare species found in creeks with rapidly flowing water over a sandy substrate. With its two external tusks and four nipple-shaped teeth on the extremity of the snout, it can feed on the scales of other fishes (Sazima 1983). Scales can be present in a fish's stomach for reasons other than lepidophagy or scale-eating, for example in fish that prey on other fish and swallow them whole. True scale-eaters like *Roeboexodon guyanensis* generally have scales in their stomach without fish bones or flesh; they have been observed removing scales in the laboratory or in the field (Roberts 1970, Sazima 1983).

Ontogenetic (developmental) diet shifts have been observed in scale-eating *Roeboides* spp.: scale consumption increased with the size of the fish, with juveniles feeding mainly on aquatic insects and microcrustaceans (Novakowski *et al.* 2004). Such ontogenetic diet shifts can be inferred for *Roeboexodon* as well.

Description: This characid is up to 15 cm long, and is striking due to its completely inferior (downward directed) mouth and a prominent pointed snout like that of a shark. The tubercles (bumps) on the upper and lower jaws do not resemble any other known characiform tooth (Géry 1977). A humeral spot and a spot at the base of the tail fin are present.

Distribution: The type locality (see Box 9) is the Marowijne River; in Suriname also known from the Suriname and Corantijn rivers. Present in French Guiana, and in Brazil in the Amazon, Tocantins, Xingu and Tapajós rivers.

Use: (No known use).

Similar species: A second Surinamese scale-eating fish with nipple-shaped external teeth on the snout is *Roeboides affinis*, which has a distinctive humpback body shape, much like *Charax* spp. (Figure 229b).

Serrasalmus rhombeus
Black Piranha, Redeye Piranha
Piren (local name used for all *Serrasalmus* spp.)
Actinopterygii, Characiformes, Serrasalmidae (pacus and piranhas)

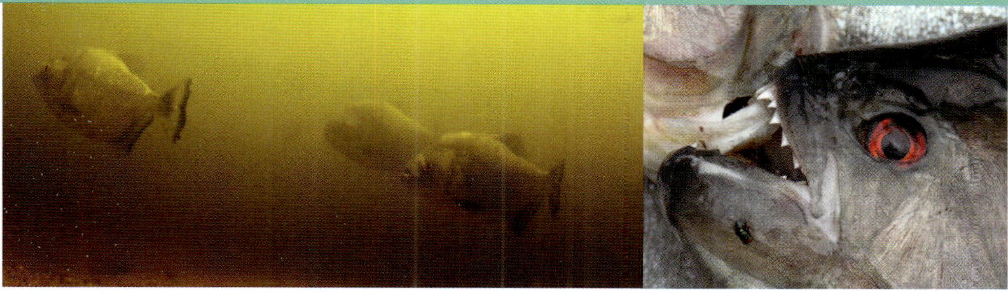

Figure 240. Black Piranhas *Serrasalmus rhombeus* in the upper Suriname River (image © Willem Kolvoort); detail of the head with razor-sharp teeth and red eye (image © Barry Chernoff).

Serrasalmus rhombeus (Figure 240) is one of the largest piranha species; in Suriname, it grows up to 45 cm long and 3 kg (Willink *et al.* 2011). In Brokopondo Reservoir, it is the dominant fish-eating species. However, in the reservoir Redeye Piranhas are 'stunted', they do not become nearly as large as they become when living in rivers (Mol *et al.* 2007a). In Venezuela, young Redeye Piranhas mostly eat aquatic insects and fish fins, adults eat entire fish, with the proportion of fish in the diet steadily increasing with length (Winemiller 1989b). Sazima and Machado (1990) presented underwater observations of the feeding behavior of three piranha species in Brazil, including fin eating in young *Serrasalmus* spp. Under certain conditions, Redeye Piranhas attack humans, such as at riverside recreation parks on the Suriname River (Mol 2006). Redeye Piranhas have a small home range (individuals do not roam widely; Cohen *et al.* 1999) and spawn in the high-water season (Vari 1982). In Guyana, Lowe-McConnell (2000) observed the eggs of *Serrasalmus rhombeus* attached to submerged tree roots, guarded by the parents.

Description: Serrasalmidae have a deep, laterally compressed body (strongly flattened sideways) with a series of saw-toothed modified scales called serrae along the underside of the body between the pectoral and anal fins and, in most genera, a front-facing spine just in front of the insertion of the dorsal fin (on the back). Piranhas possess one row of interlocking razor-sharp, pointed teeth in each jaw (Figure 240), *Serrasalmus rhombeus* has bright-red eyes and large, adult specimens are almost black in color. Juveniles are spotted with a black band along the margin of the anal fin and two vertical black bands on the caudal fin (Mol 2012b, p.113).

Distribution: The type locality (see Box 9) of this species is the Suriname River. Redeye Piranhas are widespread in tropical South America from Argentina to Venezuela. In Suriname, they are present in all rivers and large tributaries.

Use: An important food fish in the Interior of Suriname and elsewhere in tropical South America (Lowe-McConnell 1984). However, high mercury concentrations in muscle tissue (>0.5 g Hg g-1 ; Mol *et al.* 2001), related to gold mining, render piranhas from Brokopondo Reservoir unsuitable for consumption.

Similar species: The similar piranha *Pristobrycon striolatus* (30 cm) of the Marowijne River has a distinguishing black spot on the caudal peduncle (narrow part of body before tail fin) (Figure 229b). *Pygopristis denticulata* is a smaller, more colorful piranha (20 cm) found in black-water streams and swamps of the Coastal Plain. *Pristobrycon eigenmanni* (25 cm), with a distinguishing black humeral spot, lives in clear-water rivers of the Interior alongside *Serrasalmus rhombeus*; (Mol 2012b). *Pristobrycon* cf. *eigenmanni* from the Corantijn River has thin vertical lines on the body. Pacus (*pakusi*, *kumalu*, *pacu*) are serrasalmids that look like piranhas, but have two rows of molar-like teeth in the upper jaw, to grind fruits and seeds (Goulding 1980).

Hoplias aimara
Trahira, wolf fish (English names used for all Erythrinidae)
Anyumara
Actinopterygii, Characiformes, Erythrinidae

The *anyumara* (Figure 241a, Figure 229b) is a formidable ambush predator that feeds mainly on fishes, but also attacks other animals that fall into the water. Spawning is at the start of the rainy season. Young *Hoplias aimara* are not easily distinguished from other *Hoplias* spp. when they are young. They are found in a wide range of habitats and first feed mostly on tiny crustaceans and aquatic insects. Later (when >20 mm length) they switch to feeding on insect larvae and fish (Ponton & Mérigoux 2001). Fish lice (*Argulus*, *Dolops* and *Braga* spp.) can be abundant on scales and fins of *Anyumara*, but these are known to be cleaned by *Anostomus* headstanders (Lucanus 2009, p. 149). Adult *anyumara* have a small home range (Tito de Morais and Raffray 1999), meaning that individuals stay within their small stretch of water, not moving up or downstream much. So, when angling for *anyumara* – a line with baited hook is thrown with a big splash in the water – a strike does not follow, *anyumara* clearly are not present at that particular spot, so the angler will try again 10-20 m further along the stream. *Anyumara* can be dangerous to humans when entering the water at a spot where the fish is lying motionless in wait for prey (Figure 241b). People of the Interior are more afraid of the *anyumara* than of piranhas or Electric Eels. However, once you are in the water the *anyumara* does not recognize a human as potential prey; the senior author, standing waist-deep in a clear-water creek, once observed two large *anyumara* swimming slowly past him at touching distance.

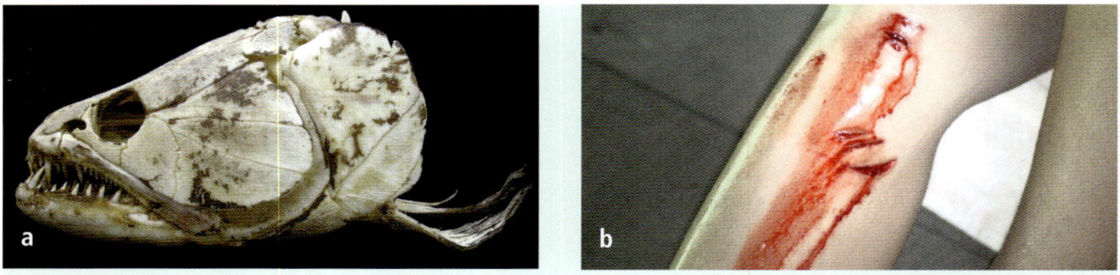

Figure 241. The *anyumara* or wolf fish *Hoplias aimara*:
a. skull from an *anyumara* of Coppename River (image © Phillip Willink), and
b. wound resulting from an attack by *Hoplias aimara* (image © Celine Artero).

Description: Like other erythrinids, this species has a cylindrical body, a rounded tail fin, a dorsal fin with 8-15 rays positioned in front of the anal fin and above the pelvic fins, and numerous sharp teeth (Figure 241a). *Hoplias aimara* is the largest erythrinid species (up to 100 cm long, 40 kg); it has large, dorsally positioned eyes and a dark spot on the hind edge of the gill cover (operculum) (Figure 229b).

Distribution: Type locality (see Box 9) for this species is Cayenne, French Guiana. Widespread in tropical South America; in the Interior of Suriname, it occurs in both large rivers and small forest creeks.

Use: *Anyumara* is an important food fish in the Interior of Suriname; it is also fished commercially and transported on ice to Paramaribo and from there even exported to the Netherlands. It is a popular target in recreational fisheries.

Similar species: Four smaller erythrinids occur in Suriname. *Hoplias malabaricus* (*pataka*, 40 cm) of coastal swamps58.274 has distinct chevron-shaped marks on the flanks. The smaller, blackish *Hoplias curupira* has a much shorter snout and large eyes. *Hoplerythrinus unitaeniatus* (*walapa*, 31 cm) has a dark lateral band, a plain caudal fin, and an ocellated (eyelike) spot on the operculum. The smallest is *Erythrinus erythrinus* (*matuli* or *stonwalapa*, 20 cm); it has a spotted caudal fin and a marbled color pattern on the flanks.

Nannostomus marginatus
Dwarf Pencilfish
Potloodvisje (name also used for other *Nannostomus* spp.)
Actinopterygii, Characiformes, Lebiasinidae (pencilfishes)

Pencilfishes change color: during the day, the color pattern consists of the distinctive longitudinal black stripes (Figure 242), but at night, the stripes disappear and two or three comma-shaped blotches materialize on the sides of the fish, extending from its back to its belly. Little is known about the biology of pencilfishes, but in Amazonian streams, they graze on algae on submerged trees and dead trunks and feed on small invertebrates (Carvalho *et al.* 2007).

Pencilfishes typically inhabit the shorelines of small creeks, streams, pools and flooded forests, and are rarely found in the open primary channels of large rivers. In these locations, they are typically found in shallow water, living among aquatic plants, leaf litter, and floating meadows. In Suriname, *Nannostomus marginatus* is found in both black-water and clear-water creeks, and particularly in swampy areas with aquatic vegetation or submerged woody debris and leaf litter. In the Para River, *Nannostomus marginatus* lives together with the related *Nannostomus beckfordi* and *Nannostomus harrisoni* (Mol 2012b). Benzaquem *et al.* (2015) found groups of closely related *Nannostomus* species that are very similar in appearance in Amazonia, among them a *Nannostomus marginatus* species complex.

Figure 242. Dwarf Pencilfish *Nannostomus marginatus* in a black-water tributary of Para River, Suriname river system (image © Willem Kolvoort).

Description: Pencilfishes (*Nannostomus* spp.) are small (<7 cm long) fishes with an elongate cylindrical body, a small mouth opening forward, and a reduced lateral line or no lateral line at all. An adipose fin may be present or not and the anal fin is short. The tiny *Nannostomus marginatus* (3.5 cm long; Figure 242) has three longitudinal black stripes, some red spots on the body and fins, and no adipose fin.

Distribution: Lower to middle Amazon River, Colombia east of the Andes, Guyana, Peru, Suriname and Venezuela; in Suriname, this species is known from the Suriname, Nickerie and Corantijn rivers.

Use: Pencifishes are popular aquarium fishes.

Similar species: Two other pencilfishes, *Nannostomus beckfordi* and *Nannostomus bifasciatus*, also lack an adipose fin, but *Nannostomus beckfordi* has only one longitudinal black band and *Nannostomus bifasciatus*, found in clear-water forest creeks in the Interior, has two longitudinal black bands. *Nannostomus harrisoni*, with an adipose fin and a long snout, was probably introduced to Suriname; it is only found in the Para River, near the international airport. The Splashing Tetra *Copella arnoldi* (Figure 229b), *Pyrrhulina filamentosa* and *Pyrrhulina stoli* have an upturned mouth, an enlarged upper lobe on the tail fin and a black spot on the dorsal fin, and they lack longitudinal black bands.

Cetopsis sp.
(no English name)
(no local name)
Actinopterygii, Siluriformes, Cetopsidae (whale catfishes), Cetopsinae

Whale catfishes of the subfamily Cetopsinae are rare in fish collections and several species are only known from the type (see Box 9). *Cetopsis* spp. are notorious for their feeding habits, feeding on carrion and attacking live fishes caught in gill nets. They are a nuisance to fishermen because with their streamlined bodies they penetrate wounds or torn flesh, becoming wedged in as they continuously chew away the flesh of the netted fish (Goulding 1980). Much like piranhas, schools of *Cetopsis* spp. can literally eat a large fish in less than a minute. Some *Cetopsis* spp. live in the deep channel of large rivers (Carvalho *et al.* 2007). *Cetopsis* spp. males have a long, thread-like extension or 'filament' in the dorsal and pectoral fins (Vari *et al.* 2005).

The only species of the genus *Cetopsis* known to occur in Suriname (Figure 243) was collected midstream in the water column of a small tributary of the Tempati Creek. Only two specimens were collected, on the same day; the largest one was 9.5 cm long. The stream in which this *Cetopsis* sp. was collected was a shallow forest creek (10 m width, 1 m depth) with medium current strength, sandy bottom and stands of the emergent macrophyte *Thurnia sphaerocephala*. The water was turbid due to upstream gold mining and had a conductivity of 24 µS/cm, pH 5.65, dissolved oxygen 7.30 mg/L and temperature 24.2°C. In captivity this *Cetopsis* sp. fed on live tetras (see *Hyphessobrycon simulatus* species account). Beyond this nothing is known about the species' biology.

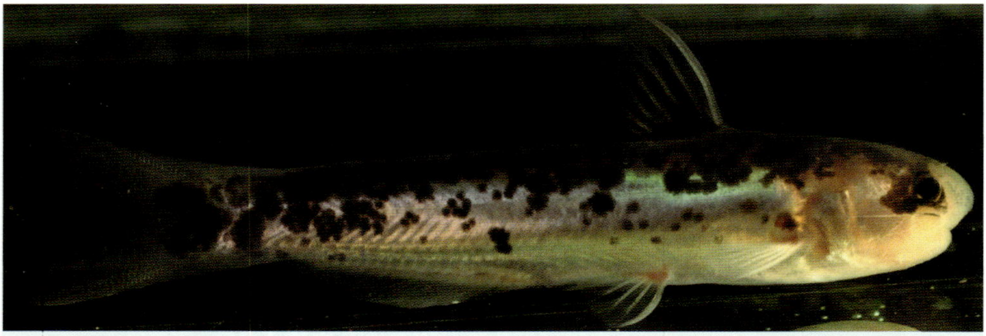

Figure 243. *Cetopsis* sp. from a tributary of the Tempati Creek, upper Commewijne River (image © Anil Gangadin).

Description: Whale catfishes are small to moderate sized with a long anal fin, no spines in the pectoral and dorsal fins and three pairs of barbels (wiry appendages on the head), one pair associated with the upper jaw and two with he lower jaw. *Cetopsis* spp. lack an adipose fin, have the eyes reduced or covered with skin, and the dorsal fin positioned far forward. The body is rounded in cross-section, the snout rounded and protruding forwards beyond the large, somewhat downward opening mouth. The *Cetopsis* sp. of Suriname resembles *Cetopsis montana* of Peru in pigmentation pattern, but it is probably an undescribed ('new') species.

Distribution: The 'new' species – if it is indeed a distinct species – is only known from a tributary of Tempati Creek, upper Commewijne river basin.

Use: (No known use).

Similar species: Two rare *Cetopsidium* spp. are known from Suriname (Mol 2012b); they are similar but smaller and with the first fin ray of both the dorsal and pectoral fin spinous. They were found hiding in woody debris and, at night, on submerged leaves in flooded forest. The most common cetopsid of Suriname, *Helogenes marmoratus* (8 cm long) differs in being laterally compressed and having an adipose fin; it occurs south of the Coastal Plain in black-water and clear-water creeks, hiding during the day in submerged leaf litter and root masses.

Bunocephalus aloikae
Camouflaged catfish, banjo catfish
Braadpan meerval (local name also used for other *Bunocephalus* spp.)
Actinopterygii, Siluriformes, Aspredinidae (banjo catfishes)

Bunocephalus spp. are bottom-dwelling catfishes of small forest creeks; they are sluggish unless disturbed and when motionless in a seine net resemble small pieces of dead wood. They are nocturnal and can bury themselves shallowly in loose substrates; they are often collected in submerged leaf litter, in which they are difficult to detect. In captivity, *Bunocephalus coracoideus* is reported to spawn 4,000-5,000 eggs in a hollow or depression (Burgess, 1989). A small group of young *Bunocephalus* was observed during the day, swimming midwater in flooded forest of the lower Kabalebo River.

Figure 244. The banjo catfish *Bunocephalus aloikae* from the Marowijne river system (image © Piotr Naskrecki).

Description: Banjo catfishes have a broad, depressed head and body followed by a slender tail (thought to resemble a banjo). They have a rough skin, small slit-like gill openings, and lack both a spine in the dorsal fin and an adipose fin on the back. The camouflaged *Bunocephalus* spp. have a short, tapering body and a short anal fin with 6-8 rays. *Bunocephalus aloikae* (Figure 244) has a very depressed body with backward positioned dorsal, anal and ventral fins and a long pectoral spine with 16 notches or 'serrae' along the inner margin.

Distribution: Type locality is the Litani River near the village of Aloikae in Suriname; this species is only known from the Marowijne river basin (Litani is a tributary thereof).

Use: *Bunocephalus* spp., especially *Bunocephalus coracoideus*, are used as aquarium fishes. They are not aggressive towards other fishes, but alsonot very active.

Similar species: Three other *Bunocephalus* species occur in Suriname: *Bunocephalus amaurus* with a shorter pectoral spine with more (18) serrae along its inner edge, *Bunocephalus coracoideus* with a longer process of the coracoid bone from the Coppename River, and *Bunocephalus verrucosus* with a 'warty' head in the Corantijn River (see Mol 2012b). Four larger and more slender banjo catfishes with a long anal fin are locally known as *trompetfisi* and live in the estuaries of Suriname (in brackish water): *Aspredo aspedo*, with plain coloration and two pairs of barbels (wiry appendages) on the upper jaw, *Platystacus cotylephorus* with mottled coloration and one pair of maxillary barbels, *Aspredinichthys filamentosus* with seven pairs of maxillary barbels and two hooked spines on the upper surface of the snout and *Aspredinichtthys tibicen* with 9-10 pairs of maxillary barbels and four hooked spines on the snout (Mol 2012b).

Stenolicmus ix
(no English name)
(no local name)
Actinopterygii, Siluriformes, Trichomycteridae (pencil or parasitic catfishes), Sarcoglanidinae

This miniature (2 cm), slender catfish was only recently described in Brazil (Wosiacki *et al.* 2011), and subsequently discovered in Suriname (Mol & Wan Tong You 2013). Most species of the subfamily Sarcoglanidinae were described based onthree or fewer known specimens and, prior to its collection in Suriname, *Stenolicmus ix* (Figure 245) was only known from a single specimen (the type; see Box 9). The low number of Sarcoglanidinae in collections can be explained by limited access to areas where the species occur, inadequate collecting methods, their translucent or pale coloration (although this is not the case with *Stenolicmus ix*), small body size and sand-dwelling behavior (Schaefer *et al.* 2005). Sand-dwelling behavior for *Stenolicmus ix* is suggested by the habitat where the Surinamese specimens were collected, a sandbank in a small creek. Vandelliine trichomycterids or *Candiru* (not present in Suriname) feed exclusively on blood of other, larger fishes (Zuanon & Sazima 2004) and have a habit of entering the human urethra, from which they are difficult to remove (Breault 1991, Spotte *et al.* 2001). However, this behavior is not known from other trichomycterid catfishes, including *Stenolicmus* spp.

Figure 245. *Stenolicmus ix* from a headwater tributary of the upper Palumeu River, Marowijne river system (image © Sandra Raredon).

Description: Trichomycterid catfishes have a patch of odontodes (teeth) on the interopercle (a bone of the gill cover), a pair of barbels (wiry appendages) at the angle of the mouth, and a backward positioned dorsal fin. They lack spines in the pectoral and dorsal fins and also lack an adipose fin. The tiny *Stenolicmus ix* has a small, depressed head, small eyes located dorsolaterally on the head, a somewhat downward opening mouth, and a snout with a flat lateral expansion along the area between the eye and the origin of the maxillary barbel.

Distribution: The type specimen (see Box 9) of *Stenolicmus ix* was collected in a small headwater creek of the Rio Curuá, Brazil (Wosiacki *et al.* 2011), in a habitat that resembled the habitat in which the 14 Surinamese specimens were collected (a sandy beach in a headwater tributary of the Upper Palumeu River, close to the watershed divide and border with Brazil; Mol & Wan Tong You 2013).

Use: (No known use).

Similar species: Other trichomycterids in Suriname include *Ituglanis amazonicus*, *Ochmacanthus* aff. *flabelliferus* (Figure 229a) and *Ochmacanthus reinhardtii*, and probably several undescribed species, two of which are known from mountain streams in the Nassau and Brownsberg mountains, respectively (at around 500 m elevation).

Hoplosternum littorale
Brown Hoplo
Hei-ede kwikwi, sokè kwikwi
Actinopterygii, Siluriformes, Callichthyidae (armored catfishes), Callichthyinae

Hoplosternum littorale (Figure 246a) lives in tropical swamps with low concentrations of dissolved oxygen (see Ecosystems chapter). They need to breathe air, and have to surface at regular intervals to inhale it. Atmospheric oxygen is taken up by the intestines (Carter & Beadle 1931). A second adaptation to the oxygen-poor swamp water is the floating bubble nest, up to 40 cm diameter and 10 cm high (Figure 246b), in which the eggs are positioned above the water surface in oxygen-rich foam produced by the guarding male (Mol 1993). In the rainy season, males construct bubble nests (Mol 1996) and develop an enlarged, ski-shaped spine in the pectoral fin (Winemiller 1987) which is used in defending the nest. Around noon, several females simultaneously spawn their eggs in the nest. Courtship and spawning (Gautier *et al.* 1988) involve 'sperm drinking' by the female. Male *hei-ede kwikwi* produce a mucous secretion that embeds sperm before release (Mazzoldi *et al.* 2007), keeping the sperm packed and protected during the passage through the hostile environment of the gut of the female. *Hei-ede kwikwi* nests contain on average 20,000 eggs (maximum 52,000) which hatch after approximately three days (Mol 1993). Two days after hatching the yolk sac is consumed and the larvae start feeding on rotifers and tiny crustaceans (Mol 1995). In three months the young *hei-ede kwikwi* grow to 10 cm (Ponton *et al.* 2001). Adults feed on detritus and associated invertebrates at the swamp bottom (Mol 1995).

Figure 246. *Hei-ede kwikwi* (images © Jan Mol): a. live *Hoplosternum littorale* in aquarium, and b. large floating bubble nest of *Hoplosternum littorale* in a herbaceous swamp at Weg naar Zee.

Description: Medium-sized (25 cm long) bottom-dweller with a forked tail fin, two pairs of long barbels (wiry appendages on the head, reaching the pectoral spine), and armor consisting of two rows of bony plates on either side of the body; uniform gray color when the fishes live in rice fields and black when they live in swamps.

Distribution: This species is widely distributed in tropical South America, including the Amazon, Orinoco and Paraguay rivers and the coastal rivers of the Guianas. In Suriname, it occurs in coastal swamps, but not in the Interior (Mol 1994).

Use: A popular, expensive food fish in Suriname that is protected by law in its 1 April – 15 July reproductive season (Rondeel 1965), but not in Brazil and Venezuela from where it is exported to Suriname and the Netherlands. A favorite of sport fishers in the Coastal Plain; commercially breeding attempts were unsuccessful (Luquet *et al.* 1989).

Similar species: Two other *kwikwi* species occur in both coastal swamps and black water streams in the Savanna Belt and clear water creeks in the Interior (Mol 1994). *Platkop kwikwi* (*Callichthys callichthys*) has proportionally smaller eyes, a more flattened head and body, and a more rounded tail fin (Figure 229a); it is an amphibious species that can 'walk' over land, and climbs out of the water into hollow tree trunks to escape shrinking pools in the dry season. *Kat(e)rina kwikwi* (*Megalechis thoracata*) differs in having spots on the body and a slightly concave caudal fin; its simple bubble nests consist of a few leaves floating on top of a thin layer of bubbles (Mol 1993).

Corydoras sipaliwini (and two similar *Corydoras* spp.)
Cory, cory cat (English name used for all *Corydoras* spp.)
Seseiguse (local name used for all *Corydoras* spp.)
Actinoptertygii, Siluriformes, Callichthyidae, Corydoradinae

Seseiguse (*Corydoras* spp.) (Figure 247) are bottom-dwellers that feed on small invertebrates; they live in small groups and are active during the day, which is exceptional for catfish. They deposit their adhesive eggs on submersed substrate, such as plants and rocks (Breder & Rosen 1966). Under low oxygen conditions in dry-season pools *seseiguse* can resort to air breathing (Kramer & McClure 1980). In the aquarium, a male *Corydoras aeneus* courts a female by presenting his abdomen to her and, before releasing eggs, the female attaches her mouth to the male's genital opening (the 'T-position') and directly drinks his sperm (Kohda *et al.* 1995). The sperm pass through the intestine of the female and are discharged together with the eggs into a 'pouch' formed by her pelvic fins (see also species account *Hoplosternum littorale*). The eggs are thus mixed with fresh, concentrated sperm in an enclosed space, ensuring effective insemination, especially in running water. A fossil *Corydoras* species is known from rock 58.5 million years old in Argentina (Cockerell 1925).

In Suriname, *Corydoras* species occur in clear-water streams in the Interior; they live in shallow water on sandy bottoms in both the river and *terra firme* forest creeks.

Figure 247. Two *Corydoras* spp. with a similar pigmentation from the Sipaliwini River, upper Corantijn river system (images © Raphael Covain):

a. *Corydoras* aff. *filamentosus* with a long snout, and

b. *Corydoras sipaliwini* with a short, rounded snout.

Description: *Corydoras* spp. are small-sized (Surinamese species up to 7 cm long) with short barbels (wiry appendages on head) and species-specific pigmentation patterns including 'masks', blotches, spots, and longitudinal and vertical bands or stripes (Nijssen 1970, Mol 2012b). *Corydoras* species with a similar pigmentation pattern can live together in the same river or forest creek, such as the short-snouted *Corydoras sipaliwini*, and the long-snouted *Corydoras filamentosus* and *Corydoras* aff. *filamentosus* in the upper Corantijn River (Figure 247).

Distribution: *Corydoras sipaliwini*, *Corydoras* aff. *filamentosus* and *Corydoras filamentosus* are all endemic to the upper Corantijn River (Suriname and Guyana).

Use: *Corydoras* spp. are among the most popular catfishes in the aquarium (see Junk 1984, Prang 2008) because they have beautiful color patterns and are active in daylight and not aggressive towards other fishes. Aquarium fish hobbyists from Japan travel all the way to Suriname to collect certain endemic *Corydoras* spp. in their natural environment.

Similar species: About 20 species of the genus *Corydoras* are known to occur in Suriname, some of them yet to be described (such as *Corydoras* aff. *filamentosus*); most Surinamese species are endemics, often restricted to a specific river system. They differ from each other in pigmentation pattern and body proportions such as size and length of pectoral spine and snout.

Parotocinclus halbothi
(no English name)
(no local name)
Actinopterygii, Siluriformes, Loricariidae (suckermouth armored catfishes), Hypoptopomatinae.

Parotocinclus halbothi (Figure 248) is another recently described catfish species (Lehman *et al.* 2014) that was subsequently collected in headwater streams in Suriname (see also species account *Stenolicmus ix*; in fact, these two species were collected in the same stream and on the same day!). Next to nothing is known about its biology. The clear headwater streams where *Parotocinclus halbothi* was collected flow through pristine rainforest where trees provide a 90 percent canopy cover over the stream. No aquatic vegetation was observed in the streams. The streams had oxygen-rich, nutrient-poor, clear water, and were shallow (<1 m water depth), 3-5 m wide, and with medium current velocity (0.44-0.76 m/s); the pH was 4.9-5.9 and the temperature 23.5-26.4°C (Mol & Wan Tong You 2013, Lehman *et al.* 2014). The bottom substrate was fine white gravel, coarse sand and leaf litter in the Brazilian streams, and coarse sand, large boulders and woody debris in the Surinamese streams.

Figure 248. *Parotocinclus halbothi* from a headwater tributary of the upper Palumeu River (image © Sandra Raredon).

Description: Suckermouth armored catfishes have a body armor consisting of 3-5 rows of bony plates on each side of the body and a ventral mouth with a lower lip that forms a sucker disc and has 'papillae' (nipple-shaped bumps). Species in the subfamily Hypoptopomatinae have a laterally compressed caudal peduncle (narrow part of body before tail fin) (see Figure 226), a tail fin with less than 14 branched rays, and the abdomen entirely covered with bony plates. *Parotocinclus halbothi* (2 cm) is one of the smallest loricariid catfishes known to date.

Distribution: *Parotocinclus halbothi* is presently known from only four localities, two small streams in the upper Trombetas river basin (Lehmann *et al.* 2014) and two forest creeks in the upper Palumeu River (Mol & Wan Tong You 2013).

Use: (No known use).

Similar species: *Parotocinclus halbothi* is differentiated from *Parotocinclus collinsae* of the Essequibo River, the most similar species in terms of coloration, by the small, circular, median abdominal plates, a poorly developed bony plate in front of the anus, and by having a rudimentary adipose fin. It differs from other *Parotocinclus* spp. from the Guianas, such as *Parotocinclus britskii* of the Coppename and Nickerie rivers, by the absence of a triangular patch of dark pigmentation on the anterior portion of the dorsal fin membrane. Its pigmentation with dark blotches resembles that of tiny *Microglanis* spp. catfishes, which lack body armor and a ventral disc-like mouth (see Mol 2012b).

Harttiella crassicauda
(no English name)
(no local name)
Actinopterygii, Siluriformes, Loricariidae (suckermouth armored catfishes), Loricariinae

Figure 249. *Harttiella crassicauda* in upper Paramaka Creek, Nassau Mountains (image © Trond Larsen).

Harttiella crassicauda (Figure 249) is a narrowly endemic species that is only found in the upper Paramaka Creek (250-500 m asl) in Nassau Mountain (Mol *et al.* 2007b, Covain *et al.* 2012). It feeds on filamentous algae, diatoms and fine detritus. In their intestines, all examined fish had horsehair worm cysts, larvae or both (Nematomorpha, Gordioidae; possibly *Gordius* sp.; A. Schmidt-Rhaesa personal communication). Fecundity is extremely low with 3-7 large, mature eggs per female. Potential predators of *Harttiella crassicauda* include Marbled Swamp Eel (*Synbranchus marmoratus*) and *platkop kwikwi* (*Callichthys callichthys*) (Figure 229a) (Mol *et al.* 2007b). In the aquarium, *Harttiella crassicauda* was mainly active during the night (Wan Tong You 2007). A male was observed to defend a territory against other small fishes.

The Paramaka Creek is a small, shallow mountain stream (3-5 m wide, <50 cm deep) with clear, nutrient-poor, well oxygenated water and a bottom substrate with boulders, gravel and sand (water temperature 22.6°C, and pH 7). Tufts of filamentous red algae (*Batrachospermum* spp.) and green algae (*Spirogyra* sp.) were found at sites where gaps in the closed canopy allowed sunlight to reach the water surface; the filamentous algae had epiphytic diatoms (*Eunotia* spp.) attached to their branches. The filamentous algae and diatoms were found in the stomach of *Harttiella crassicauda*. Stands of the emergent macrophyte *Thurnia sphaerocephala* were observed close to the edge of the plateau (Mol *et al.* 2007b).

Description: The subfamily Loricariinae has a distinct depressed caudal peduncle (narrow part of body before tail fin) (see Figure 226), lack of an adipose fin and, often, a flat snout. *Harttiella crassicauda* is a small-sized species (5 cm) with a weakly flattened caudal peduncle (narrow part of the body before tail fin), and bony plates that are covered with small teeth but lack keels projecting to the side. Due to its color, it is well camouflaged at the stream bottom (Figure 249)

Distribution: *Harttiella crassicauda* is an example of a small group of very narrowly distributed fishes, being restricted to a single mountain brook (also see *Guyanancistrus* n.sp. of Brownsberg Mountains, Figure 228).Endemic to Nassau Mountains (Covain *et al.* 2012). Illegal gold mining and a proposed bauxite mine in Nassau Mountains threaten the species with extinction.

Use: (No known use).

Similar species: An additional five narrowly endemic species of the genus *Harttiella* species were recently discovered in French Guiana (Covain *et al.* 2012), hinting at the occurrence of other, still undiscovered *Harttiella* spp. in Suriname. The *Harttiella* species in French Guiana differ from *Harttiella crassicauda* in depth or length of the caudal peduncle and coloration. Three *Harttia* spp. live in the rapids of large rivers in Suriname; they are much larger than *Harttiella* (15-20 cm) and have a strongly flattened tail (Mol 2012b).

Hemiancistrus medians
(no English name)
Goré-djab, gorenweh, gangu
Actinopterygii, Siluriformes, Loricariidae (suckermouth armored catfishes), Hypostominae

Hemiancistrus medians (Figure 250) lives in the main river channel and large tributaries, especially between boulders in and near rapids with strong currents. This species is mostly vegetarian. The sexes are markedkly different: the male has a more angular head and a more slender body than the female. The species occurs together with other fishes found in rapids such as *Leporinus granti*, *Jupiaba meunieri*, *Parodon guyanensis* (see species account), *Harttia surinamensis*, *Hypostomus gymnorhynchus*, *Guyanancistrus brevipinnis* and *Chasmocranus longior*.

Figure 250. *Hemiancistrus medians* of rapids in the Marowijne River (image © Mark Sabaj-Pérez).

Description: *Hemiancistrus medians* is a large (39 cm), robust and colorful suckermouth armored catfish in the subfamily Hypostominae with a laterally compressed caudal peduncle (narrow part of body before tail fin flattened sideways) and long extensible teeth (odontodes) in the area between gill cover bones. The body is yellowish brown with large black spots on the body and all fins. The bony plates are keeled and with many small spines(rough in touch).

Distribution: *Hemiancistrus medians* is endemic to the upper Marowijne River (Suriname and French Guiana).

Use: A food fish for local people along the Marowijne River, but obviously also a potential aquarium fish.

Similar species: The *warawara* (local name) *Hypostomus gymnorhynchus* also occurs in the upper Marowijne River, but it is more slender, smaller (17 cm SL), and lacks extensible odontodes in the interopercular area (and it is also smooth in touch). The rare *Pseudacanthicus serratus* is a large species (40 cm) with many large and strong spines and odontodes on the head and body (very rough, even painful, in touch), but this species is uniformly black. *Pseudancistrus* species have a very flat head and males have a 'beard' of long prominent odontodes along the rim of the snout (Figure 229a).

Brachyplatystoma filamentosum
Lau-lau, goliath catfish (English name used for all *Brachyplatosoma* spp.)
Lalaw
Actinopterygii, Siluriformes, Pimelodidae (long-whiskered catfishes)

Figure 251. The goliath catfish *Brachyplatystoma filamentosum* (*lalaw*) at Wonotobo Rapid, Corantijn River (image © Jan Kiel).

Goliath catfishes (*Brachyplatystoma* spp.) are freshwater species that avoid brackish water; in estuaries they move upstream during the dry season when salt water invades the estuary (Barthem & Goulding 1997). *Brachyplatystoma filamentosum* (Figure 251) is a large fish-eating catfish that lives on the bottom of the deep channel of large rivers (Barthem & Goulding 1997). In the Amazon River, the closely related *Brachyplatystoma rousseauxii* undertakes 8000 km long upriver spawning migrations from the estuary to headwaters (Barthem & Goulding. 1997, Duponchelle *et al.* 2016). In Suriname this same species may also migrate upriver as shown by large specimens caught in the Corantijn River at Wonotobo Rapids and in the Coppename River at the mouth of Wayombo River (Jan Kiel, personal communication). Such long distance migration probably also occurs in *Brachyplatystoma filamentosum* (Garcia-Davila *et al.* 2015). In Suriname, this species is collected mainly in lower freshwater reaches of large rivers (below the furthest downstream rapids), but in the Suriname and Corantijn rivers large specimens were also collected far upriver, at Brokopondo (km 185, Mees 1974) and Wonotobo Rapids (km 350; Figure 251), respectively . In the Amazon, goliath catfishes are known to negotiate rapids during the low-water season (Barthem & Goulding 1997) and in the Corantijn River *Brachyplatystoma filamentosum* was caught at the Wonotobo Rapids, upstream of the Cow Falls.

Description: *Brachyplatystoma filamentosum* is the largest freshwater fish of Suriname (length 300 cm, >200 kg). It has small eyes, large patches of small teeth on the vomer and palentine bones that are located inside the mouth, a uniform brown to gray color with a whitish belly and very long barbels on the upper jaw. The adipose fin is short and the upper lobe of the caudal fin has a long thread-like extension.

Distribution: *Lalaw* are widespread in the Amazon and Orinoco river basins and in the large rivers of the Guianas.

Use: Popular with sport fishers (Figure 251) and also caught by commercial fishermen in the lower freshwater reaches of large rivers.

Similar species: Two additional *Brachyplatystoma* species occur in Surinamese rivers: the silvery *Brachyplatystoma rousseauxii* (190 cm) with short barbels (wiry appendages on head not reaching the dorsal fin), and the relatively small (150 cm) *Brachyplatystoma vaillantii* with a long adipose fin (the base of the adipose fin is two times longer than the base of the anal fin). Other large pimelodid catfishes in Suriname include *Pseudoplatystoma* spp. (*spigrikati;* Figure 229a) and the colorful Redtail Catfish (*Phractocephalus hemioliopterus*; *mototyar* or *swit-watra yaribaka*) of the Corantijn River; these catfishes are much more colorful than the dull gray or silvery *Brachyplatystoma* spp.

Acanthodoras cataphractus
Spiny Catfish
Merkikwikwi
Actinopterygii, Siluriformes, Doradidae (thorny catfishes)

The local name *merkikwikwi* (*merki* literally means milk) of *Acanthodoras cataphractus* (Figure 252) refers to the emission of a milk-white fluid from glands at the base of the pectoral fin when disturbed. This substance is apparently toxic and, in aquarium fishkeeping, *Acanthodoras* are transported in separate containers so that they do not kill other fishes by 'fouling' the water (Mark Sabaj, personal communication). *Merkikwikwi* are omnivorous, feeding mainly on detritus and benthic invertebrates. They search for food by digging in the sediment. They are active at night and lie hidden in underwater roots and debris during the day. Doradids or thorny catfishes are also called 'talking catfishes' because, when caught, most doradids produce considerable noise either by movements of their pectoral fins or by vibrating the swimbladder (see for example Boyle *et al.* 2015).

Figure 252. *Merkikwikwi* (*Acanthodoras cataphractus*) from the Para River, Suriname river system (image © Mark Sabaj-Pérez).

Description: Doradid catfishes are distinguished by the ossified tubules of the lateral-line (set of perforated scales that allow water in the lateral line system), variously expanded to form a row of bony plates (scutes) each with a backward-directed spine or thorn in the middle. *Acanthodoras cataphractus* is a small species (13 cm long) with three pairs of barbels, a large flattened head, small eyes, a sawblade-shaped dorsal spine, a rounded tail fin, and a dark-brown color with a light longitudinal band along the lateral line.

Distribution: *Acanthodoras cataphractus* has a wide distribution in the Amazon river basin and coastal drainages of the Guianas: Bolivia, Brazil, Colombia, French Guiana, Guyana, Peru and Suriname. In Suriname, it is known from Para River, lower Tempati Creek, Wane Creek and tributaries of the lower Corantijn, Coppename and Marowijne rivers. It is best known from black-water streams.

Use: A well-known aquarium fish.

Similar species: *Amblydoras affinis* from the lower Nickerie and Corantijn rivers is about the same size as *Acanthodoras cataphractus*, but its grooved spine on the back lacks saw-like teeth and its tail fin has a truncated shape. Medium-sized to large doradids include two *Doras* spp. (*agunosu*, 30-35 cm) with fringed barbels (wiry appendages) on the upper jaw and a laterally compressed head (flattened sideways) (Figure 229a); two *Platydoras* species (*soké*, 25-30 cm) with a forked tail fin; and the large *Pterodoras* aff. *granulosus* (up to 70 cm) of the Nickerie and Corantijn rivers with very small eyes, a forked tail fin, and a lack of bony plates in the caudal peduncle (narrow part of body before tail fin) (Mol 2012b).

Ageneiosus inermis
Guyana Slopehead Catfish
Plar'plari, paypay
Actinopterygii, Siluriformes, Auchenipteridae (driftwood catfishes)

Ageneiosus inermis or *plar-plari* (Figure 253a) is an elongated catfish which prefers the middle of the water column in rivers where the current is not too strong. With its large mouth it feeds on fish and crustaceans. Very young individuals have a distinct pigmentation pattern (Figure 253b). *Plar'plari* show pronounced differences between the sexes in the breeding season, when the barbels (wiry appendages) on the lower jaw and the spine on the back of the male change, and a copulatory organ appears at the anal fin. Nuptial males can be distinguished from females by their elongated dorsal spine with hooks and ossified, fringed, stiffened barbels on the upper jaw (Figure 253c). The dorsal fin spine and stiffened barbel appear to acts as clasper organs, which hold the female near to the male during spawning; they disappear at the end of the reproductive season. Auchenipterids are unique among catfishes in that insemination is internal (inside the body) (Mazzoldi *et al.* 2007) and that the female may carry mature, unfertilized eggs and packets of sperm inside her reproductive organs for an extended period of time before triggering fertilization and deposition of the eggs.

Figure 253. ***Plar'plari Ageneiosus inermis*:**

a. adult, Marowijne River (images © Mark Sabaj-Pérez),

b. early life stage (live) from Marowijne River (images © Mark Sabaj-Pérez), and

c. detail of ossified, fringed barbel of nuptial male (image © Barry Chernoff).

Description: *Ageneiosus* spp. have a flattened head, a large mouth with short barbels, and the eyes positioned at the side of the head, at the same level as the mouth, ensuring vision above and below. *Ageneiosus inermis* is a relatively large species (47 cm long) with a tail fin that is obliquely truncate (as if cut off at the end at an angle) or weakly emarginated (slightly concave at the cut-off); its body is colored darker on the upper side and lighter on the underside ('countershaded'), with the upper half yellowish brown to gray and sometimes a longitudinal black band on the flanks.

Distribution: The type locality (see Box 9) of this species is Suriname, but it is widespread in South America from Argentina to Venezuela. In Suriname, it occurs in the main channel of rivers and large tributaries in the Interior, large black-water streams in the Zanderij Belt and the lower freshwater reaches of rivers in the Coastal Plain.

Use: It is a popular and tasty food fish. Nuptial males are difficult to release from nets because of their strong spine on the back bordered with hook-like teeth.

Similar species: *Ageneiosus marmoratus* with heavily marbled coloration and *Ageneiosus ucayalensis* with a forked tail fin are slightly smaller species (30 cm long) reported from the Corantijn and Nickerie rivers (Walsh 1990). *Trachelyopterus galeatus* (*noya*) is a shorter (22 cm), robust, round-bodied fish that is widespread in South America and, in Suriname; it is found in freshwater streams and swamps (Mol 2012b). Other auchenipterid catfishes in Suriname are also shorter, and include *Pseudauchenipterus nodosus* (*botromanki*, 22 cm long), two slender *Auchenipterus* species (12-15 cm long) and several small-sized species such as *Glanidium leopardum*, *Centromochlus* spp. and *Tatia* spp. (Mol 2012b).

Electrophorus electricus
Electric eel (English name used for all *Electrophorus* spp.)
praké, maisi, nakifisi, stroomfisi (local names use for all *Electrophorus* spp.)
Actinoptertygii, Gymnotiformes, Gymnotidae (naked-back knifefishes)

Electrophorus electricus or *stroomfisi* (Figures 254, 229a) is an obligate air breather that has to surface regularly to inhale air. In the dry season, surfacing *stroomfisi* can be observed in rapid complexes, in pools that are temporarily separated from the main river channel. During the day they hide between rocks, but at night they can be observed cruising, possibly hunting, at the water surface. It is a formidable predator of small fishes, including other gymnotiforms (Westby 1988). *Stroomfisi* first locates prey fish by weak electric signals, then stuns the prey with powerful discharges (up to 600 V / 1 A in 2 m eels), and finally swallows them down before they recover (Catania 2014). The voltage generated by the electric eel is related to the length of the specimen (Coates & Cox 1945) and large eels can be dangerous to humans, knocking a person unconscious for several minutes (Juan Montoya-Burgos, personal communication), during which he or she can drown. Other fishes, probably piranhas, attack *stroomfisi* as obvious from mutilated tails in museum specimens (Tyson Roberts, personal communication). Bubble nests of an undescribed *Electrophorus* sp. (not *Electrophorus electricus*; David de Santana, personal communication) are described from the swamps of Marajo Island in the mouth of the Amazon River (Assunção & Schwassman 1995), but such nests are not known from *Electrophorus electricus* in Suriname.

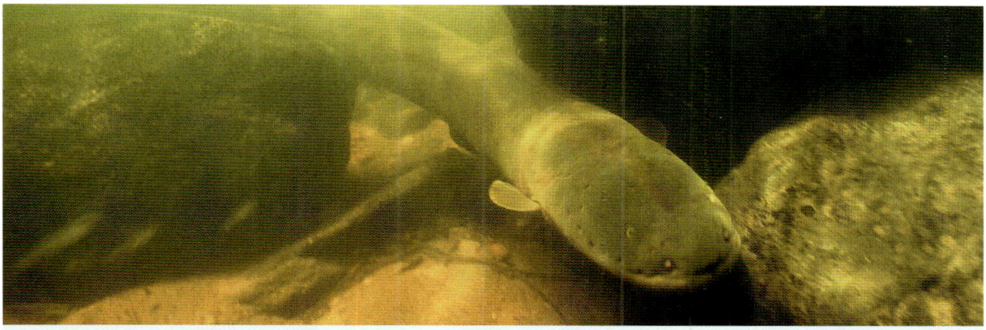

Figure 254. The Electric Eel *Electrophorus electricus* in Raleighvallen Rapid, Coppename River (image © Willem Kolvoort).

Description: A large, up to 250 cm long, eel-like fish with the upper and lower jaws equal in length; the anal fin continues up to the tail end of the body, which is rounded. Houttuyn (1764) provided an early description of *stroomfisi* under the name 'beef-aal'.

Distribution: The type specimen (see Box 9) of *Electrophorus electricus* is lost, but the type locality is Suriname (probably the Suriname River). In Suriname, it occurs in both large rivers (mostly in rapids) and small forest creeks of the Interior. A new species of the genus *Electrophorus* was recently discovered in the Palumeu River (Marowijne river system); the description of this species is in preparation by David de Santana.

Use: The electric eel is eaten by tribal people in the Interior of Suriname and collected for public aquariums (typically to demonstrate its electric power generation capacity by, for instance, lighting up a Christmas tree).

Similar species: The only other large, eel-like fish in the Interior of Suriname is the Marbled Swamp Eel (*Synbranchus marmoratus*; up to 100 cm long; Figure 229a) which can be distinguished by its single opening to the gill chamber on the underside of the body (vs. paired gill openings at the side of the body in electric eels) and incapacity to generate electricity. The related knifefish species *Gymnotus carapo* (*logologo*, 38 cm) and *Gymnotus coropinae* (8-16 cm) are much smaller and only weakly electric. *Gymnotus carapo* is known to guard its eggs and young (Crampton & Hopkins 2005).

Gymnorhamphichthys rondoni
(no English name)
Saprapi (local name used for several small knifefishes of the Interior)
Actinopterygii, Gymnotiformes, Rhamphichthyidae (sand knifefishes)

During the day, *Gymnorhamphichthys rondoni* (Figure 255) hides in the sandy bottom of small forest streams, and during the night it actively searches for prey in the sand, mainly aquatic insect larvae and small crustaceans, by probing with its long snout (Zuanon *et al.* 2006). All knifefishes, including sand knifefishes, have a species-specific electric organ discharge (EOD) signal that can be transformed to audible sound with a so-called electric fish finder apparatus (Hopkins & Heiligenberg 1978). This electric fish finder can locate and identify knifefish species by their EOD 'sound': in this way *Gymnorhamphichthys rondoni* that is hiding in the sand during the day can be found and then easily scooped out of its hiding place by sifting the sand with a dip net.

Figure 255. The sand-burrowing knifefish *Gymnorhamphichthys rondoni* (live), detail of the head, upper Nickerie River (image © Ross Smith).

Description: Rhamphichthyidae are knifefishes that are strongly laterally compressed, (like a knife, hence the name); they have a long, tubular snout and a small mouth, and they lack a tail fin. *Gymnorhamphichthys rondoni* is a small (22 cm long), sand-burrowing species with a translucent body and a long tail; the anus is located, oddly enough, in the front part of the body, only slightly behind the level of the eye.

Distribution: *Gymnorhamphichthys rondoni* has a wide Neotropical distribution in the Amazon, Parana and Orinoco river basins and coastal rivers of the Guianas. In Suriname, it is known from small forest creeks with a sandy bottom in the Corantijn, Nickerie, Coppename and Suriname river basins.

Use: (No known use)

Similar species: The related species *Rhamphichthys rostratus* of large and medium-sized rivers is much larger (90 cm) and not known to burrow in sand; its body is not translucent, but with a brownish marbled pattern and covered by small scales. It probably occurs in all rivers of Suriname (Mol 2012b). The apteronotid knifefish *Sternarchorhynchus galibi* (Figure 229a) also has a tubular snout (with a downward curve), but this species has a small tail fin (lacking in all sand knifefishes).

Apteronotus albifrons
Black Ghost Knifefish
(no local name)
Actinopterygii, Gymnotiformes, Apteronotidae (ghost knifefishes)

Black Ghost Knifefish (Figure 256) are nocturnal hunters that feed on insect larvae and small crustaceans (Winemiller & Adite 1997). They maintain a relatively rigid body and swim forwards and backwards with equal ease. They search for prey with their back forming the leading edge (head angled downwards when swimming forward), then scan the detected prey (see below) by swimming backwards, and finally capture it with a short forward lunge (Lannoo & Lannoo 1993, Nelson & MacIver 1999). The black ghost knifefish have an electric organ, and the electric field produced by it is used in electrolocation, which is to scan the environment within half a body length – both at night and in turbid water – by analyzing distortions in the electric field caused by nearby objects. Electric organ discharge (EOD) is also used for communication: males 'sing' electric courtship songs to females and engage in contests of electric one-upmanship with rivals. The EOD signals often differ between sexes, with males producing stronger EODs and females displaying higher discharge frequencies.

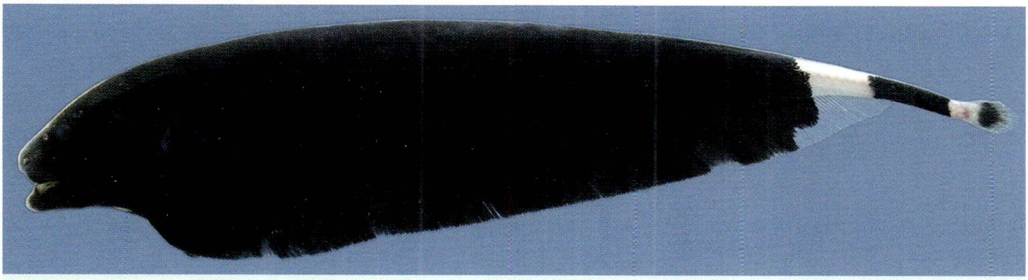

Figure 256. The Black Ghost Knifefish *Apteronotus albifrons* (image © Mark Sabaj-Pérez).

Description: Apteronid knifefishes are the only gymnotiform knifefishes with a caudal fin, albeit small. *Apteronotus albifrons* (35 cm long) has a black body with two transversal light bands on the hind part; the pectoral fins are black. A narrow white line extends from the snout along the back to the (imaginary) vertical line through the front part of the anal fin.

Distribution: The Black Ghost Knifefish occurs in northern and central South America, from Argentina to Venezuela. The type locality (see Box 9) is Suriname, where the species has been recorded from all river systems with the exception of the Coppename. It lives in forest creeks and rapids with moderate current and sandy bottoms.

Use: A beloved aquarium fish that is not aggressive towards other fishes.

Similar species: Two other apteronid knifefishes occur in Suriname; both species are only known from the Marowijne River: *Sternarchorhynchus galibi* has a downward curved tubular snout (Figure 229a), whereas *Porotergus gymnotus* has a brown-gray body and a single narrow white transversal band on the caudal peduncle (narrow part of body before tail fin) (Mol 2012b).

Laimosemion agilae (previously named *Rivulus agilae*)
(no English name)
(no local name)
Actinopterygii, Cyprinodontiformes, Rivulidae (South American annual fishes)

Figure 257. Male *Laimosemion agilae* from a small forest creek near Blanche Marie Falls, upper Nickerie River (image © Frans Vermeulen).

Held in an aquarium, reproductivey active female *Laimosemion agilae* produce some 5-10 eggs daily that hatch after 2 weeks (Vermeulen 2015; young are described by Ponton & Mérigoux 2001). Several rivulines or killifishes of the New World have eggs with long incubation times and embryos entering into diapause, a period of suspended development when environmental conditions are adverse (Vermeulen 2015). They live in habitats that seasonally fall dry; the parents die after eggs are deposited in the moist bottom substrate. Eggs can survive dry-season conditions for months, hatching in the next rainy season. Diapause of developing embryos is not known in *Laimosemion agilae*, but in Suriname it is known in the Mangrove Rivulus (*Kryptolebias marmoratus*) (Taylor 2012). Rivulids often live in ephemeral (temporary) habitats such as shallow seasonal tributaries or small seasonal pools near streams (see Ecosystems chapter, Figure 45 a,b) (Pazin *et al.* 2006). They can move over land by jumping and wriggling, to escape from drying pools. In Suriname, *Laimosemion agilae* has been collected in a pitfall trap in the forest at a distance of a few hundred meters from the nearest water body. Young *Erythrinus erythrinus* (*stonwalapa*) live in the same shallow-water habitat as *Laimosemion agilae*; they mimic female *Laimosemion agilae* in order to attract the males on which they then prey, a case of aggressive mimicry (Brosset 1997). The Mangrove Rivulus leaves the water regularly and is known to aggregate inside decaying mangrove logs that have been 'galleried' by terrestrial insects (Taylor *et al.* 2008).

Description: *Laimosemion agilae* is small (5 cm long), with <35 scales in the lateral row (longitudinal row between head and tail, tail fin, mid-way on each side of the body). Males (Figure 257) with orange and blue chevron marks on the hind part of the flanks, and tail fin with a lower black margin and yellowish submargin; females with round eye-like spot (black with light margin) at the upper side of the caudal peduncle (narrow part of body before tail fin) that faints with age and 4-5 irregular black vertical stripes on the tail fin.

Distribution: The type locality (see Box 9) of *Laimosemion agilae* is Agila Creek, Para river system, Suriname. The species occurs in the coastal plain of the Guianas, from western Guyana to Cayenne in French Guiana, mainly in small, shaded, shallow black-water creeks with slow water current and associated pools, but also in sunny swamps.

Use: Well-known and popular aquarium fish; peaceful, but a strong jumper.

Similar species: The related and similar *Laimosemion geayi* occurs in the Interior on laterite soils, whereas *Laimosemion agilae* is better known from sandy soils in the Savanna Belt. The Mangrove Rivulus occurs in brackish-water mangrove forests; this hermaphroditic species is the only fish capable of fertilizing its own eggs (producing clones of genetically identical offspring). The rivulids *Kryptolebias sepia* from the Tapanahony River, *Anablepsoides fransvermeuleni* of the Upper Nickerie River (Figure 229a), and *Rivulus amphoreus* of Tafelberg Mountain are endemic to Suriname.

Anableps anableps

Largescale Four-eye

Kutai (local name for the two *Anableps* spp. of Suriname)

Actinopterygii, Cyprinodontiformes, Anablepidae (four-eyed fishes and one-sided livebearers)

Figure 258. A school of *kutai Anableps anableps* in the lower Suriname River (image © Willem Kolvoort).

Anableps spp. or *kutai* swim with the center of the eye at the water's surface (Figure 258) and are thus capable of simultaneously seeing below and above the water surface. Schools of 10-50 *kutai* are easily spotted on tidally exposed mud flats, always swimming at the surface. They are difficult to catch as the fish can see a person approaching from a distance of 10 m or more (they are also most adept at jumping over the float line of a seine net). In Suriname, *Anableps anableps* feeds on diatoms, river-bottom invertebrates, and terrestrial insects which they occasionally capture in the air (Zahl *et al.* 1977). In the estuary of the Amazon River, *Anableps anableps* showed intertidal migration: entering the tidal creeks with the first flood rise and feeding in the inundated mangroves at high water, and then returning gradually on the ebb to ultimately converge on the subtidal parts of the main river channel at low water (Brenner & Krumme 2007). Reproduction takes place in the dry season. Courtship and mating in the aquarium were described by Mattig and Greven (1994). Four-eyes exhibit internal fertilization and are viviparous (give birth to live young) with nutrients transferred through a follicular 'placenta' during a prolonged gestation; the emproyo's weight increases 3000-fold between fertilization and birth (Knight *et al.* 1985). Female *Anableps anableps* give birth to about 12 live young of 4.5-5.2 cm length (Knight *et al.* 1985).

Description: *Anableps* spp. are called four-eyed fishes because they have eyes that are prominently raised above the top of the head; each eye is divided lengthwise forming two pupils, an upper and a lower one in each eye. *Anableps anableps* (30 cm long) has an elongated body that is cylindrical in cross-section with five conspicuous longitudinal stripes on the flanks, 50-55 scales along the lateral line (perforated row of scales on the side of the body), and a rounded tail fin.

Distribution: Brazil, French Guiana, Guyana, Suriname, Trinidad and Venezuela. In Suriname it occurs in the estuaries and lower freshwater reaches of all rivers.

Use: Four-eyes are well known in the aquarium trade (Mattig & Greven 1994).

Similar species: One related species occurs in Suriname: *Anableps microlepis* with smaller scales, 80-90 along the lateral line, and an oblique tail fin (shaped as if cut off at the end, as opposed to rounded or convex condition in *Anableps anableps*).

Pseudophallus aff. *brasiliensis*
Freshwater pipefish (English name used for all freshwater Syngnatidae)
(no local name)
Actnopterygii, Syngnathiformes, Syngnathidae (pipefishes and seahorses)

Figure 259. The freshwater pipefish *Pseudophallus* aff. *brasiliensis* in Corantijn River:
a. live male with ripe brood pouch (image © Jan Mol), and
b. female (image © Phillip Willink).

Most *Pseudophallus* aff. *brasiliensis* specimens, eight mature males (with eggs in their brood pouch; Figure 259a) and one female (Figure 259b), were collected at a site 500 m upstream of the village of Apoera (Corantijn River, 155 km upstream from ocean; Mol 2012a). This sex ratio and the difference in coloration between males and females suggests that *Pseudophallus* aff. *brasiliensis* live in a sex-role reversed harems, with territorial females competing for males that brood their eggs. In a bucket, a pregnant male gave birth to about 20 tiny offspring (7-9 mm; Mol 2012a). Pregnant pipefish males provide nutrients to the developing embryos, but under some circumstances they are also known to absorb nutrients from them (Sagebakken *et al.* 2010). *Pseudophallus* aff. *brasiliensis* (adults) accepted live *Artemia* nauplii, but the newborn did not (they probably feed on smaller food items). Most seahorses and pipefishes are near-shore marine inhabitants. However, the mature males of *Pseudophallus* aff. *brasiliensis* with eggs in their brood pouch collected at Apoera indicate that this species reproduces in fresh water (Mol *et al.* 2012a) and thus can be considered a freshwater representative of a mainly marine family, much like the freshwater stingrays.

Description: All syngnathids are encased in bony plates that form rings around the body. *Pseudophallus* aff. *brasiliensis* is a small freshwater pipefish (7.5 cm) that lacks an anal fin. The longitudinal bony ridges of the upper body and the tail are discontinuous, but the ridges on the lower body and the lateral ridges are continuous with the tail ridges. Mature males have a brood pouch below tail rings 11-22 and a mottled brown color; females have (faint) vertical stripes and a long anal papilla (a small, fleshy area near the anus from which the eggs are released).

Distribution: Collected near Apoera (km 155) in the Corantijn River in a stand of the dwarf Amazonian sword plant *Helanthium tenellum* (Mol 2012a); also recently collected in the lower Suriname River (at Carolina).They seem to live in the main channel of the river, in shallow, clear water with a slow current.

Use: (No known use).

Similar species: Recently a larger, long-snouted freshwater pipefish, the Short-tailed or Opposum Pipefish (*Microphis brachyurus*), was collected in the lower Comté River, French Guiana (Graham Short, personal communication). This very rare species is also known from Guyana and thus could also be present in Suriname. It differs from *Pseudophallus* in a longer snout and shorter tail and in a distinctive color, especially in breeding adults: the upper snout and back half of the head and body is sienna brown with a series of dark red blotches forming a red stripe between the lateral and upper trunk ridges; there is a silver stripe on the mid-side between the lateral and lower trunk ridges; the lower half of the snout is bright red with a variable number of black vertical bars, and the caudal fin is also red with a central dark stripe.

Cichla ocellaris
Butterfly Peacock Bass
Tukunari (local name used for the two *Cichla* spp. of Suriname)
Actinopterygii, Cichliformes, Cichlidae (cichlids)

Cichla ocellaris (Figure 260) feeds on shrimps and fish, including other cichlids; young *Cichla* of 4 cm feed on tiny planktonic crustacea and algal filaments (Lowe-McConnell 1969). In the Rupununi district in Guyana, *Cichla ocellaris* spawn before and during the rains which start in May (Lowe-McConnell 1969). Males develop a characteristic fatty hump on the nape just before spawning. Male and female clear a hard surface for spawning and dig several 8-20 cm diameter and 1.5-6 cm deep depressions in the bottom substrate ('nest pits') with their chin and pectoral fins. The eggs remain attached to the hard substrate until they hatch after about 3-4 days. Both male and female guard the spawn (Figures 260, 229a); little feeding takes place during spawning and incubation of the eggs. The male generally attacks intruders. The newborn (5-6 mm long) are transferred in the parent's mouth to a nest pit. The young remain at the bottom of the pit for about four days before swimming up and around. During this time they are guarded by both parents who may move them from one pit to another if they are disturbed. Parent *Cichla* spp. are not good fighters, leaving the nest if danger occurs, but returning quickly when the danger is over. Five to six days after spawning the yolk sac is resorbed and the young start to feed. The shoal of young then swims out guarded by both parents. Lowe-McConnell (1969) suggests that the caudal eye-like spot (ocellus), which becomes brilliant in breeding fish, has an important function in enabling the young to orientate themselves to the parent. Pairs of *Cichla ocellaris* with a cloud of young above them can be observed in the Interior when the water level in the river is low.

Figure 260. A pair of *tukunari* (*Cichla ocellaris*) in Brokopondo Reservoir(a); parent guarding eggs (b) (images © Willem Kolvoort).

Description: Cichlids have a single nostril on each side of the head, 7-24 spines in the dorsal fin and 3-5 spines in the anal fin. The lateral line is divided into a front portion ending below the dorsal fin (on the back) and a hind lower portion running on the caudal peduncle (narrow part of body before tail fin). *Cichla* spp. are among the largest Neotropical cichlids. *Cichla ocellaris* (50 cm long) has three dark vertical bars on a brownish-green ground color and a large eye-like spot or ocellus on the upper half of the caudal peduncle (Figure 229a). The anterior and posterior portions of the lateral line (row of perforated scales at the side of the body) are not divided, as in most cichlids, but continuous.

Distribution: Widely distributed in tropical waters of South America. Common in all rivers of Suriname and in the Brokopondo Reservoir (Mol *et al.* 2007a; Figure 260).

Use: One of the best-liked food fishes in the Interior and an important target of sport fishermen.

Similar species: In the related *Cichla monoculus* the front and back portions of the lateral line are divided (versus continuous in *Cichla ocellaris*), and the space between the front-most vertical bars under the dorsal fin is smaller than the width of the bars (versus larger in *Cichla ocellaris*).

Crenicichla saxatilis
Pike cichlid (English name used for all *Crenicichla* spp.)
Datrafisi (local name used for all *Crenicichla* spp.)
Actinopterygii, Cichliformes, Cichlidae (cichlids)

Crenicichla saxatilis (Figure 261) feeds on small fish, shrimps and aquatic insects. Breeding males are larger than females and may have iridescent scales scattered all over the body and around the humeral spot. In aquaria the eggs are deposited in a shallow pit and guarded by both parents; parents are reputed to move eggs and young up and down with the water level (Lowe-McConnell 1969). Parental care by both parents is common in cichlids, but the duration of parental care is exceptionally long among pike cichlids, which may take care of the young up to half of adult size. Young juveniles have a black band on either side and superficially resemble *Nannostomus* pencilfishes (Ponton & Mérigoux 2001).

Figure 261. ***Datrafisi* (*Crenicichla saxatilis*) in the upper Suriname River (image © Willem Kolvoort).**

Description: Pike cichlids have an elongated body with the depth more than 3.7 times in standard length and the adpressed pelvic fins not reaching the anus. *Crenicichla saxatilis* (20 cm long) has relatively few (52-72) scales in the horizontal row below the lateral line. The humeral spot has a notch pointing toward the space between the eye and the dorsal fin, and is sometimes outlined with silver (ocelled) in breeding fish; the spot on the upper part of the tail fin base is poorly visible in females.

Distribution: The type locality (see Box 9) *of Crenicichla saxatilis* is Carolina Creek, Para River, Suriname River system. The species occurs in coastal rivers of northern South America: French Guiana, Guyana, Suriname, Trinidad and Venezuela. In Suriname, it occurs in the Nickerie, Saramacca, Suriname, Commewijne and Marowijne rivers, in coastal freshwater swamps, black-water streams, and clear-water forest creeks of the Interior. In the Brokopondo Reservoir, *Crenicichla saxatilis* occurs together with *Crenicichla multispinosa*.

Use: It is a food fish in the Coastal Plain and Interior of Suriname, but commercially not very important.

Similar species: Several related species of the genus *Crenicichla* occur in Suriname and four of these are endemic (Ploeg 1987, Mol 2012b). *Crenicichla lugubris* and *Crenicichla multispinosa* (Figure 229a) have a larger number of scales (100-130) in the row below the lateral line; the latter species is endemic to the Marowijne and Suriname rivers (Figure 229a). *Crenicichla nickeriensis*, an endemic of the Nickerie and Corantijn rivers, has a series of dark blotches along the lateral line on males. *Crenicichla albopunctata* has silvery dots on the gill cover and cheeks. *Crenicichla sipaliwini* is endemic to the upper Corantijn River; it has a short upper jaw, not reaching the eye. *Crenicichla coppenamensis* is endemic to the Coppename and Sramacca rivers and has a round humeral spot (without notch) with a white margin.

Krobia guianensis
Krobia
Krobia, owruwefi (name also used for *Krobia itanyi* and *Cichlasoma bimaculatum*)
Actinopterygii, Cichliformes, Cichlidae (cichlids)

Male *Krobia guianensis* are slightly larger and more colorful than females (Figure 262). A female produces up to 500 eggs; the eggs are deposited on a broad leaf and this leaf with the eggs attached to its surface is displaced (taken into the mouth) by the parents if necessary (Keenleyside & Bietz 1981). Aquarium observations show that parents defend a territory during breeding. Ponton & Mérigoux (2001) describe early developmental stages of *Krobia guianensis*. In the Petit Saut Reservoir, French Guiana, young juveniles of *Krobia guianensis* did not switch to larger food items during their development and tiny crustaceans occurred more frequently in their diet. Ponton and Mérigoux (2000) hypothesized that the low net energy gained when eating such small prey may explain why the young were thinner at reservoir sites than in undisturbed river habitats. They also suggested that *Krobia guianensis* will develop stunted populations in the reservoir. Stunted fish populations (for example of the piranha *Serrasalmus rhombeus*) have been documented in Brokopondo Reservoir (Mol *et al.* 2007).

Figure 262. A pair of *Krobia guianensis* in the upper Suriname River (image © Willem Kolvoort).

Description: A moderately elongate, moderately compressed, medium-sized (15 cm long) cichlid with a conspicuous black spot on the upper part of the tail fin base. It has a black band extending from the base of the eye to the hind part of the cheek region ('preopercle' bone) continuing about midway the side of the body till the dorsal fin base.

Distribution: Type locality (see Box 9) is Guyana. Occurs in Guyana, Suriname and French Guiana. Present in all rivers of Suriname with the possible exception of the Marowijne River, where closely related *Krobia itanyi* occurs.

Use: A food fish in the Interior of Suriname.

Similar species: The Dolphin Cichlid *Krobia itanyi* of the Marowijne River has the black band of lateral blotches ending short of the end of the dorsal fin base (as opposed to this band continuing all the way till the end of dorsal fin base in *Krobia guianensis*). The *krobia* of coastal swamps is the Black Acara (*Cichlasoma bimaculatum*) which has four spines in the anal fin (versus three in *Krobia guianensis* and *Krobia itanyi*). *Guianacara* species (Figure 225d) lack the black spot on the caudal fin base (Mol 2012b).

Amphibians and Reptiles

Paul Ouboter

The amphibians were the first vertebrate animals or vertebrates to colonize the land. They evolved in the Devonian period from fish that possessed characteristics that pre-adapted them to a terrestrial life style: lungs and bony-limped fins. Amphibians are not fully adapted to a life on dry land, as is obvious from their characteristics (see overview in Table 8). Their skin is very permeable, easily leading to life-threatening water loss. Amphibians also need water or at least very humid surroundings to reproduce, as their eggs are gelatinous and have no shell. Nevertheless, amphibians occur in almost all terrestrial habitats, excluding extremely cold and saline environments. Many have special adaptations to cope with challenging environments, and their reproductive strategies are fascinating and varied (see below). They became a dominant animal group during the Carboniferous and Permian periods, and included giant forms. Amphibian diversity decreased later due to competition with other vertebrates. Nowadays only the so-called Lisamphibia survive, with three main groups: the Anura (frogs and toads), the Caudata (salamanders) and the Gymnophiona (caecilians) (see Duellman & Trueb 1994; Table 9).

Characteristic	Amphibians	Reptiles
Skin	Smooth, with slime and poison glands, and pigment cells	Scales and scutes, pigment cells but no glands
Webbing (skin between toes and/or fingers, partial or whole)	Present in many frog species	Only present in turtles and crocodilians
Heart	1 ventricle, 2 atria	2 ventricles, 2 atria
Breathing	Gills, skin, lungs	Lungs
Thermoregulation	Poikilothermic	Poikilothermic
Larval stage/metamorphosis	Yes	No
Senses	Eyes, ears well developed	Eyes, smell well developed. Vomeronasal organ for smell in Squamata. Heat-sensitive organ in some snakes.
Communication	Mostly sound	Mostly visual
Food	Adults predators of mostly arthropods. Larvae often algae	Mostly predators of arthropods and vertebrates. Tortoises and some adult lizards are vegetarian
Eggs	Gelatinous without shell	Parchment or calcaneus shell
Fertilization	External (no penis)	Internal, males have a (hemi)penis
Size	<1 cm to 150 cm, but usually less than 30 cm	2.5 cm to 10 m

Table 8. Characteristics of amphibians and reptiles.

Reptiles evolved from amphibians in the Carboniferous period as an adaptation to life on dry land (see overview of their characteristics in Table 8). They have scales which decrease evaporation via the skin, a very strong skeleton and musculature, and, crucially, an eggshell, which makes them independent of wetlands. What we call reptiles are vertebrate animals with four legs and scales; like amphibians, they are ectotherms, meaning that the body temperature reflects that of the external environment, and is regulated behaviorally (see below). Reptiles include descendants such as snakes which have lost the legs in the course of their evolution. Mammals and birds are also

Characteristic	Frogs and toads	Salamanders	Caecilians
Legs	4, hind legs large, elongated	4, hind legs same size as front legs	No legs
Webbing	In many species	None	None
Tail	No tail in adults	Tail in larvae and adults	Tail in larvae and adults
Metamorphosis	Much change during metamorphosis	Limited change during metamorphosis	Hardly any change during metamorphosis
Reproduction	Usually oviparous	Oviparous	Oviparous, often with direct development, and viviparous
Larvae	Exterior gills only during early development. Limbs develop in a later stage.	Exterior gills. Most species with well-developed limbs	Very similar to adult
Habitat	All habitats that are not too cold or saline	Temperate forests and grasslands; in the tropics only in high mountain ranges	Hidden underground or in rotting logs in tropical rainforests
Distribution	Temperate & tropical	Temperate	Tropical

Table 9. Characteristics of amphibian orders.

descendants of reptiles, but are endotherms without scales, although some groups do have scales on part of the body. Endotherms keep their body temperature constant by, for example, 'burning' sugars to generate heat, and panting to cool down. Most of the living reptile species on our planet are lizards, followed by snakes, turtles, amphisbaenians (worm lizards), crocodilians and a single rhynchocephalid (see Vitt & Caldwell 2013; Table 10).

As ectotherm organisms, amphibians and reptiles have a relatively low metabolism: food consumption is limited and growth slow. However, growth in reptiles continues during their entire life, and some species can grow very old and therefore reach an impressive size. In weight the Sea Crocodile of

Characteristic	Testudines	Squamata			Archosauria
		Sauria	Serpentes	Amphisbaenia	Crocodilia
Legs	4	4	No legs	No legs	4
Carapax	Yes	No	No	No	No
Scalation				Scales in rings	
Teeth	No teeth, horny beak	Mostly pleurodont teeth, all of the same size	Acrodont teeth, with specialization	Pleurodont teeth with enlarged median tooth in upper jaw	Thecodont, conical teeth, with specialization
Gizzard	No	No	No	No	Yes
Reproduction	Oviparous; egg shell calcareous; no parental care	Oviparous, ovo-viviparous & viviparous; egg shell leathery or calcareous; some guarding eggs	Oviparous, ovo-viviparous & viviparous; egg shell leathery some guarding eggs	Mostly oviparous; egg shell leathery; no parental care	Oviparous; egg shell calcareous; all guarding eggs and juveniles
Habitat	Wet to dry	Wet to dry	Wet to dry	Dry	Wet
Distribution	Temperate to tropical	Temperate to tropical	Temperate to tropical	Tropical	Subtropical to tropical

Table 10. Characteristics of the orders and suborders of the reptiles (Sphenodontia are not included in this table).

the Pacific Ocean is the largest living reptile on earth: it may weigh almost 1400 kg and reach 6.3 m total length. Some snakes may reach or even exceed this length; the Green Anaconda (*Eunectes murinus*; see species account) grows to almost 6 m, and is the heaviest snake. The largest living lizard is the Komodo Dragon of the Indonesian island Komodo, at 3 m; in Suriname the largest lizard is the Green Iguana (*Iguana iguana*; see species account), which grows up to 2 m. It should be noted that in species descriptions – including many species accounts in the present chapter – the body length is often expressed as 'snout-vent length', which is a standardized measurement from the tip of the snout till the cloaca (opening via which excrement leaves the body, as well as eggs and sperm).

Both amphibians and reptiles have well-functioning eyes; the second best sense is hearing for amphibians, but for reptiles it is smelling. The lizards and snakes have, apart from their nose as first nasal organ, a well-functioning second nasal organ, the vomeronasal organ, for chemical detection. It is also present in some other animals, like mammals. This organ is located in the roof of the mouth; the tongue transfers chemical stimuli to it. The system of communication among most amphibians, in particular frogs, is based on sound; with most reptiles it is based on vision.

The first information on the amphibians and reptiles of Suriname can be traced back to Maria Sibylla Merian (1705) and to Carolus Linnaeus (1758), who described 8 amphibian and 35 reptile species from the country. Many species were collected during subsequent expeditions to the Interior of Suriname. Marinus Hoogmoed, who visited the country many times between 1968 and 1990, gathered much knowledge on the amphibians and reptiles of Suriname (see for example Hoogmoed 1973, 1979). Hoogmoed's publication of 1973 is an extensive overview of the lizards of Suriname. Holthuis (1959) and Hoogmoed (1973) list a total of 54 collectors of herpetological specimens. Between 1988 and 2017 the National Zoological Collection of Suriname (NZCS) organized 12 expeditions to the Interior. Together with four rapid assessment programme expeditions by Conservation International, baseline studies and environmental impact assessments, these expeditions contributed considerably to our knowledge of the fauna of the country (see for example Ouboter *et al.* 2007, 2011). In 2012 Ouboter and Jairam published a monograph on the amphibians of Suriname. So, at present there is a fairly comprehensive overview of the lizards and amphibians of Suriname, but not yet of the snakes and turtles. All the research mentioned above was taxonomic or biogeographic. Other research in Suriname has focused on ecology or behavior: studies on sea turtles (Hill & Green 1971, Schultz 1975, Hilterman & Goverse 2007), the studies of Glastra (1983), Ouboter and Nanhoe (1987, 1988, 1989) and Ouboter (1996) on caimans, and recent ethological work on *Atelopus hoogmoedi* (Luger *et al.* 2009) and *Ameerega trivittata* (d'Orgeix *et al.* 2015).

Diversity of amphibians and reptiles in Suriname

Species richness. In June 2015, a global list of 7,395 species of living amphibians was compiled; 88 percent of these were of the Anura group, which are frogs and toads (Frost 2015). The number of species new to science being described is accelerating, mainly due to molecular (DNA) analysis. On the other hand, many species have recently gone extinct, and many are on the verge of extinction (IUCN 2016). The tropics are particularly rich in amphibian species, except salamanders, which occur where the climate is temperate. South America, most of which is tropical, has the greatest diversity of amphibian species. For Suriname, a relatively small country, 104 species were listed in 2012

(Ouboter & Jairam 2012). Since then, 19 additional species have been recorded, most in the Sipaliwini area in southern Suriname (Fouquet *et al.* 2015), bringing the total to 123. Figure 263 shows a comparison with neighboring territories and entire regions.

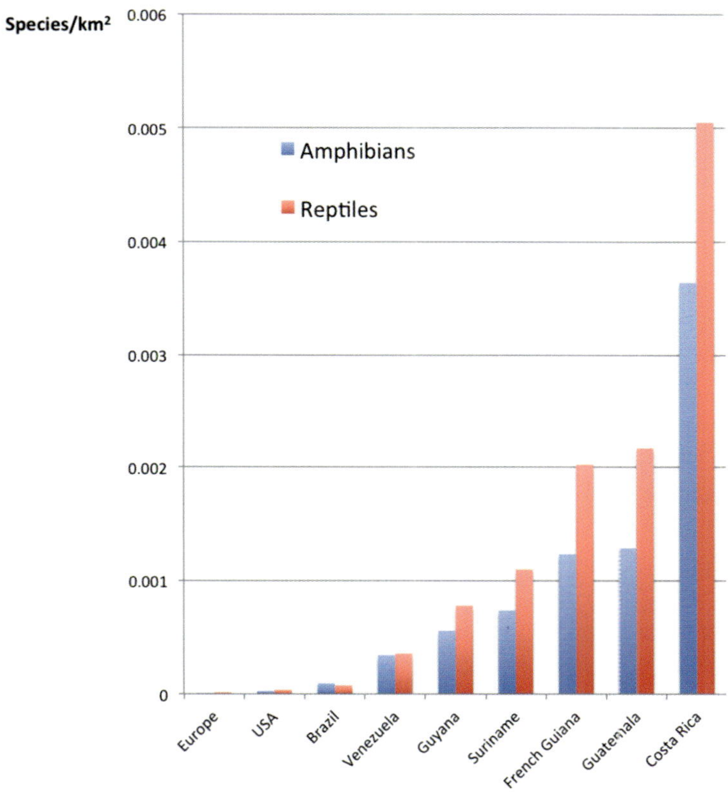

Figure 263. Comparison of the number of species of amphibians and reptiles per square kilometer between Suriname, countries of the region, USA and Europe.

Over 10,000 species of living reptiles are known to science (Uetz & Hošek 2015). Suriname has 175 species of reptiles listed: 47 lizards, 106 snakes, 4 amphisbaenians, 15 turtles and 3 crocodilians (Avila-Pires 2005). Recently one additional lizard and two additional snakes have been recorded by NZCS staff. Also, two introduced lizards are missing from the list of Avila-Pires (2005). This brings the total number of reptiles of Suriname to 180 (see also Figure 263). The rate at which new reptile species are being described is much lower than for amphibians.

Endemism. Suriname is centrally located in the Guiana Shield, a region with a high level of amphibian endemism (species unique to the region): 42 percent (Hoogmoed 1979). This is an underestimate, because many species with a wide distribution have in recent times been shown to actually be complexes of species with narrower distributions. Some species are even limited to a single mountain; this is called micro-endemism. Ouboter and Jairam (2012) calculated amphibian endemism for Suriname at only 5 percent; however, given the rate at which new species are discovered, this number will likely surpass 9 percent. The level of endemism among Suriname reptiles is much lower, with only one species being endemic; most species having a wide Amazonian distribution.

Shaped by the environment and enemies

Water. The characteristics of amphibians and reptiles are mostly determined by their dependence or independence of water. Contrary to reptiles, amphibians are strongly dependent on water. They have a permeable, smooth skin with many glands, and desiccation (drying out) is a serious risk. Their eggs have no shell, so can be fertilized externally, but have to be deposited in water or a very humid environment. Most amphibians have a free-swiming aquatic larval stage, better known as tadpoles (Duellman & Trueb 1994). Reptiles have a dry skin with scales or scutes that minimize evaporation. Scutes resemble scales but have a bony component in the skin and are often larger; they are common in crocodilians and turtles. Reptilian eggs have a shell and can be laid on land. Fertilization has to take place before shell formation, and thus has to be internal (inside the female body). Reptilian hatchlings are not larvae; they are essentially shaped like and feed like an adult, but are smaller and not yet reproducing.

Beware of predators. Some large amphibians and many large reptiles were top predators in distant, prehistoric times. Of the living amphibians and reptiles in South America, only the crocodilians may be considered top predators, alongside some large boid snakes or boas. The rest are prey for a wide range of predators such as other amphibians and reptiles, birds and mammals, and even spiders and insects (Figure 264). Amphibians and reptiles are, however, masters in defense and protection. Many species are well camouflaged or can even change color to enhance camouflage. Several species are poisonous for defensive purposes, and many are cryptic (see below).

Camouflage and related tricks. Among Suriname frogs, impressive examples of dead leaf camouflage are *Rhinella martyi*, *Ceratophrys cornuta* (Figure 281), *Pristimantis chiastonotus* and *Pipa pipa* (*pipatodo*; Figure 291). The first three species live among the leaf litter on the forest

Figure 264. ***Hypsiboas cinerascens*** **being eaten by a spider, Aug. 2010 (image © Raffael Ernst).**

floor, while *Pipa pipa* is an aquatic frog living in quiet streams and lakes, often where dead leaves have accumulated. Frogs living on tree trunks, like many species of the genera *Osteocephalus* and *Trachycephalus*, are typically well camouflaged, as are frogs living on tree leaves, like glass-frogs (Centrolenidae). The same applies to many reptiles; among lizards most species that inhabit leaf

Figure 265. Confusing coloration pattern of *Arthrosaura kockii*, Brownsberg, Nov. 2014 (image © Vanessa Kadosoe).

litter are brown. Good examples are *Anolis chrysolepis* (Figure 296), *Chatogekko amazonicus*, and *Arthrosaura* spp.; *Arthrosaura kockii* in addition has an orange dorsal stripe that resembles a leaf vein, and makes it difficult to recognize the body shape (Figure 265). Other leaf litter lizards (*Tretioscincus agilis* and juvenile *Mabuya nigropunctata*) have bright blue tails: this is to concentrate the attention of predators on the tail, which can break off when seized (autotomy). *Chatogekko amazonicus* (Figure 301), as well as being well camouflaged, is one of the smallest lizards in the world, growing to no more than 24 mm (Figure 301).

Like frogs, lizards living on tree trunks are mostly brown, often with a pattern of mosses, like *Plica* spp. and *Uranoscodon superciliosus*, while lizards living in the forest canopy are green, such as *Iguana iguana* (Figure 297), *Anolis punctatus, Polychrus marmoratus* and *Uracentron azureum* (Figure 300). Many of these species have additional defense mechanisms. *Uranoscodon superciliosus* always lives on trees along the edge of streams. In case of danger it runs to the water and walks on the water to another tree. *Iguana iguana* is a large lizard often living in trees and bushes above the water (Figure 266). When alarmed it jumps in the water and stays underwater

Figure 266. *Iguana iguana*, Raleighvallen, Aug. 2012 (image © Paul Ouboter).

for some time. *Polychrus marmoratus*, by contrast, stays immobile and completely relies on its camouflage, being able to change color quickly and adapt to its background. Most other reptiles are well camouflaged too.

Snakes on the forest floor are usually brown, like the poisonous viperid snakes or vipers *Bothrops atrox* (*owrukuku*; Figure 312) and *Lachesis muta* (*makasneki*; Figure 313), while tree snakes are green and slender, like *Oxybelis fulgidus* (*busiswipi*; Figure 310). However, snakes have many other defense mechanisms when discovered by a predator. Most will try to escape, but when confronted many will hiss and bite in the direction of the attacker. Some species are able to enlarge their neck region to look bigger, such as *Pseustes* spp. (Figure 267), or show bright colors on the underside of the body, such as *Helicops* and *Oxyrhopus* spp., to startle predators. Water turtles often look like pieces of wood or dead leaves, such as *Platemys platycephala* (Figure 293), and caimans are very difficult to find in daytime because they resemble logs floating in the water (Figure 296).

Poison. Another line of defense is being poisonous. Many frogs and toads have poison glands in their skin. Frogs of the family Dendrobatidae usually advertise their poisonousness by bright warning colors (Santos *et al.* 2003a); they forage in plain sight on the forest floor. In Suriname this family is represented by the black-and-fluorescent-green *Ameerega trivittata* and black-blue-yellow *Dendrobates tinctorius* (Figure 282). Snakes use poisons for defense but also to capture and digest prey. Although in Suriname many poisonous snakes, like the vipers, are camouflaged, the coral snakes of the family Elapidae (*Micrurus* spp.; relatives of cobras) have bright warning colors of black, yellow and in most species also red, such as *Micrurus surinamensis* (Figure 311). Viper poison is hemotoxic, causing the destruction of blood proteins and body tissues; elapid poison is neurotoxic, disabling parts of the nervous system. Apart from the front-fanged poisonous snakes of the latter two families, many Colubridae have poison fangs in the back of their mouth (referred to as an opisthoglyph arrangement). Some of these colubrids, such as *Clelia clelia*, can cause poisoning in people who are bitten, but in most species the poison merely contains enzymes that help with the digestion of prey. Some non-poisonous snakes look very similar in coloration to coral snakes, such as *Erythrolamprus aesculapii* (Figure 309), or vipers, such as *Xenodon* spp. and some boas.

Figure 267. Defense behavior of *Pseustes poecilonotus*, Brownsberg, Jun. 2015 (image © Paul Ouboter).

Figure 268. *Anolis chrysolepis* sleeping on a leaf 1 m above the forest floor, Gros-Rosebel, Jun. 2014 (image © Paul Ouboter).

CRYPSIS. Despite camouflage or bright coloration, many amphibians and reptiles do not sleep on the forest floor because of the many predators foraging there at night. They climb up a bush and sleep on a leaf one to two meters above the forest floor (Figure 268). Amphibians and reptiles are experts in crypsis, which essentially means hiding. It is even a way of life for many species. Most leaf-litter lizards forage between and under the leaves, and caecilians, amphisbaenians and blind snakes may live most of their time underground or in rotting logs. Lizards and snakes may hide in tiny holes and crevices during rest periods. *Uracentron azureum* (Figure 300) lives in the forest canopy and can run very fast on the branches. When resting or in danger it hides in crevices in the wood. It has a spiny tail that prevents it from being pulled out by predators. Caimans hide underwater or in accumulated mud or leaves.

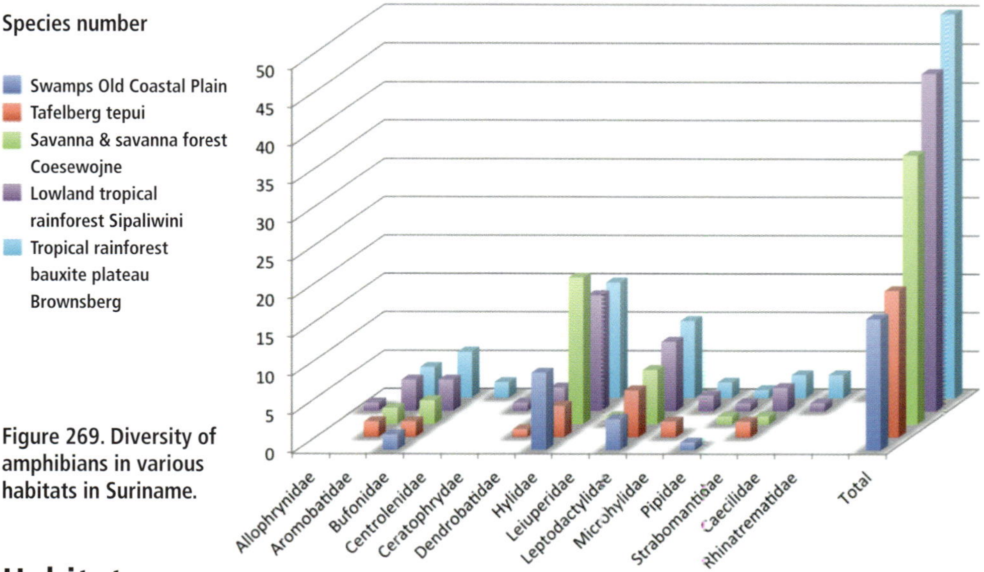

Figure 269. Diversity of amphibians in various habitats in Suriname.

Habitat

SUITABLE ECOSYSTEMS. Amphibians and reptiles occur in all terrestrial and freshwater ecosystems of Suriname; unlike reptiles, no amphibians occur in marine habitats. The brackish Estuarine Zone and Coastal Zone in general has a few anthropophilic species, meaning species that like to live near people, like the large toad *Rhinella marina* (*krastodo*; Figure 278), common on yards, and the small tree frog *Scinax ruber*, common in houses (especially in bathrooms). The tropical rainforest of the

Interior is the richest in species, especially the bauxite-capped mountains (Figure 269). In rainforests all potential microhabitats seem occupied: the forest floor (soil, leaf litter and understory vegetation), small creeks and temporary pools, tree trunks and crowns, and so on (see below).

RAINFOREST FLOOR. Many small lizards and snakes, as well as small to large amphibians, live in the leaf litter of the forest floor. A large frog like *Ceratophrys cornuta* (Figure 281) digs itself in between the dead leaves to camouflage itself from predators and prey. Large leptodactylid frogs like *Leptodactylus pentadactylus* (Figure 289) burrow in soil covered with leaf litter. The toad *Rhinella martyi* relies on its color pattern that makes it look exactly like a dead leaf. Many snakes also forage on the forest floor, like *Chironius* spp. during the day, and *Bothrops atrox* (known in Suriname as *owrukuku* or *labaria*) and *Lachesis muta* (*makasneki*) at night. During the day *Bothrops atrox* camouflages itself between the leaf litter, often even partly hiding under the leaves, or on or between tree trunks. Some tree snakes occupy the lower vegetation in the forest, like *Dipsas catesbyi* (Figure 307). In and around tree gaps more bushes are available because of the increase in light intensity. This is the habitat of many other tree snakes, like *Oxybelis fulgidus* (*busiswipi*). In the open areas of the tree gaps, heliotherm or sun-loving lizards of the family Teidae thrive, like the large *Tupinambis teguixin* (*sapakara*; Figure 304) and the smaller *Ameiva ameiva*.

Figure 270. *Plica umbra*, a species living on tree trunks, Rosebel, Jun. 2014 (image © Paul Ouboter).

TREE TRUNKS AND BRANCHES. Several species of lizards and a few frogs use tree trunks as their main micro-habitat. The lizard *Plica plica* is usually found on big tree trunks and *Plica umbra* (Figure 270) on smaller ones. *Uracentron azureum* is well at home up in the canopy. Some snakes may also live all the way up in the canopy, but as far as we know most tree snakes prefer edge situations (edge of tree falls or rivers). The lizard *Mabuya nigropunctata* is often seen on fallen trees, but may climb up into the canopy to forage (Hoogmoed & Avila-Pires 1991). *Iguana iguana* prefers to hide and forage in the lower canopy leaf masses along streams.

RAINFOREST CREEKS. Small creeks draining the forest are especially rich in species. Tree frogs and leptodactylid frogs thrive here, such as *Phyllomedusa bicolor* (Figure 287), *Dendropsophus minutus*, *Hyalinobatrachium cappellei* (Figure 279) and *Leptodactylus validus*. It is also the habitat of *Neusticurus bicarinatus* (Figure 302), a small lizard with the appearance of a caiman. Of the three species of caiman that are found in Suriname, only *Paleosuchus trigonatus* (*blakakaiman*) is found in creeks in closed forest, the other two are more sun-loving. *Paleosuchus trigonatus* is found in larger rivers in the Interior, but reproduces along small tributaries. Its newborns migrate upstream after a few weeks and may be found in swampy areas, small creeks and pools in the forest.

Temporary rainforest pools. Creeks are not the only water bodies in the forest. During the rainy season, depressions in the forest inundate after the first heavy rains. This often causes breeding frenzies of frogs. *Ceratophrys cornuta* (Figure 281 is one of the first to arrive; its tadpoles are carnivores that feed on other tadpoles. A temporary pool may attract more than ten species of frogs simultaneously. Microhylid frogs often breed in small temporary ponds; *Chiasmocleis shudikarensis* for example can be found in these ponds in large numbers, but only for a few nights each year. After that, it is as if this species has completely disappeared from the forest, probably hiding under the leaf litter.

Exposed bedrock. Areas with exposed bedrock occur dispersed in the Interior of Suriname. These inselbergs and rock plates carry the same rock savanna vegetation. At least one frog and one lizard are typical of this habitat: the frog *Leptodactylus myersi*, which hides in the crevices between the rocks during the day and comes out at night, and *Tropidurus torquatus,* which looks like lizards of the genus *Plica*, but climbs rocks instead of trees. The latter species can also be found in some savannas and along the wide, rocky shores of rapids complexes in large rivers.

Large water bodies. Larger rivers are home to only a few species of amphibians and reptiles. Few tadpoles can survive in rivers because of the numerous predatory fish. Reptiles associated with rivers are usually large, like the three species of caimans (*Caiman crocodilus*, *Paleosuchus palpebrosus* (Figure 296) and *Paleosuchus trigonatus*), the Green Anaconda *Eunectes murinus* (*aboma*) (Figure 305), the giant water snake *Hydrodynastes gigas* and some species of turtles, the largest being *Podocnemis unifilis*.
Treeless seasonal swamps are abundant in the Coastal Plain and the Zanderij Belt, but are also found dispersed and in smaller size in the Interior. These open swamps are usually flooded in the long rainy season, sometimes also in the short rainy season, but not during the rest of the year. The swamps of the Coastal Plain and Zanderij Belt have a characteristic fauna, although some of the species are also found in the swamps of the Interior. During the rainy season the choir of frogs in these swamps can be deafening. The choir may consist of *Rhinella marina*, *Dendropsophus minutus*, *Hypsiboas crepitans*, *Hypsiboas punctatus*, *Sphaenorhynchus lacteus*, *Elachistocleis surinamensis* (Figure 290), and *Scinax* spp. The swamps are also the habitat of several species of water snakes feeding on these frogs and on fishes: juvenile Green Anaconda, *Helicops angulatus* (Figure 308) and *Erythrolamprus* spp. Other predators in the swamps are the Spectacled Caiman (*Caiman crocodilus*; *wetberekaiman*) and the Dwarf Caiman *Paleosuchus palpebrosus* (*redikaiman*; Figure 296). Even some savanna lizards occur here: *Anolis auratus* and *Kentropyx striata* (Hoogmoed 1993).

Savannas and urban areas. Gardens in Paramaribo resemble savannas in general appearance, so it is not surprising that many savanna species can also be found in the capital of Suriname, such as the lizards *Anolis auratus* and *Cnemidophorus lemniscatus*, and the frogs *Leptodactylus fuscus* and *Leptodactylus longirostris*. Typical savanna species not yet found in Paramaribo are the frog

Figure 271. Male *Rhinella merianae* calling at Brownsberg, Apr. 2015 (image © Paul Ouboter).

Figure 272. *Phimophis guianensis*, Coesewijne Savanna, Jul. 2007 (image © Paul Ouboter).

Leptodactylus ocellatus, the toad *Rhinella merianae* (Figure 271; also occurring in some forested localities in the Interior), the lizard *Kentropyx striata* and the snakes *Epicrates maurus*, *Phimophis guianensis* and the rattlesnake *Crotalus durissus* (*sakasneki*). *Phimophis guianensis* has a turned-up nose to dig in loose sandy soil (Figure 272).

Because large areas in and around Paramaribo may become inundated after heavy rain showers, many swamp species mentioned earlier inhabit Paramaribo as well. Very typical are the calls of *Scinax ruber* and the very high pitched, long call of *Elachistocleis surinamensis*. *Scinax ruber* is highly anthropophilic and can be found in virtually every bathroom.
Paramaribo also harbors at least four introduced species of amphibians and reptiles: the frog *Eleutherodactylus planirostris*, the lizard *Anolis roquet,* and the gekkonid lizards or geckos *Lepidodactylus lugubris* and *Hemidactylus mabouia* (as to this species, see Hoogmoed 1981).

Thermoregulation and seasonality

Temperature. Amphibians and reptiles are ectotherms. For aquatic species and amphibians this means that their body temperature is equal to the temperature of the environment. Many terrestrial reptiles, especially lizards and snakes, actively raise or lower their body temperature by thermoregulation (Avery 1979). After a cold night they slowly appear from their hiding places. When cold, their metabolism is low and moving quickly to escape predators is impossible. So reaching an optimum body temperature as soon as possible is critical. Many species therefore start their morning basking in the sun, enlarging the surface area of their back by flattening themselves. Some species may even change color, being dark when still cold and becoming much lighter when warm. In Suriname temperature is nowhere too cold for the occurrence of amphibians and reptiles and most lizards and snakes occur in the shaded forest, such as *Anolis chrysolepis* (Figure 298) and *Bothrops atrox*. Teid lizards, however, are heliotherm (sun-loving) and have body temperatures that are higher than lizards living in the forest. The Green Ameiva Lizard (*Ameiva ameiva*) and the Rainbow Whiptail Lizard (*Cnemidophorus lemniscatus*) can often be seen basking in the sun, even in gardens in Paramaribo. Spectacled Caimans (*Caiman crocodilus*) thermoregulate by changing their position between the sunny shore and the much cooler water. They may also dig themselves into the cool mud.

While ectothermy can be disadvantageous because it may limit activity, it can also be advantageous. In periods of sub-optimal environmental conditions an individual can hide in a cool place, thereby lowering its metabolism and food requirements. Caimans, turtles and many frogs do this by digging themselves into the mud during the dry season. In Suriname the seasonal differences in temperature are limited, but temperatures between January and March are perceptibly lower than between September and November. This could explain why many leaf-litter lizards, such as *Chatogekko amazonicus,* are hardly seen at the beginning of the calendar year.

Precipitation. Seasonality is mostly determined by precipitation (see Climate and Hydrology chapter). It has an enormous impact on the reproductive strategy of amphibians and caimans (more on this below). It is not clear if other groups of reptiles also have a reproductive season linked to precipitation. However, most snakes are seen during rainy periods, and most lizards are seen during dry periods. Spells of extreme dryness occur from time to time, especially in El Niño years. During a dry spell at the Brownsberg Plateau, Ouboter *et al.* (2016) observed numerous frog individuals of 14 species flattening themselves in contact with humid tire tracks in the mud which were all that remained of pools in the road (Figure 273). In this way they seemed to be able to absorb at least some water.

Food

Adults. Apart from the larval stages of most amphibians, over 90 percent of amphibians and reptiles are opportunistic predators (Vitt & Caldwell 2013). Their prey selection is mainly based on size. Small frogs and lizards eat small insects and spiders, while large frogs and lizards can even handle small mammalian prey. Baby caimans eat insects and small fish; as they grow, they switch to larger fish, frogs and reptiles, and finally to birds and mammals (Ouboter 1996). Only *Chelonoidis* spp. tortoises and some large lizards like the Green Iguana are vegetarians as adults.

Figure 273. *Leptodactylus pentadactylus* flattening itself on humid mud of a tire track at Brownsberg, Nov. 2012 (image © Vanessa Kadosoe).

All snakes are predators, but they are often more specialized in prey type than lizards. Many tree snakes are specialized in lizards (such as *Oxybelis* spp.), birds (such as *Corallus* spp.) or snails (*Dipsas* spp.), while terrestrial species may specialize in insects and earthworms (such as *Tantilla melanocephala*), frogs and toads (such as *Xenodon* spp.), mammals (such as *Chironius* spp. and *Lachesis muta*), or even poisonous snakes (*Clelia clelia*). Snakes living in aquatic environments will usually prey on fish (*Helicops* spp., *Micrurus surinamensis*) or amphibians (*Erythrolamprus* spp.) (see Starace 2013). Poisonous vipers are one of the most advanced predators in the world. The members of the viper subfamily Crotalinae have so-called loreal pits between the nostril and the eye on either side of the head which serve as very accurate temperature-sensing organs: they form a realistic thermal image of their environment in complete darkness! Their very potent poison kills prey in minutes. After the bite the snake tracks its prey – if it ran away after the bite – based on chemical information collected by the tongue and brought to the vomeronasal organ.

Adult amphibians are sit-and-wait predators, or move slowly through their home range to look for food. Active foraging for food is characteristic of dendrobatid frogs like *Ameerega trivittata*. Most species use visual clues to locate potential prey items and only see them when they are moving, they lose interest when the prey stays motionless. Some of the larger species, with more brain mass are exceptions to this rule, such as *Rhinella marina* (see species account). Anurans (frogs and toads) use their sticky extendable tongue to capture prey, although large anurans may just grasp large prey with the mouth. Species of the genus *Pipa* are the exception; they are strictly aquatic and have no use for a tongue (prey is sucked in with the water when the mouth is opened). Caecilians – legless snake-like amphibians – slowly move to their prey, often earthworms, and seize it with a powerful bite; they have long recurved teeth to hold on to their slippery prey.

Larvae. While reptiles have no larval stage, most amphibians do; their larvae usually are herbivorous tadpoles, living off algae. The teeth of these tadpoles are specially adapted to scrape algae off a substrate such as stones, logs and leaves. However, several species have carnivorous tadpoles, preying on the tadpoles of other species and on invertebrate animals that live in the water (such as *Ceratophrys cornuta*) (Figure 281) or frog eggs (such as *Osteocephalus taurinus* and *Osteocephalus oophagus*; Figure 286). *Phyllomedusa* spp. larvae have a funnel-shaped mouth, a special adaptation to feeding on small, floating particles.

Figure 274. Male *Hyalinobatrachium cappellei* with eggs Brownsberg, Jan. 2015 (image © Vanessa Kadosoe).

Reproduction

Reproduction in amphibians. In frogs potential reproductive partners usually find each other acoustically. The males position themselves at exposed or concealed localities and make calling sounds (Figure 271). For the call to carry far, high exposed localities are favorable, but on the other hand these sites increase the risk of being caught by predators. Even within the same genus, the choice of exposed or concealed locations varies between species. The call of all frogs is species-specific, and even humans can learn to distinguish frog calls. Not all frog calls serve to attract a female, some calls are territorial (or serve both purposes).

The reproductive strategies of amphibians are varied and intriguing (see for example Duellman & Trueb 1994). Fertilization is external. In frogs and toads the pair stays in amplexus for some time, which means that the male is on the back of the female, waiting for her to deposit eggs. He will fertilize the eggs with his sperm as soon as the eggs are deposited. Amphibians have to cope with the risk of desiccation of themselves and their eggs. The simplest solution is to return to an aquatic environment to deposit the eggs, and most species do this. However, there is much variation in exactly how they do it. In Suriname most amphibians reproduce in the rainy season and deposit their eggs in pools after the first heavy rains. Toads (Bufonidae) and many tree frogs (Hylidae) do this. One of the problems with this is the high risk of egg predation. In some species the eggs are poisonous (for example *Rhinella marina*). Male and female frogs of the families Leptodactylidae and Leiuperidae produce a foam nest. The foam is produced by the male while the pair is in amplexus. The foam probably has the main purpose of protecting the eggs against predators, but also prevents desiccation (Heyer 1969), exposure to extreme temperatures (Gorzula 1977) and growth inhibition (Duellman & Trueb 1994). Species of the genera *Osteocephalus* and *Trachycephalus* deposit the eggs in water-filled tree holes (phytotelmata), high up in the trees, thereby at least avoiding the many predators that roam on the forest floor. *Dendrobates tinctorius* deposits its eggs in fallen palm bracts. After hatching, the tadpoles are transported to bromeliads and tree holes high up the tree on the back of the male (Born *et al.* 2010). Many tree frogs deposit their eggs above water on or under a leaf. After hatching the tadpoles fall into the water. Males of the glass frog *Hyalinobatrachium cappellei* call from the underside of a leaf to attract females to deposit their eggs there (Figure 274). Since the males keep calling atop the eggs, it seems that other females may be persuaded to deposit in the same location. After hatching the tadpoles fall in the brooklet below. Here too predation is a problem. In *Phyllomedusa* spp. the males therefore fold the leaf together to keep the eggs out of sight. Despite this, these nests are often raided, especially by snakes (Figure 288). A very special method of reproduction is seen in *Pipa pipa*. In this species, the eggs are lodged on the back of the female and hatch there (Figure 292). The genus *Pristimantis* (family Strabomantidae) has direct development of the embryo: eggs are deposited between wet leaves or the moss on trees, and after an incubation period froglets (not tadpoles) hatch from the eggs.

Reproduction in reptiles. The reptiles are a heterogeneous group, so it is not surprising that their reproduction is quite varied (see Vitt & Caldwell, 2013). Potential partners find each other by vision in lizards and turtles, by chemical detection in snakes, and by sound in crocodilians. When the potential partners get closer to each other the visual clues become important in all groups. Lizards have a specific, well-defined courtship behavior in which especially head movements, and a dewlap – a skin flap below the throat – are important. The lizards of the genus *Anolis* have a brightly colored dewlap that is displayed in both territorial encounters and courtship. In the introduced *Anolis roquet* this dewlap is very large and bright yellow. If the female allows copulation, male lizards hold on with a bite in the neck or body of the female. Head movements are important in most groups of reptiles: this has been observed in snakes, tortoises and crocodilians. In water turtles the male swims above or in front of the female and touches her head with the fore limbs. Acoustic communication in courtship is only important in gekkonid lizards and crocodilians.

Most lizards and snakes lay eggs with a leathery shell. These eggshells are permeable and absorb water during incubation. The eggs also grow in size. This means that the female has to be very selective in the choice of nest locality. Nests can be made in slightly humid soil, decomposing logs, tree holes, termite nests and other hidden localities. Gekkonid lizards have calcareous egg shells that can be deposited under less strict environmental conditions, such as in crevices in tree bark or rocks. The introduced gekkonid lizards lay eggs behind the books in the bookcase, in carton boxes, bags and elsewhere. This explains the ease of dispersal to other human settlements, once introduced. Crocodilians have calcareous eggs as well, but these are deposited with great care by the female. Two nesting strategies exist: nesting in a hole in the sand, usually during the dry season, and nesting in a mound of decomposing plant matter, usually during the rainy season (Ouboter & Nanhoe 1987, Ouboter 1996). Caimans are mound-type nesters, although a limited amount of plant material may force them to use an intermediate strategy. The decomposing plant matter produces warmth that helps keep the incubation temperature at the optimum level. In crocodilians (Ferguson 1985) and sea turtles (see for example Mrosovsky & Yntema 1980), the incubation temperature is critical because it determines the sex of the hatchlings.

Anomaloglossus surinamensis
Suriname Rocket Frog
Todo (local name used for all frogs and toads)
Amphibia, Anura, Aromobatidae

Figure 275. *Anomaloglossus surinamensis*, Brownsberg, Dec. 2015 (image © Vanessa Kadosoe).

The recently established family Aromobatidae (Grant *et al.* 2006) is represented in Suriname by at least nine species which were formerly considered to belong to the Dendrobatidae. Like male dendrobatids, male aromobatids carry the tadpoles to water reservoirs. The general build of aromobatids also resembles that of dendrobatids, but they are usually smaller, have no skin poison and have a cryptic coloration (are camouflaged). In recent times, several new species belonging to this Neotropical frog family have been described each year.

Anomaloglossus surinamensis (Figure 275) is a habitat specialist that only occurs in leaf litter near tiny headwater streams of forested mountainous areas in Suriname and French Guiana (Ouboter & Jairam 2012; Vacher *et al.* 2017). Here it may reach a high population density with individuals spaced less than five meters apart. At Brownsberg it occurs alongside an as yet undescribed species of the genus *Anomaloglossus* that occupies the forest litter further away from streams. *Anomaloglossus surinamensis* is only seen and heard during the rainy season when the headwater streams carry some water. It is assumed that they aestivate, which means that they undergo a prolonged period of inactivity during drier periods. Reproduction occurs only in the rainy seasons, and during these periods males can be seen carrying two or three tadpoles. The male call to attract females consists of a single high note repeated every half ≠second.

Description: Tiny (17 mm body length) but stocky, mostly brown frog with large eyes. First and second toes only are webbed. On either side of the body, there is a pair of contrasting light brown bands, with a much darker band just below each, stretching from behind the eyes till the cloaca; a pair of contrasting dark bands stretches from the snout tip to the eyes, narrowing and continuing downward behind the eye.

Distribution: As yet only known with certainty from Brownsberg and Nassau mountains in Suriname. Similar small frogs occur at the Tafelberg and Bakhuis mountains and in mountainous parts of French Guiana; their exact taxonomic status is being investigated.

Use: (No known use).

Similar species: This recently described species (Ouboter & Jairam 2012) has been confused with *Anomaloglossus degranvillei* which is only known from French Guiana. It is very similar to other species of the genus *Anomaloglossus,* such as *Anomaloglossus stepheni* and the undescribed species mentioned above, but these are larger and less stocky, and have more distinct contrasting dark and light bands from eyes to cloaca.

Atelopus hoogmoedi
Hoogmoed Harlequin Toad
Todo (local name used for all frogs and toads)
Amphibia, Anura, Bufonidae

Nine species of the family Bufonidae are known from Suriname, but just a single one of the genus *Atelopus*. The Bufonidae are the 'real' toads. They have a skin that looks dry, lay their eggs in strings, and have a poison gland behind the ear (paratoid gland). The skin of Bufonidae has many poison glands, often concentrated in very distinct bumps, yet species of the genus *Atelopus* lack these bumps.

Atelopus hoogmoedi (Figure 276, 277) is a species from hilly and mountainous areas covered with tropical rainforest. The males are usually found near small streams; the females seem to stray further away from water (Luger *et al.* 2009). Like all members of the genus, this species is active during the day. At night it sleeps on top of small bushes or seedlings. During the rainy season the males call from perches under dense vegetation near streams. At the Brownsberg Mountain, a high density of the males was recorded during the reproductive season, of almost 0.5/m2 (Luger *et al.* 2009). Between 1982 and 1985 the species was quite abundant at this site; after 1992 no individuals could be found there until 1998, when the population seemed to begin a full recovery (Paul Ouboter, pers. obs.). Although a link was assumed with a possible epidemic of the lethal fungus *Batrachochytrium dendrobatidis*, later research showed that this fungus is absent from Brownsberg (Luger *et al.* 2008; Christian D'Orgeix, personal communication).

Figure 277. *Atelopus hoogmoedi nassaui*, Merian, Dec. 2008 (image © Paul Ouboter).

Figure 276. *Atelopus hoogmoedi* pair in amplexus, Brownsberg, Aug. 2014 (image © Vanessa Kadosoe).

Description: Small (26-38 mm body length) with slender head and legs; paratoid gland not visible externally. Upper side of the body and flanks black with a series of interconnected yellow rings; belly and throat largely yellow. Color is variable: populations are known with individuals that are largely yellow, such as at the upper Tapanahony River, or with pink rather than yellow rings, such as at Nassau Mountain (Ouboter & Jairam 2012) (Figure 277).

Distribution: Guianas and bordering Brazilian states. In Suriname it only occurs in the Interior; the distribution is patchy, and no populations are known yet from the southwestern part of the Interior.

Use: Because of its beauty, and because it is active during the day, it is a popular species in the pet trade, and caught in Suriname for export.

Similar species: Its coloration is very distinct, and therefore confusion with other frog species is unlikely.

Rhinella marina
Cane Toad, Giant Toad
Krastodo
Amphibia, Anura, Bufonidae

This species and four others of the genus *Rhinella* in Suriname are also Bufonidae (see preceding account). *Rhinella marina* (Figure 278), formerly named *Bufo marinus*, has the plump and warty look of a toad. It is known as the Giant Toad, since individuals can become very large. Giant Toads are most abundant near human settlements, where they feed on insects like cockroaches, and young rodents. They are often seen eating dog and cat food, as well as dog excrement, and have been observed to show up like clockwork at the feeding bowl just ahead of the dog's regular feeding time. This shows their intelligence. While most amphibians need the stimulus of movement to recognize potential prey, *Rhinella marina* does not. These toads are quite poisonous; dogs and cats that try to eat them often die. Their eggs and larvae are also poisonous (Lescure & Marty 2000). Reproduction occurs in the rainy season, when long strings of thousands of eggs are laid in pools with stagnant water. The tadpoles are small (10-12 mm) and black. They grow quickly to adulthood and are mature at a size of 90-100 mm (Zug & Zug 1979). *Rhinella marina* protects itself against predators by its poison and by inflating itself to look bigger. Despite its toxins, it has many predators in its natural distribution, especially snakes and raptors.

This species was introduced in many countries, such as in Australia and the Caribbean (Lever 2001), to control insect and rodent pests in sugarcane plantations. These introductions were mostly unsuccessful. Worse, the toad became invasive, exterminating local fauna (Easteal 1981) due to the lack of effective native predators, and the poisoning of 'naïve' predators (Doody *et al.* 2009).

Figure 278. *Rhinella marina*, Brownsberg, Feb. 2014 (image © Vanessa Kadosoe).

Description: Females can grow to 250 mm in body length, males to 140 mm. Large paratoid glands (see previous species account) behind the ears, and many bumps on the back; head with bony ridges; toes partly webbed. Predominant color brown to yellow-brown and gray; the underside of the entire body is lighter. Young individuals usually have a mosaic of different brownish colors.

Distribution: From Texas south through most of Central and tropical South America; in Suriname occurring all over the country, especially where people live and in disturbed areas; much rarer in pristine rainforest.

Use: Some are wild caught in Suriname and exported for the pet trade.

Similar species: Resembles other toads that occur in Suriname, especially *Rhinella merianae* and *Rhaebo guttatus*. The latter has a smoother head and back, flatter warts, and a contrasting dark flank. The former is much smaller, with smaller paratoid gland and bumps. All other toads in Suriname have small paratoid glands.

Hyalinobatrachium cappellei
Banded-limp Glass Frog
Todo (local name used for all frogs and toads)
Amphibia, Anura, Centrolenidae

The family Centrolenidae, the glass frogs, are represented in Suriname by ten species, eight of which belong to the genus *Hyalinobatrachium*. It's a family of relatively small frogs with a transparent skin on the underside of the body. Consequently the internal organs and sometimes also the bones are visible. The bone at the tip of each digit is T-shaped.

Hyalinobatrachium cappellei (Figure 279, 280) lives along small mountain streams in primary rainforest. The males call during or just after heavy rain showers from the underside of leaves, 2 to 5 m above the stream. The cream-colored eggs are also deposited under the leaves, and the males remain calling on top of the eggs, probably to attract other females to deposit too. When within a few days the eggs hatch, the tadpoles fall into the small stream underneath. According to Myers and Donelly (1997) the tadpoles are tan with small black dots. Outside the rainy season and after dry days during the rainy season, these frogs cannot be heard or seen. They probably undergo a prolonged period of rest during long dry periods (aestivation), presumably higher up in the trees.

Figure 280. *Hyalinobatrachium cappellei* underside, Brownsberg, Jan. 2015 (image © Paul Ouboter).

Figure 279. *Hyalinobatrachium cappellei*, Brownsberg, Jan. 2015 (image © Paul Ouboter).

Description: Small green frog, males up to 24 mm body length. Head relatively large and rounded; eyes protruding; iris golden with brown dots; toes fully webbed. Underside of the body transparent, upper side light green with large yellow spots and many small black dots; the digits have white tips; the heart, intestines, liver and stomach are well visible through the skin on the underside of the body; the liver is covered by white pigment.

Distribution: Guianas; in Suriname known from seven localities spread over the Interior.

Use: (No known use).

Similar species: Quite similar to other *Hyalinobatrachium* spp., but differs from these in the size, density and/or color of the spots on the upper side of the body, color of and spots in the iris, and color of the tips of the digits.

Ceratophrys cornuta
Suriname Horned Frog
Todo (local name used for all frogs and toads)
Amphibia, Anura, Ceratophryidae

Figure 281. ***Ceratophrys cornuta*, Brownsberg, Jun. 2014 (image © Paul Ouboter).**

The horn frog family (Ceratophryidae) is represented by just a single species in Suriname. Horn frogs are rather large frogs with a bulky head and a wide mouth. They are ambush predators with carnivorous larvae.

Ceratophrys cornuta (Figure 281) is a very well camouflaged, cryptic species, usually hiding itself by partly burrowing its body in the rainforest soil, with only the top of the head protruding. The adult frogs prey on large arthropods, such as insects, and small vertebrates, like other frogs and small mice. Reproduction is very 'explosive': immediately after the first heavy rains of the rainy season, essentially all mature frogs appear simultaneously to congregate in pools and mate. Eggs are deposited in depressions in the forest that temporarily fill with water. The tadpoles are predators, and while they also feed on invertebrate animals, they feed mostly on other tadpoles. They thus need a developmental head start before the onset of reproduction by other frog species (Duellman & Lizana 1994).

Description: Large frog – up to 120 mm body length – with big head and mouth, but relatively short, tapering body; has a striking, horn-like skin growth on the eyelids. The upper side of the entire body is well camouflaged with different shades of brown, sometimes also green, in a pattern that is very variable; belly whitish, and throat black; the iris is silvery with a horizontal black band.

Distribution: Guianas, Venezuela and Amazon Basin. In Suriname known from the Interior; however, the species is usually difficult to find, and therefore known only from a limited number of locations. According to indigenous people, it is very easy to find this frog in large numbers at the start of the long rainy season.

Use: Some are caught in the wild in Suriname and exported for the pet trade.

Similar species: No other frog or toad species in Suriname has the disproportionately large head and the horn-like structures on the eyelids.

Dendrobates tinctorius
Dyeing Poison Frog
Okopipi
Amphibia, Anura, Dendrobatidae

Figure 282. ***Dendrobates tinctorius*** **with tadpoles, Raleighvallen, Feb. 2010 (image © Paul Ouboter).**

Figure 283. ***Dendrobates tinctorius azureus*****, Vier Gebroeders (forested hills in Sipaliwini Savanna), Apr. 2014 (image © Paul Ouboter).**

The poison frogs or Dendrobatidae are represented by only three species in Suriname. Dendrobatids are characterized by bright coloration and divided scutes (scale-like structures) at the end of the digits. Toxic alkaloid chemicals in the skin are derived from insect prey (Caldwell 1996).

Dendrobates tinctorius (Figure 282, 283) is only found in pristine rainforests in or near mountain ranges, specifically along rocky streams. It is day-active and typically not shy. It has no advertisement call, and only vocalizes when courting a female or in aggressive encounters between males (Born *et al.* 2010). During the rainy season, females follow males to sites for depositing eggs, usually in leaf litter. Here the male guards the 8-15 eggs for 14-18 days (Walls 1994). When the tadpoles hatch, the male carries them to small above-ground water reservoirs like a hole in a tree trunk, bromeliads or palm bracts, and leaves them to fend for themselves. The male may even climb trees to reach these reservoirs. The tadpoles develop into small froglets in about two months. During the rainy season the frogs prefer recently formed tree gaps and seem to benefit from the variety of insect food provided by the recently fallen trees; during the dry season the frogs mainly stay in retreat sites, are less active, and have less diversity in their diet (Born *et al.* 2010). Although the lethal fungus *Batrachochytrium dendrobatidis* is known to infect populations of *Dendrobates tinctorius*, it was absent from populations at the Sipaliwini Savanna in 2015 (Christian D'Orgeix, personal communication).

Description: Large for a poison frog: up to 50 mm body length. Skin smooth; digits without webbing; distinct, large divided discs at the finger tips, those on the toes slightly smaller. Extremely variable in color; typically predominantly black, with bright yellow designs on head and back, and with legs partly or completely blue with black spots (Figure 282). Males usually have more yellow than the females, which is related to sexual selection (Rojas & Endler 2013). At the Tafelberg mountain, these frogs are entirely black with three yellow spots on the head and some narrow yellow stripes on the back; in parts of southeastern Suriname they are mostly yellow, and at the Vier Gebroeders mountains in the Sipaliwini Savanna they are mostly blue (no yellow; see Figure 283).

Distribution: Occurs in the Guianas and adjacent parts of Brazil; in Suriname known from the mountainous parts of the Interior, but not from the more isolated Brownsberg, Nassau and Lely mountains in the east.

Use: Exported for the pet trade; often used as a model organism for the transmission of *Batrachochytrium dendrobatidis* (see Nichols *et al.* 2001).

Similar species: None. The other poison frogs occurring in Suriname have a different color pattern, having longitudinal bright stripes.

Dendropsophus marmoratus
Marbled Tree Frog
Todo (local name used for all frogs and toads)
Amphibia, Anura, Hylidae

Figure 284. ***Dendropsophus marmoratus*****, Coesewijne, Mar. 2014 (image © Paul Ouboter).**

The Hylidae or tree frog family is the largest frog family in Suriname, with 46 species. Tree frogs are usually characterized by adhesive pads on the digits and large, protruding eyes for binocular vision. Most species are tree-inhabiting and are brown to green in coloration.

Dendropsophus marmoratus (Figure 284) is a well-camouflaged frog looking like tree bark or bird droppings (Toledo & Haddad 2009). It is most commonly found at forest edges, for instance in bushes between closed forest and an open area like savanna, rock outcrop or airstrip. The males call from bushes after heavy rain showers at a height of 1 to 4 m. This happens most during the rainy season, but is also triggered by rain in the dry season. For instance during a severe drought at Brownsberg Plateau, in October 2014, fourteen species of frogs were seen flattening themselves at humid mud patches on the road (Ouboter *et al.* 2016), but *Dendropsophus marmoratus* was not seen or heard at all. A rain shower in the afternoon of November 1st caused all mud-patch associated frogs to disperse in the forest again, and that night males of *Dendropsophus marmoratus* were calling from the bushes near what used to be a large pool in the road. Several individuals were sitting near tire tracks that started filling with water again. The female deposits eggs in shallow pools in or bordering the forest.

Description: A medium-sized tree frog of up to 56 mm body length. Along the lateral edge of the front and hind limbs skin folds are present; the webbing between most fingers and all toes is complete. The coloration of the upper side of the body is very cryptic and similar to tree bark, with brown, yellow-brown and gray colors; the underside is white to orange on the hind part, with black spots, except the underside of the upper arms and legs, which is black, as well as the webbing. The tadpole is brown-green with dark spots (Duellman 1978).

Distribution: The Amazon Basin, Venezuela and the Guianas. In Suriname occurring in both the Zanderij Belt and Interior, but not recorded from the southern part of the country.

Use: (No known use).

Similar species: Characterized by its coloration, especially the underside of the body which is in part orange with black spots. The skin folds along the limbs are straight; in *Dendropsophus melanargyreus* these are serrated (edge not smooth, but more like that of a saw blade).

Boana boans
Giant Gladiator Frog, Rusty Tree Frog
Todo (local name used for all frogs and toads)
Amphibia, Anura, Hylidae

Figure 285. ***Boana boans*, Blanche Marie Falls, Aug. 2009 (image © Paul Ouboter).**

Boana boans (Figure 285), formerly *Hypsiboas boans*, occurs near streams in both pristine rainforests and secondary forests. During most of the year they are high up in the trees. In the reproductive season these tree frogs come down to engage in courtship near streams, and to prepare the nest. The call of the males is very loud and resembles that of a blow on an anvil. Contrary to most other frogs, reproduction is usually in the dry season, because the frogs need shallow water over sand banks to make their nests. The nest is made at the border of the stream by the male with its hind legs, and is a small circular basin in the sand with a sand dike around it. The average basin diameter is 47 cm. The eggs are deposited in it, and hatch in two to three days. Through the movement of the tadpoles, the walls of the basin may collapse in a couple of days. If not, the tadpoles hide in the humid sand. The tadpoles seem to be unpalatable to fish (Lima *et al*. 2006). Occasionally *Boana boans* nests with tadpoles are found outside the dry season.

Description: One of the largest tree frogs of the Guianas, up to 120 mm body length. Large head with a depression in the middle; the last three fingers and all toes are fully webbed. The upper parts of the body are brown to greenish brown, to orange brown; the underside is white; the discs at the tip of the digits are light green to brown. The tadpoles hatch with external gills that are covered later in their development. They are light brown to whitish.

Distribution: Eastern Panama, Pacific lowlands of Colombia and Equador, eastern Amazon Basin and the Guianas. In Suriname, common throughout the Zanderij Belt and Interior, but only known from certain localities in the Coastal Plain.

Use: (No known use).

Similar species: Other large tree frog species in Suriname, such as *Hypsiboas geographicus,* have less or no webbing between the fingers; all other tree frog species are considerably smaller.

Osteocephalus oophagus
Oophagus Slender-legged Treefrog
Todo (local name used for all frogs and toads)
Amphibia, Anura, Hylidae

Figure 286. *Osteocephalus oophagus* at egg-laying tree hole, Sep. 2011 (image © Raffael Ernst).

Osteocephalus oophagus (Figure 286) lives high up in the trees in pristine and secondary rainforests. Reproduction also takes place in the trees. Although the full life cycle can be completed in the trees, in October 2014 several specimens were found near ground level in the Gros-Rosebel area and at Brownsberg Plateau in Suriname. October 2014 was very dry, so the most likely explanation for their descent was that they were in search of more humid locations to prevent desiccation.

The males call during rainy periods to attract females. The characteristic call is very deep, starting with a number of notes shortly after each other and then further apart. Mating pairs remain in amplexus for a long period of time. The eggs are deposited in small water-filled tree holes and bromeliads. After hatching, the tadpoles have to be fed, so the female has to remember the precise locations where eggs were laid. The tadpoles are fed with fertilized eggs that are deposited by the female in the breeding reservoirs every 5 to 7 days. The tadpoles metamorphose after about 34 days (Jungfer & Weygoldt 1995, 1999).

Description: A medium-sized tree frog of up to 53 mm body length Nostrils projecting; skin fold present over the upper and hind margin of the ear drum; skin of the back is rough; toes and part of the fingers are three-quarters webbed. The upper side of the body is brown to grey-brown with darker brown areas; the legs have dark brown bands; the iris is golden with fine black lines radiating from the center.

Distribution: Most of the Guiana Shield. In Suriname widespread in the Zanderij Belt and Interior. Because it lives high up in the trees, this frog is not often seen, but the characteristic call gives it away.

Use: (No known use).

Similar species: *Osteocephalus buckleyi* and *Osteocephalus taurinus* may be confused with this species. The former has a rougher skin on the back, and has a bright orange groin. The latter has a skin that is usually less rough than that of *Osteocephalus oophagus*; it is a larger frog with a more elongate body.

Phyllomedusa bicolor
Giant Monkey Frog
Todo (local name used for all frogs and toads)
Amphibia, Anura, Hylidae

Phyllomedusa bicolor (Figure 287) lives along pools and small streams in savanna forests and rainforests. During most of the year they are high up in the trees, but descend during the rainy season. Then males call from bushes and trees at 1 to 4 m above the ground. The call is a loud 'bok', often with a couple of notes following. A batch of 600 to 2000 white eggs (Lescure & Marty 2000) is deposited by the female on a large leaf that the male folds to form a cylinder. This may protect the eggs against predators. Despite this protection, snakes are quite good at finding the eggs to feed on them. In a forest in the Boven Coesewijne Nature Reserve, two snake species were observed eating *Phyllomedusa bicolor* eggs: *Leptodeira annulata* and *Chironius carinatus* (Figure 288). The eggs hatch in 8 to 10 days (Lima *et al.* 2006) and the tadpoles fall into the water below, where they form a group. Ouboter and Jairam (2012) reported observations that indicate that eggs may be deposited directly in pools as well.

Figure 287. *Phyllomedusa bicolor*, Bakhuis Mountains, Apr. 2015 (image © Paul Ouboter).

Figure 288. *Phyllomedusa bicolor* nest raided by *Chironius carinatus*, Coesewijne Savanna, Apr. 2013 (image © Vanessa Kadosoe).

Description: A large tree frog, up to 135 mm body length. Head large and slightly angular, with a large poison gland above and behind the eardrum; the pupils are positioned vertically; the digits are flattened, flesh-colored with green discs at the tip; the first finger and toe are opposable to hold on to branches. The predominant color of the upper side of the body is grass green; the flanks and the sides of the legs have white spots with a black border; the iris is silvery to gray. The tadpoles are brown gray oranges (Lescure & Marty 2000).

Distribution: The Guiana Shield and Amazon Basin. In Suriname occurring throughout the Zanderij Belt and most of the Interior, but not yet recorded from the central part of Suriname (see Ouboter & Jairam 2012).

Use: The poison of this species is being investigated for medical use (see for example Daly *et al.* 1992, Erspamer *et al.* 1993, Chen *et al.* 2005).

Similar species: *Phyllomedusa bicolor* resembles the three other species of *Phyllomedusa* that occur in Suriname, however these species are considerably smaller and two of them have orange markings on the flanks. *Phyllomedusa bicolor* is most similar to *Phyllomedusa vaillantii*, but this species is slightly smaller and has rougher skin on the upper side of the body.

Leptodactylus pentadactylus South American Bullfrog *Todo* (local name used for all frogs and toads) Amphibia, Anura, Leptodactylidae

Figure 289. *Leptodactylus pentadactylus*, Brownsberg, Jul. 2012 (image © Vanessa Kadosoe).

The Leptodactylidae are a large family of ground-dwelling frogs, represented by at least 18 species in Suriname. They are usually brownish colored frogs, with skin folds between the upper parts of the trunk and the flanks, but without webbing between toes or fingers. Most species make foam nests.

Leptodactylus pentadactylus (Figure 289) is a species of pristine forests, living on the forest floor, often near streams and pools and always near holes in which they escape when disturbed. Males call during the rainy season, a call that can be heard over a long distance. The pair in amplexus produces a foam nest in pre-existing or self-made cavities (Muedeking & Heyer 1976), which may be on dry land or near water. In case the hatched tadpoles cannot get to water, they complete their entire development in the foam (Muedeking & Heyer 1976, Hero & Galatti 1990). Near Manaus, in Central Amazonia, the frogs reach sexual maturity in two years (Galatti 1992).

Description: One of the largest leptodactylids, up to 185 mm body length. Two dorsolateral skin folds – from behind the eyes to the groin – are present. 'Supratympanic' skin folds – above the eardrum – bend to the flanks and are bordered by a black band below. The upper part of the body is bronze brown with more or less distinct transversal darker brown band that may be bordered by a narrow black line; the flanks and hind legs are often more grayish; on either side of the head there is a black band from the snout to the eye, and there are several black triangular spots on the upper lip; the iris is orange.

Distribution: Honduras to northwestern Ecuador, Amazon Basin and the Guianas. Occurs in Suriname throughout the Zanderij Belt and the Interior.

Use: This frog is eaten by some Indigenous tribespeople. Its skin toxins are being investigated for antibiotic properties (Sousa *et al.* 2009).

Similar species: In *Leptodactylus pentadactylus* the dorsolateral folds continue all the way to the groin uninterrupted; in *Leptodactylus knudseni* they are incomplete, or continue to the groin as a series of bumps. The supratympanic fold in *Leptodactylus pentadactylus* bends towards and continues on the flanks, while in *Leptodactylus knudseni* it bends towards the shoulder.

Elachistocleis surinamensis
Suriname Oval Frog
Papitodo
Amphibia, Anura, Microhylidae

Nine species of the family Microhylidae are known to occur in Suriname. Most microhylids have a small, pointed head and much wider oval body. Usually they live a secretive life under leaf litter or in holes in the soil.

Elachistocleis surinamensis (Figure 290) is a species from lowland areas with open vegetation, like pastures, gardens, swamps and savannas. In Trinidad, however, this species seems to occur in forests, while the closely related *Elachistocleis ovalis* is found in open areas (Murphy 1997). It may be that in Trinidad *Elachistocleis ovalis* forces *Elachistocleis surinamensis* into another habitat. During and after heavy rains, the long, high-pitched calls of the males of *Elachistocleis surinamensis* can be heard all over Paramaribo. The males call from under grass or leaves and are very difficult to find. In Trinidad, the floating eggs are deposited during the first three months of the rainy season, and hatch in 48 hours. The tadpoles become froglets after eight weeks (Kenny 1969). In Suriname, reproduction occurs in inundated meadows, gardens and swamps. Mebs *et al*. (2010) reported that *Elachistocleis* spp. mainly eat ants, but poisonous substances obtained from the ants are not stored in the skin as in dendrobatid frogs (see species account *Dendrobates tinctorius*).

Figure 290. *Elachistocleis surinamensis*, Paramaribo, May 2013 (image © Vanessa Kadosoe).

Description: A small frog, up to 43 mm body length. Head small with pointed snout; eyes small; body oval in shape; fingers and toes have no webbing. The predominant color is grayish black; the underside of the trunk has irregular yellow to orange spots (orange spots on the groin). The tadpole is black with white spots, and a cream-colored stripe at the base of the tail (Lescure & Marty 2000)

Distribution: Venezuela, Trinidad and Guianas. A concealed but quite common species in the Coastal Plain; not recorded elsewhere in Suriname, except at one locality in the Savanna Belt, and at Nassau Mountain (Ouboter & Jairam 2012), but the latter record is in doubt.

Use: (No known use).

Similar species: This species has been mistaken for *Elachistocleis ovalis*, which has a yellow underside without spots (also no spots on the groin), a longitudinal white stripe on the back, and a white stripe on the thigh (Lavilla *et al*. 2003, Caramaschi 2010, Ouboter & Jairam 2012). *Elachistocleis ovalis* may not occur in Suriname. In size and shape, *Elachistocleis surinamensis* also resembles other species of the family Microhylidae, but these lack yellow and orange spots on the underside.

Pipa pipa
Suriname Toad
Pipatodo
Amphibia, Anura, Pipidae

The Pipidae or tongueless frogs are a small family with only two species in Suriname. All species of this family are fully aquatic. They lack a tongue. The toes are fully webbed. The body is flattened and the eyes are small and positioned on the top of the head (not the sides). They possess a lateral line system like fish (see Freshwater Fishes chapter) to sense changes in water pressure.

Pipa pipa (Figure 291) is a fully aquatic species that lives in stagnant or slow moving creeks, pools and swamps. Most individuals have been seen in Suriname during night surveys. They were floating just under the water surface with their legs spread out. In this position they wait till a fish comes nearby to swallow it by opening the mouth. The water displacement of the opening mouth directs the prey inside. Potential prey may be attracted to the skin appendages at the sides of the mouth or the long, thin fingers with star-shaped tips.

The male attracts the female under water with clicking sounds. The male and female in amplexus describe vertical circles in the water while the male releases sperm to fertilize the eggs. The eggs fall onto the back of the female; the male helps to get them in the right position (Rabb & Rabb 1960). Here the eggs adhere to the back of the female. Her skin swells around the eggs, so that every egg is embedded. There is no free-living larval stage, and after about 15 weeks froglets hatch from the embedded eggs (Figure 292).

Figure 292. *Pipa pipa* froglet hatching from back of mother, Jan. 2011 (image © Raffael Ernst).

Figure 291. *Pipa pipa*, Bigi Kroetoe, Jun. 2015 (image © Fleur van der Sterren).

Description: Female *Pipa pipa* can reach 200 mm body length, the males are much smaller (80 mm). Entire body flattened; the skin on the upper side of the body is rough; mouth is large with skin appendages in the corners; eyes small with a round pupil; toes are fully webbed; the finger tips are star shaped. The color of the upper part of the body is dark grayish brown; the underside is white with dark spots and a T-shaped dark line.

Distribution: The Guianas, Amazon Basin and Trinidad. In Suriname common in swamps and creeks in the Zanderij Belt and Coastal Plain. Only two records from the Interior are known, one from the upper Kabalebo River and one from the Gros-Rosebel area. Based on the *Pipa pipa* distribution map of Lescure and Marty (2000), the species has also been found along the upper Marowijne River.

Use: (No known use).

Similar species: A closely related species, *Pipa aspera*, occurs in Suriname as well; it is much smaller, and lacks the skin appendages at the corners of the mouth, as well as the star-like finger tips. This species is mostly known from the Interior.

Platemys platycephala
Grooved Sideneck, Twist-necked Turtle
Sekrepatu (local name used for all turtles and tortoises)
Reptilia, Testudines, Chelidae

Figure 293. ***Platemys platycephala*, Brownsberg, Jun. 2014 (image © Vanessa Kadosoe).**

The Chelidae is a family of mostly aquatic turtles. Three chelid species occur in Suriname. The family is mostly characterized by the structure of the skull. Chelids hide their head in the shell by folding the neck sideways. Their toes are completely webbed.

Platemys platycephala (Figure 293) lives in small pools, creeks, swamps and inundated areas in tropical rainforests. It is often encountered walking on land, probably in search of other water bodies or food. Its coloration provides a good camouflage between the dead leaves of the forest floor. Emmons (1989) reports predation by jaguars. Hardly anything is known about its food choice in the wild. *Platemys platycephala* eats all kinds of animal food in captivity, so will probably eat worms, arthropods, small fish, and small frogs in nature. After mating, which in Colombia takes place in the rainy season (Medem 1983), a single egg is deposited. This egg is huge compared to the size of the female turtle, averaging 5.5 x 2.8 cm. No nest is excavated; the egg is laid between rotting leaves, often partly or fully exposed (Medem 1983).

Description: A small species with a maximum shell length of 17 cm. The shell is quite flat (not distinctly dome-shaped, like in many tortoises), and has two ridges on the upper side, with a groove in between. The shell is brownish above, the ridges may be slightly darker; the underside is blackish with a yellow border; the head is yellow above, orange in the center and black on the sides; the legs are dark gray.

Distribution: Amazon Basin, Guianas and eastern Venezuela. In Suriname found throughout the Interior.

Use: Usually considered too small to eat.

Similar species: This species is much smaller than other related species of aquatic turtles, and the only species with a pair of longitudinal ridges on the upper side of the carapace. The color of the head is distinctive.

Chelonoidis carbonaria
Red-footed Tortoise
Redifutu sekrepatu
Reptilia, Testudines, Testudinidae

Figure 294. ***Chelonoidis carbonaria*, Coesewijne Savanna, Jul. 2014 (image © Vanessa Kadosoe).**

The Testudinidae or tortoises are purely terrestrial species. In Suriname two species occur. The shell is usually high and dome-shaped. The feet and digits are short. The hind legs are elephant-like, and have only four toes. Most species are herbivores.

Chelonoidis carbonaria (Figure 294) lives mostly in savannas and rock savannas and the bordering forest. In many areas, especially near savanna-forest transition zones, this species and its close relative *Chelonoidis denticulata* occur together. Both species seem to prefer fruits in their diets. In Suriname they were observed to eat fallen fruits of *Spondias mombin* trees and *Astrocaryum vulgare* palms. Moskovits (1981, in Pritchard & Trebbau 1984) reported fruits as the principal component of their diet in Roraima (Brazil); in addition, leaves, vine shoots, mushrooms, soil, arthropods and carrion were recorded as food. *Chelonoidis carbonaria* excavates a nest hole in the soil in which two to seven eggs are deposited (Medem *et al.* 1979). It is unclear if there is a well-defined nesting season. Although adults are well protected against most predators, shells opened up by the bite of a jaguar are found frequently. Making it easier to collect tortoises for food is one of the reasons that Indigenous people living in the Zanderij Belt and Interior frequently burn savannas. Tortoises often gather below *Astrocaryum vulgare* palms to eat the fallen fruits and can be easily collected then.

Description: Medium-sized tortoise with a maximum shell length of 40 cm. Pritchard and Trebbau (1984) reported two larger specimens (one of 50.5 cm held at the Paramaribo Zoo, and a 51.2 cm male from Brazil). In adult males the shell may be constricted in the middle. The upper side of the shell is black with yellow in the center of each scute (individual plate that is part of the shell); the head and legs are black with contrasting red scales on the legs and yellow to red scales on the head and neck.

Distribution: Northern and eastern Colombia, Venezuela, Guianas, and from eastern Brazil to northern Argentina. In Suriname occurring in and around the northern Zanderij Belt; also known from many isolated rock outcrops, small savannas in the Interior, as well as the Sipaliwini Savanna.

Use: This species is threatened by over-exploitation, both for the international pet trade, and for food (see above). In many areas where the species was common in Suriname in the 1980s, it is now rare. Alas, it is not yet included on the IUCN Red List (2015) (*Chelonoidis denticulata* is listed as vulnerable; see Use, Threats and Conservation chapter).

Similar species: *Chelonoidis denticulata* can be distinguished from *Chelonoidis carbonaria* by its brown shell, becoming lighter brown near the center of the scutes. The scales on its legs and head are yellow.

Dermochelys coriacea
Leatherback Turtle
Aitkanti
Reptilia, Testudines, Dermochelyidae

The Dermochelyidae are large, and have a shell with considerable gaps between the bony plates. The shell has distinct longitudinal ridges, and no scales nor scutes. The only living species is the Leatherback, *Dermochelys coriacea* (Figure 295), but more species are known from fossil records.

Leatherbacks travel long distances in the open ocean. To reproduce they return to shallow waters and make nests on sandy or shell beaches, typically the very beaches where they hatched. They may dive to a depth of 1280 m to feed on jellyfish (Doyle *et al.* 2008); they also eat other soft-bodied marine organisms (Pritchard & Trebbau 1984). The nesting biology of this species was studied in Suriname by Hilterman and Goverse (2007). Females arrived at beaches during high tide, and dug nest cavities in the sediment with their flippers, typically above spring tide level. Between 80 and 88 eggs were deposited in a nest cavity, after which the cavity was filled again with sediment. The female then moved in circles over and near the nest, stirring the sediment over a large area to obscure the nest's exact location. This 'camouflage' may confuse egg predators. Some 4-5 nests were laid by a female during a single nesting season (April-July). The hatching success in the Galibi area (at/west of Marowijne river mouth) where many beaches are stable was low (10.6-25.8%), compared to that in the Matapica area where beaches moved westward each year (52.7-56.0%). This was likely due to disease-causing organisms and predators lingering at stable beaches. The Leatherback population in the area around the Marowijne river mouth is one of the largest in the world (IUCN 2015), with 6,600 to 31,000 nests made annually on the Surinamese side of the border by an estimated 1,545 to 5,500 females (Hilterman & Goverse 2007). Due to westward beach movement, the population previously nesting at Matapica currently nests close to Paramaribo (at/east of the Suriname river mouth).

Figure 295. *Dermochelys coriacea*, Galibi Nature Reserve, Jul. 2006 (image © Jerrel Pinas, WWF Guianas).

Description: The largest living sea turtle with a shell length up to 180 cm. The shell has five longitudinal ridges above and one on each side; it is wide in front, tapering behind, covered with black skin with white spots. The flippers have no claws; the upper jaw has two sharp points in front.

Distribution: In all oceans in tropical, subtropical and temperate regions; nesting occurs in the tropics, such as along the Suriname coast.

Use: Leatherback eggs are considered a delicacy by many people. The species is fully protected in Suriname, but it is threatened by egg poaching and the drowning of adults in gill and trawl nets. Internationally, it is listed as vulnerable (IUCN 2015) and its trade is prohibited (CITES 2015).

Similar species: None. Other sea turtles are smaller and lack the ridges on the shell; they have shells with scutes and flippers with claws.

Paleosuchus palpebrosus
Cuvier's Smooth-fronted Caiman, Dwarf Caiman
Redikaiman
Reptilia, Crocodylia, Alligatoridae

Of the family Alligatoridae (alligators and caimans) three species of caimans occur in Suriname. They have toes that are only half webbed; being freshwater species they lack a salt gland. All alligators and caimans make nest mounds of vegetation, usually in the rainy season. Alligatoridae have generally wider, shorter snouts than Crocodilidae (true crocodiles).

Medem (1958) reports *Paleosuchus palpebrosus* (Figure 296) from swift streams with rapids in Colombia, often near savannas. In the Amazon Basin it occurs in the floodplains (Magnusson 1985). In Suriname it is abundant in the black-water rivers that drain the Zanderij Belt and in the Sipaliwini Savanna. The species seems to have an affinity with savanna landscapes. It often occurs together with the caimans *Caiman crocodilus* or *Paleosuchus trigonatus*. Campos and Magnusson (2013) show that *Paleosuchus palpebrosus* can function at low body temperatures, yet it does not enter the shaded rainforest streams that are the habitat of *Paleosuchus trigonatus*. Medem (1981) describes its nests: mounds of vegetation with between 13 and 18 eggs. In Suriname a nest was located on a small vegetation island with no mound, where the eggs were instead dug into rotting leaves (Ouboter 1996). Nesting occurs in the rainy season and the eggs hatch early in the dry season (Ouboter 1996). The incubation period is approximately 90 days. Parental care seems to be limited since most hatchlings are found alone. However, in October 1988 a pod of six hatchlings was found; when one of the hatchlings emitted a distress call an aggressive adult caiman appeared (Ouboter 1996). Campos *et al.* (2012) recorded parental care for as long as 21 months in three areas in Brazil.

Figure 296. ***Paleosuchus palpebrosus*, Coesewijne River, Mar. 2014 (image © Vanessa Kadosoe).**

Description: One of the smallest species of crocodilians (body length typically <1.5 m; up to 1.82 m in Suriname; see also Campos *et al.* 2010). Both *Paleosuchus* spp. of Suriname lack the bony ridge between the eyes characteristic of the genus *Caiman*. Snout of *Paleosuchus palpebrosus* relatively short and turned up; skin heavily armored. The predominant color of the body's upper parts is reddish-brown; the underside is white with dark blotches; such blotches also occur on the sides of the lower jaw.

Distribution: Venezuela, Guianas, the Amazon Basin, southern Brazil and Paraguay. In Suriname it occurs in the Zanderij Belt, but not in the Interior, except at the Sipaliwini Savanna and the upper Kabalebo River (a single individual collected there by Marinus Hoogmoed; Ouboter 1996).

Use: Its heavy armor makes it useless for the leather industry. It is eaten by Indigenous tribespeople. It is a fully protected species in Suriname.

Similar species: *Paleosuchus trigonatus* has a longer snout, and has a darker, gray color; it has a shorter tail, and double crest on the tail is pointing outwards (as opposed to upwards in *Paleosuchus palpebrosus*).

Iguana iguana
Green Iguana
Legu, legwana
Reptilia, Squamata, Sauria, Iguanidae

Figure 297. *Iguana iguana*, Paramaribo, Apr. 2012 (image © Vanessa Kadosoe).

Most members of the family Iguanidae are adapted to dry habitats and are of medium to large size. Suriname has just a single species of iguanid, *Iguana iguana* (Figure 297), a tree-inhabiting, sun-loving species. It is often seen basking in the vegetation at 3 to 6 m above the ground, especially above medium to large streams. It is abundant in forested areas as well as in gardens and agricultural areas, where it is often seen walking on the ground. Green Iguanas are primarily vegetarians feeding on forest plants and agricultural crops; they supplement their diet with small living animals and carrion. In the forest they only come down to soil level to nest or escape from danger; they often escape by diving into a stream. *Iguana iguana* is a very good swimmer and may stay submerged for up to four and a half hours (Moberly 1968). Nesting mainly takes place in the dry season. In the Interior the females often nest on exposed sandbanks in the riverbed. Nesting in isolated sandbanks provides some protection against terrestrial predators. The eggs are laid at the end of a long tunnel the female digs in the sand (Hoogmoed 1973). Clutch size varies from 9 to 71 eggs (Avila-Pires 1995). The eggs have an incubation period of three months (Hoogmoed 1973). When attacked and unable to flee, the Green Iguana defends itself by biting, scratching and whipping with its long tail.

Description: The largest lizard of Suriname; males up to 2 m long, which includes a tail twice as long as trunk and head combined. Head relatively small with a blunt snout; body cylindrical; adults with spiny crest lengthwise along the back; head with dewlap (skin flap hanging down from the throat) and large round scale below the ear; legs well developed. Body mostly grass green, grayish green in old adults; tail with grey to black bands.

Distribution: From Mexico to Paraguay. Introduced in many Caribbean islands, Hawai and Florida, and there typically invasive, dispersing rapidly and negatively impacting related species. In Suriname occurring all over the country in forested areas and forest margins, especially near streams; also in mangrove forests and cultivated areas.

Use: Both eggs and adults are eaten by an ethnically diverse array of people, yet the Green Iguana remains common in Suriname. In some countries it is bred for its eggs and meat. It is exported for the pet trade and listed by CITES (2015; on Appendix II); it is not listed on the Red List by IUCN (2016), but based on recent IUCN workshops it's likely a species of 'Least concern' (see also chapter Use, Threats and Conservation).

Similar species: Large adults are unmistakable, but young *Iguana iguana* resemble tree-inhabiting lizards such as *Anolis punctatus*, which is green with white spots, has a more pointed snout and a shorter tail, and *Polychrus marmoratus*, which may be completely green but can change color rapidly, and has 2-3 distinctive black lines that radiate from the eyes.

Anolis chrysolepis
Goldenscale Anole
Agama or *lagadisya* (local names used for all lizard species)
Reptilia, Squamata, Sauria, Dactyloidae

Figure 298. *Anolis chrysolepis*, Nassau, Jul. 2006 (image © Paul Ouboter).

Of the Neotropical family Dactyloidae or anoles five species occur in Suriname. They are mostly arboreal lizards, with digital lamellae – pads of microscopic hairs under fingers and toes to improve climbing ability. For territorial and courtship display they use a usually brightly colored dewlap (a skin flap, typically folded when not used, hanging down from the throat).

Anolis chrysolepis (Figure 298, 299), in contrast to most other anoles, stays on the forest floor during most of the time that it is active. It occurs in undisturbed savanna forest and rainforest. Its coloration is an effective camouflage that makes this species difficult to see in the forest floor leaf litter. When disturbed it makes short runs and stops, the stops to re-evaluate the level of threat. Like many other amphibians and reptiles, they often sleep on leaves 0.5 to 1.5 m above the forest floor (see Figure 268). Hoogmoed (1973) studied the stomach contents of several individuals and found various arthropods and small spiders. Reproduction seems to occur during most of the year; females lay one to two eggs (Hoogmoed 1973).

Figure 299. *Anolis chrysolepis* juvenile, Brownsberg, Jun. 2014 (image © Vanessa Kadosoe).

Description: A small anole with a maximum snout-vent length of 74 mm; total length (including tail) is 180 mm. Dewlap small, not reaching the forelegs; legs relatively long; digital lamellae narrow; back with a row of enlarged scales along the spine. The upper side of the entire body is brown, often with light brown spots and a light brown stripe lengthwise along the back. The dewlap of males is bright blue, of females yellow to orange with some gray or cream scales in between.

Distribution: Southern Guyana, Suriname, French Guiana and northern Brazil; in Suriname present from the Zanderij Belt to the southernmost parts of the Interior, including Tafelberg (at 700 m elevation). *Anolis chrysolepis* has been one of the model organisms for the development of the 'refugium hypothesis' that during the Pleistocene the Amazonian forest shrunk, leading to the formation of sister species in (parts of) the Guiana Shield and other supposedly isolated large forest areas (Vanzolini & Williams 1970; see Biogeography chapter).

Use: (No known use).

Similar species: In older publications this same species is referred to as *Anolis nitens*, a name that is not valid anymore. *Anolis chrysolepis* differs from other anoles in Suriname by its ground-dwelling habits, long legs, small dewlap and the color of the dewlap.

Uracentron azureum
Green Thornytail Iguana
Agama (local name used for many lizard species)
Reptilia, Squamata, Sauria, Tropiduridae

The Neotropical lizard family Tropiduridae has five species in Suriname. It is a family of small to medium-sized lizards with spiny scales and bodies that are usually cylindrical or dorso-ventrally flattened.

Uracentron azureum (Figure 300) is a tree-inhabiting species that is rarely observed, most likely because it spends most of its time up in the canopy of the forest. According to Avila-Pires (1995) individuals may also be found at sites with more open vegetation. The author of this chapter observed two individuals in trees in swamp forest, one on a tree trunk in dryland rainforest, and one on an isolated tree in open savanna. Hoogmoed (1973) studied the stomach contents of several individuals and concluded that this species is a food specialist that prefers ants, although some other insects were found as well. An individual kept in captivity readily ate different species of ants, but also accepted grasshoppers (Vanessa Kadosoe, personal communication); it was out and active during daytime, but hiding at night in tree holes. Two pregnant females studied by Hoogmoed (1973) had respectively one and two eggs.

Figure 300. *Uracentron azureum*, Rosebel, Nov. 2013 (image © Paul Ouboter).

Description: A small lizard with a maximum snout-vent length of 8.7 cm; full body length up to 15 cm (with tail). Tail short and spiny; snout short and rounded; rump and tail moderately flattened; legs well developed. It is a beautiful lizard, bright green with black to gray bands; the digits are more yellow with indistinct gray bands.

Distribution: The Guianas and eastern Amazon Basin. In Suriname this species occurs in the Coastal Plain, Zanderij Belt as well as the Interior, but only few individuals have been caught or observed in the country.

Use: (No known use).

Similar species: The general body shape of *Uracentron azureum* is like that of the lizard *Tropidurus hispidus*, but the latter lives on rocky substrates, is larger and grayish-brown, and lacks spines on the tail.

Chatogekko amazonicus
Proposed English name: Amazonian Dwarf Gecko
Lagadisya (local name used for all lizards)
Reptilia, Squamata, Sauria, Sphaerodactylidae

Five species of the family Sphaerodactylidae or dwarf geckos occur in Suriname. The Sphaerodactylidae were formerly considered a subfamily of the Gekkonidae (geckos), but have recently been given family rank by Gamble *et al.* (2011). They are mostly very small to small gecko-like lizards that are active during the day (whereas gekkonids are mostly nocturnal).

Chatogekko amazonicus (Figure 301) is a day-active species of the leaf litter of secondary and primary forests. It seems to avoid wet locations and is seen much less during the rainy season. It can be very abundant locally. *Chatogekko amazonicus* feeds mostly on springtails (Collembola; tiny arthropods) (Avila-Pires 1995). Females can often be seen with a single egg in the body, visible through the skin. The skin is water repellent; this probably helps to prevent this extremely small lizard from drowning during heavy rainstorms of the rainy season.

Figure 301. *Chatogekko amazonicus*, Rosebel, Jun. 2014 (image © Paul Ouboter).

Description: A very small gecko-like lizard with a maximum snout-vent length of 2.4 cm and a total length of only 4.8 cm (including tail). Trunk is cylindrical; scales on the back keeled; claws are enclosed by four asymmetrical scales. Upper side of the body brown, usually with a few lighter spots; short longitudinal line present low on each side of the base of the tail, beginning behind the hind legs.

Distribution: The eastern Guianas and Amazon Basin. In Suriname present in dryland forests of the Zanderij Belt and Interior.

Use: (No known use).

Similar species: Formerly named *Coleodactylus amazonicus*, *Chatogekko amazonicus* was placed in a genus of its own in 2011. *Coleodactylus septentrionalis* looks similar but usually has a white line on the back of the head and more obvious white spots on the body. *Pseudogonatodes guianensis*, another small, day-active leaf litter inhabitant also looks similar but is more elongated. Both these similar species have claws enclosed by five asymmetrical scales (not four as in *Chatogekko amazonicus*).

Neusticurus bicarinatus
Two-faced Neusticurus
Lagadisya (local name used for all lizards)
Reptilia, Squamata, Sauria, Gymnophthalmidae

There are at least 15 species of Gymnophthalmidae (spectacled lizards) in Suriname. Gymnophthalmidae are characterized by small size, an usually elongated body and often a reduction of the limbs. A transparent disc is present in the lower eyelid, to accommodate vision while the eyelids are closed. This protects the eyes while the lizard is active in leaf litter and soil. In some species the eyelids have completely grown together.

Neusticurus bicarinatus (Figure 302) is a day-active, semi-aquatic lizard that lives near creeks in the forest. It forages on the edge between water and land. Based on the analysis of stomach contents, they eat mostly small terrestrial animals (land snail, millipedes, insects, spiders), but also aquatic and semi-aquatic ones (water bug, small catfish) (Hoogmoed 1973). When disturbed this lizard flees into the water and swims away with the head emerging, or it dives underwater into a hiding place, for instance under a rock. Pregnant females that have been collected carried two eggs (Hoogmoed 1973).

Figure 302. *Neusticurus bicarinatus*. Brownsberg, Sep. 2009 (image © Paul Ouboter).

Description: Large for a gymnophthalmid with a maximum snout-vent length of 11 cm; total length 34 cm (including tail). Snout slightly elongated and pointed; head wider than the neck; ear drum located deep inside the ear; legs well developed. Upper side light to dark brown, usually with lighter and darker spots; red spots, sometimes forming a circle, around the eyes; two black stripes above the eye; a white, black-edged eye-spot in front of each foreleg. Resembles a tiny caiman, due to large keeled scales on the back, a double row of large scales forming a double crest on the base of the otherwise laterally flattened tail.

Distribution: The Guianas and western Amazon Basin. In Suriname only known from the Zanderij Belt and Interior, along creeks. Small-scale gold mining in creeks (see chapter Use, Threats and Conservation) destroys this species' habitat and may lead to its local extinction.

Use: (No known use).

Similar species: Resembles the only other species of the genus *Neusticurus* known from Suriname, *Neusticurus rudis*. In *Neusticurus bicarinatus,* the enlarged scales on the tail form two continuous rows; in *Neusticurus rudis* the rows are not continuous due to the presence of smaller scales between the enlarged ones.

Tretioscincus agilis
Smooth Tegu
Lagadisya (local name used for all lizards)
Reptilia, Squamata, Sauria, Gymnophthalmidae

Figure 303. *Tretioscincus agilis*, Brownsberg, Dec. 2016 (image © Vanessa Kadosoe).

Tretioscincus agilis (Figure 303) is an inhabitant of the primary rainforest, where it lives on dead and living tree trunks and the surrounding leaf litter (Hoogmoed 1973; Avila-Pires 1995; Paul Ouboter, pers. obs.). It is a sun-loving species that is not seen active before about 10 am, nor after 4 pm, and not at all on cloudy days. When disturbed, it usually tries to escape by climbing a tree, although Hoogmoed (1973) reported it circling a tree when pursued. Almost anywhere *Tretioscincus agilis* is rare and researchers may do years of fieldwork without seeing one. However, recently at Brownsberg Nature Park, seven specimens were seen in two years. Hoogmoed (1973) reported spiders, crickets and bugs in the stomach contents of two individuals and concluded that this species finds its prey under objects on the soil, such as logs.

Description: A relatively small lizard with a maximum snout-vent length of 6 cm; total length 15 cm (including tail). Rump cylindrical, covered by large, smooth scales; legs well developed, but the inner finger reduced in size; lower eyelid movable, with a transparent disc. The upper parts of the head and the back are bronze brown, the flanks and limbs are gray to black; between the upper parts and the flanks is a longitudinal light yellow band that transitions into a bright blue tail.

Distribution: The Guianas and the Brazilian states of Para and Amazonas. Occurs in Suriname throughout the Interior.

Use: (No known use).

Similar species: Several other gymnophthalmids that occur in Suriname have the same size, shape and smooth scales, but none of these has a blue tail. Young *Mabuya nigropunctata* (of the family Scincidae) may have a blue tail, but their scales are much smaller and the inner finger is not reduced in size.

Tupinambis teguixin
Black Tegu, Golden Tegu
Sapakara
Reptiles, Squamata, Sauria, Teidae

Six species of Teidae occur in Suriname. They are medium-sized to large lizards, with the scales on the underside of the body large and rectangular. Their limbs are well developed, and they have a snake-like tongue. Some species have all-female populations and reproduce by parthenogenesis (females lay unfertilized eggs that hatch).

Tupinambis teguixin (Figure 304) is a sun-loving species of more open areas, though it requires these areas to have hiding places such as bushes, or to be near the forest edge. Suitable areas include beaches, gardens, agricultural areas, roadsides and clearings in the forest, but not dense, pristine forest. Natural treefall gaps in dense forest will only be colonized when these are close to already occupied areas with more open vegetation. When disturbed it flees into the cover of bushes, under logs or into a hole. *Tupinambis teguixin* is a predator of large invertebrates and small vertebrates like lizards, frogs and small rodents and birds. It also feeds on eggs and carrion. It has been observed eating the eggs of birds, turtles, snakes and caimans. According to Avila-Pires (1995) it also eats leaves and fruits. A summary of its reproductive habits is presented by Avila-Pires (1995): the eggs are often deposited in termite nests, even in nests high up in trees; the number of eggs in nests varied between 4 and 32. The nests with high numbers of eggs may be communal ones. In captivity the incubation period of the eggs was 5-6 months.

Figure 304. *Tupinambis teguixin*, Brownsberg, Aug. 2007 (image © Paul Ouboter).

Description: A large lizard with a maximum snout-vent length of 34.5 cm; total length almost 1 m (including tail). Head slightly tapered; rump cylindrical; strong legs and long tail. The upper side of the body is black with predominantly white to cream colored transverse bands that are quite irregular and spotty. Adult males can be recognized by their swollen throat.

Distribution: Northern South America, south till central Brazil. In Suriname occurring all over the country, and common in more open, disturbed habitats where some cover is available.

Use: In Suriname incidentally hunted for its meat; in rural areas, the Black Tegu is often regarded as a pest and killed, as it devours young chickens and eggs. It is listed in Appendix II of CITES (2015) (see Use, Threats and Conservation chapter); its skin is used by the leather industry (not in Suriname), and live individuals are caught and exported as pets.

Similar species: Most similar to the lizard *Ameiva ameiva*, but the latter species is much smaller and green-brown colored.

Eunectes murinus
Green Anaconda, Common Anaconda, Water Boa
Aboma
Reptilia, Squamata, Serpentes, Boidae

Of the snake family Boidae seven species occur in Suriname. They are large non-venomous snakes that kill their prey by constriction and suffocation ('strangulation'). Vestigial limbs (spurs) and remnants of the pelvis bones testify of their descent from four-legged reptiles.

Eunectes murinus (Figure 305) is semi-aquatic, typically found with its body submerged and its head at the water surface. It is occasionally seen ashore, especially when digesting large prey. In Suriname, it is abundant in swampy areas, but also occurs in rivers and creeks. Rivas (2000) reports that it prefers shallow water in the Llanos (Venezuela) and migrates, because of seasonal water level changes. It is often considered strictly nocturnal, but day-active individuals have often been seen in Suriname. Rivas (2000) observed it to be active at all times of day and night. Green Anacondas stalk their aquatic or terrestrial prey from the water, mostly birds when young, when older mammals and reptiles, including turtles and caimans. In the Llanos the most important anaconda predator is the Spectacled Caiman (Rivas 2000). In Suriname a caiman has been seen eating an anaconda, and anacondas killing caimans. Green Anacondas also prey on their own kind: the smaller males are eaten by females (Rivas 2000). In the Llanos mating occurs in the dry season; the female attracts numerous males that coil themselves around the female (Rivas 2000). Like all boas, this species is viviparous (lays no eggs, but gives birth to live young); clutch size varies from 5 to 75 and is related to female size.

Figure 305. *Eunectes murinus*, Kabalebo, Aug. 2009 (image © Paul Ouboter).

Description: Heaviest snake in the world (102 kg individual caught in French Guiana; Starace 2013) and second longest, up to 6 m (but the largest of 780 individuals measured in the Llanos was 521 cm; Rivas 2000). Males are usually 3 m or less. Head small compared to much of the rest of the body; eyes on upper part of the head; distinct spurs on both sides of the cloaca (spurs longer in males); scales very smooth. Body olive green; black oval spots on the back; black oval spots with a pink to orange center on the sides; head on each side has a black stripe from the eye to the corner of the mouth, with an orange band above.

Distribution: Northern South America, south till Paraguay; also Trinidad. In Suriname present throughout the country.

Use: Anaconda fat is used to alleviate rheumatic pain and inflammations. Anacondas are killed as vermin in Suriname, and have become rare in some areas where they were once abundant.

Similar species: *Eunectes deschauenseei* is known from the Lower Marowijne River. It is much smaller than *Eunectes murinus*, is grayish-green to yellowish with black-only markings.

Chironius carinatus
Amazon Whipsnake, Sipo
Reditere
Reptilia, Squamata, Serpentes, Colubridae

The Colubridae are the largest family of snakes in Suriname (and almost everywhere else) with at least 68 species. They lack obvious pelvic bone remnants and have no functioning left lung. The scutes on the head are large. While most species are not venomous, many are opisthoglyph, meaning that they do secrete venom, and in the back of the mouth have either poison fangs (teeth with venom canal) or 'normal' enlarged teeth (see for example Kardong 1982).

Chironius carinatus (Figure 306) is very active during the day on the ground and in the vegetation; according to Henderson *et al.* (1976) it inhabits trees in Peru. It occurs in primary and secondary forests, savannas and on cultivated land. It is well known for its bold behavior when it encounters people: it raises the front part of its body, comes closer and may even follow someone walking or running away. It never really attacks, however. When threatened it vibrates its tail, expands the front part of its body, and may even open its mouth. It will bite when harassed. This species feeds on amphibians, lizards, birds and small mammals (Starace 2013). At Coesewijne Nature Reserve, this snake was observed eating the eggs from a nest of the frog *Phyllomedusa bicolor* (Figure 288). It is oviparous, meaning that it lays eggs; clutch size is between 5 and 12 eggs (Dixon *et al.* 1993).

Figure 306. *Chironius carinatus*, Brownsberg, Dec. 2012 (image © Vanessa Kadosoe).

Description: Maximum body length is 180 cm; the eyes are relatively large. The upper part of the body is olive-green to gray to brownish-gray, brown in young individuals; the flanks are gray to bluish-gray; the underside of the body is yellow to orange, and color transition towards the flanks is abrupt, in a zigzag pattern (where the lateral gray and the yellow or orange scales below meet); the head is often of a lighter brown color.

Distribution: From Costa Rica south to northern Argentina; also on Trinidad. Common throughout Suriname.

Use: (No known use).

Similar species: Quite similar to other species of the genus *Chironius*, but can usually be distinguished from these by its uniformly gray upper body and yellowish underside. *Chironius exoletus* looks very similar, but has no zigzag pattern where the contrasting colors of the flanks and the underside of the body transition; instead the transition zone is straight (some scales thus being part gray, part yellowish).

Dipsas catesbyi
Catesby's Snail-eating Snake
Sneki (local name used for all snakes)
Reptilia, Squamata, Colubridae

Figure 307. ***Dipsas catesbyi*, Rosebel, Jun. 2014 (image © Paul Ouboter).**

Dipsas catesbyi (Figure 307) is a nocturnal snake that lives in primary and secondary forests. It is mostly seen in bushes at a height of 1 to 4 m above soil level. It is a diet specialist, eating slugs and snails; its teeth are adapted to remove the snails from their shells. When disturbed this snake does not try to bite but releases a foul-smelling fluid from glands located at the cloaca. It is oviparous (egg-laying) and lays 2 to 6 eggs (Starace 2013).

Description: Maximum body length is 70 cm. Slender snake with wide head and short snout; eyes large. The predominant body color is brown to reddish-brown, but the brown of the front part of the body, behind the head, is very pale, almost white; the brown of the upper part of the body and flanks is interrupted by very large spots with a white border; the head is black with a white band in front of the eyes (on the snout, but not its tip) and a broad white and/or brownish or orange band at the back of the head.

Distribution: Venezuela, the Guianas and the Amazon Basin. Occurs throughout Suriname.

Use: (No known use).

Similar species: The relatively wide head and short snout are characteristic for the genus *Dipsas*. Differs from other *Dipsas* spp. in its color pattern, especially the head region. The closest resemblance is with *Dipsas pavonina* and *Dipsas copei*. *Dipsas pavonina* has a much narrower white band on snout and a narrow white band on the back of the head; *Dipsas copei* has wide orange-brown bands (the band on the snout continues to the tip of the snout).

Helicops angulatus
Brown-banded Water Snake
Watrasneki (local name used for many aquatic colubrid species)
Reptilia, Squamata, Serpentes, Colubridae

Figure 308. *Helicops angulatus* eating a *Rhinella* sp., Wanica, Dec. 2013 (image © Aniel Gangadin).

Helicops angulatus (Figure 308) is an aquatic snake living in swamps, ponds and ditches. It is one of the most common snakes in cultivated land. When harassed it will bite and may release a musky, foul-smelling substance from the cloaca. This snake's rear teeth are enlarged and its saliva is mildly toxic, affecting the nervous system (Estrelia *et al*. 2011). *Helicops angulatus* mainly feeds on fish and amphibians. Most reports mention this snake to be oviparous, laying 8 to 11 eggs (Murphy 1997); however Rossman (1973, 1984) reported females that laid eggs with fully developed embryos, meaning that these eggs were ready to hatch when leaving the body of the female (this is known as ovovivipary). Whether the embryos require a substantial amount of time to continue developing inside eggs that have already been laid (ovipary) or not (ovovivipary) may depend on environmental conditions affecting the female.

Description: Maximum body length is 100 cm. Relatively small eyes, located high on the upper part of the head; scales on the back keeled. The predominant color is yellowish to grayish brown on the upper side of the body and flanks, interrupted by a total of 20-27 wide, grayish to black transverse bands that are narrower on the flanks; the brown transitions to anything between pale yellow to bright red on the underside of the body, where the gray to black bands continue, many asymmetrically, to produce a checkered pattern on much of the underside of the body.

Distribution: Venezuela, Guianas and Amazon Basin, as well as Trinidad.

In Suriname known mostly from the Coastal Plain and Zanderij Belt, but a few individuals have been collected in the Interior, indicating that this species may occur in lowland areas all over the country.

Use: (No known use).

Similar species: The only other aquatic snake in Suriname with the eyes high up on the head is the close relative *Helicops leopardinus*. This species can be distinguished based on its smooth temporal scales (rows of scales behind the scales that encircle the eye; as opposed to the keeled ones of *Helicops angulatus*), and its brown-grayish upper body with rows of black spots (as opposed to black bands).

Erythrolamprus aesculapii
Common False Coral Snake
Krarasneki (local name use for a variety of similarly colored snakes)
Reptilia, Squamata, Serpentes, Colubridae

Figure 309. ***Erythrolamprus aesculapii*****, Brownsberg, Sep. 2010 (image © Paul Ouboter).**

Erythrolamprus aesculapii (Figure 309) is a snake of the forest floor in primary and secondary rainforests, active by day and night. It often hides under the leaf litter or in holes in the ground. It feeds mainly on other snakes, including poisonous ones, but may occasionally eat lizards, eels and insects. It is opisthoglyph, having poison fangs in the back of the mouth, and its bite may cause local swelling and pain in people. It is considered part of a mimicry complex that includes extremely poisonous coral snakes of the genus *Micrurus* with very distinct color patterns, and similarly colored non-poisonous false coral snakes of the genus *Atractus*. In such mimicry complexes, predators learn to avoid poisonous or distasteful species, as well as innocuous species resembling them. Unless *Erythrolamprus aesculapii* serves as a model, it is difficult to imagine how this particular mimicry complex works, since predators bitten by an actual coral snake cannot learn to avoid it, as they will die (but see Greene & MacDiarmid 1981). Most of the body of *Erythrolamprus aesculapii* typically has a repeated banding pattern of red, followed by black, white, and black. The pattern resembles the slightly more complex banding pattern of some coral snakes of the genus *Micrurus*: red, followed by black, white or yellow, black, white or yellow, black, and finally red again.

Description: Maximum body length is 95 cm. The head is white with a transverse black band over the eyes and another one in the neck; the rest of the body has the above-mentioned banding pattern. However, there is much variation in the pattern: bands vary in width, and part of the black and white bands may be missing, especially on the front part of the body (including the head) which then is entirely red. Individuals may lack red bards altogether; they have black and white bands only.

Distribution: Venezuela, Trinidad, Guianas, Amazon Basin and south till Paraguay and northern Argentina. Known from the Brownsberg and Tafelberg mountains, and given its wide distribution in French Guiana (Starace 2013), probably occurs throughout Suriname's Interior.

Use: (No known use).

Similar species: *Atractus badius* has duller colors, a black head with just one white band in the neck, and narrower white bands as part of the repeated banding pattern. Other false coral snakes have different repeated banding patterns: in *Rhinobothryum lentiginosum* it is a black band, followed by a narrow white one, a red one, and again a narrow white one; *Oxyrhopus petola* has only red and black bands. *Micrurus diutius* and *Micrurus lemniscatus* look very similar but always have the extra white bands (see above). Atypically colored *Erythrolamprus aesculapii* can be difficult to distinguish from real coral snakes (*Micrurus* spp.).

Oxybelis fulgidus
Green Vine Snake
Busiswipi (local name also used for other *Oxybelis* spp.)
Reptilia, Squamata, Serpentes, Colubridae

Figure 310. ***Oxybelis fulgidus*, Berg en Dal, Sep. 2009 (image © Paul Ouboter).**

Oxybelis fulgidus (Figure 310) is a species from primary and secondary rainforests and savanna forests, living in bushes mostly between 1 and 3 m above ground level, but also traveling on the forest floor. It is also found on cultivated land. Its diet consists mainly of birds, but lizards and small mammals are also eaten. *Oxybelis fulgidus* is opisthoglyph (with poison fangs in the back of the mouth), and kills captured prey by moving it further down the mouth where the fangs are then used to inject venom. It is an aggressive snake that will bite quickly when harassed. Its bite may cause local swelling and pain in people bitten, and in some cases more severe reactions. Its poison contains toxins that affect the nervous systems; the poison immobilizes lizards immediately, but has no effect on mice (Heyborne & Mckessy 2013). In Brazil, mating occurs at the end of the rainy season and eggs are laid in the dry season; clutch size varied between 6 and 12 eggs (Scartozzoni *et al.* 2009).

Description: A very long and slender snake; maximum body size is 180 cm. According to Scartozzoni *et al.* (2009), females are larger than males. Elongated, very pointed snout, and a large mouth. The predominant body color is grass green; a narrow yellowish-white stripe is present along each side of the belly and tail; the underside of the body is light green to yellow; the head is light green below a longitudinal blackish stripe on either side at the level of the eye, above the stripe the head is darker green.

Distribution: From Mexico through Central and South America, south till northern Argentina. This species has been observed in lowland and upland areas in the Interior of Suriname; it probably occurs throughout the Zanderij Belt and Interior.

Use: Its toxins are the subject of research for possible use in medicine.

Similar species: Two other snakes also live in bushes, and are very slender and elongated: *Oxybelis aeneus* and *Philodryas argenteus*. *Oxybelis aeneus* is not green but bronze-brown. *Philodryas argenteus* is grass green but does not have a pointed snout.

Micrurus surinamensis
Surinam Coral Snake, Aquatic Coral Snake
Krarasneki (local name use for a variety of similarly colored snakes)
Reptilia, Squamata, Serpentes, Elapidae

Figure 311. ***Micrurus surinamensis*, Coesewijne, Jun. 2015 (image © Christian D'Orgeix).**

Suriname is home to at least seven species of Elapidae, a family that includes brightly colored coral snakes in the New World and cobras in the Old World. In general appearance many species look quite similar to colubrid snakes, with large head scales, but elapids differ in having poison fangs in the front part of the mouth that are fixed, meaning that they do not fold back when the mouth is closed. Elapid poison mostly affects the nervous system of prey and people bitten.

Micrurus surinamensis (Figure 311) is an aquatic snake living in forested areas of the Zanderij Belt and Interior. Although wandering on land occasionally, it is mostly active completely under water. Most literature reports it to be a nocturnal species, but the specimens observed under water in Suriname were active during the day. It feeds almost exclusively on fish. An oviparous species, it deposits its 5 to 11 eggs during the long dry season (Starace 2013). It is quite aggressive when harassed, and has one of the strongest poisons of all coral snakes (Roze 1996).

Description: The largest coral snake of the Guianas, with a maximum body length of 140 cm, and a robust build. The head is red with all scales having a black border; the neck has a yellowish white band; the rest of the body has a banding pattern that is repeated 5-8 times: a black band, followed by a white, black, white, black, and finally a broad red one; the scales of the red bands are entirely red except at the hind tip.

Distribution: Venezuela, Guianas and Amazon Basin. In Suriname most records are from black water rivers in the Zanderij Belt; while in Suriname a few individuals have been observed in the Interior, in French Guiana it is only known from the coastal zone (Starace 2013).

Use: The poison of *Micrurus surinamensis* is a popular subject of research (for example Olamendi-Portugal *et al.* 2008) to find medical applications.

Similar species: Differs from all other coral snakes (*Micrurus* spp.) by its size, the banding pattern, and the very distinct black borders of the red head scales. The eyes and nostrils of *Micrurus surinamensis* are located higher up the head than in other coral snakes; only the fourth supralabial plate (part of a row of scales on the upper lip) is in contact with the eye, while in other coral snakes two supralabials border the eye.

Bothrops atrox
Fer-de-lance, Common Lancehead
Labaria, owrukuku, rasper
Reptilia, Squamata, Serpentes, Viperidae

Six species of vipers (family Viperidae) are found in Suriname. These poisonous snakes look sluggish and stocky. They have a short tail and a triangular head (seen from above). The scales on the head are small and numerous; in most species the scales on the upper side of the body and flanks are keeled. The long poison fangs are attached to short, moveable upper jawbones; when the mouth is closed, the fangs are folded backward against the roof of the mouth, but when it opens, the fangs are erected. The poison mostly affects the tissue and blood of prey and people that are bitten. All New World vipers are pit vipers (subfamily Crotalinae); they possess a pair of heat-sensitive organs in pits between the eyes and the nostrils.

Bothrops atrox (Figure 312) forages on the ground and sometimes in low bushes from the late afternoon till the early morning, but during daytime it is immobile, coiled up between dead leaves where it is well camouflaged. When disturbed it may quickly vibrate its tail, produce a rattling sound when stirring the dry leaf litter, and will strike when approached. It occurs in many habitats: primary and secondary forests, savannas, seasonal swamps and cultivated land, including gardens. In Suriname, it is mostly encountered during the rainy season near water, hardly at all during the dry season; this is in agreement with observations from Central Amazonia (Oliveira & Martins 2001). Young individuals mostly feed on frogs; adults feed on frogs, lizards and small mammals. It is viviparous (lays no egs, but bears live young). According to Starace (2013) in French Guiana copulation takes place at the end of the rainy season, and females subsequently produce between 11 and 43 live young.

Figure 312. *Bothrops atrox*, Nassau, Mar. 2006 (image © Paul Ouboter).

Description: Maximum body length is 162 cm (Campbell & Lamar 2004), but typically less than 100 cm. The broad triangular head makes for a distinct neck; the heat sensitive pits are clearly visible. The body color is quite variable, predominantly grayish to yellowish brown; often there are triangular brown blotches present with contrasting pale borders; behind each eye is a dark longitudinal stripe; the end of the tail of most individuals, especially young ones, is light colored.

Distribution: Eastern Venezuela, Guianas and the Amazon Basin; one of the commonest snakes of Suriname, and cause of most snake bites.

Use: *Bothrops atrox* venom is well studied; some of its components are used in medical science (see for example Niewiarowski *et al.* 1979)

Similar species: *Bothrops brazili* looks similar but has a more regular and more contrasting pattern of blotches; it is much rarer than *Bothrops atrox* and lives in primary forests only, especially in mountainous areas, and seems to be intolerant to habitat alterations.

Lachesis muta
Bushmaster, South American Bushmaster
Makasneki, kapasisneki
Reptilia, Squamata, Serpentes, Viperidae

Figure 313. ***Lachesis muta*, Kleine Saramacca, May 2015 (image © Dick Lock).**

Lachesis muta (Figure 313) is a large night-active viper (see preceding species account) of primary forests. According to Campbell and Lamar (1989) it is mostly found in relatively wet rainforests and montane forests with an annual precipitation above 2000 mm. It seems to be intolerant to changes in its habitat. During the daytime this snake hides in holes, between tree buttresses or under roots or fallen branches in the forest. It is often reported to live in the burrows of armadillos, explaining its local name *kapasisneki* (*kapasi* = armadillo; *sneki* = snake). On Trinidad it also seems to be associated with paca burrows (Murphy 1997). It has been reported to be aggressive at night. When threatened it inflates it neck and vibrates its tail; between dead leaves this produces a rattling sound. Because this is a large snake, it produces a lot of venom, and when a person is bitten, this often has very serious consequences; several people in the Interior of Suriname have lost a leg due to a bite from this species. *Lachesis muta* feeds mostly on small mammals, especially rodents and opossums (Starace 2013). It is the only oviparous pit viper in the Americas, laying 7 to 20 eggs in the burrow of an armadillo or paca (Campbell & Lamar 2004; Starace 2013). The female protects the eggs and may even have a role in regulating their temperature (Starace 2013).

Description: This is the largest viper in the world, reaching 360 cm body length (Campbell & Lamar 1989). Head more oval than triangular; scales on the upper side of the body strongly keeled, in the middle (along the spine) forming a longitudinal ridge; scales at the end of the tail spiny and irregular, the very tip of the tail is a spine. The predominant body color is yellowish to reddish brown, with the underside of the body white; behind each eye there is a dark stripe till the corner of the mouth, usually with a pale stripe above it; on the upper side of the rest of the body there are 20-37 diamond-shaped dark brown to black blotches that form triangles on the flanks; the center of these blotches is paler.

Distribution: Northern South America: Venezuela, Guianas, Amazon Basin and Trinidad; an isolated population occurs in the southeast of Brazil (Atlantic Forest region) that is considered by some as a subspecies (an isolated, distinct group of populations in that region). Occurs throughout Suriname in undisturbed forested areas.

Use: Much research has focused and is still focusing on the venom (see for example Fuly *et al.* 1997).

Similar species: *Lachesis muta* is much larger than the other vipers in Suriname; smaller specimens can be distinguished from *Bothrops* spp. by the shape of the head, the strongly contrasting diamond-shaped dark blotches and the distinct spine at the end of the tail.

Birds

Brian O'Shea

The lowland Neotropics is home to more bird species than anywhere else on Earth, a fact reflected in Suriname's spectacular diversity of birds. From the forested hills of the Interior to the mangroves of the Coastal Zone, Suriname is blessed with largely intact natural landscapes supporting more than 700 bird species. Suriname lies on the ancient Guiana Shield (Hammond 2005), long recognized as an area of endemism for South American birds (Haffer 1969, Cracraft 1985), and indeed many of its resident forest birds occur only in the region north of the Amazon River and east of the Rio Branco (Naka 2011). Savannas, mountains, and coastal regions support distinctive bird communities to complement the hundreds of species found in Suriname's extensive Interior forests. The result is outstanding bird diversity for a country of its size.

Figure 314. Suriname's coastline and estuaries support large concentrations of wading birds, such as these American Flamingos (*Phoenicopterus ruber*) taking flight at Bigi Pan Nature Reserve in Nickerie (image © Brian O'Shea).

The birdlife zones of Suriname can be envisioned as a series of east-west-oriented bands progressing inland from the coast (Ottema *et al.* 2009). Mudflats, mangrove forests and herbaceous swamps along the coast support one of the highest concentrations of migratory shorebirds in the world, as well as a diverse assemblage of resident species, many of them restricted to the Coastal Zone (Figure 314). Moving inland, the coastal mangroves quickly give way to a varied landscape of tall swamp and marsh forest, agricultural lands, and, near Paramaribo, urban development. This diversity of habitats is largely responsible for the abundant and diverse birdlife of the Coastal Plain. Still further inland is the Zanderij Belt, a region with sandy soils featuring savanna vegetation and stunted forests. Many of the same generalist species found in secondary habitats to the north also occur here, in addition to more specialized savanna birds.

South of the Zanderij Belt lies the most extensive ecosystem in Suriname – the lowland rain forest of the Interior (the Precambrian Shield), which stretches unbroken to the country's southern border and beyond. The lowland rain forest of the Interior is by far the most species-rich habitat in Suriname – roughly 500 of the 726 bird species known from the country can be found here. Although mostly below 300 m in elevation, the Interior also features several ranges of forested hills and plateaus, some of which approach or exceed 1000 m. The larger of these formations, such as the Tafelberg in central Suriname, support a limited suite of highland birds in addition to more widespread lowland species (Zyskowski *et al.* 2011). The bulk of these highland species' ranges are hundreds of kilometers to the west in the tepui highlands of Guyana, Venezuela and northern Brazil.

At the far southern end of Suriname is a unique rolling grassland peppered with trees – the Sipaliwini Savanna (see Ecosystems chapter), with an avifauna (bird fauna) more similar to that of the central Brazilian Cerrado than savanna regions elsewhere in the Guianas (Robbins *et al.* 2004, O'Shea 2005, Mittermeier *et al.* 2010). Many species, such as the Campo Flicker (*Colaptes campestris*) and Gray Monjita (*Xolmis cinerea*), are only known in Suriname from this fascinating region.

Figure 315. The Tropical Kingbird (*Tyrannus melancholicus*) is one of 91 species of tyrant flycatchers occurring in Suriname (image © Kester Clarke).

Under current taxonomy (Remsen *et al.* 2017), the avifauna of Suriname comprises 726 species, in 80 families and 28 orders. The 620 resident species are supplemented by 82 migratory species that visit Suriname during their non-breeding seasons, 21 species that do not regularly occur in the country but have been observed once or twice, and two introduced species. One additional species, the Horned Screamer (*Anhima cornuta*), is known from historical accounts, but no longer occurs in the country. With an increasing number of birdwatchers in Suriname documenting their observations, the number of species known from the country continues to rise.

Suriname shares the majority of its lowland forest avifauna with adjacent Guyana and French Guiana. The most diverse family is the Tyrannidae, or tyrant flycatchers (Figure 315), with 91 species occurring in the country. Bird diversity is highest in the Interior, where some of the sites most visited by birdwatchers, such as Brownsberg Nature Park, boast lists in excess of 400 species. In the forested Interior, approximately 10 to 15 percent of the bird species at any locality are endemic to the Guiana Shield, meaning that they are not found outside this region (Naka 2011). Suriname's avifauna reflects the high bird diversity of the Neotropics in general, and also the largely intact natural landscapes found throughout the country. What makes Suriname special is the healthy population status of many large and spectacular birds – cotingas, parrots, guans and curassows, the Harpy Eagle, and other birds of prey – that have declined in other parts of the Neotropics where habitat has been fragmented and lost.

Bird habitats in Suriname

A diverse Coastal Zone. The Coastal Zone of Suriname features a mosaic of natural vegetation types interspersed with extensive urban and agricultural habitats. In contrast to the ancient forests of the Interior, the landscape of the Coastal Zone is dynamic at multiple timescales. In geological time (from the Miocene to the present), the distribution of habitats has been shaped and reshaped by deposition and reworking of sediments associated with sea level changes, fluvial processes, and ocean currents. On an ecological time scale, the vegetation of the Coastal Zone has been transformed by human activities over the past several centuries, including rice and sugarcane farming, bauxite mining, and urban development. There is no doubt that human activities have strongly influenced the distribution and composition of Suriname's Coastal Plain avifauna. In particular, recent habitat changes have benefited generalist species and those of open habitats (Figure 316). With 90 percent of Suriname's growing human population living in the Coastal Zone, the extent of non-forest habitat continues to increase.

Figure 316. The *kawfutuboi*, or Smooth-billed Ani (*Crotophaga ani*), is one of many ecological generalists that thrive in developed areas of Suriname's Coastal Plain (image © Meshach Pierre).

Estuarine areas of the Coastal Plain. Suriname's coastline is protected by extensive areas of mangroves, which also grow in tidal zones along major rivers. These forests offer protection to nesting waders such as the Boat-billed Heron (*Cochlearius cochlearius; arapapa*) and Scarlet Ibis (*Eudocimus ruber; korikori*). At low tide, crabs emerge to forage on the exposed mud, only to become prey for Rufous Crab-Hawks (*Buteogallus aequinoctialis*; *krabu-aka*) lurking above. These forests also offer a safe haven for roosting Orange-winged Parrots (*Amazona amazonica; kulekule*), thousands of which can sometimes be seen winging their way northward from the coastal plain in the early evening. The mangrove forests support an abundant and diverse resident avifauna. Common resident species include Spotted Tody-Flycatcher (*Todirostrum maculatum*), Yellow-breasted Flycatcher (*Tolmomyias flaviventris*), Great Kiskadee (*Pitangus sulphuratus*) and Bicolored Conebill (*Conirostrum bicolor*). In Suriname, this last species, along with the elusive Mangrove Cuckoo (*Coccyzus minor*), is essentially restricted to coastal mangroves.

The coastline also features vast river estuaries that empty into the Atlantic along Suriname's shores. These swamps, lagoons and mudflats are home to a spectacular diversity of birds, including some

Figure 317. The Arrowhead Piculet (*Picumnus minutissimus*), Suriname's only endemic bird species, is readily found in and around Paramaribo (image © Meshach Pierre).

of the largest concentrations of migrant shorebirds in the Western Hemisphere (Spaans 2003). This habitat also supports substantial populations of waterfowl and wading birds, many of which also are migratory.

EXTENSIVE FRESHWATER SWAMP AND MARSH FORESTS. South of the Estuarine Zone, the freshwater herbaceous swamps and swamp forests of the Coastal Plain support an avifauna that is even more diverse than that of the mangroves. Typical species of this habitat include Cinereous Tinamou (*Crypturellus cinereus*), Cream-colored Woodpecker (*Celeus flavus*), Black-crested Antshrike (*Sakesphorus canadensis*), Blackish Antbird (*Cercomacroides nigrescens*), Straight-billed Woodcreeper (*Dendroplex picus*), Yellow-crowned Elaenia (*Myiopagis flavivertex*), Cinnamon Attila (*Attila cinnamomeus*), and Buff-breasted Wren (*Cantorchilus leucotis*). Many forest species that are widespread in the Coastal Zone are restricted to seasonally flooded, riverine forests and swamps in the Interior. Coastal swamp and marsh forests also support a few unique species, including Suriname's only endemic bird, the Arrowhead Piculet (*Picumnus minutissimus*; Figure 317), a tiny woodpecker that forages on small dead twigs and vines in the mid-story and canopy; the Blood-colored Woodpecker (*Veniliornis sanguineus*), which is restricted to the coastal forests of Suriname and Guyana; and the subspecies *fumosus* of Fuscous Flycatcher (*Cnemotriccus fuscatus*), which is common along the Guianan coast and is quite different in appearance and voice from other populations of that species.

Throughout the Coastal Plain, many widespread ecological generalists such as House Wren (*Troglodytes aedon*), Tropical Mockingbird (*Mimus gilvus*), Pale-breasted Thrush (*Turdus leucomelas*), and Blue-gray Tanager (*Thraupis episcopus*) thrive in secondary habitats created by humans, and are likely much more common in Suriname today than they were a few hundred years ago. These species benefit directly from humans – for example, by using buildings for nesting and roosting sites, or by feeding on crop plants and fruit trees planted by people. Finally, some species more characteristic of Suriname's Interior, including Ruddy Quail-Dove (*Geotrygon montana*), Black-and-white Hawk-Eagle (*Spizaetus melanoleucus*), and White-throated Toucan (*Ramphastos tucanus; bigikuyake*), can be found surprisingly close to the coast where suitable corridors and

Figure 318 The Rufous-capped Antthrush (*Formicarius colma*) is a species of Interior forests that still occurs in parts of the Coastal Plain where sufficient forest cover remains. This individual was marked as part of an ongoing study of birds in Peperpot Nature Park (image © Meshach Pierre).

large patches of forest remain, such as along the east bank of the Suriname River south of Paramaribo (Figure 318). These interior species mix with more typical species of secondary habitats, creating a diverse avifauna in settled areas of the Coastal Plain.

The Zanderij Belt. South of the Coastal Plain is the complex landscape of the Zanderij Belt, a region characterized by diverse soils and vegetation, ranging from open savanna to tall mesic forest (see Ecosystems chapter). Changes in soil and vegetation are often abrupt, creating a fine-scale mosaic of habitats that supports a diverse avifauna. Indeed, this is a true transition zone – many bird species of the Interior forests reach their northern limits here, whereas species typical of secondary and savanna habitats of the Coastal Plain do not occur in the continuous forests to the south. The Zanderij Belt also harbors its own distinct species, many of which occur in Suriname only within this narrow zone, or are far more common here than elsewhere. These include Red-legged Tinamou (*Crypturellus erythropus*), Rufous Nightjar (*Antrostomus rufus*), Rufous-crowned Elaenia (*Elaenia ruficeps*), Pale-bellied Mourner (*Rhytipterna immunda*), Black Manakin (*Xenopipo atronitens*), Saffron-crested Tyrant Manakin (*Neopelma chrysocephalum*), and Black-faced Tanager (*Schistochlamys melanopis*). *Mauritia* palm swamps are characteristic features of this region. They support specialized birds such as the Fork-tailed Palm-Swift (*Tachornis squamata*) and Point-tailed Palmcreeper (*Berlepschia rikeri*), and provide nesting and feeding resources for parrots, especially Red-bellied Macaw (*Orthopsittaca manilata*) and Blue-and-yellow Macaw (*Ara ararauna*).

Lowland forests of the Interior. Traveling south from the coastal savanna belt, open landscapes and their resident generalist species begin to disappear as the forest closes in and reveals a different suite of birds. Groups of macaws and solitary King Vultures (*Sarcoramphus papa*) now fly overhead, while that unmistakable sentinel of lowland rain forest, the Screaming Piha (*Lipaugus vociferans*), throws its voice through the trees. Traveling by boat, you quickly become familiar with the sight of toucans flying across the river ahead, while White-winged Swallows (*Tachycineta albiventer*) and White-banded Swallows (*Atticora fasciata*) skim the surface of the water. This is the lowland forest for which Suriname is famous, stretching unbroken from the Corantijn to the

Marowijne, and from the Zanderij Belt to the Grensgebergte – the mountains on the southern border with Brazil – and beyond. In this continuous forest landscape, mixed-species flocks composed of antbirds, ovenbirds, flycatchers, and tanagers work the canopy and understory, partitioning the foraging space in their constant quest for insect prey. On the forest floor, tinamous, curassows, and trumpeters forage in the leaf litter, while bands of antbirds and woodcreepers follow army ant swarms to gorge on arthropods dislodged from their hiding places by the ants. Fruiting trees and shrubs attract guans, toucans, manakins, cotingas, thrushes, and tanagers. Flocks of parrots course overhead, commuting between roosting and feeding areas. Cormorants, herons, and kingfishers patrol the creeks and rivers. It is possible to record upwards of 200 bird species in a single day in this extraordinarily rich ecosystem.

Although the forested Interior may seem like a monotonous landscape, particularly from above, that is far from the case. Subtle shifts in topography and soils create different types of forests, marked mainly by differences in tree species composition (see Ecosystems chapter). Most bird species of lowland forest in Suriname occur in multiple forest types. The main influence on bird species composition is the amount of flooding a forest experiences during the annual rainy seasons. Forests growing along rivers, where flooding is common, share roughly 90 percent of their bird species with adjacent 'terra firme' forest (forest on dry land; see Ecosystems chapter) (O'Shea & Ottema 2007), but distinct avifaunal changes do occur across this ecotone (zone of transition between ecosystems of different type). For example, as you cross from terra firme to seasonally flooded riverine forest in the Interior, White-necked Thrush (*Turdus albicollis*) is replaced by Cocoa Thrush (*Turdus fumigatus*) and Olive-backed Foliage-gleaner (*Automolus infuscatus*) by Chestnut-crowned Foliage-gleaner (*Automolus rufipileatus*). Other species indicative of riverine forests include Guianan Streaked-Antwren (*Myrmotherula surinamensis*), Black-chinned Antbird (*Hypocnemoides melanopogon*), Blackish Antbird (*Cercomacroides nigrescens*), and Striped Woodcreeper (*Xiphorhynchus obsoletus*). Many of these riverine forest species are also common in swamp and marsh forests of the Coastal Plain.

Upland forests. Suriname's numerous bauxite plateaus, few of which exceed 500 m, are not high enough to support a distinct highland avifauna. Virtually all bird species of the surrounding lowlands also occur up the slopes and across the tops of these plateaus. However, there are several species widespread in the lowlands of southern Suriname that, in the northern half of the country, occur primarily or only on plateau tops. These include Great-billed Hermit (*Phaethornis malaris*), Collared Trogon (*Trogon collaris*), Golden-olive Woodpecker (*Colaptes rubiginosus*) White-throated Pewee (*Contopus albogularis*), White-fronted Manakin (*Lepidothrix serena*), and White-breasted Wood-Wren (*Henicorhina leucosticta*).

There are a few areas in Suriname high enough to support a limited suite of montane bird species. The largest of these is the Tafelberg (Zyskowski *et al.* 2011), which reaches just over 1000 m and, importantly, has a significant amount of forest at that elevation. Smaller patches of upland forest can be found in the Wilhelmina, Eilerts de Haan, and Grensgebergte mountains in central and southern Suriname (O'Shea & Ramcharan 2013), but these have proven difficult to survey as access is challenging. As a consequence, the distribution of highland bird species in Suriname is incompletely known. Most true highland species occur around 800 m and above. The upland forests of Suriname are an eastern outpost for species whose ranges lie primarily in the 'tepui' or table-top mountain region of western Guyana, southeastern Venezuela, and northern Brazil. Because Suriname's uplands are so limited in extent and isolated from the tepuis, and because birds differ in their ability to disperse across unsuitable habitats, only a subset of the tepui avifauna is found here. The most widespread of these species in Suriname is the Tepui Vireo (*Vireo sclateri*), which occurs as low as 500 m across the southern borders of the Guianas and is fairly common on the Tafelberg (Robbins *et al.* 2007, Zyskowski *et al.* 2011). Other species of the tepui highlands whose ranges reach Suriname are Velvet-browed Brilliant (*Heliodoxa xanthogonys*), Tepui Toucanet (*Aulacorhynchus whitelianus*), Pale-eyed Thrush (*Turdus leucops*), Black-hooded Thrush (*Turdus olivater*), and Rufous-collared Sparrow (*Zonotrichia capensis*).

The Sipaliwini Savanna. In extreme southern Suriname, a vast savanna appears abruptly out of the forested landscape. Although the open grasslands, patches of forest, and *Mauritia* palm swamps are superficially similar to savannas of the Coastal Plain, this region is floristically very different. This is the Sipaliwini Savanna, and it contains an avifauna unique within Suriname. In the Sipaliwini, Swallow-tailed Hummingbirds (*Eupetomena macroura*) and Horned Sungems (*Heliactin bilophus*) probe flowers on the open savanna, while Scissor-tailed Nightjars (*Hydropsalis torquata*) roost on the ground. Isolated savanna trees offer commanding perches for Suiriri Flycatchers (*Suiriri suiriri*), Gray Monjitas (*Xolmis cinerea*) and Chalk-browed Mockingbirds (*Mimus saturninus*), as Narrow-billed Woodcreepers (*Lepidocolaptes angustirostris*) explore crevices in their trunks and branches. Patches of long grass are occupied by Rusty-backed Antwrens (*Formicivora rufa*) and Rufous-sided Pygmy-Tyrants (*Euscarthmus rufomarginatus*). Overhead fly flocks of Peach-fronted Parakeets (*Eupsittula aurea*) and the spectacular Sulphur-breasted Parakeet (*Aratinga maculata*). Many of the species here occur nowhere else in the country, nor are they known from Guyana or French Guiana. Indeed, the avifauna of the Sipaliwini savanna shows stronger affinities with the Cerrado of central Brazil than it does with other savanna regions of northern South America (Robbins *et al.* 2004, O'Shea 2005, Mittermeier *et al.* 2010). The unique avifauna of the Sipaliwini Savanna is a major contributor to avian (bird) diversity in Suriname.

Ecology and species interactions

Guilds. A staple of ecology is the concept of guilds, which are groups of organisms that exploit similar classes of resources in similar ways (Simberloff & Dayan 1991). Guilds provide a logical way to break down a complex fauna into groups of ecologically similar species – for example, birds that eat mainly insects and those that eat fish comprise separate guilds. In addition to diet, guilds are often defined by foraging height and substrate, or a combination of the two. Guilds tend to contain more species in diverse and structurally complex environments such as tropical rain forest. A summary of typical rain forest bird guilds is given in Table 1.

Mixed-species flocks. The guild concept is most clearly illustrated by the mixed-species flocks that are a characteristic feature of tropical forests. These flocks, which may contain dozens of species, form daily and forage over large areas of forest, moving as a cohesive unit. Flocks provide much excitement for birdwatchers as the forest comes alive with the sound and activity of birds darting after insects and following each other along their foraging routes. Studies of flocks have revealed that each species is usually represented by a mated pair or family group, and that pairs tend to stay with the same flock for their entire lives (Munn & Terborgh 1979). Within each flock, the constituent

Guild	Description	Example families/species
Sallying insectivore	Fly from perch to catch insects	*Onychorhynchus coronatus, Thamnomanes caesius*
Ant-following insectivore	Forage on arthropods flushed by army ants	*Gymnopithys rufigula*
Arboreal frugivore	Feed on fruit in forest canopy	*Trogon melanurus, Rupicola rupicola, Perissocephalus tricolor*
Nectarivore	Probe into flowers for nectar	*Phaethornis superciliosus*
Terrestrial insectivore	Feed on insects at ground level, usually in leaf litter	*Formicarius colma*
Frugivore/ insectivore/nectarivore	Opportunistic; eat wide range of foods	*Psarocolius viridis*
Aerial insectivore	Catch insects during sustained foraging flights	Apodidae, Hirundinidae
Woodpecker	Bore into wood for insects	*Veniliornis sanguineus*

Table 11. A summary of typical rain forest bird guilds in Suriname, with representative species (see species accounts for details) (adapted from Terborgh 1992).

species typically differ in foraging behavior. Many species occur only in such flocks and are never found away from them. The main types of mixed species flocks in Suriname are reviewed here.

Understory flocks of birds that primarily eat insects (insectivores) are a common feature of lowland forest and tend to occupy the same home ranges throughout the year. They almost always include one or two pairs of antshrikes of the genus *Thamnomanes*, which function as the flock leaders, using their loud calls to summon the flock at dawn and to keep it moving through the forest during the day. They are also the ones to sound the alarm whenever a predator (or human) appears nearby. In Suriname, understory flocks typically consist of three to ten core species, rarely more. In addition to *Thamnomanes* antshrikes, these flocks usually contain up to four species of antwrens of the

Figure 319. The Brown-bellied Antwren (*Epinecrophylla gutturalis*) is a common member of mixed-species flocks in rain forest understory, where it probes hanging dead leaves for insects hiding within (image © Leon Moore).

genera *Myrmotherula* and *Epinecrophylla* (Figure 319), which are frequently joined by Chestnut-rumped Woodcreeper (*Xiphorhynchus pardalotus*), Olive-backed Foliage-gleaner (*Automolus infuscatus*), Plain Xenops (*Xenops minutus*), Tawny-crowned Greenlet (*Tunchiornis ochraceiceps*), and one or two species of flycatchers such as Sulphur-rumped Flycatcher (*Myiobius barbatus*) or Olivaceous Flatbill (*Rhynchocyclus olivaceus*). Certain other understory species, known as flock 'dropouts' (Stouffer & Bierregaard 1995), maintain territories separate from flocks but will join flocks when they pass through, presumably to take advantage of the extra anti-predator vigilance provided by other members of the flock.

Canopy insectivore flocks exploit the wide array of foraging niches in the tropical rain forest canopy. They tend to be more variable and diverse than understory flocks, and have larger home ranges. Like understory flocks, canopy flocks have 'leader' species that are highly vocal and maintain flock cohesion. In Suriname, typical canopy flock species include Slaty-capped Shrike-Vireo (*Vireolanius leucotis*) and Fulvous Shrike-Tanager (*Lanio fulvus*), both of which serve as flock leaders, as well as Todd's Antwren (*Herpsilochmus stictocephalus*), Yellow-margined Flycatcher (*Tolmomyias assimilis*), and Buff-cheeked Greenlet (*Pachysylvia muscicapina*). These core species are joined by various woodpeckers, antbirds, woodcreepers, flycatchers, becards, and tanagers. Many maintain close associations with the flock throughout the day, whereas others may be 'dropouts' (see above) that join the flocks temporarily. A few species – notably Yellow-throated Flycatcher (*Conopias parvus*), Blue-backed Tanager (*Cyanicterus cyanicterus*) and Red-billed Pied Tanager (*Lamprospiza melanoleuca*) – maintain positions at the periphery of canopy flocks, often flying off for long distances and returning.

Army ant flocks consist of birds that attend swarms of army ants to devour arthropods and small vertebrates dislodged from their hiding places by the marauding ants. Two species – White-plumed Antbird (*Pithys albifrons*) and Rufous-throated Antbird (*Gymnopithys rufigula*) – forage exclusively at these swarms and cannot survive in forest patches lacking army ants. The sight of an ant swarm with these birds darting above, clinging to vertical saplings and jostling for position at the leading edge of the swarm, is one of the great spectacles of the Neotropical rain forest. Ant swarms are attended by other species as well, including Black-headed Antbird (*Percnostola rufifrons*), Common Scale-backed Antbird (*Willisornis poecilinotus*), Amazonian Barred-Woodcreeper (*Dendrocolaptes certhia*), Black-banded Woodcreeper (*Dendrocolaptes picumnus*), and the relatively rare White-chinned Woodcreeper (*Dendrocincla merula*), which is seldom seen away from ants. Other, more sedentary bird species may follow army ant flocks as the swarms pass through their territories.

Flocks of birds that primarily feed on nectar (nectarivores) occur in most forest types, but are especially conspicuous in so-called savanna forest in the Zanderij Belt that has a low stature and high density of thin-stemmed trees, as well as structurally similar mountain savanna forest in the Interior. These flocks are dominated by small tanagers, chiefly Green Honeycreeper (*Chlorophanes spiza*), Red-legged Honeycreeper (*Cyanerpes cyaneus*), Purple Honeycreeper (*Cyanerpes caeruleus*), and Blue Dacnis (*Dacnis cayana*), along with at least one species of the genus *Tangara*. In taller lowland forest, most of these species are also found with canopy insectivore flocks, but they dominate in lower forests, which are structurally less complex, and where many insectivorous canopy flock species tend to be rare or absent.

Aerial flocks are composed entirely of swifts (Apodidae), which spend all day in the air in pursuit of flying insects. Aerial flocks range widely and can appear seemingly out of nowhere. A large emergence of flying ants or termites is often accompanied by the dramatic appearance of a flock of swifts, zooming just above the canopy to feast on the bounty. Much as members of canopy and understory flocks forage at different levels in the forest, swifts tend to partition the airspace (Collins 2015). Upper altitudes are the domain of the large White-collared Swift (*Streptoprocne zonaris*)

Figure 320. Many Neotropical birds consume fruit and are important seed dispersers. Here, an Opal-rumped Tanager (*Tangara velia*) swallows melastome fruits (image © Kester Clarke).

and the elegant Lesser Swallow-tailed Swift (*Panyptila cayennensis*), whereas small swifts of the genus *Chaetura* forage mostly at lower altitudes.

Frugivory. Although most Neotropical birds feed on insects to some extent, fruit is a large component of the diets of many species (Moermond & Denslow 1985; Figure 320). Fruit can be regarded as the way a plant attractively packages its seeds to promote their dispersal by animals (see Plants chapter). Birds exert tremendous influence on the structure and floristic composition of the forest through their roles as primary seed dispersers for a wide variety of plants. Indeed, birds are important agents of reforestation in degraded landscapes in the tropics through their dispersal of tree seeds from adjacent forested areas (Pejchar *et al.* 2008).

In some cases, frugivory – feeding predominantly on fruits – influences the seasonal movement patterns of birds. Powerful fliers like parrots are a good example: they often range over many kilometers to track their preferred food resources, even within a single day, and their populations at any given location may fluctuate dramatically over the course of a year (Karubian *et al.* 2005). Large cotingas, many of which are primarily frugivorous, are known to move up and down mountains in response to food availability in some parts of the Neotropics (Chaves-Campos 2004). Other species, notably flycatchers of the genus *Tyrannus*, shift the proportion of fruit in their diet while moving seasonally over distances of thousands of kilometers (Morton 1971).

The remaining species of tropical fruit-eating birds are generally sedentary, meaning they spend most of their lives within a limited home range. Individual trees can be critical resources for these birds when food is scarce. Chief among them in the tropics are figs (*Ficus* spp.; Leighton & Leighton 1983). A single fruiting fig tree can provide hours of fascination for an observer as multitudes of birds and mammals visit to take advantage of the crop. This is perhaps the best way to appreciate the importance of frugivory in Neotropical forests.

The widespread availability of fruit in tropical rain forest allows for specialization on this resource. Birds that consume mostly fruit generally need to spend less time foraging than their insectivorous counterparts, because fruit is a conspicuous and easily exploited resource. This frees frugivorous birds to focus on other activities – such as mating. Males gather into groups, called leks, to display for females, and spend considerable time doing so. Lekking species – which include virtually all species of manakins (Pipridae) and cotingas (Cotingidae), many hummingbirds (Trochilidae), and some flycatchers (Tyrannidae) – employ a variety of methods to woo females. Examples include the Guianan Cock-of-the-rock (*Rupicola rupicola*), which has brilliant plumage with bizarre ornaments; the Pompadour Cotinga (*Xipholena punicea*), which uses modified flight feathers to create mechanical sounds during flight displays; the Blue-backed Manakin (*Chiroxiphia pareola*), which performs coordinated group displays; and the famous voice of the rain forest, the Screaming Piha (*Lipaugus vociferans*), whose males forego bright plumage in favor of sheer vocal power. The beautiful plumages and bizarre displays of these species have been shaped by millions of years of evolution, based on females' preferences for increasingly outlandish male traits, to produce the diversity that we see today.

Migration. Migration is another important feature of the natural history of many birds in the Neotropics (Levey & Stiles 1992). In Suriname, migrants occur as non-breeding visitors, usually in response to environmental changes on their breeding grounds. Three broad classes of migrants are reviewed here.

Nearctic migrants are species that breed in North America and spend the northern winter in the tropics. In Suriname, they are present primarily between October and April, although some individuals are present during all months of the year. Shorebirds are perhaps the most conspicuous Nearctic migrants in Suriname; roughly two dozen species of sandpipers (Scolopacidae) and plovers (Charadriidae) either winter along the Suriname coast or use it as a stopover on their way to wintering grounds in southern South America. Non-breeding individuals of many shorebird species sometimes remain in Suriname during the May-July breeding season, whereas the first migrants begin to arrive back in Suriname as early as mid-July. Passerine (birds in the order Passeriformes) migrants are more strictly seasonal in occurrence, and also less diverse, as the majority of migrant species breeding in North America spend the winter in the Caribbean, Central America, and northwestern South America; relatively few venture as far east as Suriname. Nevertheless, certain species, particularly Barn Swallow (*Hirundo rustica*) and Yellow Warbler (*Setophaga petechia*), are common in Suriname during the northern winter.

Austral migrants are species that breed in southern South America. They are present in Suriname during the southern winter, from roughly May to October. The majority of austral migrants that reach Suriname are in the families Tyrannidae (tyrant-flycatchers; Chesser 1997) and Hirundinidae (swallows). In several of these species, such as Tropical Kingbird (*Tyrannus melancholicus*) and Gray-breasted Martin (*Progne chalybea*), austral migrants mix with resident populations of the same species in Suriname; the interactions between these populations are poorly studied.

Birds from the Caribbean and other nearby tropical areas are not subject to annual climatic fluctuations as drastic as those faced by Nearctic and austral migrants, but nevertheless some tropical species migrate seasonally to and from Suriname. The most common Caribbean migrant in Suriname is the Gray Kingbird (*Tyrannus dominicensis*), which occurs along the coast during the northern winter. Migration within the tropics remains a poorly known phenomenon and deserves further study, which is becoming more possible with the development of geolocator and satellite tracking technology (Kays *et al.* 2015).

BOX 10 The legacy of François Haverschmidt

Figure 321. François Haverschmidt at Mariënburg in 1967 (image courtesy of VACO Publishers N.V., publisher of Haverschmidt and Mees' Birds of Suriname (1994).

No single person has contributed more to our knowledge of the birds of Suriname than François Haverschmidt (1906-1987; Figure 321). Born in the Netherlands, Haverschmidt took a post as a judge in Paramaribo in 1946, eventually becoming acting governor of colonial Suriname before his retirement in 1968. In this latter position, he played a key role in the establishment of the first nature reserves in Suriname. During his long residency in the country, and for a number of years after, he published six books and more than 100 scientific papers, a number of which remain the only published information on the natural history of some Guianan bird species. These publications are essential reading for anyone interested in the avifauna of Suriname. In his writings, Haverschmidt reveals himself to be a consummate naturalist with a special interest in the feeding and nesting habits of the birds of the Coastal Zone. His oft-stated expressions of surprise at his own discoveries suggest an almost limitless fascination with the rich natural world of Suriname, so different from what he had known in his youth in Europe. His book 'Birds of Suriname', first published in 1968 and revised and expanded by Gerlof Mees in 1994, was produced before any guidebooks to South American birds existed, and was thus a truly pioneering work. His productivity is all the more remarkable considering that it took place on the sidelines of a highly successful legal career.

Haverschmidt realized that many species of South American birds were poorly known, especially compared to those of Europe, and he supplemented his behavioral studies with an outstanding collection of almost 10,000 bird specimens representing some 540 species. This material, most of which is now housed in the Naturalis Biodiversity Center in Leiden, Netherlands, forms the core of Haverschmidt's ornithological legacy. Specimens are the gold standard for documenting a species' occurrence at a particular place and time, and they also provide valuable body data and vouchers for genetic material that are impossible to obtain by other means. When properly cared for, specimens last for hundreds of years. Haverschmidt's collection provides a snapshot of the avifauna of Suriname during the mid-20th century, and its scientific value will only grow in the coming era of rapid technological and environmental change.

For a full account of Haverschmidt's life and career, including a full list of his publications, see Mees (1988).

Tachornis squamata
Fork-tailed Palm-Swift
(No specific local name)
Apodiformes, Apodidae

Brian O'Shea

The swifts, close relatives of hummingbirds, are abundant but seldom-noticed birds of the open skies. They spend the entire day aloft, feeding on insects which they scoop from the air with their wide bills. Swifts are especially common over the canopy of lowland rain forest and along rivers, where large flocks may appear suddenly in response to hatches of winged ants and termites, a favorite food.

The Fork-tailed Palm-Swift (Figure 322) is one of the more specialized and unusual species found in Suriname. It shuns rain forest, but is common wherever palms, especially those of the genus *Mauritia*, dominate the landscape, such as in wet savannas, swamps, and even urban areas, where it can often be seen flying low overhead in zigzagging flight with distinctive flicking wingbeats. Following one of these birds may lead you to its nest, which is invariably attached to a dead palm frond close to the trunk of the tree (Sick 1993; Cooper 2015). As in most swifts, the nest is affixed to the frond and held together with the birds' own saliva; unlike them, the palm-swift uses many feathers in nest-building, which it obtains in a remarkable way. The swift aggressively steals feathers directly from other birds by chasing them down and pulling out beakfuls of feathers, either in flight or by alighting briefly on the birds' backs. Whitney (2007) observed palm-swifts stealing feathers from 50 species representing 21 families, although pigeons (Columbidae) and parrots (Psittacidae) seem to be the favorite targets.

Figure 322. ***Tachornis squamata*** **(image © Ian Davies/eBird).**

Description: A small swift (9-12 g), mostly grayish, paler below, with a long, forked tail that is usually held in a single point.

Distribution: Found in palm swamps, savannas and along rivers from Venezuela, Colombia, and Trinidad to southern Brazil. In Suriname, found mainly in the Coastal Plain and the Sipaliwini Savanna, where it breeds mainly in the rainy season (Haverschmidt 1968); it is common in Paramaribo.

Use: (No known use)

Similar species: Distinguished from swallows (Hirundinidae) by shallow, stiff wingbeats. Other swifts tend to fly higher, appear mostly blackish rather than gray, and lack the palm-swift's distinctive point-tailed profile.

Phaethornis superciliosus
Long-tailed Hermit
Korke (local name used for all Trochilidae)
Apodiformes, Trochilidae

Brian O'Shea

The Long-tailed Hermit (Figure 323) is not what comes to most peoples' minds when they think of a hummingbird. Large, impish, and mostly brown with no iridescent sheen, this hermit is one of the most common hummingbirds in the forest understory, where it is an important pollinator, especially of Heliconiaceae, the flowers of which it can access easily with its curved bill. Hermits have a different foraging strategy from many other hummingbirds. Rather than defend patches of flowers, hermits 'trapline', visiting widely dispersed flowers at regular intervals to harvest the accumulated nectar. In doing this, they face a trade-off: a longer wait between visits means more nectar, but also increases the chance that another bird will harvest it first (Gill 1988). Traplining hummingbirds must therefore make continuous adjustments to their foraging routine to minimize losses to competitors (Garrison & Gass 1999). Perhaps for this reason, they are eager to explore potential new food sources, even when that turns out to be a piece of clothing or flagging tape! Hermits are familiar to many naturalists because of their tendency to investigate brightly colored objects, and they will often approach humans to within a meter or two, their long central tail feathers twitching constantly, before darting off to their next patch of flowers. In Suriname, the Long-tailed Hermit shares its habitat with up to three other species of *Phaethornis*, all of which differ in either body mass or bill shape and presumably exploit slightly different nectar resources.

Because hermits do not need to expend energy defending their nectar sources, they have evolved a lek display system much like that of manakins and cotingas. These leks are a common sound during the dry season in Suriname, when many male hermits convene on their song perches in the understory and give an endless series of squeaks while twitching their tails up and down.

Figure 323. *Phaethornis superciliosus* (image © Frédéric Royer).

Description: A medium-sized (5-8 g) hummingbird. Plumage mostly brownish. The distinctive wedge-shaped tail is tipped white, with two long white central feathers. Bill long and curved.

Distribution: Found in lowland forest from the Orinoco River southeast to east-central Brazil. Occurs throughout Suriname.

Use: (No known use)

Similar species: Straight-billed Hermit (*Phaethornis bourcieri*) is often sympatric (sharing the same geographical range) with Long-tailed, but is smaller, with straight bill (not curved downward). Reddish Hermit (*Phaethornis ruber*) is considerably smaller and bright reddish-brown. Great-billed Hermit (*Phaethornis malaris*), restricted to plateau tops in eastern Suriname, is substantially larger and darker, with noticeably longer bill.

Psophia crepitans
Gray-winged Trumpeter
Kamikami
Gruiformes, Psophiidae

Serano Ramcharan

Trumpeters are an unusual family of birds found only in lowland rain forests of Amazonia and the Guiana Shield. North of the Amazon River, the Gray-winged Trumpeter (Figure 324) occurs in regions of undisturbed primary forest, often in flocks of up to 50 individuals. They are quite noisy birds that walk on the forest floor in search of insects, fruit, and small vertebrates. Despite their noisy tendencies, they are shy and difficult to see, often running away at the first sign of danger. Although they are weak fliers, trumpeters ascend to the forest canopy to roost at night. Trumpeters are good indicators of intact forest ecosystems, as they are highly vulnerable to forest fragmentation and do not occur in secondary forests or cultivated landscapes (Thiollay 1992).

Trumpeters have a polyandrous mating system in which a female mates with several males (Potter 2011). Mating groups of up to five individuals will splinter off the main group and make a nest containing 3-4 eggs, often in a hollow tree (Haverschmidt & Mees 1994). Members of the group share incubation duties. Young birds are brought back to the main group and defended vigorously by the adults. The trumpeter's strong defensive instincts are well known to indigenous people, who often keep these birds as pets, both to warn of intruders and for their prowess in hunting snakes.

With its vast expanses of intact forest, Suriname supports a globally significant population of trumpeters, and also happens to be one of the best places to see them due to the semi-tame flock that resides around the main compound at Brownsberg Nature Park. Up to 60 trumpeters come daily to the restaurant at Brownsberg to feed on food scraps set out for them, providing great opportunities to study these fascinating birds.

Figure 324. *Psophia crepitans* (image © Leon Moore).

Description: A large (1.2-1.5 kg), oddly shaped bird, with a long neck and legs, small head, and practically no tail. The soft, fluffy plumage is mostly black, with gray wings and back, and some iridescence on the neck. More often heard than seen, trumpeters make a variety of grunting, cooing and rasping noises, especially when disturbed.

Distribution: Found in lowland rain forest north of the Amazon. In Suriname, it is confined to primary forest in the Interior.

Use: Trumpeters are occasionally hunted for their meat and feathers. They are often kept as pets by indigenous people.

Similar species: Not easily confused. Black Curassow (*Crax alector*) is much larger, with a conspicuous white belly. Tinamous (Tinamidae) are brown, with shorter legs, and do not occur in flocks.

Calidris pusilla
Semipalmated Sandpiper
Snepi (local name used for most Charadriiformes)
Charadriiformes, Scolopacidae

Arie L. Spaans

For most of the year, the exposed tidal mudflats, shallow lagoons, and swamps along the Suriname coast are teeming with large numbers of North American shorebirds. Of the 26 species regularly migrating through the country, or spending the northern winter here, the Semipalmated Sandpiper (Figure 325) is the most common. In the 1970s and 1980s, hundreds of thousands were often present during the southward and northward migration periods, but recent surveys indicate a decline of roughly 80 percent since that time (Morrison *et al.* 2012). Semipalmated Sandpipers have small webs between their toes which enable them to walk on the soft tidal mudflats. They forage by probing the upper mud layer for tanaids (Tanaidacea), small crustaceans that occur here at densities of several thousand per square meter (Swennen *et al.* 1982). In the lagoons and swamps, the birds' diet consists mainly of insects and plant seeds (Spaans 2003).

Like many shorebirds, the Semipalmated Sandpiper undertakes an epic migration between its breeding and wintering grounds. Birds put on tremendous amounts of fat just before the start of their migration. The migration route of most central Canadian breeding birds is elliptical, with a 4000 km-long flight across the ocean to northeastern South America during late summer, returning across the interior of North America in spring (Gratto-Trevor *et al.* 2012). Migration from Suriname starts in early May and takes place fast: by the end of May, the first birds have arrived on the Arctic breeding grounds. Summer in the Arctic is short, and birds begin leaving the breeding grounds by early July. By mid-August, the first adults have returned to Suriname; that year's young follow some weeks later (Spaans 2003).

Figure 325. *Calidris pusilla* (image © Lloyd Spitalnik).

Description. A small brownish sandpiper (more grayish brown in non-breeding plumage) with a black bill and black legs. Birds of western origin (Alaska) have much shorter bills than those from eastern Canada.

Distribution. Semipalmated Sandpipers nest in the sub- and mid-Arctic areas of North America and spend the northern winter mostly along the northern and central parts of the Atlantic and Pacific coasts of South America.

Use. Semipalmated Sandpipers are legally protected, yet each year tens of thousands are netted and shot for food (Ottema and Spaans 2008).

Similar species. Long-billed Semipalmated Sandpipers in non-breeding plumage can be mistaken for Western Sandpipers (*Calidris mauri*), which are much less common in Suriname. Least Sandpiper (*Calidris minutilla*) is smaller and has yellow legs, a more slender bill, and a sharp demarcation between breast and belly. Slightly larger White-rumped Sandpiper (*Calidris fuscicollis*) has a white rump, and wings which extend beyond the tail.

Eurypyga helias
Sunbittern
Sonfowru
Eurypygiformes, Eurypygidae

Brian O'Shea

Broad, shallow creeks in lowland rain forest are the domain of the Sunbittern (Figure 326), a spectacular but shy and seldom-seen inhabitant of the forests of Suriname's Interior. The Sunbittern is named for the the exquisite sunburst pattern of colors on its wings. When a Sunbittern is disturbed, particularly in territorial encounters or when defending its young, it faces its antagonist, crouches forward, and spreads its wings and tail in an effort to ward off the intruder (Frith 1978). You need a bit of luck to witness this amazing display.

Sunbitterns are typically seen as they fly up- or downstream on stiffly bowed wings, away from an observer, pausing after each short flight to assess the danger behind them. Too quickly, they vanish around a stream bend or duck away out of sight. Occasionally they may be observed along the edges of smaller rivers, where their orange legs, odd shape and flight style, and shy behavior quickly distinguish them from other wading birds, such as herons and sandpipers. Sunbitterns forage by walking slowly along creek and river edges, lunging and gleaning for invertebrates and small frogs and fish (Stiles & Skutch 1989, Hilty 2003, MacLean 2010).

The Sunbittern builds a bulky, globular nest of leaves, sticks, and mud that may be situated on a low branch or on the ground, often near a creek (Haverschmidt 1968, Thomas & Strahl 1990, Dubs 1992). Nesting has been documented in Suriname from May to August (Haverschmidt 1968), but it likely breeds in other months of the year as well. Chicks remain in the nest for about three weeks before fledging (Thomas & Strahl 1990).

Figure 326. *Eurypyga helias* (image © Nick Athanas – antpitta.com).

Description. A medium-sized (200-300 g), oddly proportioned wading bird: thin-necked, small-headed and long-tailed, often holding its body horizontally. Plumage mostly gray and buff, with contrasting yellow-orange legs.

Distribution. Found in Neotropical lowland moist forests from southern Mexico through the Guianas and the Amazon basin. Occurs throughout Suriname, except in coastal mangroves and urban and agricultural areas.

Use. Occasionally hunted for food.

Similar species. Herons (Ardeidae) have an upright posture and a spearing bill. Distinguished from wood-rails (*Aramides* spp.) by larger size, different shape and coloration, and longer neck.

Cochlearius cochlearius
Boat-billed Heron
Arapapa
Pelecaniformes, Ardeidae

Otte Ottema and Brian O'Shea

The Boat-billed Heron (Figure 327) gets its name from its unusually broad, boat-shaped bill, which imparts such a strange appearance that the species was formerly considered to be in its own family (Meyer de Schauensee 1970). Like the night-herons with which it often occurs, the Boat-billed Heron is nocturnal, hence its very large eyes. Whereas most herons capture their prey by spearing with the bill, this is not a practical strategy for the *arapapa*. Instead, this bird uses its broad bill to scoop fish and invertebrates from the water, either from an overhanging branch or while wading in the shallows (Kushlan 2009).

Paramaribo is blessed to have a heron colony in the center of town at the mouth of the Sommelsdijckse Creek, where six species of herons are breeding. In 2013 this colony contained more than 1000 pairs of birds, of which 300 pairs were Boat-billed Herons. Observations of the colony have revealed that males and females share nest-building duties, and that birds begin breeding in their second year but are less successful than older birds. This species' extraordinary bill is often used in aggressive encounters, along with the expressive crest, which is longer in males than in females. Birds may respond to intruders by rushing toward them with crest raised and bill open; they then close the bill forcefully, producing a loud snapping sound. Boat-bills will attack other herons that invade their space in the nest colony, as well as iguanas and potential nest predators such as the Great Black Hawk (*Buteogallus urubitinga*).

Figure 327. ***Cochlearius cochlearius*** **(image © Jean-Louis Rousselle).**

Description. A medium-sized, stocky heron (500-800 g) with an exceptionally broad, keel-shaped bill and large, dark eyes. Adult plumage mostly pale gray above and white below, with a chestnut belly and black flanks. The head is mostly black with a long crest during the breeding season. Immature birds are darker and browner than adults, transitioning to adult plumage over approximately five years.

Distribution. Found in tropical lowlands from northern Mexico to southern Brazil. Occurs throughout Suriname, where it is most common in mangroves and flooded forest.

Use. (No known use)

Similar species. Unlikely to be confused in its coastal habitat. Most similar to Black-crowned Night-Heron (*Nycticorax nycticorax*) which has a very different bill and a black back.

Eudocimus ruber
Scarlet Ibis
Flamingo, korikori
Pelecaniformes, Threskiornithidae

Arie L. Spaans

Adult Scarlet Ibises (Figure 328), with their bright red plumage, are among the most colorful wading birds that can be seen along the coast of Suriname. The species is almost entirely confined to the soft tidal mudflats, the mangrove forest bordering the mudflats, and the brackish lagoons and swamps located just behind the mangrove zone. But during extremely dry seasons, Scarlet Ibises can be seen as far as 10 km upstream of the estuaries of large rivers. Scarlet Ibises are very social birds. They roost and nest communally, often together with herons and egrets. On the feeding grounds the birds are often more dispersed. They feed mainly on crustaceans, fishes, and insects. Fishes are especially important as a food source for nestlings. Some of the crustaceans eaten cause the red color in adult birds.

Suriname is located in the center of the species' distribution. Each year, 2-4 rookeries are present with thousands to tens of thousands of breeding pairs. In 1986, some 35,000 pairs (almost one-third of the world population) were found nesting here (Spaans 2003). There is no relationship between the number of birds nesting in Suriname and Guyana or French Guiana, suggesting that local circumstances determine how many birds start breeding each year (Spaans 1990). Recoveries of birds banded in the Coppenamepunt Nature Reserve during the 1960s revealed that after the breeding season, birds dispersed from the nesting area at least as far as the Nickerie District in the west and eastern French Guiana in the east (ffrench & Haverschmidt 1970).

Figure 328. ***Eudocimus ruber*** **(image © Hervé Breton).**

Description. A medium-sized wading bird. The adult is scarlet red with black wing tips. The bill and legs are red as well. During the nesting season, however, the bill is black. The juvenile bird is grayish brown with a white lower back and belly, and white central underwings. Immature birds show an increasing number of red feathers when maturing. It takes more than one year before they are in full adult plumage.

Distribution. Caribbean and Atlantic coasts of northern South America, with a small outlying population on the coast of southeastern Brazil. Inland nesting is known only from the Venezuelan llanos.

Use. Kappler (1881) vividly describes how Scarlet Ibis rookeries in Suriname were raided for their eggs and young. Although they have been protected by law since 1954, hundreds of birds are still poached each year.

Similar species. One of the few wading birds with a strongly downward-curved bill. The bright red adult is unmistakable. The juvenile bird somewhat resembles Tricolored Heron (*Egretta tricolor*), which has a straight bill.

Harpia harpyja
Harpy Eagle
Gonini
Accipitriformes, Accipitridae

Brian O'Shea

The Harpy Eagle's immense size and power are without equal among the world's birds of prey. In Suriname's expansive forests, the Harpy (Figure 329) thrives on the abundance of arboreal mammals that constitute the majority of its prey (Schulenberg 2009). Seldom seen, it is nevertheless a powerful presence. Experienced naturalists can often surmise that a *gonini* is near by listening to the calls of monkeys, which take on an urgent quality when this predator is spotted.

Harpy Eagles feed primarily on sloths and monkeys, which they take by surprise flying through or just above the canopy, quickly and easily pulling even the strongest animals from the trees with their massive talons (Rettig 1978; Peres 1990). Harpies also prey on other arboreal mammals such as coatis, opossums and porcupines, as well as agoutis, young brocket deer, curassows and macaws (Schulenberg 2009). Top predators like this require fully intact and functional forest expanses – not remnant patches of forest from which many mammal species have disappeared.

The Harpy chooses a massive emergent tree for nesting, often a *kankantri* (*Ceiba pentandra*). The bulky stick nest is built where the huge branches radiate from the tree bole. Typically, only one young survives to fledge, at 5-6 months. The young eagle uses the branches of the nest tree to practice running and flapping. It remains near the nest for up to a year after fledging (Schulenberg 2009). Pairs of Harpy Eagles can only reproduce every 2-3 years.

Figure 329. *Harpia harpyja* (image © Leon Moore).

Description. A huge (males 4-5 kg; females 7.5-9 kg), broad-winged, long-tailed eagle with a characteristic double crest and massive beak and talons. Plumage is generally gray above and white below, with a broad black chest band. Immatures are mostly gray and white.

Distribution. Very rare in lowland rain forests of Mexico and Central America and SE Brazil; more common in the Guianas and the Amazon basin. Found throughout the Interior of Suriname; occasionally seen in the Coastal Plain.

Use. Harpy Eagle feathers are kept and worn by indigenous people in many parts of the Amazon basin. Due to their unsuspecting nature, Harpies are frequently shot when they appear near human settlements, and are often eaten.

Similar species. Could be confused only with the rarer Crested Eagle (*Morphnus guiananensis*), which is smaller and slimmer, has a single (not double) crest, and lacks the Harpy's black chest band.

Rostrhamus sociabilis
Snail Kite
Pakro-aka
Accipitriformes, Accipitridae

Brian O'Shea and Serano Ramcharan

The Snail Kite (Figure 330) is one of Suriname's most visible raptors, and is unusual among birds of prey in that its diet is composed almost entirely of apple snails (*Pomacea* spp.). These large snails are abundant in the freshwater habitats of Suriname's coastal lowlands. The *pakro-aka* is often seen watching for snails while perched on wires or fence posts, or flying slowly along roadsides. It is especially numerous in rice-growing areas. Haverschmidt (1954a, 1962) described roosts with hundreds of Snail Kites in Nickerie, and noted that some of them carried snails in their bill or talons when flying to the roost site. Murphy (1955) provided a colorful description of the kite's method for extracting snails from their shells. By his account, the birds would take live snails to a perch and wait for them to emerge before piercing them at a specific point behind the operculum, paralyzing the snails and causing them to eventually release from their shells. This account was believed by many ornithologists – including Haverschmidt, who likely thought that the kites were carrying snails to the roost for this purpose – before subsequent studies confirmed that the Snail Kite uses its extraordinary bill to reach far inside the snail's shell to sever the columellar muscle attachment (Haverschmidt 1970, Snyder & Kale 1983). The kites perform this procedure quite efficiently, taking around 90 seconds to dispatch and consume each snail (Beissinger 1990, Bergmann *et al.* 2013).

In the 1970s, the Snail Kite population in Suriname's Nickerie rice district declined following the use of pesticides to control apple snails (Haverschmidt & Mees 1994). Many birds were found dead during this time. Fortunately, populations today appear to be stable, although local populations can fluctuate as birds wander in search of food. When snails are scarce, Snail Kites may consume alternative prey including small turtles and crustaceans; in Venezuela, they also routinely feed on crabs (Beissinger 1990).

Figure 330. *Rostrhamus sociabilis* (image © Meshach Pierre).

Description. A medium-sized raptor (350-500 g), adult male mostly slate gray with white rump and base of tail; legs, feet, and bill orange-red, upper mandible extremely long and curved. Females and immatures are mottled brown and white.

Distribution. Occurs from the southern USA (Florida), Cuba, and southeastern Mexico to Argentina; absent from large areas of Central America and northern South America, where confined to savanna and coastal regions. In Suriname, it is most abundant in rice-growing areas (e.g. Nickerie).

Use: (No known use)

Similar species. Adult male unlikely to be confused except with Slender-billed Kite (*Rostrhamus hamatus*), which does not typically forage in open habitats and has a much shorter tail with no white. Females and immatures may superficially resemble other birds of prey, but are quickly distinguished by their distinctive bill and slow, buoyant flight.

Buteogallus aequinoctialis
Rufous Crab Hawk
Krabu-aka
Accipitriformes, Accipitridae

Brian O'Shea

Suriname's ecosystems support robust populations of birds of prey. No fewer than 33 species of hawks and eagles occur in forest and savanna habitats throughout the country. Like many top predators, these birds tend to have large home ranges encompassing multiple habitats and prey types. However, a few of them are true ecological specialists. The Rufous Crab Hawk (Figure 331) is one of these – it is found exclusively in the coastal zone of northeastern South America, where it tends to be the most numerous bird of prey in mangrove forest (Thiollay 2007, Da Silva 2014). As its name implies, this species feeds almost exclusively on crabs, which it captures by pouncing on them from a low perch as they meander over exposed mud at low tide. Rufous Crab Hawks are notably unwary and can sometimes be approached to within a few meters. This species is quite vocal and is easily found at Weg naar Zee and other coastal locations near Paramaribo.

Mangrove habitats are under threat worldwide, both from rising sea levels and human encroachment. While the mangrove forests of the Guianan coasts have largely been spared the destruction seen in more densely settled regions of the world, climate change is a more insidious threat to this vital ecosystem. The Rufous Crab Hawk, although locally common, has nonetheless recently been upgraded to Near Threatened by the IUCN (IUCN 2015; see also chapter on Use, Threats and Conservation). Although this fine bird appears to tolerate some degree of disturbance – no doubt because its main prey is so abundant – it remains vulnerable to large-scale habitat changes and environmental pollution, as well as direct persecution by people.

Figure 331. *Buteogallus aequinoctialis* (image © Hervé Breton).

Description: A medium-to-large (600-800 g), stocky hawk with broad wings and a short tail. Adult plumage mostly dark gray and chestnut; immatures are mottled brown and white. Long legs are yellow but are often caked with mud.

Distribution: Occurs exclusively in coastal forests from eastern Venezeula (Orinoco Delta) to eastern Brazil.

Use: (No known use)

Similar species: Adult is unlikely to be confused in its mangrove habitat. Immature birds are variable and may appear similar to immature Great Black-Hawk (*Buteogallus urubitinga*), but that species is slightly larger, with a more clean-cut appearance, and is less common in mangroves.

Trogon viridis
Green-backed Trogon
Uduloso-fowru
Trogoniformes, Trogonidae

Brian O'Shea

Trogons are colorful, slow-moving birds found in tropical regions worldwide, though they are most diverse in the New World (Collar 2001). The five species found in Suriname are all birds of lowland forest, and at least four species typically occur at any given site in the Interior. The Green-backed Trogon (Figure 332) is generally the most common and widespread trogon in Suriname. Trogons feed on both fruit and arthropods, although larger species tend to be more frugivorous than smaller ones (Remsen *et al.* 1993). Trogons are quite vocal and sometimes gather in communal sessions to call and chase one another; the function of these gatherings remains unknown (Riehl 2008).

Trogons excavate nest cavities using their serrated bills (Skutch 1962), and this could partially explain their fondness for nesting in arboreal termite nests, or termitaria, which are used for nesting by many bird species throughout the tropics (Hindwood 1959, Brightsmith 2004). In many tropical forests, including those of Suriname's Interior, termite nests are an abundant resource, easier to excavate and offering better support than decayed wood. But there may be an additional benefit to nesting in termitaria – they appear to have lower rates of nest predation than old tree cavities, perhaps because they are comparatively short-lived, so predators do not learn to associate them with food (Brightsmith 2005a). It is not easy to be a baby bird in the tropics. Helpless nestlings are defenseless and nutritious, and even animals that consume a mostly vegetarian diet will seldom refuse the opportunity to raid a nest. Consequently, predation rates on tropical bird nests can be as high as 80-90% (Stutchbury & Morton 2001). Some birds that nest in termitaria have evolved longer nestling periods (Brightsmith 2005b), suggesting that these are safe sites during this vulnerable time.

Figure 332. *Trogon viridis* (image © Harold Greeney).

Description: A medium-sized (80 g), chunky, lethargic bird of the forest canopy. Male is mostly iridescent blue-green, especially on the back, with a conspicuous pale bluish bill and eye-ring, yellow belly, and black tail with large white patches underneath. Female similar but has less white in tail and is mostly gray where the male is blue-green.

Distribution: Found in humid lowland forest and second-growth forest throughout Amazonia. Occurs throughout Suriname in forested habitats.

Use: (No known use)

Similar species: Most similar to Guianan Trogon (*Trogon violaceus*), which is slightly smaller, has finer barring on the underside of the tail, and has a yellow (not pale bluish) eye-ring. Females of the two species can be especially difficult to separate.

Monasa atra
Black Nunbird
Redimofo, woyopita
Galbuliformes, Bucconidae

Brian O'Shea

The Black Nunbird (Figure 333) stands out among other puffbirds for a variety of reasons. Puffbirds, as a whole, are lethargic birds that spend much of their day perched quietly in the forest, looking around slowly before darting out to capture large insects and small vertebrates. They dispatch these by beating them against a perch in much the same way that kingfishers, who are fairly closely related, handle fishes. Most puffbirds live in pairs and are relatively inconspicuous by virtue of their quiet and sluggish behavior.

The Black Nunbird is perhaps the most flamboyant member of the family in this regard. The rollicking chorus of a group of nunbirds is a familiar sound in the rain forests of Suriname's interior, and is the source of the local name *woyopita*. Unlike other puffbirds, nunbirds are fairly easy to see as they perch along roadsides and river edges, often in groups of 3-5 birds, their bright red bills contrasting sharply with the green foliage. These groups are composed of close relatives, as nunbirds live and breed cooperatively (Skutch 1972), nesting in long burrows which they excavate in level to gently sloping terrain (Greeney & Port 2010). Nunbirds have a more flexible foraging strategy than other puffbirds, and it is not unusual to see them following swarms of army ants (Chaves-Campos 2005) or groups of monkeys (Thiollay & Jullien 1998) to take prey flushed out by these animals as they move through the forest.

Figure 333. ***Monasa atra*** **(image © Sean McCann).**

Description: A medium-sized (70-90 g), long-tailed bird that perches upright in the mid-story and along forest edges. Plumage mostly black with a conspicuous white wing patch. Bill bright coral-red and somewhat curved.

Distribution: Endemic to the Guiana Shield. Replaced by other species of *Monasa* elsewhere in Amazonia and in southern Central America. Common in the interior of Suriname, rarely venturing north of the Zanderij Belt.

Use: (No known use)

Similar species: Combination of black plumage and bright red bill is unique among birds in Suriname.

Ramphastos tucanus
White-throated Toucan
Bigikuyake
Piciformes, Ramphastidae

Brian O'Shea

The genus *Ramphastos* includes two species that are among the most recognizable and conspicuous birds of interior Suriname. The White-throated Toucan (Figure 334) is the larger of the two, and its distinctive yapping call – much like the barking of a small dog – is familiar to anyone who has spent much time in Amazonian rain forest. The smaller Channel-billed Toucan (*Ramphastos vitellinus*) is virtually identical to its larger relative in habits, distribution and ecology, and often occurs with it in groups that can contain dozens of individuals, although smaller groups are more common. *Ramphastos* toucans are noisy and boisterous birds that tend to congregate around fruiting trees or on bare snags above the canopy, which they often use as song perches, delivering their far-carrying calls with great swagger as they swing their bills and tails back and forth. The toucan chorus is a characteristic sound after rains and during the last hour of daylight in lowland forest landscapes.

The toucans' defining feature is their enormous bill, which may be up to one-third the length of the body. These bills are well suited for grabbing a wide variety of fruits, and although toucans occasionally take bird eggs, nestlings, and other small vertebrates, they appear to be mostly frugivorous (Skutch 1971; Remsen *et al.* 1993). Accordingly, toucans are among the most important avian seed dispersers of tropical rain forest canopy trees (Snow 1981). Studies of their ranging behavior reveal that they may carry seeds hundreds of meters away from the parent trees (Holbrook 2011, Kays *et al.* 2011).

Figure 334. *Ramphastos tucanus* (image © Brian O'Shea).

Description: 450-800 g. Males larger than females. Plumage mostly black, with face and chest white, and base of tail red; huge bill mostly deep red with a yellow base. Skin around eye bright blue.

Distribution: Lowland forest of Amazonia. Occurs in Suriname wherever patches of tall forest remain, but is most common in the Interior.

Use: Toucans are hunted for food throughout Amazonia, and their feathers and bills are often used by indigenous people for ceremonial purposes.

Similar species: Channel-billed Toucan is slightly smaller, with a strong yellow wash across the breast and a black bill; its voice is a repetitive croak, much different than that of White-throated. The two species often occur together and call simultaneously.

Veniliornis sanguineus
Blood-colored Woodpecker
Temreman (local name used for all Picidae)
Piciformes, Picidae

Brian O'Shea

The gardens and old plantations of Suriname's Coastal Plain support an avifauna dominated by widespread ecological generalists. But it is here that one of Suriname's most range-restricted species also makes its home. The Blood-colored Woodpecker (Figure 335) is one of the most colorful members of its family. Although similar in ecology to other members of the genus *Veniliornis* and small woodpeckers in general, this bird provides one of the highlights of a visiting birdwatcher's trip to Suriname. Often found in pairs or small family groups, the Blood-colored Woodpecker is a familiar garden bird in and around Paramaribo. It is especially common in old plantations such as Peperpot Nature Park in Commewijne, where it forages by boring into bark along branches at lower and middle strata, often alongside Suriname's only endemic bird species, the Arrowhead Piculet (*Picumnus minutissimus*). These two species highlight the conservation value of old plantations, which provide habitat for a wide variety of birds as they regenerate into marsh and swamp forest.

Like most woodpeckers, the Blood-colored Woodpecker excavates holes for roosting and nesting (Haverschmidt 1953). These holes are in turn used for the same purposes by numerous other species of birds, mammals, and even insects. In this sense, woodpeckers perform a vital ecological role as they create habitat for other species.

Figure 335. ***Veniliornis sanguineus*** **(image © Kester Clarke).**

Description: A small woodpecker (20-30 g), mostly bright red above, barred brown and white below. Males have more red on the head than females.

Distribution: Found only in the Coastal Plain of Guyana and Suriname, where it occurs in mangroves, swamp and marsh forests, and urban areas.

Use: (No known use)

Similar species: Unmistakable in its limited range. Replaced in the Interior by Golden-collared Woodpecker (*Veniliornis cassini*), and in French Guiana by Little Woodpecker (*Veniliornis passerinus*).

Ibycter americanus
Red-throated Caracara
Busikaka
Falconiformes, Falconidae

Brian O'Shea

The social insects – ants, bees, wasps and termites – are abundant in tropical rain forest, and few biologists pass a single day afield without being bitten or stung by them at least once. Some of these insects are notoriously aggressive in defending their nests, and it is a common strategy among rain forest birds to build nests close to those of wasps to reduce their risk of predation. It defies reason that a predator may actually attack a wasp nest and risk the wrath of these insects to get access to their larvae – yet that is precisely what the *busikaka* (Figure 336) does. A member of the falcon family, this unusual bird is a specialist predator on wasp nests. For decades, ornithologists have marveled at this bird's ability to attack vicious species of wasps while apparently avoiding their stings, prompting some to speculate that these birds possess a chemical defense that thwarts the insects' attacks (Lowery & Dalquest 1951, Thiollay 1991, Weldon & Rappole 1997). Recent studies have found little evidence for this – instead, the birds appear to take advantage of the tendency of wasps to abandon their nests following disturbance. The birds employ a 'hit and run' technique to damage the wasps' nests, and then feast on the contents once the wasps have absconded (McCann *et al.* 2013).

Red-throated Caracaras are cooperative breeders and spend their lives in noisy, territorial groups (Thiollay 1991). Their raucous calls are a distinctive sound of the forests of Suriname's Interior. This species frequently hunts in the forest understory, but sentinels in the canopy warn other group members of approaching danger and may even follow potential predators while calling loudly. For this reason, Red-throated Caracaras are despised by hunters, and have been found to be less common where hunting and deforestation are widespread (Thiollay 1991).

Figure 336. ***Ibycter americanus*** **(image © Sean McCann).**

Description: A medium-sized (500-700 g), long-tailed raptor, mostly bluish-black with a sharply defined white belly and red facial skin.

Distribution: Found in lowlands from Central America (where now very rare) through the Amazon basin.

Use: Reputedly avoided by indigenous hunters due to their odor and distasteful meat (Lowery & Dalquest 1951, Thiollay 1991), though they may be shot by hunters who resent their mobbing behavior.

Similar species: Unlikely to be confused. Black Caracara (*Daptrius ater*) is much slimmer, occurs mainly as singles and pairs along rivers, has different vocalizations, and is almost entirely black with orange-yellow facial skin. Vocalizations of Red-throated Caracaras may be mistaken for those of macaws (*Ara* spp.), but are more varied and cacophonous.

Amazona amazonica
Orange-winged Parrot
Kulekule
Psittaciformes, Psittacidae

Serano Ramcharan

Suriname is a fantastic place to see parrots. Of the 26 species that occur in the country, the *kulekule* (Figure 337) is perhaps the most familiar and, on the Coastal Plain, the most abundant. Thousands may be seen each evening winging their way northward to spend the night in coastal mangroves after a day of foraging. Roosting concentrations also occur in the middle of Paramaribo, for example at the Palmentuin and near the Academic Hospital. This adaptable species is also found throughout the Interior of Suriname, though it is most common near the coast and can be scarce at higher elevations. As is true for many parrots, this species wanders widely in search of its preferred foods, although the extent of its daily and seasonal movements is not fully known. Because it is so common in the Coastal Plain, it frequently visits groves of fruit trees but is not considered a pest by most people.

The Orange-winged Parrot accounts for a large percentage of the parrots trapped in Suriname for the pet trade. It is commonly seen for sale on the street and in markets, in part because it is considered to be among the best imitators of all the parrots. Trappers often poach young birds from breeding areas because it is believed that birds raised from a young age make better pets and develop a wider range of vocal skills. Efforts by game wardens to reduce trapping have met with mixed success, though they are important to maintain parrot populations, since parrots have fairly low reproductive rates.

Figure 337. ***Amazona amazonica*** **(image © Kester Clarke).**

Description: A medium-sized (300-400 g), stocky parrot with a chunky shape and short tail. Plumage mostly green, with an orange patch on the flight feathers and yellow cheeks.

Distribution: Found widely in South American lowlands east of the Andes, with isolated populations in southeastern Brazil and on Trinidad and Tobago. Occurs throughout Suriname, but most numerous in the Coastal Plain.

Use: Widely trapped for the domestic and international pet trade. Occasionally hunted for food; feathers used for ceremonial purposes by indigenous people.

Similar species: Most similar to Blue-cheeked Parrot (*Amazona dufresniana*), which has blue (not yellow) cheeks and is not nearly as numerous, especially in the Coastal Plain; and Mealy Parrot (*Amazona farinosa*), which is larger and entirely green with a white eye-ring, and is more restricted to the Interior.

Brian O'Shea

Ara ararauna
Blue-and-yellow Macaw
Tyambarafru
Psittaciformes, Psittacidae

Brian O'Shea

Palms of the genus *Mauritia* are an important component of forest ecology in Amazonia, not least because they provide a rich food source for both humans and wildlife (Bodmer 1990, Manzi & Coomes 2009, Rosa *et al.* 2014). These palms occur in poorly drained areas, where they often form dense stands, especially in savanna regions. A number of parrots are highly dependent on *Mauritia* swamps, including the spectacular Blue-and-yellow Macaw (Figure 338), which nests primarily in large cavities in standing dead *Mauritia* trees (Brightsmith & Bravo 2006). Large tree cavities tend to occur only in areas that have remained undisturbed long enough for trees to develop them through natural processes. Other large Amazonian macaws, like Red-and-green (*Ara chloropterus*) and Scarlet Macaw (*Ara macao*), avoid competition with the Blue-and-yellow Macaw by nesting in large rain forest tree cavities (Renton & Brightsmith 2009), which fortunately are abundant in the ancient forests of Suriname's Interior.

The Blue-and-yellow Macaw shares its affinity for palms with two smaller species, the Red-shouldered Macaw (*Diopsittaca nobilis*) and especially the Red-bellied Macaw (*Orthopsittaca manilata*), both of which utilize smaller cavities. The nutritious fruit of the *Mauritia* draws numerous other species of parrots. Unfortunately, as palm savannas are generally more accessible and limited in extent than lowland rain forest, and because nest cavities in isolated palm snags are more visible than those in rain forest canopy, palm-nesting macaws are relatively easy to trap for the pet trade, and their numbers have dropped sharply in areas near human settlements (Juniper & Parr 1998). However, parrots do seem to respond well to management and protection of nest sites (Barré *et al.* 2010), providing hope that Suriname will remain a stronghold for these spectacular birds.

Figure 338. ***Ara ararauna*** **(image © Kester Clarke).**

Description: A large (1.0-1.2 kg), long-tailed parrot, brilliant blue above and yellow below.

Distribution: Found in humid lowlands from eastern Panama to Bolivia and southern Brazil. Occurs throughout Suriname.

Use: Indigenous people across Amazonia hunt macaws for their meat and use their feathers for ceremonial purposes. Macaws are also widely trapped and sold for the pet trade, both within Suriname and internationally.

Similar species: Unmistakable. The only other macaws of similar size (Red-and-green and Scarlet Macaws) are mostly red.

Thamnomanes caesius
Cinereous Antshrike
Mehatere (name used by Trio indigenous people)
Passeriformes, Thamnophilidae

Brian O'Shea

The Trio people of southwestern Suriname have a song about the *mehatere* (Figure 339), who invites other forest birds to come and feast with him. It is easy to see why. Follow the rattling calls of the *mehatere* to their source, and you will find yourself at the center of an understory mixed-species flock.

Throughout Amazonia, *Thamnomanes* antshrikes play a key role in maintaining the integrity of mixed-species flocks (Munn & Terborgh 1979, Jullien & Thiollay 1998). These flocks, which in Suriname typically contain 6-10 species, move through the forest within a defined home range, searching for insect prey while subdividing foraging space to minimize competition among species. *Thamnomanes* antshrikes often sing at dawn to summon flock members, and then lead the flock's movements throughout the day. They are the first to sound the alarm when danger threatens, and are generally easy to see as they take exposed perches, sometimes up into the mid-story, and dart out after insects – often nabbing prey dislodged by their flock-mates. A controversial idea that *Thamnomanes* may use false alarm calls to deceive other birds, allowing them to steal insects flushed out of hiding (Munn 1986), hints at the complex web of associations within a mixed-species flock; despite appearing chaotic, the flock's movements and behaviors are as interdependent as might be expected from birds that spend their entire lives foraging together. Flocks are known to occupy the same areas not only day to day, but over generations – flocks at a site in French Guiana were using the same foraging territories and roosting sites almost two decades after they were first mapped (Martínez & Gomez 2013). When forest is fragmented or otherwise altered on a large scale, however, this stability begins to break down (Mokross *et al.* 2014). Forest patches lacking *Thamnomanes* are virtually always missing other flock species as well.

Figure 339. *Thamnomanes caesius* (image © Harold Greeney).

Description: A small (18-21 g), nondescript bird of the forest understory. Males are entirely dark gray, females mostly brown with a rufous belly.

Distribution: Most of northern and eastern Amazonia; absent from SW Amazonia, where replaced by Bluish-slate Antshrike (*Thamnomanes schistogynus*). In Suriname, common in continuous lowland forest of the Interior; does not occur in the Coastal Plain. Always found in the presence of mixed-species flocks in the forest understory.

Use: (No known use)

Similar species: Most likely to be confused with Dusky-throated Antshrike (*Thamnomanes ardesiacus*), which is also common in Suriname and often occurs in the same flocks. Male Dusky-throated has a well-defined black throat; female is paler and duller than female Cinereous; the two species have different vocalizations. Other gray-plumaged antbirds have different shapes, behaviors, and vocalizations.

Gymnopithys rufigula
Rufous-throated Antbird
Mirafowru (local name used for all Thamnophilidae)
Passeriformes, Thamnophilidae

Brian O'Shea

One of the great ecological spectacles of Neotropical rain forest is the sight of a swarm of army ants moving through the forest understory, dislodging any organism that happens to be in their path. Any animal too slow to escape is consumed by the swarm. Those that do manage to evade the ants face another threat – a flock of sharp-eyed avian predators lurking above.

Antbirds are a large and exclusively Neotropical family, so named because some of them – like the Rufous-throated Antbird (Figure 340) – obtain all of their prey by following army ants. In Suriname, this species and the White-plumed Antbird (*Pithys albifrons*) are the only two 'professional' ant-followers in the Thamnophilidae. These two species only occur in forests of the Interior where army ants are abundant; so dependent are they on the ants that they do not forage when ants are inactive. Some woodcreepers (Furnariidae) forage primarily near army ants as well, and many other bird species take advantage of the swarms when they move through their territories. Competition is intense for prime foraging locations, usually at the center and leading edge of the swarm, and the hierarchy is determined primarily by body size. Rufous-throated Antbirds are larger than White-plumed Antbirds, and are thus dominant over them at swarms, although the latter species is more agile; a third, smaller species, the Common Scale-backed Antbird (*Willisornis poecilinotus*), is often relegated to the periphery of the swarm (Willis & Oniki 1978). Antbirds are quite noisy as they jostle for position over the swarm, and their churring calls are a sure indicator that army ants are nearby.

Figure 340. *Gymnopithys rufigula* (image © Marc Chrétien).

Description: Plumage mostly warm brown, paler below, with rufous-orange throat; bare skin around eye forms distinctive whitish ring, legs rather long and pale. Found in forest understory, almost always in the vicinity of army ants.

Distribution: Endemic to the Guiana Shield. In Suriname, confined to understory of lowland rain forest in the Interior; does not occur in fragmented forest. Replaced elsewhere in Amazonia by other species of *Gymnopithys*.

Use: (No known use)

Similar species: Whitish eye ring and habit of foraging over army ants should separate this species from other forest understory birds. White-plumed Antbird, which often occurs with it, is very different, with mostly chestnut plumage, orange legs, and a prominent white plume on the forehead.

Polystictus pectoralis
Bearded Tachuri
(No specific local name)
Passeriformes, Tyrannidae

Brian O'Shea

In the far south of Suriname, the Sipaliwini Savanna provides an otherworldly contrast with the surrounding lowland forest. This perception is heightened by the fact that many bird species here occur nowhere else in Suriname, and the avifauna overall is more reminiscent of savannas in central Brazil, far to the south and east (Renssen 1974, O'Shea 2005). This region has thus far escaped the fate of numerous other savannas in the Neotropics (Parker & Willis 1997) and remains unaffected by large-scale agriculture and cattle ranching, which has allowed ecological specialists such as the Bearded Tachuri (Figure 341) to prosper.

Although many savanna bird species can be found in deforested areas (for example, in the Coastal Plain of Suriname), a handful of them are true specialists requiring particular habitat features that are often destroyed when savanna is burned too frequently, grazed by cattle, or converted to agricultural use. The Bearded Tachuri only occurs in areas of savanna with long grass and scattered shrubs, usually near water; it is dependent on long grass for both foraging and nesting (Collar & Wege 1995). Such habitat cannot regenerate under a regime of frequent disturbance, and as a consequence this species is highly local in occurrence and vulnerable to local extinction. The Bearded Tachuri shares its habitat with another grassland specialist, the Rufous-sided Pygmy-Tyrant (*Euscarthmus rufomarginatus*), which maintains a small and isolated population in the Sipaliwini Savanna; both species are listed as Near Threatened by the IUCN (2015; see also chapter on Use, Threats and Conservation).

Figure 341. *Polystictus pectoralis* (image © Leon Moore).

Description: A tiny flycatcher, mainly brown above and bright buffy below, males with a contrasting black-and-white throat and blackish head.

Distribution: Found locally in appropriate savanna habitat from eastern Colombia through the Guianas, and south of the Amazon from southern Brazil into Argentina. In Suriname, it is confined to the Sipaliwini Savanna.

Use: (No known use)

Similar species: Unlikely to be confused in long-grass habitat. Rufous-sided Pygmy-Tyrant is similar in size but is much duller overall, lacking any buffy tones.

Onychorhynchus coronatus
Royal Flycatcher
(No specific local name)
Passeriformes, Tyrannidae

Brian O'Shea and Serano Ramcharan

The Royal Flycatcher (Figure 342) is an outlandish member of the most diverse bird family in the world, the Tyrannidae or tyrant-flycatchers. Restricted to the New World, tyrant-flycatchers can be found in almost any terrestrial habitat in North and South America, though they are most numerous and diverse in the tropics. Of the more than 400 species of tyrannids, 91 occur in Suriname; in many places, particularly in the Coastal Plain, they are the most common and conspicuous birds. As befits their name, many tyrannids are fiercely territorial and sport some type of distinctive crown patch, which they raise during confrontations with other birds. But the Royal Flycatcher takes this plumage feature to an absurd extreme. Although its spectacular crest is usually held flat along the top of the head, when raised to its full extent it is a sight to behold. Few people have seen it in free-flying birds, since this species normally lives quietly in pairs along creeks and forest edges, where it is rather difficult to see. But when handled, this bird puts on a mesmerizing performance, raising the crest and constantly turning its head almost completely upside down and from side to side while opening the bill to reveal the bright orange mouth lining. Although some authors have hypothesized that this display may function to startle predators (Dick & Mitchell 1979), the available evidence suggests that it is used in both courtship and territorial encounters (Skutch 1960, Graves 1990).

Royal Flycatcher nests are distinctive hanging structures, sometimes up to two meters long, usually placed over streams or narrow roads in the forest. Although Royal Flycatchers occur in primary forest, they can be quite common in secondary forest and abandoned plantations as well, a habitat they share with scores of other flycatcher species. Flycatchers avoid competition with each other by specializing on different prey types and by using different foraging heights and substrates.

Figure 342. ***Onychorhynchus coronatus*** **(image © Meshach Pierre).**

Description: Small (12-16 g), mostly yellowish-brown songbird with a long, flat bill and buffy-orange rump and tail. Spectacular crest is red in males, more yellowish in females, and often folded flat along the head, giving this bird a distinctive profile.

Distribution: Found in Amazonian lowlands with a separate population in Central America. Occurs throughout Suriname, often in secondary forests. Royal Flycatchers have recently been found to be quite common in Peperpot Nature Park near Paramaribo.

Use: (No known use)

Similar species: Distinguished from other forest understory flycatchers by orange tail, long bill, and long head profile.

Tyrannus savana
Fork-tailed Flycatcher
Seseigrikibi
Passeriformes, Tyrannidae

Brian O'Shea

At dusk during the rainy season in Paramaribo, it is not unusual to see large flocks of Fork-tailed Flycatchers (Figure 343) settling into communal roosts, sometimes in the middle of town. These are austral migrants – birds that breed in the Southern Hemisphere and migrate north during the southern winter, roughly between May and October. This habit is shared by a number of species that occur in Suriname, including several other flycatchers, martins and the Lined Seedeater (*Sporophila lineola*), but the Fork-tailed Flycatcher is the most conspicuous. It is also one of the most thoroughly studied austral migrants in South America (Jahn *et al.* 2013, 2016).

The Fork-tailed Flycatcher's communal roosts are a remarkable spectacle – they can contain thousands of birds and tend to grow through the rainy season, attaining peak size from July to September (Haverschmidt 1954b). Birds often begin streaming into roosts just before sunset and disperse to forage in the morning prior to sunrise. Like other migratory species of the genus *Tyrannus* (Morton 1971), Fork-tailed Flycatchers shift from being mostly insectivorous when breeding to frugivorous in migration and during the non-breeding season. Although they consume a wide variety of fruits, they have been observed in Guyana to preferentially forage on *Trema micrantha* (Ulmaceae) and certain species of Loranthaceae (M. MacPherson, pers. comm). A permanent resident population of Fork-tailed Flycatchers (*Tyrannus savana monachus*) occurs in the Sipaliwini Savanna in the far south of Suriname (Ottema *et al.* 2009).

Figure 343. ***Tyrannus savana* (image © Leon Moore).**

Description: A medium-sized bird (~30g) with greatly elongated outer tail feathers. Generally gray above and white below with a black cap and tail. Males and females are similar but males have much longer tails.

Distribution: Several breeding populations (subspecies) occur from southern Mexico to southern Argentina. The southernmost populations migrate into Amazonia and northern South America during the austral winter. Can occur throughout Suriname from May to October, but most abundant in the Coastal Plain, where scattered individuals may be seen year-round.

Use: (No known use)

Similar species: Unlikely to be confused with other species due to its long, forked tail. Short-tailed individuals may be mistaken for Gray Kingbird (*Tyrannus dominicensis*), but are distinguished by slimmer shape, smaller bill and black cap.

Rupicola rupicola
Guianan Cock-of-the-rock
Rotshaan
Passeriformes, Cotingidae

Brian O'Shea and Serano Ramcharan

There are few birds as emblematic of the great forests of the Guiana Shield as the Cock-of-the-rock (Figure 344). Its outlandish plumage, curious habitat preference, and general wariness ensure that this bird remains mysterious enough to capture the imagination. It is electrifying to see one streak through the forest, a virtual comet of orange, gone almost instantly. To see them in their full glory, you have to visit them on their display grounds.

Male Cocks-of-the-rock owe their stunning plumage to the process of sexual selection, which is common among Neotropical manakins and cotingas. In the Cock-of-the-rock, males possess courts within a display ground, or lek. Females visit the lek and assess each male's court before deciding with whom to mate. While a female is present, the males drop to their courts and crouch motionless, showing off their splendid plumage (Snow 1971, Trail 1987). Over evolutionary time, this has created pressure for more brilliant adornment – but it comes at a cost, since the showy males increase their risk of predation. Indeed, the Cock-of-the-rock seems to be pushing the limits, as available evidence suggests that displaying males are frequently attacked by predators, especially the Ornate Hawk-Eagle (*Spizaetus ornatus*; Trail 1987, 1990). Cocks-of-the-rock require large boulders for nesting. Females build nest cups of mud and affix them directly to the boulders using saliva (Gilliard 1962, Snow 1971). Leks are often located near rock outcrops and inselbergs in the forest. For this reason, the distribution of the Cock-of-the-rock is somewhat patchy. In Suriname, the most famous lek is near the Voltzberg in the Central Suriname Nature Reserve (Trail 1987), which has been active for at least 70 years and probably much longer.

Figure 344. *Rupicola rupicola* (image © Kester Clarke).

Description: A fairly large (200-230 g), stocky bird. Adult males are mostly brilliant orange, with a semicircular crest extending from the bill to the back of the head; the wings have black and white markings. Females and immature males are very different, mostly dull brown with a less developed crest.

Distribution: Endemic to the Guiana Shield. In Suriname, confined to interior forests, around inselbergs and in hilly regions with large boulders. Readily seen at the Voltzberg and near Kasikasima.

Use: Cock-of-the-rock feathers have been found in indigenous ceremonial garb in Amazonia and the Andes (where a related species occurs). In Suriname, these birds were hunted for their feathers prior to the 1950s (Haverschmidt & Mees 1994). This almost wiped out the lek near the Voltzberg, which has since recovered.

Similar species: Nothing remotely similar occurs in Suriname. Females are recognizable by their chunky shape and crest.

Perissocephalus tricolor
Capuchinbird
Busikaw
Passeriformes, Cotingidae

Brian O'Shea

One can only imagine the confusion that early explorers felt as they ascended the rivers of interior Suriname and heard their first Capuchinbirds (Figure 345). This bizarre cotinga has one of the most unusual vocalizations in the bird world – it is most frequently likened to a distant chainsaw or a lowing cow, or some hybrid of the two. Equally strange is the Capuchinbird's visual display. As they deliver their calls, males adopt an erect posture and fluff their body plumage to exaggerate their hunchbacked, bare-headed appearance (Snow 1961, 1972). A Capuchinbird lek is a telltale sound of the Guianan lowlands, as the species is restricted to the Guiana Shield, where it is a rather uncommon but widely distributed inhabitant of primary rain forest.

Undoubtedly, those early explorers lost no time in pursuing the source of the new sound they were hearing – perhaps expecting to find a group of cows in the forest. Instead, they would have stumbled onto a chaotic scene. Capuchinbird leks feature intense competition for a single perch, which is controlled by a dominant male who performs all of the copulations with visiting females (Trail 1990). Up to eight males may vie for this coveted spot, chasing each other furiously and frequently interrupting copulation attempts by the dominant male. Unlike many lekking species such as the Cock-of-the-rock, in which the sexes have drastically different plumages, male and female Capuchinbirds are virtually identical. Trail (1990), who studied the lekking behavior of Capuchinbirds at Brownsberg Nature Park, observed multiple cases of sexual mimicry at leks, in which males would imitate females to gain access to the mating perch, and females would imitate males to drive other females from it.

Figure 345. *Perissocephalus tricolor* (image © Kester Clarke).

Description: A crow-sized cotinga (300-400 g), with body plumage mostly rich reddish-brown. Head is unfeathered and blue-gray. Oddly proportioned with a unique hunchbacked appearance.

Distribution: Endemic to the Guiana Shield. Found in tall forest throughout interior Suriname; does not occur in the Coastal Plain.

Use: This species is likely hunted for food occasionally, though no ceremonial or commercial uses are known.

Similar species: Size, shape, color and bare head render this species unmistakable.

Ceratopipra erythrocephala
Golden-headed Manakin
(No specific local name)
Passeriformes, Pipridae

Brian O'Shea

In the understory of most Neotropical lowland forests, manakins are the most numerous avian frugivores (e.g. Karr *et al.* 1990, Blake & Loiselle 2009). They are known for the elaborate displays of the males, who typically spend considerable amounts of time on their leks, advertising for females. The flashy Golden-headed Manakin (Figure 346) is a common and widespread species in the Interior of Suriname, where its buzzing, crackling display is a frequent sound, especially during the dry season. Although these birds are generally difficult to see, they play a critical role in rain forest ecology through their dispersal of understory plants.

When a tree falls in the forest, the resulting canopy gap allows light to penetrate to the forest floor, triggering a riot of new growth. In the race to reach the light, certain plant species tend to dominate, among which the family Melastomataceae is prominent. Melastomes produce copious amounts of fruit, which are eagerly devoured by manakins – indeed, they comprise a substantial proportion of the birds' diet (Krijger *et al.* 1997). The sheer abundance of manakins ensures that melastome seeds are dispersed widely through the forest, and behavioral differences among age and sex classes promote the distribution of melastome seeds across a variety of microhabitats (Théry 1992). In French Guiana, melastome seeds were found to be especially abundant in the seed bank beneath manakin leks (Krijger *et al.* 1997). This system ensures that melastomes are positioned to take advantage of random disturbances virtually anywhere in the forest (Blendinger *et al.* 2011).

Figure 346. ***Ceratopipra erythrocephala*** **(image © Kester Clarke).**

Description: A small (11-13 g), chunky bird of tall forest understory. Older males are entirely black with a golden-yellow head and white eyes. Females and young males are plain grayish-green with yellowish bills and dark eyes.

Distribution: Found in humid lowland forest from eastern Panama across South America north of the Amazon River. Occurs throughout Suriname, but is rare to absent in the lower Coastal Plain.

Use: (No known use)

Similar species: Showy males are unmistakable. Females are similar to other female manakins, but are distinguished by small size, plain facial expression and yellowish bill.

Vireo sclateri
Tepui Vireo
(No specific local name)
Passeriformes, Vireonidae

Brian O'Shea

Rising out of the middle of the vast lowland forest of the Central Suriname Nature Reserve, the Tafelberg is the eastern outpost of the tepuis, flat-topped mountains that dominate the landscape of southeastern Venezuela, western Guyana, and adjacent northern Brazil. The tepui region is renowned for its high endemism, particularly of plants (Kelloff & Funk 2004). The unique flora and fauna of these mountains has captivated biologists for centuries, no doubt because many remain so inaccessible.

Although the summit of the Tafelberg barely exceeds 1000 m, the mountain supports Suriname's most extensive area of upland forest. Here can be found a handful of endemic tepui bird species (Zyskowski *et al.* 2011). The most common is the Tepui Vireo (Figure 347), whose persistent *'pe-teeer'* is a frequent sound in these otherwise rather quiet forests. Like many tepui endemics, virtually nothing is known of the Tepui Vireo's natural history. It is known to be insectivorous and to follow mixed species flocks, much as other vireos do. What is less clear is how this and other tepui endemics maintain such widely scattered populations, often on tiny 'islands' of upland forest in a sea of lowland rain forest. Are these birds relics of formerly widespread populations? Is the Tafelberg population descended from a single dispersal event, or is the population supported by steady immigration from the core tepui region, hundreds of kilometers away? Do birds from Tafelberg cross lowland forest to colonize other upland patches in the mountains of the Grensgebergte and Eilerts de Haan? Many tepui endemic bird species are absent from Tafelberg, and the Tepui Vireo has been found as low as 500 m in other parts of the Guiana Shield (Robbins *et al.* 2007, O'Shea & Ramcharan 2013), so it seems plausible that some bird species may wander through the lowland forest to colonize new areas, whereas others are less likely to do so. But this crucial aspect of the natural history of highland forest birds in the Guiana Shield remains as mysterious as the birds themselves.

Figure 347. ***Vireo sclateri*** **(image © Foek Chin-Joe).**

Description: A small bird of the forest canopy. Greenish above, white below, washed with yellow across the chest, with a grayish head. The eye is conspicuously pale.

Distribution: Endemic to the tepui region of eastern Venezuela, western Guyana, and northern Brazil. Also occurs eastward on the Guiana Shield, as low as 500 m in southern Guyana and Suriname, where it is known from Tafelberg and the Grensgebergte. Extent of distribution incompletely known.

Use: (No known use)

Similar species: Unlikely to be mistaken in upland forests where it occurs. Similar to the related Lemon-chested Greenlet (*Hylophilus thoracicus*), a widespread lowland species, but distinguished by very different voice and pale eye.

Hirundinidae
Swallows
(No specific local name)
Passeriformes, Hirundinidae

Brian O'Shea

Travelers on Suriname's rivers are likely to become familiar with a host of birds skimming gracefully just over the water's surface, often flying alongside boats before wheeling up into the air and looping around again. These are swallows, the most conspicuous members of the guild of aerial insectivores in Suriname. They spend much of the day aloft, where they feed on a variety of insects (Haverschmidt 1968) in much the same manner as the unrelated swifts; unlike swifts, swallows spend considerable time perched between foraging bouts. Swallows occur just about everywhere, but are easiest to see along rivers, where they offer passing entertainment during long boat trips.

Both permanent resident and migrant swallows occur in Suriname. The resident White-winged Swallow (*Tachycineta albiventer*) and White-banded Swallow (*Atticora fasciata;* Figure 348) are tied closely to rivers. White-banded Swallow is particularly common in upstream areas of the Interior, where the rivers are narrower and rocky; White-winged tends to be more conspicuous along broader rivers, hundreds of meters wide, although the two species frequently occur together. White-banded Swallow is clearly the more social of the two species. Groups of 10-15 or more are frequently encountered around boulders and snags, often near a steep bank, where they nest in burrows excavated by other birds, such as kingfishers. White-winged Swallows, on the other hand, typically occur in pairs and nest in crevices in stumps, between rocks, or in old woodpecker holes (Sedgwick 2010). Swallows are agile flyers and have relatively few predators, but the tendency of some species to crouch low on river boulders when resting may be a strategy to avoid detection by birds of prey, particularly the Bat Falcon (*Falco rufigularis*), a nimble aerialist common along rivers in lowland rain forest.

Figure 348. *Tachycineta albiventer* (above); *Atticora fasciata* (below) (image © Carlos Alcantara).

Description: White-winged Swallow is blue-green above and white below with a large white wing patch. White-banded Swallow is entirely blue-black with a sharply contrasting white chest band. White-winged averages slightly larger than White-banded (14-20 g vs.12-16 g).

Distribution: Both species occur throughout Amazonia. In Suriname, White-winged occurs throughout, whereas White-banded is more confined to the Interior. Both species are most common along rivers.

Use: (No known use)

Similar species: Swallows are distinguished from swifts (Apodidae) by their more graceful, fluid flight and habit of perching frequently. Both White-winged and White-banded are quite distinctive. White-banded may be confused with Black-collared Swallow (Pygochelidon melanoleuca), but is longer-tailed and has a different plumage pattern (white below with a black band).

Tangara mexicana
Turquoise Tanager
(No specific local name)
Passeriformes, Thraupidae

Brian O'Shea

The remarkably diverse tanagers comprise the second largest bird family in the world, and include some of the most abundant birds in the Neotropics. *Tangara* is the largest and arguably the most spectacular genus of Neotropical birds. Although the diversity of *Tangara* is highest in the cloud forests of the Andes, several species occur in the lowland forests of Suriname. They are rather difficult to see, as they spend much of their lives high in the canopy, moving in mixed-species flocks over large areas in search of fruit and insects. The Turquoise Tanager (Figure 349) is one of the more frequently observed species, in part because it is common on the Coastal Plain as well as in the Interior. Pairs or small groups can often be seen foraging in fruit trees in gardens in Paramaribo.

The tanager family is thought to have diversified extremely recently relative to other lineages of birds in the Neotropics (Burns *et al.* 2014). Bill shapes, in particular, have evolved quickly to enable birds to respond to new foraging niches. Until recently, this rapid evolution had puzzled taxonomists, who had relied on morphology and plumage patterns to classify birds; many members of the tanager family seemed to defy categorization, and some were originally classified in other families. The advent of molecular data shed considerable light on this family's relationships (Sedano & Burns 2010, Barker *et al.* 2013, Burns *et al.* 2014) and yielded some surprises – for example, that Darwin's famous Galapagos 'finches' are actually tanagers!

In contrast to the family as a whole, the 49 species in the genus *Tangara* are all quite similar in both morphology and ecology. However, they have evolved dramatic differences in plumage, many with patches of brilliant yellow, red and blue. This is likely driven by the high light intensity in the canopy where these birds forage, particularly in the ultraviolet spectrum (Gomez & Théry 2004). Bright plumage serves as a social signal among species foraging in the same flocks, as well as a sexually selected ornament within species.

Figure 349. ***Tangara mexicana*** **(image © Kester Clarke).**

Description: A small, blue-and-black bird with a pale yellow belly.

Distribution: Found in humid lowland forest and forest edges throughout Amazonia; tolerant of disturbed and secondary habitats. Occurs throughout Suriname.

Use: (No known use)

Similar species: Plumage color and pattern are unique. Other species of *Tangara* are very differently colored and tend to occur in smaller groups or mixed-species flocks. This is the only species of *Tangara* likely to be seen in secondary habitats around Paramaribo.

Sporophila crassirostris
Large-billed Seed-Finch
Twatwa
Passeriformes, Thraupidae

Brian O'Shea

Each Sunday morning, the calm of Paramaribo's *onafhankelijkheidsplein* (the grassy square in front of the presidential palace) is broken by the melodious whistling of cage birds. In a time-honored tradition, men congregate here to mill about the open space, assessing each other's birds before positioning two cages in close proximity to spur a singing contest between their captives. Most of these birds are seed-finches and seedeaters in the genus *Sporophila*, sparrow-sized birds of open grassland and disturbed habitats. Although each species has a characteristic and generally pleasing song, the *twatwa* (Figure 350) is widely considered to be the finest singer of them all. A well-trained and battle-tested bird can fetch a high price indeed.

Historical accounts indicate that the *twatwa* was once widespread in the coastal marshes and interior savannas of Suriname (Haverschmidt & Mees 1994). Although it has a large geographic range, it appears to be more specialized than many other species of *Sporophila* – it is closely tied to wet savannas, where it breeds during the rainy season (Lill 1974). Perhaps this is one reason the species has mostly disappeared from Suriname and elsewhere in the Guianas – its patchy distribution may render it more vulnerable to trapping for the cage bird trade. Trapping is widely believed to have decimated this species across the Guianas, where sightings of free-flying *twatwa* are now rare.

Figure 350. *Sporophila crassirostris* (image © Meshach Pierre).

Description: A small (20 g) bird with a very thick, conical bill. Adult males are black with a small white wing patch, white underwings, and a bluish-white bill; females and young males are entirely dark brown.

Distribution: Occurs in marshy, open habitats in lowlands of northern South America (north of the Amazon River). In Suriname, it was already quite rare in the Coastal Plain by the 1960s, but is now virtually extirpated from the country, with a few individuals possibly remaining in the Sipaliwini Savanna.

Use: Trapped for the cage bird trade throughout its range.

Similar species: Chestnut-bellied Seed-Finch or *pikolet* (*Sporophila angolensis*) is smaller, and males have a chestnut lower breast and belly and a black bill.

Psarocolius viridis
Green Oropendola
Busiponpon
Passeriformes, Icteridae

Brian O'Shea

One of the characteristic sights of the lowland Neotropics is a tree with oropendola nests. These large members of the blackbird family build hanging pouches roughly one meter long, and up to 30 may be found in a single tree. Although highly visible, the nests are woven onto the very tips of canopy branches, so are quite inaccessible to predators. The nesting tree itself is chosen for its isolation to discourage visits by monkeys and other arboreal mammals. Oropendolas (Figure 351) are noisy at their nests. The constant comings and goings of the parent birds provide a background for the remarkable gurgling calls of the males. These calls, from which the species' Surinamese name derives, are integrated in a ritualized display during which the bird pitches forward on his perch, vibrating his wings and turning almost completely upside-down while delivering an extraordinary sound that can be heard a kilometer away.

The activity around an oropendola colony is further augmented by other bird species that take advantage of the relative safety of the oropendolas' nest sites. One of these is the Piratic Flycatcher (*Legatus leucophaius*), a feisty little bird that takes over other birds' nests by driving out the occupants. Oropendolas and their close relatives, the caciques, are favorite targets of this bird, whose distinctive buzzing call is often heard around their colonies. The Giant Cowbird (*Molothrus oryzivorus*) also lurks around oropendola colonies, slipping into nests to lay its eggs, which are incubated and reared by the oropendolas. Research in Central America suggested that oropendolas will sometimes tolerate this parasitism, because the young cowbirds remove botfly larvae from their oropendola nest-mates, thereby increasing their survival (Smith 1968). However, subsequent studies of other species indicate that oropendolas almost never tolerate the presence of cowbirds around their colonies (Robinson 1988; Webster 1994).

Figure 351. ***Psarocolius viridis*** **(image © Mathias Fernandez).**

Description. A large (200-400 g) songbird with mostly olive-green and chestnut body plumage, a long, conical bill, and bright yellow outer tail feathers.

Distribution. Found in lowland forest throughout the Amazon basin. In Suriname, confined to the Interior, where it occurs in both dryland and seasonally flooded forest; wanders into the Coastal Plain only rarely, if ever.

Use. (No known use)

Similar species. Distinguished from the related caciques (*Cacicus spp.*) by much larger size, distinctive rowing flight, and green-and-chestnut plumage with conspicuous yellow outer tail feathers. Crested Oropendola (*Psarocolius decumanus*), which can co-occur with it in seasonally flooded forests and is common in the Coastal Plain, is predominantly black.

Mammals

Burton Lim, Marilyn Norconk

Mammals are undoubtedly one of the most identifiable and familiar groups of animals in Suriname and throughout South America. Among the most iconic of these is the Jaguar (*Panthera onca*; *penitigri*) – the largest carnivore in the New World. Such top-level predators are important to the ecosystem and have a huge impact on the diversity and abundance of prey species including other mammals and vertebrates.

However, the overwhelming majority (80%) of the 196 species of mammalian biodiversity in Suriname (Lim 2016) are small, weighing about 1 kg or less. Smaller mammals include bats, rodents and opossums (Lim *et al.* 2005a), many of which are prey to predator species. Because of their small size, they are also among the most poorly known groups with species new to science being discovered surprisingly often. For example, the Guianan Broad-headed Bat (*Platyrrhinus guianensis*) was recently described from the lowland regions of Guyana and Suriname (Velazco & Lim 2014). Furthermore, DNA analysis suggests that the biodiversity of mammals in the Neotropical region is underestimated by at least a quarter (Lim 2012b).

Although not as species-rich as small mammals, large mammals are the most recognizable to people. Next to the wild cats, primates are popular and relatively easy to identify by sight and sound. Anyone who hears the eerily haunting roar of the Guianan Red Howler Monkey (*Alouatta macconnelli*; *babun*) will never forget it. Visually distinctive is the White-faced Saki (*Pithecia pithecia*; *wanaku*), though the white face is only seen in adult males, contrasting with the gray-brown color of females.

Upon its publication in 1978, 'The Mammals of Suriname' by A.M. Husson instantly became the most comprehensive monograph of the mammalian fauna in any country of South America (Pine 1982). However, at that time most scientific studies dealt with taxonomy and distribution with ecological and behavioral information primarily derived from naturalists' anecdotal observations. Since then, mammalogical research in Suriname has lagged in comparison to other countries on the continent (Lim & Catzeflis 2014, Lim 2016). Furthermore, much of the information on many species of mammals found in South America has been derived from studies on captive animals in zoos. More recent field research is providing descriptive information on wild species, but work in Suriname is still relatively minimal with perhaps the exception of primates.

In addition to Husson (1978), long out of print but now available free online (**repository.naturalis.nl/document/572868**), general information on the biology of the mammals in this chapter has been gleaned from various sources. These include 'Neotropical Rainforest Mammals' by Emmons and Feer (1997), the International Union for the Conservation of Nature (IUCN) Red List of Threatened Species (**iucnredlist.org**), Animal Diversity Web (**animaldiversity.org**), the scientific publication series Mammalian Species (**science.smith.edu/msi**), and 'All the World's Primates' (**alltheworldsprimates.org/Home.aspx**).

Mammal groups

The classical scientific groupings (taxa) within mammals are hierarchical and based on anatomy, molecular biology and behavior. Mammals include 29 taxa at the rank of 'order'; 12 of these occur in Suriname. The country's mammal biodiversity (excluding strictly marine species) ranges from the order of bats with 108 species, to four orders with only a single species (Table 12). There are alternative methods of grouping these animals; one of the more comprehensive is based on 'guild', which groups together species according to their adaptation to similar ecological pressures, such as feeding behavior associated with available food resources. Suriname's mammal species can thus be categorized into food-related or trophic guilds according to whether their primary foraging strategies rely on meat (carnivore guild), fruit (frugivore guild), seeds (granivore guild), leaves (herbivore guild), insects (insectivore guild), nectar (nectarivore guild), a combination of plants and animals (omnivore guild), or blood (sanguinivore guild) (Table 12). Most species (80) of non-marine mammals in Suriname are insect-feeders, whereas blood-feeding occurs in only two species of bats. Bats are also notable in having the most diverse range of feeding behaviors, with only granivory and herbivory absent. Bats are also the only order of mammals to include nectarivores and sanguinivores.

Taxonomic Order	Common Name	Trophic Guild[1]								Total
		C	F	G	H	I	N	O	S	
Didelphimorphia	Opossums	2	2			4		6		14
Pilosa	Anteaters and Sloths				2	3				5
Cingulata	Armadillos					5				5
Chiroptera	Bats	4	26			63	8	5	2	108
Primates	Monkeys		2	2	1	1		2		8
Carnivora	Carnivores	12						3		15
Cetacea	Whales and Dolphins	1								1
Sirenia	Manatees				1					1
Perissodactyla	Odd-toed Ungulates				1					1
Artiodactyla	Even-toed Ungulates		2		3					5
Rodentia	Rodents		13	4	4	4		7		32
Lagomorpha	Rabbits				1					1
Total		19	45	6	13	80	8	23	2	196

Table 12. The number of species of mammals found in Suriname grouped by natural scientific classification (taxonomic order) and categories based on diet1 (trophic guild).

[1] C=carnivore, F=frugivore, G=granivore, H=herbivore, I=insectivore, N=nectarivore, O=omnivore, and S=sanguinivore (terms explained in text).

Body size, in addition to the general habitat where the animal lives, is perhaps the most intuitive way to group species. For convenience of interpretation, this is how we present the mammals of Suriname (Table 13). Some are primarily terrestrial (ground-dwelling), such as the Giant Armadillo (*Priodontes maximus*; *granmankapasi*), while others are arboreal (tree-dwelling) including monkeys that spend essentially all of their time up in the forest canopy. Primates have anatomical and behavioral adaptations that help them live in the trees. For example, Brown Capuchins, Guianan Red Howler Monkeys and the Guiana Spider Monkey (*Ateles paniscus*; *kwata*) have prehensile tails that act as a fifth appendage to grab branches as they forage for food. The Guiana Spider Monkey is also notorious for hanging by its tail and violently shaking debris from the canopy when disturbed by human intruders.

Ecological Guild	Common Name	Species
Small non-flying mammals	Guianan Terrestrial Spiny Rat	*Proechimys guyannensis*
	Broad-headed Rice Rat	*Hylaeamys megacephalus*
	Common Opossum	*Didelphis marsupialis*
	Delicate Slender Mouse Opossum	*Marmosops parvidens*
Small flying mammals	Common Vampire Bat	*Desmodus rotundus*
	Seba's Short-tailed Fruit Bat	*Carollia perspicillata*
	Pallas's Long-tongued Bat	*Glossophaga soricina*
	Proboscis River Bat	*Rhynchonycteris naso*
Large arboreal mammals	Golden-handed Tamarin	*Saguinus midas*
	Guianan Red Howler Monkey	*Alouatta macconnelli*
	Guiana Spider Monkey	*Ateles paniscus*
	Brown Capuchin	*Sapajus apella*
	South American Squirrel Monkey	*Saimiri sciureus*
	White-faced Saki	*Pithecia pithecia*
	Kinkajou	*Potos flavus*
	Linnaeus's Two-toed Sloth	*Choloepus didactylus*
Large terrestrial mammals	Paca	*Cuniculus paca*
	Red-rumped Agouti	*Dasyprocta leporina*
	Capybara	*Hydrochoerus hydrochaeris*
	Red Brocket Deer	*Mazama americana*
	White-lipped Peccary	*Tayassu pecari*
	South American Tapir	*Tapirus terrestris*
	Jaguar	*Panthera onca*
	Forest Rabbit or Tapeti	*Sylvilagus brasiliensis*
	Bush Dog	*Speothos venaticus*
	Giant Anteater	*Myrmecophaga tridactyla*
	Giant Armadillo	*Priodontes maximus*
Aquatic mammals	Guiana Dolphin	*Sotalia guianensis*
	West Indian Manatee	*Trichechus manatus*
	Giant Otter	*Pteronura brasiliensis*

Table 13. General ecological grouping of mammals with species accounts in the present chapter.

However, habitat separation between terrestrial and arboreal is not always clearly delineated. The smaller cats, such as the Ocelot (*Leopardus pardalis*) are as comfortable on the ground as they are climbing and resting in trees. Some groups of mammals such as anteaters have species that live exclusively on the ground (Giant Anteater; *Myrmecophaga tridactyla*; *tamanwa*) and others that use both the ground and the trees (such as the Southern Tamandua; *Tamandua tetradactyla*; *mirafroiti*).

Other large-bodied arboreal mammals include species of sloths (*loiri*) and the lesser-known Kinkajou (*Potos flavus).* Its Surinamese name, *netikeskesi*, translates into English as the 'night monkey': as the name suggests, it is nocturnal (active at night) and resembles a primate, with a rounded head and prehensile tail, though it is in fact a member of the raccoon family (Procyonidae). The majority (80%) of mammals in Suriname sleep in the heat of the day and search for food during the cool of night. Primates, including humans, are unusual mammals because they are primarily diurnal – active during the day. Guianan Red Howler Monkeys can give loud calls or roars during the night, but do not move from their sleeping tree until dawn (Vercauteren & Gautier 1993).

A few species of large mammals are fully aquatic in freshwater or estuarine environments along the coast. The West Indian Manatee (*Trichechus manatus*; *seku*) and Guiana Dolphin (*Sotalia guianensis*; *profosu*) spend all of their time in the water and their front limbs are modified as flippers for maneuvering. Over evolutionary time, the hind limbs have been lost resulting in a more streamlined body for swimming and the tail fluke provides forward propulsion. Manatees are a little more secretive and difficult to spot, but dolphins can be readily seen in the Suriname River not too far from Paramaribo.

Some mammals are better categorized as semi-aquatic with adaptations for foraging in water while sleeping on land. The Giant Otter (*Pteronura brasiliensis*; *bigiwatradagu*) has a horizontally flattened tail and webbed feet for swimming, but is adept on land where it constructs extensive living quarters up to 50 square meters in area. Less aquatically adapted and spending more time out of the water is the largest land mammal in the Neotropics, the South American Tapir (*Tapirus terrestris*; *bofru*).

Bats (*fremusu*) are specialized mammals and the only ones that fly. In Suriname, they account for more than half (108 of 196 species) of the mammalian species diversity. But bats are perhaps the most mysterious to people because they are usually seen as fleeting silhouettes against the evening sky in the fading light of dusk. Many people think that bats are bloodsuckers, but only 2 of the 108 species in Suriname are vampires, with one feeding on mammals (*Desmodus rotundus*) and another on birds (*Diaemus youngi*). Most other bats provide important ecosystem services such as seed dispersal and flower pollination, and control herbivorous insect populations that may otherwise do damage to agricultural crops or trees in naturally forested areas.

Rodents and opossums (*awari*) comprise almost one-quarter of the diversity of mammals in Suriname. They are mostly small in size such as mice (*moismoisi*), rats (*alata*), squirrels (*letyan, bonboni*), and mouse opossums (*busmoismoisi*), but a few are large, including the Capybara (*kapuwa*), which is the largest species of living rodent in the world, averaging about 50 kg. Typically, rats and mice weigh less than 500 g, including the relatively common Broad-headed Rice Rat (*Hylaeamys megacephalus*) which is found throughout the country.

All opossums in Suriname belong to the family Didelphidae, which is unique to the New World, and more closely related to the marsupials found in Australia than to any other mammals in Suriname. In contrast to these placental mammals, which give birth to relatively well-developed young after a longer period of gestation, opossums and other marsupials give birth to underdeveloped young after a short pregnancy. To compensate, young marsupials spend a longer time with the mother and many, but not all, species in this group have a pouch (or marsupium) for their offspring. The Delicate Slender Mouse Opossum (*Marmosops parvidens*) is one species in a group of small arboreal marsupials that does not have a pouch.

Diet and foraging

Top-level predators. The classic top-level predator in the Neotropics is the Jaguar: a carnivorous species at the apex of the food chain with no other animals preying on it. The presence of large predators is an indicator of a high density of prey species, including other mammals (Duplaix 2002). In aquatic environments, the Giant Otter fills this niche as an efficient predator of fish. Not to be outdone, the night sky is dominated by the Greater False-vampire Bat (*Vampyrum spectrum*), the largest bat in the New World with a wingspan reaching up to 1 meter from tip to tip. They feed on small vertebrates such as birds, mice, and other bats.

Fruit eaters. Suriname's eight species of monkeys are fruit eaters, although they vary in the amount of fruit in their diet. They are among the most important seed dispersers in Suriname. Guiana Spider Monkeys (*kwata*) are particularly well known for their seed-dispersing capabilities (Mittermeier & van Roosmalen 1981, Russo *et al.* 2005, Link & DiFiore 2006). Guianan Bearded Sakis (*bisa*) and White-faced Sakis (*wanaku)* are also fruit-eaters, but they eat more seeds than fruit pulp; thus they benefit some tree species by dispersing seeds, but eat or destroy seeds of other plant species (Norconk *et al.* 1998).

Monkeys tend to be high-quality seed dispersers because they can eat large quantities of fruit at a single sitting, feed and move in groups that may exceed 50 individuals, and defecate seeds away from the parent plant. Most monkeys have more than 80 plant species on their menu over an annual cycle. When these primates are uncommon, scatter-hoarding rodents such as the Red-rumped Agouti become important secondary seed dispersers (Forget & Cuijpers 2008). Their buried caches of food have better chances of survival than seeds that are left above ground.

Many other species of mammals also eat primarily fruit including leaf-nosed bats in the family Phyllostomidae (26 of 60 species) and rodents (13 of 32 species); even some species in the order Carnivora such as the Tayra (*Eira barbara*; *aira*), a member of the weasel family, consume fruit so they are better categorized as omnivorous.

Leaf eaters. Sloths (*loiri*) primarily subsist on leaves and have many adaptations, such as microbial fermentation in the foregut of the stomach, for consuming this low-energy food source (Dill-McFarland *et al.* 2016). Their low metabolic rate has resulted in the English common name being co-opted as a synonym for slow or lazy, but this physiological adaptation reduces the need for energy. Other mammals consume leaves as a secondary supplement in their diets, such as fruit-eating bats that get important proteins by chewing and spitting out the fibrous leaf pellets (Kunz & Diaz 1995).

The larger-bodied monkeys in Suriname (*babun* and *kwata*) eat leaves, preferably young leaves. Leaves make up about 50 percent of the diet for Guianan Red Howler Monkeys (*babun*), but that percentage can increase to 80-90 percent when fruit is rare (Di Fiore et al., 2011). Guiana Spider Monkeys (*kwata*) add young leaves to their diet as their primary protein source (about 8 percent of annual diet). White-faced Sakis (*wanaku*) also eat leaves almost every day (Norconk 1996), but always at a lower percentage than fruit and seeds. Interestingly, the other seed-eating monkeys in Suriname, the Guianan Bearded Sakis (*bisa*), rarely eat leaves and may get most of their protein from insect prey (Norconk & Conklin-Brittain 2016).

Insect eaters. Anteaters require a diet of ants and termites. They have evolved many adaptations for this specialized insect diet, including anatomical changes such as an elongated mouth, absence of teeth, and a long tongue (Naples 1999). Their behavior is also modified to minimize counter-attacks from prey by quickly feeding at many different ant and termite nests.

Not as well known, but with the highest species diversity of any mammalian feeding guild (Table 12), bats have also mastered the consumption of insects. They have evolved two main strategies: aerial bats use echolocation to catch insects on the wing in flight, whereas gleaning bats use hearing to snatch stationary insects that are making calls to attract mates.

Small- and medium-sized monkeys in Suriname eat insects opportunistically or intentionally search for them. Insects provide dietary fat and protein and are important energy sources for these active primates. Golden-handed Tamarins (*saguwenke*) and South American Squirrel Monkeys (*monkimonki*) are opportunistic insect eaters – moving rapidly through the forest, scaring up insects that they grab with their hands. They have a special relationship with insect-eating birds, such as Double-toothed Kites, that may benefit the birds, but do not affect the ability of the monkeys to find food. In Brownsberg Nature Park, which has a high density of *saguwenke*, it is common to see kites flying in and among (but not preying on) a foraging group of small monkeys (M. Norconk, pers. obs.). Brown Capuchins (*keskesi)* feed both opportunistically on walking sticks, katydids and praying mantis, and seek colonial insects – wasps, bees, termites, ants and caterpillars. White-faced Sakis also eat insects opportunistically, particularly termites and ants. All of these monkeys also eat fruit.

Granivory. Two closely related species of monkeys in Suriname, White-faced Sakis (*wanaku*) and Guianan Bearded Sakis (*bisa*), are primarily seed eaters (granivores), which from the plant perspective could also be called seed "predators." Seeds of dozens of species make up 50 to 75 percent of the annual Guianan Bearded Saki diet and about 60 percent for White-faced Sakis (Norconk 2011). Even though both species are seed eaters, Gregory & Norconk (2013) found less than 20 percent overlap in plant species used in the same years at Brownsberg Nature Park. The two species have very similar teeth, adapted for prying open fruit to gain access to seeds, but Guianan Bearded Sakis are larger-bodied, more specialized in terms of their jaw anatomy (Kinzey 1992), and heavily reliant on seeds. Interestingly, a recent study by Shafler (2013) in Guyana suggests that we may have underestimated the frequency with which Guianan Bearded Saki eat insects and flowers when seeds are not available.

All rodents consume seeds to some degree, but few rely on this food as their primary source. This is also true for mammals in general (Table 12). Although agoutis are well known as scatter hoarders of seeds, they are primarily fruit-eaters, with pulp comprising most of their stomach contents (Dubost & Henry 2006).

Seed predation and dispersal. The Red-rumped Agouti and Red Acouchy are both seed dispersers and seed predators in that they eat seeds, but also bury them in the ground to hedge against a possible shortage of food during the dry season. If the cache is not needed or its location is forgotten, then the seeds can potentially germinate into seedlings and grow as part of natural forest regeneration. One primary example is the tree species *Carapa procera* (*wetikrapa* or *krapa*). A study in French Guiana and Suriname found that the harvesting of seeds by humans was usually associated with the subsistence hunting of the seed dispersing animals. which negatively impacted the trees by loss of both seed and disperser (Forget & Jansen 2007).

Bats are good seed dispersers because they typically do not eat the fruit at the parent tree. They either discard large seeds after eating the fleshy parts or defecate small seeds after consuming the whole fruit. One of the largest seeds (60 by 30 mm) carried in the mouth of a bat is from the emergent tree species *Dipteryx odorata* (*tonka*). The large fruit-eating bats, *Artibeus lituratus, Artibeus obscurus* and *Artibeus planirostris,* have been reported as seed dispersers of this tree (Lobova *et al*. 2009). These three species are usually among the most common bats flying in the understory of lowland rainforest in Suriname (Lim *et al*. 2005b, Lim & Joemratie 2011, Lim & Banda 2013).

Solanum spp., belonging to Solanaceae or tomato family, produce small-seeded fruits eaten whole by bats. These shrubs are usually found in disturbed or open areas, and bats in the genera *Carollia* and *Sturnira* are their primary seed dispersers (Lobova *et al*. 2009). *Solanum* is an important pioneer genus that provides shade in forest gaps for canopy and emergent trees to take hold resulting in natural forest regeneration.

Tapirs are also considered good dispersers because seeds are undamaged by either being spat out or passing through the digestive tract intact (Henry *et al*. 2000). The palm tree *Mauritia flexuosa* (*morisi*) has fruit with large, hard seeds up to 30 mm in length that tapirs do not swallow after eating the fleshy pulp. By contrast, the small seeds of *Geissospermum laevis* are swallowed when the fruit is eaten but not damaged by the teeth of tapir; seeds can germinate after defecation.

Other species of hoofed mammals such as deer and peccaries are seed predators. Brocket deer digest seeds using gut microbes and the powerful jaws of peccaries easily crack seeds, although some small seeds may evade the digestive process (Bodmer 1991). And, all primates are seed dispersers to varying degrees (see above).

Nectarivory and pollination. Although other species of mammals have been documented to occasionally feed on nectar and pollinate flowers (Carthew & Goldingay 1997), bats are the only taxonomic order to have nectarivory as a primary feeding behavior. In the Neotropics, there are nectar-feeders in two subfamilies of the leaf-nosed bat family Phyllostomidae: Glossophaginae and Lonchophyllinae. All have obvious morphological adaptations including an elongated snout and a long tongue that illustrate their co-evolution with flowers (Heithaus *et al*. 1975). A study in French Guiana found that the common canopy tree *Eperua falcata* (*walaba*) is pollinated by the nectar-feeding bats, *Anoura geoffroyi, Lionycteris spurrelli*, and *Lonchophylla thomasi*, which are species that also occur in Suriname (Delaval *et al*. 2005).

Many other species of bats also secondarily feed on nectar. Omnivorous bats, including the spear-nosed bats *Phyllostomus hastatus* and *Phyllostomus discolor*, are important pollinators of the emergent kapok tree (*Ceiba pentandra*; *kankantri*). Various animals ranging from diurnal bees, wasps and hummingbirds to nocturnal bats, opossums and moths feed on the nectar, but the spear-nosed bats are the main pollinators because they forage in large groups and visit other kapok trees over long distances (Gribel *et al*. 1999).

Keystone plant species. The tree genus *Cecropia* is considered a keystone species because many different animals rely on it as a food source, such as birds and mammals that eat the fruit (see also Plants and Birds chapters). It is also an important pioneer plant that provides shade in gaps for emergent trees to grow during natural forest succession. Although not high on their food list, primates and rats are known at times to eat unripe fruits with undeveloped seeds, making them ineffective seed dispersers. By contrast, some species of *Cecropia* rely on bats as a primary disperser as the seeds pass through the gut intact: they include the large fruit-eating bats in the genus *Artibeus* and, to a lesser extent, the smaller Short-tailed Fruit Bat (*Carollia perspicillata;* Lobova et al 2009).

Both free-standing and strangling figs (*Ficus* spp.) are also known to be keystone species for birds and mammals (Terborgh 1983), and are important in the reproductive cycle of some wasp species that form a mutualistic (pollinating) relationship with fig trees (Cook & Rasplus 2003). Fig trees fruit irregularly throughout the year. If they fruit in the dry season they are a highly productive, but short-term resource, providing a dramatic fruiting event for about 14 days (Janzen 1979). During that time, fruit-eating bats, birds and monkeys may return to the tree many times to ingest the fruit and disperse tiny seeds.

Mixed-species foraging. Similar to the mixed-species foraging flocks of birds (see Birds chapter), some primates form temporary associations between species. One of the better-studied associations, between Brown Capuchins and South American Squirrel Monkeys (Podolsky 1990), provides evidence for mutual benefits to both species. While large groups of meandering South American Squirrel Monkeys benefit from the Brown Capuchins' knowledge of available feeding trees and their natural alertness to predators, Brown Capuchins benefit from faster food processing time when visiting small-crowned feeding trees. Thus, both species benefit from the larger search area afforded by adding more individuals to the group, but Brown Capuchins benefit under certain circumstances given their larger body size and more assertive behavior.

Seasonality. Most species of mammals in the Neotropics have a varied diet and can shift to other food resources during the dry season. For example, peccaries will eat fruit when in season, but rely on roots and seeds during drier times. Similarly, nectar-feeding bats rely on insects and fruits when flowers are not in bloom.

Suriname has two wet seasons and two dry seasons each year (see Climate and Hydrology chapter). Vreedzaam (unpublished data) found less than 15 percent of trees – those monitored for fruiting and flowering at Brownsberg Nature Park – had fruit in the long dry season months of September and October, compared with 20 to 25 percent of trees fruiting from November to May. He also found substantial variation between years (2003-2004). The dry season was found to be low in fruit availability, but was peak season for flower production – forecasting the next fruiting season.

In Suriname, no mammals hibernate or migrate, but there are movements within home ranges that may be very large, and which depend on resource availability – such as food and roosting or denning sites – when some lowland habitats flood or food becomes scarce. Paca typically den in hollow logs or burrows in the ground near water, but 30 percent of forested habitat is susceptible to flooding during the rainy season. This requires them to search for alternative sites (Beck-King *et al.* 1999). White-lipped Peccaries were originally thought to be migratory, but the first radio-tracking study in the northern Amazon of Brazil documented a core non-flooded home range that expanded during the rainy season (April to September) (Fragoso 1998).

Seasonality also influences habitat selection for Giant Otters that prefer shallow waters to feed on fish at the bottom of slow-flowing, clear black-water creeks and rivers (Duplaix 2002). Large groups consisting of up to 14 individuals have been observed traveling between dry and wet season areas.

Territorial behavior. Most mammals exhibit some type of territorial behavior. Although the stealthy Jaguar is rarely seen, they leave many signs of their presence such as scrapes on the ground, sometimes in combination with feces, urine markings, and scratches on tree trunks as a message to neighboring individuals (Rabinowitz & Nottingham 1990). Guianan Red Howler monkeys are famous for their loud roars. While not specifically defending an area, males advertise their presence and space themselves relative to neighboring males (da Cunha & Byrnes 2006). Sheath-tailed bats in the family Emballonuridae show both territorial and mating aggression or defense (Bradbury & Vehrencamp 1977). Males of Proboscis Bats (*Rhynchonycteris naso*) defend a harem of females, whereas males of Greater Two-lined Sac-winged Bat (*Saccopteryx bilineata*) defend a feeding territory for their females.

Some of the monkey species in Suriname, such as the Golden-handed Tamarins and White-faced Sakis, defend territories and actively engage neighbors in aggressive territorial bouts. These may be physical in nature, but are more often displays of dominance that entail vocalizations and exaggerated movements. These species have relatively small and defendable home ranges of about 30 ha. Other species, the capuchins (*bergikeskesi* and *keskesi*), Guianan Bearded Sakis (*bisa*), spider monkeys (*kwata*) and squirrel monkeys (*monkimonki*), have large home ranges of more than 150 ha. Thus, they rarely encounter neighbors although they may be aggressive when they do. Guianan Red Howler Monkeys (*babun*) use a third strategy: avoidance. They have relatively small home ranges and rely on loud vocalizations to space groups; direct encounters are infrequent but can be physically aggressive (Di Fiore *et al.* 2011).

In addition to defending space, interactions between groups of the same species may have social or reproductive benefits. For example, frequent encounters with neighboring groups may provide information about group composition to individuals that are old enough to transfer between groups. Or intermingling of individuals from different groups may be an important opportunity for out-breeding, particularly for Golden-handed Tamarins (Digby *et al.* 2011). Finally, males and females may behave differently during aggressive encounters between groups. For White-faced Sakis, only adult males and their adult-aged sons take part in territorial behavior (Thompson *et al.* 2012).

Reproduction. In general, larger-bodied species of placental mammals in tropical areas have longer pregnancies, give birth to one young once every year (or longer), and are longer-lived than smaller-bodied species. For example, tapirs have gestation periods of over one year and can live more than 35 years, and Guiana Spider Monkeys give birth every three-to-four years. By contrast, smaller-bodied species usually have short gestation periods and/or multiple births per year with many offspring and have short lifespans. For example, spiny terrestrial rats have pregnancies that last two months, can have up to four litters per year producing six young per litter, and a lifespan ranging from one-to-four years. Small mammals are typically at the bottom of the food chain and have evolved a strategy of a high reproductive rate and early reproduction to maintain population sizes. An exception to this rule are bats, which have one young twice a year, but they have few natural predators because of their nocturnal, flying behavior. Reproduction is usually correlated to the primary rainy season (May-July) and the secondary one (Dec.-Jan.) when food availability is highest.

Reproductive rate in primates is often measured using the inter-birth interval (IBI), the duration between giving birth to one infant and the next one. Suriname's primates straddle the IBI range in all of the Neotropical monkeys: Golden-handed Tamarins have the shortest (minimum 8 months), Guiana Spider Monkeys the longest (approximately 48 months) (Di Fiore *et al.* 2011). The shortest

IBI is found in the smallest monkey, but the difference between howler monkeys and spider monkeys, which have a similar body mass, is striking. Guianan Red Howler monkey females mature faster than Guiana Spider Monkeys and have shorter IBIs so that populations of the formar can grow twice as fast as those of the latter (Di Fiore *et al.* 2011). This has a number of profound implications for the conservation of these two species. The other primate species have IBIs of approximately 24 months. At Brownsberg Nature Park, monkey infants can be observed in any month, but are most common in November and December (M. Norconk, pers. obs.). Mothers are primary care givers in all species except Golden-handed Tamarins, in which males and other group members regularly carry infants.

Bush meat

Mammals comprise the largest land animals in the world and are a main source of meat to humans, even when domesticated livestock is excluded. Suriname's Indigenous people, such as the Trio in the southern part of the country, and Maroon's still hunt species such as Paca (*Cuniculus paca*; *hei*), a large rodent. Another large mammal that forms part of their diet is the White-lipped Peccary (*Tayassu pecari*; *pingo*), but it is usually hunted irregularly because of its large territorial range and fewer chances of encounters. However, herds may number into the hundreds so when found several wild hogs may be taken at once and can feed the entire village

According to Hofwijks & Madhar (2006), 'the commercial bush meat trade is a burgeoning industry in Suriname with professional hunters working year-round in the interior.' Hunting pressure on primates appears to be particularly high for Guiana Spider Monkeys and Guianan Red Howler Monkeys (Ouboter 2001, Duplaix 2001). Ouboter (2001:7) noted that the bush meat trade is 'almost completely uncontrolled'. No recent estimates are available of species and volume of primates and other mammals entering the bush meat trade, but (Neme 2015) suggests that governmental monitoring is insufficient to curtail export from Guyana and Suriname to international ports of entry, such as the Netherlands. All large-sized rodent species are hunted for food in the Interior, particularly agoutis and Capybaras (Ouboter 2001).

Natural ecosystem services

Not only are mammals amazing and captivating animals to see in the wild, they are also a key component in the sustainability of the environment – or in other words, they provide important ecosystem functions to maintain the intricate balance of the natural world. In the complex web of life, seen in its full glory in the tropical rainforests of Suriname, mammalian diversity is low compared to other groups of organisms such as plants and insects, but no less important in keeping the ecosystem healthy and productive for natural forest succession and regeneration.

Take for example the much-maligned bats; their high species diversity and wide range of feeding habits makes them one of the more important groups of mammals. The different foraging guilds of bats include: fruit-eaters, such as Short-tailed Fruit Bats (*Carollia perspicillata*), that function as excellent seed dispersers because they always pluck the fruit from the parent tree and eat it somewhere else, ensuring that the seeds are spread throughout the forest; nectar-feeders including the Long-tongued Bat (*Glossophaga soricina*) that serendipitously act as pollinators when they steal the sweet juice of flowers; and insect-feeders such as the Proboscis River Bat (*Rhynchonycteris naso*) that by their sheer numbers and volume of consumption – about half their body weight in food every night – have an impact on insect populations that could damage agriculturally important crops or natural forest trees if left unchecked.

Although no group of non-flying mammals covers all of these functions thoroughly, especially pollination, they complement the ecological benefits that bats provide. For example, the Common Opossum (*Didelphis marsupialis*; *dagu-awari*) is an omnivorous evening forager that also takes nectar from flowers not regularly visited by bats. Similarly, monkeys such as the Brown Capuchin (*Sapajus apella*; *keskesi*) are known to visit flowers during the day; they may drink nectar or even eat the flowers, but also transport pollen on their hair or faces (Garber 1988). Other primates, like the South American Squirrel Monkey (*Saimiri sciureus; monkimonki*) are more specialized and insects may make up 75 percent of their diet. And as for frugivory, Golden-handed Tamarins, Guiana Spider Monkeys and Guianan Red Howler Monkeys (*babun*) ingest fruits with larger seeds than those dispersed by bats. Although not as species rich or abundant as bats, the ecological impact of monkeys is substantial because of their higher biomass effect – larger animals consume more food than smaller ones.

In a similar vein, top-level predators such as the carnivorous Jaguar can hunt large animals including the Red Brocket Deer (*Mazama americana; redidia*); smaller wild cats, such as the Ocelot, rely on smaller prey such as the Guianan Terrestrial Spiny Rat (*Proechimys guyannensis*). In fact, most of the carnivores would probably consume any of the small mammals if confronted with the opportunity so they have a significant impact on prey species of mammals and other vertebrates.

Some species may have a negative impact on the ecosystem, in some respects, including rodents that eat seeds so they can be considered as predators of plants. This is true for the Red-rumped Agouti *(Dasyprocta leporina; konkoni*), but is offset by their scatter-hoarding behavior of burying seeds away from the parent tree as a reserve food cache. Many times these seeds are not re-encountered, never eaten and end up germinating into seedlings.

Indicator species

The wild cat species have typically been used as indicator species of the health of the rainforest. If habitat disturbance such as deforestation or overhunting occur, their absence from biological surveys can signal a local extirpation that requires measures to reverse this trend in adjacent areas. Likewise, other large mammals such as primates (particularly Guiana Spider Monkeys or *kwata*) can indicate the degree of forest loss because of their dependence on fruit. All primates in the New World are arboreal, so loss of forest also predicts loss of primate species. Some species of bats can also be indicator species of modified habitats. An environmental assessment of the Bakhuis Mountains in western Suriname found that the Short-tailed Fruit Bat can adapt to moderate levels of human disturbance including roosting in buildings (Lim 2009). By contrast, in pristine forests their relative abundance is substantially lower than in secondary habitats where their primary food source, pioneering shrubs of the genus *Piper*, have taken hold.

Conservation

Although mammals are not one of the most species-rich groups of animals, they have the highest proportion of threatened species after amphibians – 20 percent according to the IUCN. In Suriname, 11 of the 12 mammal species listed as endangered, vulnerable, or near threatened by IUCN are large mammals. The exception – the Greater False-vampire Bat – is the largest species of bat in the New World and is carnivorous. Similarly, the 25 species of mammals from Suriname that are listed by the Convention on International Trade in Endangered Species (CITES) are large in size because these are the animals typically exploited for the pet trade and hunted for food or as souvenirs or trophies. This includes all species of primates and wild cats, and the 11 large mammals listed as threatened by the IUCN, which should be considered the most vulnerable mammals in Suriname. It also includes the Bush Dog (*Speothos venaticus*; *busidagu*), one of the more secretive and elusive mammals in the Neotropics. But perhaps the rarest species of Surinamese mammal is the Forest Rabbit (*Sylvilagus brasiliensis*; Tapeti) that is known in the country by only one museum specimen.

Historical Perspective

Suriname was, at one time, a hotbed of biological exploration and for the study of mammals in South America (Lim & Catzeflis 2014, Lim 2016). In the early 18th century, an apothecary based in Amsterdam named Albertus Seba acquired natural history specimens from crew members returning on ships that had voyaged overseas, including Paramaribo as a port of call in the Dutch colony. In 1734, Seba published the first of four volumes of illustrated plates of animals. In turn, these were used by the Swedish biologist Carl Linnaeus, in his 1758 monograph that marks the beginning of modern binomial nomenclature, the system of scientific classification used for animals and plants. Many unknown explorers collected mammals from Suriname that Seba then examined and used to illustrate his book. The plates were, in turn, used by Linnaeus as the holotype or reference specimen for 13 species, including typical Neotropical animals such as Linnaeus's Two-toed Sloth (*Choloepus didactylus*; *skapuloiri*).

There have also been anecdotal reports of wildlife, by amateur naturalists, since Christopher Columbus skirted the Guianan coastline in 1498. Although none concentrated solely on mammals, select publications over the centuries include the following: Warren (1667) documented early colonial life and the animals, Bancroft (1769) focused primarily on natural history, Kappler (1881) reported the first authoritative accounts of wildlife (Husson 1978), and Walsh and Gannon (1967) chronicled an animal rescue operation during the completion of Afobaka dam near Brownsweg. Many of these older publications are readily available at **archive.org** for free download.

Unfortunately, there was a relative lull in research on Suriname's mammals after Linnaeus that lasted for over two centuries until Antonius Husson at the Natural History Museum in Leiden, The Netherlands, completed his monograph on the mammals of Suriname (Husson 1978). He set a new standard for other related publications in South America. At this time, there was a resurgence of interest, including primate studies at the Central Suriname Nature Reserve (e.g. Mittermeier & van Roosmalen 1981, Boinski 1987, Kinzey & Norconk 1990, Boinski & Cropp 1999, Boinski 2002, Gunst *et al*. 2010) and faunal surveys of small mammals throughout the country (Genoways *et al*. 1982). Similar work has continued at other sites including Brownsberg Nature Park with research on monkeys (Norconk *et al*. 2003, Thompson *et al*. 2012, Gregory & Norconk 2013) and more generally on mammals (Lim *et al*. 2005b, Ouboter & Kadosoe 2016). Recently, the non-governmental organization, Conservation International, has also conducted biodiversity assessments, including on mammals, in the remote southern Interior in the vicinity of the Indigenous communities of Kwamalasamutu (O Shea *et al*. 2011) and Palumeu (Alonso & Larsen 2013)

Panthera onca
Jaguar
Penitigri, tigri
Carnivora, Felidae

As the quintessential top-level predator in the Neotropics, Jaguars (Figure 352) feed on large prey including peccaries, deer, Capybaras, agoutis, porcupines, and turtles. Near human habitation, Jaguars also prey on domesticated animals such as dogs, cattle and pigs. They usually stalk their quarry and then bite at the nape for the kill; prey is typically dragged to a more concealed area (Seymour 1989). Humans are the only natural predator of adult Jaguars, but cubs are preyed upon by caiman and anacondas; there are few reports of Jaguars attacking healthy adult Humans (Seymour 1989).

Jaguars are good climbers and typically sleep in a tree during the day. They hunt primarily at night and on the ground, but are also active at dawn and dusk. Except when breeding and rearing their young, Jaguars are solitary animals. Average size of territory is 30 km^2 for females and about double this area for males (Seymour 1989). They mark territories by urinating and scratching on trees. They are also good swimmers. Jaguar and Puma have broadly overlapping ranges across the Neotropics, but they seem to avoid each other behaviorally and ecologically in moist and dry habitats, respectively. The average gestation period is just over three months and typically two young are born (Seymour 1989). They are weaned by six months but remain with the mother up to two years of age. Females reach sexual maturity by their third year and males by their fourth year.

Figure 352. *Panthera onca* captured by a camera trap walking on Mazaroni Road in Brownsberg Nature Park during the early afternoon (image © Paul Ouboter, July 2015).

Description: The largest carnivore in the Americas and readily identifiable by its distinctively yellow fur with black spots; however, there is a melanistic (black) form. Average head-body length is 1.5 m and weight is 75 kg.

Distribution: Widely but patchily distributed from southwestern USA to southeastern Brazil, typically associated with forest ecosystems but variable (also occurring in rural, agricultural areas). Listed as 'near threatened' by IUCN because of habitat loss and hunting, the Jaguar has disappeared from more than half of its historical range, yet is considered common throughout Suriname.

Use: Hunted for trade of body parts including the use of meat as food, teeth in jewelry, bones for medicinal purposes and skin as souvenirs (Kerman 2010); hence its appearance on Appendix I of CITES.

Similar species: There are six species of cats in Suriname, of which the Jaguar is the largest. The Puma (*Puma concolor*) is a similarly large wild cat, but is a uniform brown color.

Speothos venaticus
Bush Dog
Busdagu
Carnivora, Canidae

One of the most enigmatic of large mammals in the Neotropics, Bush Dogs (Figure 353) are rarely seen and difficult to study in the wild, despite being primarily diurnal. They have been observed at night, but are typically crepuscular (active at dusk and dawn). Because of habitat fragmentation and hunting by people of their prey species, such as large rodents including agoutis, populations of this already uncommon species are estimated to be declining. Bush Dogs are listed as near threatened by the IUCN, and their trade is highly regulated.

Bush Dogs are good swimmers and have webbed toes that also seem to be useful for walking on soft ground around water. Bush Dogs den in burrows in the ground and in trunks of large fallen trees. There are also reports that they use burrows originally dug by Giant Armadillos (Beisiegel & Zuercher 2005). They usually hunt in packs of up to six individuals to take larger prey such as peccaries and Capybaras. However, solitary hunters prey on smaller rodents. Bush Dogs probably have no regular predators, but Jaguar or Puma have been reported to prey on them. Bush Dogs are reproductively active by 12 months of age. Gestation period is at least 65 days after which 3-6 pups are born. In Suriname, females are reported to give birth during the rainy season (Husson 1978). Pups are weaned within one month. Bush Dogs are monogamous; social groups include a pair of adults with extended family of offspring. Whines and barks are used for social communication, but more study is needed. Home ranges can reach up to 10 square kilometers.

Figure 353. A pair of *Speothos venaticus* caught in the late morning on a camera trap set on the South Plateau of Brownsberg Nature Park (image © Paul Ouboter, December 2015).

Description: Low-slung animal with short legs and an average weight of 5.5 kg. The head-body length is about 660 mm and there is a short bushy tail that is about 20 percent of the head-body length. Overall color of the body is usually brownish black with a paler head.

Distribution: Found from Panama to southeastern Brazil, but considered rare throughout its range. Considered uncommon in Suriname. Bush Dogs have been historically documented in both forest and savanna ecosystems.

Use: There are no regular human uses of the Bush Dog.

Similar species: The Crab-eating Fox *(Cerdocyon thous)* is the only other canid or dog-like carnivore occurring in Suriname. It has a longer, bushier tail that typically touches the ground, longer legs and neck, and grizzled gray fur.

Cuniculus paca
Paca
Hei, ei
Rodentia, Cuniculidae

Paca (Figure 354) are monogamous with an adult pair mating for life (about 10 years), and defending a territory of about three hectares. Females have a single young after a gestation period of up to four months and typically two litters per year. Offspring are weaned after about six weeks and independent by the first year. Other than for reproduction, the adults are solitary with separate dens in hollow logs or burrows in the ground, and usually near water.

They forage at night on a diet that consists primarily of fruit, when in season, but also other vegetative matter such as roots, seeds and leaves. Paca readily take to water as an escape from predators, which include most wild cats, especially the larger species, Jaguar and Puma. Bush Dogs are also known to prey on Paca (Deutsch 1983). Pacas can be considered a pest near human habitation because they may eat fruit and tubers in farms or gardens. It has been suggested that Pacas could be raised on farms for food (Smythe & Brown de Guanti 1995), but this has not been done in Suriname. Population densities of Paca from several localities throughout the Neotropics have a wide range of variation from 3.5 individuals/km^2 in Peru to 93 in Colombia (Beck-King *et al.* 1999). These large differences may be due to several factors such as flooding of burrows during the rainy season, availability of food, and density of predators. During a biological assessment of the Kwamalasamutu region of southern Suriname, paca was the next most frequently photographed large mammal after the Red Acouchy (*Myoprocta acouchy*) during camera trap surveys (Gajapersad *et al.* 2011).

Figure 354. *Cuniculus paca* photographed by camera trap during the late evening on the South Plateau of Brownsberg Nature Park (image © National Zoological Collection of Suriname, July 2013).

Description: Fur on the body is brown with white spots and lines on the sides, but belly fur is white. Pacas are the size of a piglet; weight is about 9 kg; and head-body length is about 650 mm.

Distribution: Widespread from Mexico to Uruguay. Common in Suriname in forested areas, but nocturnal, and rarely seen during daytime.

Use: This is prized game meat in Suriname and throughout the Neotropics. 'Their meat, which is white and tender, forms an important source of proteins' (Husson 1978: 474).

Similar species: The Capybara (*Hydrochoerus hydrochaeris*) is larger with a squarish snout and solid brown fur, whereas the Red-rumped Agouti (*Dasyprocta leporina*) is smaller and without spots or stripes on its flanks.

Dasyprocta leporina
Red-rumped Agouti
Konkoni
Rodentia, Dasyproctidae

Brazil Nuts (*Bertholletia excelsa*) are primarily dispersed by Red-rumped Agoutis (Figure 355), a large rodent species with strong front teeth (incisors) that can break through the hard outer coating of the fruit. Although the agouti is predominantly frugivorous, its association with the Brazil Nut is one of the better-known examples of plant-animal interactions. Red-rumped Agoutis are seed predators – they destroy many of the Brazil Nut seeds – but they are also scatter hoarders, meaning they bury intact seeds up to 25 meters from the parent tree. If these cached seeds are not rediscovered and eaten during times of plentiful food, they may eventually germinate and become a sapling. Brazil Nut trees are tall, emergent canopy species that are usually found clustered together in non-flooded lowland forest; they are harvested by the indigenous Trio people, primarily at Alalaparu, Coeroenie and Kutari in southern Suriname (Amazon Conservation Team 2004).In spite of the multimillion dollar business, the cultivation of Brazil Nut has not been successful so must rely on manual harvesting of wild populations that rely on natural seed dispersal by Red-rumped Agoutis. A study at Brownsberg Nature Park found that Red-rumped Agoutis were also important seed cachers of *Virola kwatae*, a tall emergent canopy tree species (Forget & Cuijpers 2008).

Reproductively, agouti females usually give birth to two young born twice a year. Red-rumped Agoutis live in pairs with offspring and can be seen running around the forest during the day. Population densities have not been estimated for Suriname, but in other parts of South America they range from 5 to 44 individuals per km^2 (Peres *et al.* 1997).

Figure 355. *Dasyprocta leporina* in a clearing around the headquarters of Brownsberg Nature Park during the late afternoon (image © Bart De Dijn, July 2016).

Description: Glossy golden-reddish rump contrasts sharply with the more brownish remainder of the body. Adults weigh > 3 kg and the head-body length is 500 mm. Red-rumped Agoutis sit on their haunches, much like a rabbit, hence the Surinamese common name.

Distribution: Found primarily in forested areas in Suriname and elsewhere in the Guiana Shield; may be the most common rodent seen during daytime

Use: Hunted for its meat, although Pacas are preferred by hunters.

Similar species: Red Acouchy (*Myoprocta acouchy*) is smaller with a short tail 50 mm in length and weighs < 2 kg. The back is usually black with red flanks, but the color of the rump does not contrast sharply from the rest of the body as seen in the Red-rumped Agouti.

Hydrochoerus hydrochaeris
Capybara
Kapuwa
Rodentia, Caviidae

Capybaras (Figure 356) are semi-aquatic with adaptations for living in water, including slight webbing between the toes to help in swimming. They are capable of staying submersed underwater for as long as five minutes. Capybaras are primarily grazers that eat grasses, but they also feed on roots. Like rabbits, they eat their feces during the morning when protein content is high from microbial fermentation of grass consumed the previous day (Moreira *et al.* 2013).

Herds range in size from a single pair of adults to up to 14 individuals or more with usually a single dominant male. Copulation occurs in the water and gestation of five months results in the birth of about five offspring averaging 1.5 kg each. The young are usually weaned by four months of age and become independent by the end of their first year. Females are reproductively active at two years of age. They are primarily active during the day, but not exclusively. Morning is typically for resting on land, early afternoon for regulating body temperature in the water, and late afternoon into the evening for feeding (Barreto & Quintana 2013). Their home range averages 10 hectares, and are never too far from water, but the core area can be as small as one hectare. They have vocal communication, but this is poorly studied and understood. They mark their territory using glands on top of the snout and around the anus. Capybaras are hunted by large predators, such as Jaguar, Puma, Caiman, and anaconda, as well as people.

Figure 356. *Hydrochoerus hydrochaeris* on a sandy bank (image © Arioene Vreedzaam).

Description: The largest living species of rodent in the world, weighing on average 50 kg with a head-body length of 120 cm and height at shoulders of 60 cm. The fur is a uniform brown color and the snout is wide, heavy set.

Distribution: Most of tropical South America east of the Andes. In Suriname, found throughout the lowland regions, in or near water bodies.

Use: The Capybara is hunted for its meat.

Similar species: Paca (*Cuniculus paca*) is less than half the size and has white spots or stripes on the sides of the body.

Mazama americana
Red Brocket Deer
Redidia, prasaradia
Artiodactyla, Cervidae

The Red Brocket Deer (Figure 357) is a browser, eating primarily fruit, seeds and flowers when plant reproductive parts are available during the wet season. It shifts to plant vegetative parts such as leaves, stems and bark as a secondary food source during the dry season. Fungi are also commonly eaten, peaking in July, during the longer wet season in Suriname (Branan *et al.* 1985). Jaguar and Puma are their main predators, as well as people, and Red Brocket Deer will dart away at full speed and frequently take to water to escape predation.

They are usually found solitary, but also in pairs during the mating season (April to October in Suriname; Branan & Marchinton 1987). The average density in Suriname is 1 individual/km^2, but densities of up to 7/km^2 have been documented in Bolivia. Red Brocket Deer are primarily nocturnal and typically active throughout the night. They are considered territorial and use facial scent glands, urination and defecation for marking. Red Brocket Deer were more commonly recorded in Suriname than closely related Gray Brocket Deer during transect surveys at Brownsberg Nature Park (Lim *et al.* 2005) and during biodiversity surveys of the Kwamalasamutu region in southern Suriname using camera traps (Gajapersad *et al.* 2011). The gestation period of the Red Brocket Deer is about seven months with the birth of a single young, occasionally twins. In Suriname, fawning (birth of young) peaks during the short wet season in December and January (Branan & Marchinton 1987). Offspring are weaned by six months and reach sexual maturity after the first year.

Figure 357. *Mazama americana* mother and fawn documented by camera trap during the mid-morning on Mazaroni Road in Brownsberg Nature Park (image © Paul Ouboter, December 2015).

Description: Uniform reddish brown body with not much contrast on the belly. Short tail is dark on top and white underneath with a small white tip. Head-body length averages 125 cm and weight 35 kg. Fawns have white spots. Only males have antlers; these are short, not branched.

Distribution: Found primarily in forested habitat of South America, from Colombia to northern Argentina; widespread and common in Suriname.

Use: Hunted for its meat.

Similar species: There are two other species of deer in Suriname. The Gray Brocket Deer (*Mazama gouazoubira*) also occurs in forest, but is about half the size and has grayish fur. The White-tailed Deer occurs in savanna or the forest edge, but is larger (mature males with larger, branched antlers). Both of these species have more contrasting white bellies than the red brocket.

Tayassu pecari
White-lipped Peccary
Pingo
Artiodactyla, Tayassuidae

White-lipped Peccary (Figure 358) form large herds of up to 200 individuals (on average 40), which typically stay together throughout the year. Their home range is 60-200 km^2 (larger when herd size is large) and increases during the wet season if part of their territory floods. They are diurnal, and spend about one-third of the day eating, one-third moving, and one-quarter resting (Altrichter *et al.* 2002). In the dry season, they spend more time searching for and eating food, and less time resting. This peccary species eats fallen fruits from the forest floor when available in the wet season, but otherwise resorts to food sources such as seeds and tubers. It is a predator of large seeds, but disperses smaller seeds less than 10 mm in length, many of which may survive transit through the gut.

Typically, they travel in a single file with the alpha male in the lead. Clacking of tusks is a defensive warning against signs of potential danger, such as humans. White-lipped Peccaries have relatively low tolerance to human presence and will migrate to other areas when disturbed. Other than Humans, Jaguar, Puma and anaconda are natural predators. After a gestation period of about five months, females typically give birth to two offspring per litter and may have two litters per year. In Costa Rica, this is correlated to the wet season when fruit is plentiful (Altrichter *et al.* 2001). Young are weaned by about six months of age and become independent in their second year.

Figure 358. *Tayassu pecari* herd walking during the late morning along Mazaroni Road in Brownsberg Nature Park, captured by a camera trap (image © Vanessa Kadosoe, November 2016).

Description: Resembles its much larger relative the pig (from Eurasia), with the characteristic flat nose but long face. The hairs below the nose to the chin and cheeks are white; the rest of the body has blackish brown hairs that are long and bristly, forming a mane along the midline of the back. Average body weight is 38 kg and head-body length 115 cm.

Distribution: Found from Mexico to Argentina; common away from villages in the forested south of Suriname, but largely absent elsewhere due to hunting.

Use: Avidly hunted for its meat; tens of individuals from a single herd may be killed during a single hunt. Hunters remove the scent gland on the animal's back immediately after a kill to avoid it spoiling the taste of the meat.

Similar species: The Collared Peccary is almost 50 percent smaller in weight than the White-lipped Peccary, and has a white stripe on the neck (not a white patch near the mouth).

Tapirus terrestris
South American Tapir
Bofru, tapir
Artiodactyla, Tapiridae

The South American Tapir (Figure 359) is the largest indigenous herbivore in the continent and is typically seen foraging in natural forest gaps and along the sides of quiet dirt roads. When available, fruit accounts for about one-third of their diet, otherwise they feed on vegetation. In the daytime, they are usually sleeping in dense bushy areas. They are primarily nocturnal or crepuscular (active at dusk and dawn), but not exclusively, and can sometimes be heard during the day making a whistling noise. Their locomotion was described by Husson (1978: 343) as 'walks like a horse, jumps like a deer'. Outside of mating and caring for young, tapirs are solitary animals. Camera trapping in French Guiana estimated density as 0.32 individuals/km^2.

Tapirs are seed predators and seed dispersers. Both large and small seeds can pass through the gut intact and germinate. In northern Amazonian Brazil, tapir eat whole the fruit of the large-seeded palm *Attalea maripa* and defecate seeds up to 2 km away (Fragoso 1997). In French Guiana, the fruit is consumed but the large seed of the palm *Mauritia flexuosa* is spat out (Henry *et al*. 2000). Some small seeds are secondarily eaten or dispersed by large rodents, such as Red-rumped Agouti. Defecation occurs on both land and in water so rivers can facilitate seed dispersal as well. Their large, fibrous dung heaps betray their presence, as does extensive browsing damage to *Cecropia* spp. (*busipaya*) saplings. Predators include large animals such as Jaguar and caiman, as well as people. Tapirs often take refuge in water to escape their hunters. After a gestation period of about 400 days, females give birth to a single young approximately every other year. Newborns can weigh up to 6 kg and become weaned as early as six months. They are independent after 1.5 years of age and become sexually active a year after that. Longevity is reported to be at least 35 years.

Figure 359. *Tapirus terrestris* at Sipaliwini Airstrip just before midnight (image © Vanessa Kadosoe, April 2014).

Description: The largest Neotropical land mammal; snout with a projecting nose (proboscis); average head-body length of 2.2 m and a weight of 225 kg. Adults uniformly grayish black, with a contrasting darker mane along the midline from forehead to shoulders; young with white spots and lines on the side of the body.

Distribution: Primarily forested areas east of the Andes in South America, but in Suriname also known from the Sipaliwini Savanna. The tapir is equally at home swimming in lowland rivers to climbing mountains.

Use: They are hunted for their prized meat, and as a result are often rare or at least rarely observed near towns and villages unless they are kept as pets.

Similar species: No other animal in Suriname can be confused with the tapir because of its large size.

Myrmecophaga tridactyla
Giant anteater
Tamanwa
Pilosa, Myrmecophagidae

Giant Anteaters (Figure 360) are terrestrial and essentially solitary animals except when breeding (Reid 1997). They appear to walk awkwardly on the outside of clenched fists. Jaguar and Puma may prey on them, but Giant Anteaters also have formidable defenses with the slashing action of their strong sharp claws. Giant Anteaters have been seen active during both the day and night. When resting in a shallow depression with some type of vegetational cover, they drape their bushy tail over the body like a blanket.

The Giant Anteater uses large powerful claws on the front legs to rip into termite mounds and ant nests. The long tongue, which can be extended at least 45 cm or more beyond the mouth (greater than the length of the head), has small backward-facing protrusions or papillae that are covered in sticky saliva that helps it slurp up insects (Naples 1999). The tongue can be flicked in and out more than twice per second. Ants or termites are initially ground up against the palate before being ingested for further mastication in the stomach. Giant anteaters don't destroy the whole mound or nest to minimize attacks from the insects, so food is available for later comsumption.

The other two species of anteaters that occur in Suriname are arboreal, as are the two living species of sloths, which are each other's closest living relatives. This suggests that the terrestrial behavior of the Giant Anteater may be, in an evolutionary sense, a new or derived adaption within anteaters or that arboreal behavior arose at least twice within the anteaters and sloths. More research is needed to confirm the traits of the common ancestor of these unusual mammals.

Figure 360. *Myrmecophaga tridactyla* photographed by camera trap on the South Plateau of Brownsberg Nature Park in the early afternoon (image © Paul Ouboter, April 2014).

Description: Has an elongated snout (up to 45 cm) and lacks teeth; a black stripe bordered with white runs across the shoulder on an otherwise grayish brown body. Tail is about as long as the body (averaging 75 cm), bushy with stiff hairs hanging to the ground; weight averages 28 kg.

Distribution: Found from Honduras to Argentina; occurs throughout Suriname in lowland and upland rainforest and savanna.

Use: Is hunted for food, but the meat is not considered the best (Husson, 1978).

Similar species: The Giant Anteater is unmistakable. Other species of anteaters lack the fringed tail and characteristic black and white striping on the body. Tamanduas and sloths are smaller and arboreal, living primarily in trees.

Priodontes maximus
Giant armadillo
Granmankapasi
Cingulata, Dasypodidae

Giant Armadillos (Figure 361) are nocturnal, solitary and terrestrial. They spend most of their time in large, self-made burrows deep in the ground with an entrance that is typically domed and larger (>30 cm) than that of any other species of armadillo (Carter & Encarnaçao 1983). New burrows are typically dug every few days and different ones are used on a regular basis. Giant Armadillos can be heard rustling through the forest understory as they forage for food at night. They use large, strong front claws to break into nests and mounds to gain access to ants and termites. They also dig around for other foods, such as fruits, roots and insects. Giant Armadillos have up to 100 permanent peg-like teeth, the most of any terrestrial mammal.

They are considered ecosystem engineers (Desbiez & Kluyber 2013) because they alter the environment in a way that affects many other organisms. In the Pantanal of southern Brazil, their burrows are known to be used by other armadillo species, anteaters, Collared Peccaries, agoutis, Tayras, Ocelots, rats, Crab-eating Raccoons, coatis, and the Crab-eating Fox. The excavations or sand mounds at the burrow entrance are also used by animals such as Giant Anteaters for taking a sand bath, and South American Tapirs and Pumas for resting on the sand.

Figure 361. *Priodontes maximus* on Mazaroni Road in Brownsberg Nature Park just after midnight (image © Paul Ouboter, October 2015).

Description: The largest living species of armadillo, head-body length averaging 90 cm, tail about half as long; weighs up to 45 kg. It has massive claws, and a carapace or shell that is brownish gray on the back and paler on the edges, covering the back and sides but not the legs. Therefore, the carapace does not enable the Giant Armadillo to roll completely into a protective ball as is possible in other species of armadillos. However, its powerful claws are a good defense against potential predators, such as Jaguar and Puma.

Distribution: This species is rare but widespread in tropical areas east of the Andes; it has disappeared from much of southern South America due to habitat loss and overhunting (listed as vulnerable on the IUCN Red List). In Suriname, it occurs in lowland and upland forest and savanna.

Use: Hunted as game, but not considered the choicest meat.

Similar species: All other armadillos are less than half the size.

Sylvilagus brasiliensis
Tapeti, Forest Rabbit
Konkoni, konijn
Lagomorpha, Leporidae

The Tapeti (Figure 362) is probably the least-known mammal in Suriname with only a single museum specimen ever collected (Husson 1978). Not surprisingly, most information on the biology of this species is based on studies from other parts of its distribution.

Reproductive activity can occur at any time throughout the year. Gestation ranges from 28-44 days followed by the birth of 1-8 young. But the Tapeti is considered the least fecund of rabbit species with an average of only 1.2 offspring per litter. Nests are usually multi-chambered structures built on the ground in dry vegetation. Tayra and wild cats, such as Margay and Oncilla, have been reported to be predators of Tapeti (Calouro 2000). They often freeze and stay motionless when alarmed, but when fleeing will use a bounding, erratic zig-zag escape pattern. Tapeti are usually solitary and nocturnal, but may be found in pairs and foraging at dawn or dusk for shrubs and grasses. They have characteristic droppings that consist of many round pellets in a pile.

Figure 362. *Sylvilagus brasiliensis* (image © Kevin Schafer, WWF)

Description: Typical rabbit characteristics of long ears closely set on the top of the head, stubby furred tail, and large hairy hindfeet; its fur is brown with specks of black from longer, sparsely distributed and coarser guard hairs. The total length averages about 340 mm with weight of 910 g.

Distribution: Found in forests from Mexico to Argentina, and common in some parts of its range; patchily distributed in Amazonia, where it is considered a species of the forest edges or open (non-forest) ecosystems (De Sousa e Silva Júnior *et al.* 2005). In Suriname, its presence is only confirmed for the upper Corantijn River area (Husson 1978). There are unconfirmed sightings from other parts of Suriname, but none from Guyana or French Guiana.

Use: Hunted for food, but is not common and is found in sparsely inhabited parts of Suriname.

Similar species: May be confused with the more abundant Red-rumped Agouti (*Dasyprocta leporina*), which has shorter ears and red hindquarters that are distinct from an otherwise uniform brown body. Both rabbits and agoutis have a habit of sitting on their haunches. In fact, there may be confusion in the use of the Surinamese names since Konkoni and Hei are used for both Red-rumped Agoutis and Forest Rabbits.

Saguinus midas
Golden-handed Tamarin
Saguwenke
Primates, Platyrrhini, Callitrichidae

Golden-handed Tamarins (Figure 363) form small social groups of about six individuals consisting of a single breeding female, usually two cooperatively breeding males and dependent offspring. All females in the primate family Callitrichidae have high reproductive rates. In addition to their small body size and ability to inhabit secondary or degraded forests, both sons and daughters disperse from the natal group, enabling them to readily colonize new habitats. The dominant female in the group gives birth to twins every 8 to 12 months and females are capable of lactating and simultaneously gestating the next set of young (Löttker *et al.* 2004). Infants are carried by adult males and other group members until they are weaned starting at about four months of age. Offspring are able to reproduce by 16-18 months (French *et al.* 2003), but daughters do not always reproduce if they are in their mother's group. The behavior is regulated by scent cues from the dominant female (French *et al.* 2003). These characteristics apply to many tamarin species, although information specific to *Saguinus midas* is limited.

Tamarin groups are strongly territorial and their small body size puts them at risk of predation. They seem to counter high predation risk by being silent in the afternoon and stealthily entering a tree with dense shelter before dusk to sleep. Golden-handed Tamarins are fruit eaters and supplement their diets with insects (Ramdarshan *et al.* 2011) and tree gum. Trees respond to injury by producing sap or gum. High in water and sugar content, gum is eaten opportunistically by tamarins particularly in the dry season. Tamarins prefer forest edges (Sussman & Kinzey 1984) and disperse small seeds by defecation (Oliviera & Ferrari 2000).

Figure 363. *Saguinus midas* peeking around a branch at Brownsberg Nature Park (image © Marilyn Norconk, June 2008).

Description: Small-bodied, black, 'clawed' monkeys with golden hands and feet; bare faces, mottled 'saddle' of brown, ochre and black hairs on the back, and a long non-prehensile tail. Females are slightly larger than males (575 g versus 515 g) (Smith & Jungers 1997).

Distribution: Guianas and Brazil north of the Amazon River.

Use: Species of the genus *Saguinus* have been used extensively in biomedical research. In captivity they have low survival rates due to their dietary specializations and disease (e.g., cockroach-borne pathogens).

Similar species: In Suriname, tamarins have a distinctive pattern of hair coloration from other primates. Their ability to negotiate vertical tree trunks is similar to squirrels, but tamarins are found in social groups and their tails are not bushy.

Alouatta macconnelli
Guianan Red Howler Monkey
Babun
Primates, Platyrrhini, Atelidae

The haunting roars of these howler monkeys are loud and travel far (> 1 km). Typically sitting high above the ground, in trees near the center of their home range, the group's adult male begins and ends the lengthy chorus joined by adult females and juveniles (Da Cunha & Bryne 2006). While not specifically defending an area, the males use these calls to advertise their presence and space themselves relative to neighboring males.

Males are larger than females and have the largest hyoid bone of any modern primate (Youlatos *et al.* 2015). The cup-like hyoid bone in the throat increases the resonance of vocalizations, and thus their loudness. In humans this bone is slender, but the hyoid of howler monkey males makes their call much louder than any human call. Despite their relatively large body size, orange coloration and loud calls, Guianan Red Howler Monkeys (Figure 364) can be difficult to spot, as they spend a lot of time resting high in the forest canopy.

Their diet consists of approximately equal amounts of young leaves and unripe fruit. Resting for long periods of time (up to 75 percent of the day) complements digestion of tough plant foods (Di Fiore *et al.* 2011) and by hanging from prehensile tails they are able to reach preferred foods below them (Organ 2010). Groups ranged from 2-7 individuals at Brownsberg Nature Park (Norconk *et al.* 2003). Males and females begin to breed at about 3.5 years of age and females give birth to single infants every 17 months on average (Di Fiore *et al.* 2011). In these one-male social units, male takeovers are known to occur that may result in infanticide of unweaned infants.

Figure 364. ***Alouatta macconnelli*** **feeding on young leaves at Brownsberg Nature Park (image © Marilyn Norconk, June 2013).**

Description: Large monkeys with long prehensile tails; males weigh 5.6 to 7.5 kg, females 4.0 to 6.2 kg (Di Fiore *et al.* 2011); hair color ranges from blond to dark brown with an often strong golden red or orange shine.

Distribution: Guianan Red Howler Monkeys are found from eastern Venezuela to the Guianas. They are entirely arboreal, and common in lowland and upland forests in Suriname, from the coast to the Interior. Their diverse diet allows them to survive in forest fragments, even in the city limits of Paramaribo.

Use: They are hunted, but are not preferred bush meat (Koster *et al.* 2010).

Similar species: Guiana Spider Monkeys and Brouwn Capuchins also have prehensile tails, but Guianan Red Howler Monkeys are reddish, have a stockier build and enlarged throats. They move slower, cautiously, not swinging from one branch to the next.

Ateles paniscus
Guiana Spider Monkey
Kwata
Primates, Platyrrhini, Atelidae

Guiana Spider Monkeys or *kwata* (Figure 365) are the largest primates in Suriname. They live in communities of about 25 individuals that combine and divide into larger or smaller groups. Contact is maintained either visually or vocally. This flexible group structure may limit competition for their preferred food: ripe fruits (van Roosmalen 1985b) that are often rare. They are also very slow breeders, having the longest inter-birth interval (3-4 years) of Neotropical monkeys (Fedigan & Rose 1995). And, they start breeding late. Females first give birth at about seven years, twice the age of similarly-sized Guiana Red Howler Monkeys (Di Fiore *et al.* 2011). Guiana Spider Monkeys are large-brained and long-lived. Adult females have a precise memory of the location of fruit trees and the timing of fruiting. They are also known to deceive other group members about the location of food (Amici *et al.* 2009). Guiana Spider Monkey communities inhabit large areas (> 250 ha; van Roosmalen 1985b) and are at high risk of extinction in disturbed forests when feeding trees are lost.

A number of characteristics of Guiana Spider Monkeys are similar to chimpanzees. Both species are ripe fruit specialists, long-lived, slow breeders and very intelligent. In addition, females of both species tend to migrate from their natal group at maturity – a contrast with most other primate species. Related males form coalitions to defend territories; unrelated females use small areas repeatedly to search for food (van Roosmalen 1985b).

Figure 365. Adult female Guiana Spider Monkey (named 'Lola') at the Paramaribo Zoo. Note her prehensile tail grasping the support behind her (image © Marilyn Norconk, July 2008).

Description: Large, black-haired with red face and prehensile tail; males on average larger (9.1 kg) than females (8.4 kg) (Smith & Jungers 1997).

Distribution: Found in tall rainforest north of the Amazon River and east of the Rio Branco, but locally extinct due to hunting, as in most of the Suriname Coastal Zone. 'Vulnerable' according to the IUCN with 30 percent population decline since the 1960s, especially in Brazil.

Use: They are avidly hunted for their meat, and, being slow breeders and habitat specialists, they have gone extinct locally.

Similar species: Closely related and similar to Guianan Red Howler Monkeys, but with completely black body hair and red face, and arm-swinging locomotion.

Sapajus apella
Brown Capuchin
Keskesi
Primates, Platyrrhini, Cebidae

Of the two capuchin species in Suriname (Brown Capuchins and Wedge-capped Capuchins), Brown Capuchins (Figure 366) have been better studied. Both capuchin species are medium-sized, prehensile-tailed monkeys known for their large brains (Fragaszy *et al*. 2004), social intelligence, and ability to use natural objects as tools to extract resources. Brown Capuchins are more robust (they have deeper jaws and greater manual strength) and live in smaller groups than Wedge-capped Capuchins. The former extract insect hives hidden in trees or palm fronds; pound large, gourd-like *Lecythis zabucajo* ('monkey pot') fruit on branches until they crack open; and forage in dead bark for insects. Boinski (2002) characterized them as specialists in eating 'food that fights back' – eating poisonous insects and wasp larvae, also bird eggs, bamboo piths, palm nuts and small mammals.

Brown Capuchin group sizes range from 11 to 18 individuals (Norconk *et al.* 2003; Jack 2011) and include multiple males and females. Females begin reproduction at about seven years of age. They give birth to one infant every 1.5 to 2 years and mothers are exclusive infant caregivers. Juveniles have a long period of development that is associated with brain and motor skill development. Adult-aged males join a new group before breeding and may disperse several times in their lifetime (Jack 2011). Adult females form close social ties with related females.

Figure 366. Group leader male *Sapajus apella* outside the Paramaribo Zoo (image © Marilyn Norconk, July 2008).

Description: Brown Capuchins range from darker to lighter brown hair and have square faces outlining robust mandibles. Adult females have tufts of hair on their head. Arboreal, but may come low to the ground to access food.
Distribution: From the Guianas southwest to Bolivia; in rainforest, dryer forest types and more open vegetation (Lynch Alfaro *et al.* 2012). Occurs throughout Suriname, also in forest fragments, e.g. in Paramaribo and Wanica District.
Use: Brown Capuchins are hunted for meat and are used in biomedical research (Suriname exports more than 100 Brown Capuchins a year, according to CITES figures). They have been trained (with variable success) to aid disabled people (Fragaszy *et al*. 2004).
Similar species: Deceptively similar, especially juveniles, to closely related Wedge-capped Capuchin (*bergikeskesi*); both species have brownish-gray pelage that exhibits considerable variation among individuals. Wedge-capped Capuchins have typically lighter hairs on the scalp, are slenderer and live in larger groups (typically > 20 individuals) than Brown Capuchins (< 15).

Saimiri sciureus
South American Squirrel Monkey
Monkimonki
Primates, Platyrrhini, Cebidae

South American Squirrel Monkeys (Figure 367) are arboreal, small-bodied, active monkeys that travel in groups of 15-20 individuals in Suriname (Boinski 2002). Each group has multiple adult males and females, as well as many immature individuals (Jack 2011). Reproduction is strongly seasonal and unusual (among primates) in its synchronicity. From age 2.5 years onward, females become sexually receptive for only two days during a one-week period each year (Boinski 1987). Males begin to reproduce at about five years of age and undergo a dramatic increase in size during the short breeding season. Competition among males is manifested as physical aggression that results in scars and debilitating injuries (Boinski 2002). During the non-breeding season, adult males occupy the edges of social groups and are readily displaced by females and even juveniles (Izar *et al*. 2009). Young males are cooperative and may be closely related since they tend to disperse to new groups together (Boinski & Cropp 1999).

South American Squirrel monkeys are active from dawn to dusk, maintaining constant vocal communication, moving slowly, and hunting mainly for insects (70 percent of their diet). Squirrel monkeys and Brown Capuchins often form mixed-species groups – a mutually beneficial strategy in which both species may be safer from predators by virtue of an increase in overall group size. However, there may be separate feeding advantages. *Saimiri* benefit from Brown Capuchin knowledge of resource locations and Brown Capuchins benefit from higher feeding rates in small-crowned trees and reduced vigilance when in association (Podolsky 1990).

Figure 367. Young male *Saimiri sciureus* photographed at the Paramaribo Zoo. He was a member of the 'outside' group of free-ranging monkeys taking advantage of food set out for captive monkeys (image © Marilyn Norconk, July 2008).

Description: Small-bodied (< 1 kg body mass), yellowish-brown with a high white arch above the eyes. They walk on top of branches or leap between branches. They have long, non-prehensile tails.

Distribution: Found in the Guianas and eastern Brazil, north of the Amazon River; prefers secondary forests and river-edge habitats, including heavily disturbed forest habitats and gardens, such as those in Paramaribo.

Use: Used for biomedical research since 1949 (Abee 1989); they are difficult to breed in captivity, yet many are thought to be kept as pets in Suriname. Often caught in nets, up to 1,000 may be exported from Suriname per year (CITES national export quotas: **cites.org**).

Similar species: May be mistaken for squirrels, but *Saimiri* travel in large groups and tails are not bushy.

Pithecia pithecia
White-faced Saki
Wanaku
Primates, Platyrrhini, Pitheciidae

There are two genera of sakis in Suriname, White-faced Sakis (Figure 368) and Guianan Bearded Sakis (*Chiropotes sagulatus*; *bisa*). The two species differ in body size and color, group size and home range size, but have similar diets (seed eaters), robust jaws and bushy tails (Norconk 2011). White-faced Sakis generally chew seeds up and disperse only very small seeds that escape mastication. Males and females of both species have dental adaptations to extract seeds from hard-husked fruits that are ignored by other monkeys (Kinzey 1992; Norconk & Veres 2011). Other fruit-eating monkeys in Suriname prefer yellow and red (ripe) fruits, compared to green and brown fruits preferred by White-faced Sakis. Their strong teeth provide them with aseasonal access to seeds – a relatively stable and nutritious food supply.

White-faced Sakis live in small groups of 5-7 individuals with one adult male and one or two females, in small home ranges (ca. 30 ha). Males are black with a contrasting white facial mask; females are more uniformly brown. Regardless of sex, infants are born looking like their mothers. Males gradually develop a white mask and black body hair during their first three years of life. White-faced Sakis are strongly territorial with the dominant male and adult-aged sons engaging in vocal and physical displays at territorial boundaries (Norconk 2006; Thompson *et al.* 2012). Adult males are larger bodied and less shy than females.

Figure 368. 'Junco,' an adult male *Pithecia pithecia*, feeding on *Garcinia madruno* (Clusiaceae) at Brownsberg Nature Park. Note his prominent canine tooth used for prying open tough fruit (image © Marilyn Norconk, July 2008).

Description: Males have a broad white face and uniform long black body hair; females are grayish-brown with whitish stripes down either side of the face. They are fully arboreal.

Distribution: From the Guianas south to the Amazon River in both undisturbed and secondary forests.

Use: (No known use).

Similar species: Closely related to Guianan Bearded Sakis (*Chiropotes sagulatus* in Suriname), but White-faced Sakis form small groups led by a black-and-white adult male in the understory compared with large, rapidly moving groups of Guianan Bearded Sakis in the canopy.

Potos flavus
Kinkajou
Netikeskesi
Carnivora, Procyonidae

As alluded by the Sranan name, Kinkajous (Figure 369) are often locally referred to as 'night monkeys'. However, despite superficially resembling a primate, they belong to the raccoon family. Kinkajous are arboreal where they have an omnivorous diet feeding primarily on fruit, but also insects and the nectar from flowers. It is a beneficial species to the ecosystem as a seed disperser and flower pollinator, thus an important species for forest regeneration. The nocturnal feeding behavior of the Kinkajou is considered similar to that of the diurnal Guiana Spider Monkey (Kays 1999).

Kinkajous have a prehensile tail that is used as a fifth appendage to move through the canopy and to hang from while grabbing food with their hands. Kinkjaous are usually solitary animals, but can be found in pairs and they typically travel in small groups (Kays & Gittleman 2001). Although difficult to see in trees during the night, their presence can be divulged by vocalizations including whistling calls and barks. They rarely come to the ground.

Harpy Eagles have been documented as predators of Kinkajous in Guyana (Rettig 1978). Jaguars are also known to prey on them. The average home range in French Guiana of females is 5.5 hectares and about double the area for males, who also spend more of their time along the periphery (Julien-Laferriere 1993). Similarly, males may travel almost twice as far as the 1.5 km that females usually cover during the night. In the day, they typically sleep in tree hollows, but during the active period, they will rest at several different spots with males preferring the edge of their range.

Figure 369. *Potos flavus* at Brownsberg Nature Park (owner of image unknown).

Description: Slender animal with a head-body length of about 400 mm and a tail that is slightly longer with a prehensile end for grasping branches. Fur on the back is a uniform brown and the belly is paler. Their eyes are large and the ears are rounded.

Distribution: Widely distributed from southern Mexico to southeastern Brazil and occurs in rainforest throughout Suriname.

Use: There are no regular uses of this species, but they are at times hunted for meat and may be considered an exotic pet.

Similar species: There are two other species in the raccoon family Procyonidae that occur in Suriname, but they have alternating bands of black and brown around the tail; the Crab-eating Raccoon (*Procyon cancrivorus*) and South American Coati (*Nasua nasua*) are also primarily terrestrial animals.

Choloepus didactylus
Linnaeus's Two-toed Sloth
Skapuloiri
Pilosa, Megalonychidae

Using specimens from Suriname, Linnaeus (1758) described Two-toed and Three-toed Sloths as being in the same genus. In the next century, Two-toed Sloths were placed in a new genus but both genera were still considered closely related in the same family (Bradypodidae). More recently, Two-toed Sloths were put in the family Megalonychidae that previously had only extinct species of primarily large ground sloths (Webb 1985). Unlike the extinct ground sloths, Linnaeus's Two-toed Sloth (Figure 370) spends its time in the forest canopy except when descending to the ground to change trees or to defecate and urinate, about once every four days.

As suggested by their common name, sloths are slow moving in trees with an average speed of 0.5 km/h in captivity. On the ground, they move awkwardly at about half that speed (Adams 1999). But they are apparently good swimmers!

Not surprisingly, sloths are most vulnerable on the ground, preyed on by Jaguars and anacondas. In the canopy, Harpy Eagles prey on them during the day (Rettig 1978). Although they are difficult to study in the wild because of their nocturnal and arboreal behavior, Two-toed Sloths are known to feed primarily on leaves, but probably also eat other foods such as fruits and insects.

Figure 370. *Choloepus didactylus* in a backyard garden in Paramaribo-Noord during the late afternoon (image © Vanessa Kadosoe, May 2017).

Description: The common name refers to the number of toes on the hand, as opposed to the three toes that are common to both genera of sloths on the feet. Fur is relatively long, uniformly pale brown in coloration, and falls from the stomach to the back (opposite of other mammals). Hair may have a greenish tinge from algae that live in the fur and may also act as camouflage in the forest canopy.

Distribution: Occurs in rainforest throughout the Guianas and most of the Amazon.

Use: They are opportunistically hunted for their meat, but are difficult to see as they live high up in trees and are often motionless.

Similar species: The Pale-throated Three-toed Sloth (*Bradypus tridactylus*) is smaller, has shorter fur with a mottled coloration, and relatively longer arms with three toes on the upper extremities. It is the only similar species found in Suriname and is more common. During the Operation Gwamba animal rescue project in 1964, Three-toed Sloths outnumbered the Two-toed Sloth by 2,104 to 840 individuals (Walsh and Gannon 1967).

Sotalia guianensis
Guiana Dolphin
Profosu, dolfijn
Cetacea, Delphinidae

Guiana Dolphins (Figure 371) are considered relatively common throughout their range. Group size usually numbers about 10 individuals, but their composition can change regularly by merging with neighboring groups. Fish are usually caught cooperatively by the group with feeding tactics that include herding prey (Oliveira *et al.* 2013). They communicate with a series of ultrasonic tonal sounds or whistles at a frequency of about 48 kHz that are used for behavioral activity and social interaction. Their home range has been documented in Brazil to range from 15 to 135 km^2.

The Guiana Dolphin is quite acrobatic, regularly breaching out of the water. It typically dives for up to one minute at a time, but usually stays within 30 meters of the surface. They feed primarily on fish, but also eat squid, shrimp and crab (Di Beneditto & Ramos 2004). At times, the Guiana Dolphin gets caught in fishing nets, their greatest threat to survival, and may drown. Human pollutants in rivers are also of conservation concern.

Females reach sexual maturity at 5-8 years of age. Gestation lasts for about 12 months and females produce one litter every two or three years. After birth, mothers lactate for approximately nine months. Typically, young dolphins swim behind adults, a behavior thought to reduce drag by at least 50 percent and which allows easier access to mothers for nursing (Tardin *et al.* 2013). Guiana Dolphins can live as long as 30 years.

Figure 371. *Sotalia guianensis* near the mouth of the Suriname River during the late afternoon (image © Burton Lim, August 2013).

Description: About 1.8 m in length and averaging 60 kg in weight with a bluish-gray body. The Guiana dolphin has a low dorsal fin midway along the back, triangular in shape with usually a slight hook on the tip. It has a roundish head with a blowhole on top and a well-defined narrow snout that tapers to a pointed beak. There is no sexual dimorphism or difference in size between males and females.

Distribution: Found along coastal estuaries of Central and South America from Nicaragua to Brazil where the brackish waters of rivers and oceans intermingle. The Guiana Dolphin occurs in the larger rivers of Suriname, including the Suriname River to Paramaribo and beyond depending on tidal currents.

Use: Although the Guiana Dolphin is caught in fishing nets, it is not typically consumed by people.

Similar species: The only other fully aquatic mammal within the borders of Suriname is the West Indian Manatee (*Trichechus manatus*), which is differentiated by having a blunt face, no dorsal fin and nostrils on the front of the face

Trichechus manatus
West Indian Manatee
Seku
Sirenia, Trichechidae

Legend has it that manatees were the original mermaids, misidentified by early explorers to the New World after long voyages across the Atlantic Ocean. But as hinted in the Surinamese (*seku*) and Dutch (*zeekoe*) common names, sea cow in the unofficial English version, the sailors were sadly mistaken.

The West Indian Manatee (Figure 372) is a herbivore that eats primarily aquatic plants, grabbed with the dexterous upper lip. They typically feed from 6-8 hours per day at depths of at least one meter below the surface of the water. Manatees usually stay under water for about four minutes but have been documented to hold their breath for up to 16 minutes. There are no natural predators of manatees; Humans are their major threat because of hunting and habitat alteration (de Thoisy *et al.* 2003).

Manatees do not form pair bonds and breeding can occur throughout the year with a gestation period up to 13 months (Husar 1978). Usually one calf is born, but twins have been reported, with an average weight of 20 kg and length of one meter. Young may not be fully weaned until two years of age with sexual maturity up to six years old. Except for mating activities and caring for young, manatees are essentially solitary animals.

A census of the manatee population in Suriname is urgently needed as the subspecies *Trichechus manatus manatus* that is found in the country is listed as endangered and probably consists of fewer than 600 individuals (Pool 2013).

Figure 372. *Trichechus manatus* in the manatee pond at the Guyana Zoo during the late afternoon in Georgetown (image © Meshach Pierre, September 2016).

Description: Obvious aquatic adaptations include heavy-set cylindrical body with an average length of 3 m and weight of 400 kg. The snout is squarish, forelimbs are modified as flippers, hindlimbs are absent, and the tail is horizontally flattened into a broadly rounded paddle. Color of the nearly hairless body is grayish.

Distribution: The northern subspecies is common around Florida but ranges from North Carolina to Louisiana. The southern subspecies extends from the Caribbean to Central America from Mexico to Brazil in South America. In Suriname, found in the coastal region to about 60 km upstream on most rivers.

Use: Hunted for its meat.

Similar species: The only other completely aquatic mammal is the Guiana Dolphin (*Sotalia guianensis*), which has a pointed beak, blowhole on the top of a rounded head, more stream-lined body and dorsal fin.

Pteronura brasiliensis
Giant Otter
Bigiwatradagu
Carnivora, Mustelidae

Groups form around a monogamous breeding pair and their offspring with occasionally unrelated subadults. Giant Otters (Figure 373) are diurnal and feed primarily on fish, such as catfish and piranha (Duplaix 2002), consuming at least 3 kg per day. They are quite noisy, using vocalizations to communicate with their extended family that typically may number up to eight individuals. Their home range covers about 12 km^2 and they can be seen on river trips playing or floating on their backs while eating fish.

Dens are usually built in the sides of riverbanks with multiple openings and chambers (Duplaix *et al.* 2001). Females give birth in the den to usually 2-3 young and the offspring emerge after a couple of months. They are weaned at about nine months and are sexually mature in two years. By the third year, they are independent and disperse from the natal den to establish their own home. The life span is estimated at eight years in the wild.

Figure 373. *Pteronura brasiliensis* (image © Mary Ann McDonald, WWF).

Description: Dense water-repellent fur is uniformly brown except for the chin and throat that have irregular white spots or stripes, which are individually unique and useful for identification. Giant otters can be seen in a characteristic periscoping position with their head and neck exposed above the waterline. They have an average weight of 30 kg and a streamlined form with a head-body length of about 1 m and a horizontally flattened tail just over half this length. The tail is used for propulsion by moving it in an undulating fashion. Webbed feet provide another aquatic adaptation.

Distribution: Primarily the Amazon basin and adjacent drainages including the Orinoco and Guiana watersheds, but its distribution has become fragmented due to loss of habitat and overhunting in many areas of South America. Populations in Suriname are doing better than in other parts of its range; it can be relatively common in some rivers, although increased human encroachment can quickly change this situation (Duplaix *et al.* 2001). Giant Otters tend to favor black-water streams in Suriname, such as Kaboeri Creek, Tibiti River and the upper Coesewijne River (Duplaix 1980).

Use: The Giant Otter is hunted for its plush fur and is on the CITES Appendix I list, which affords it the most restriction on trade. They are believed to be a threat to fish stocks.

Similar species: The smaller, by about half the size and uncommon Neotropical River Otter (*Lontra longicaudis*) does not have white on the chin or throat and has a cylindrical tail.

Proechimys guyannensis
Guianan Terrestrial Spiny Rat
Maka-alata (general term for spiny rat)
Rodentia, Echimyidae

The Guianan Terrestrial Spiny Rat (Figure 374) is one of the most abundant and largest rats in Suriname, weighing up to 500 grams. It is common fodder for medium-sized predators such as Ocelots, and is also regularly taken by the larger Puma. Other nocturnal hunters of this rat species include snakes and owls (Husson 1978).

Although considered seed predators, *Proechimys guyannensis* are also beneficial as seed dispersers. They are primary scatter-hoarders of palm seeds, caching the seeds under forest litter for later consumption (Forget 1991). However, some seeds are not eaten so can eventually germinate and grow. Other studies in French Guiana have documented fruit as being more prevalent in stomach contents than seeds (Dubost & Henry 2006)

There is not much published data on reproductive activity for this species, but information from specimens trapped for museums indicates that most females are pregnant from January to April and not pregnant from August to November. However, this area needs more detailed study.

During evening walks in the forest, this spiny rat can be observed foraging for food on the ground. When scared, it will run off for a short distance before stopping so can be easily seen and tracked with a strong light. Dens are usually found at the base of trees with hollows and roots where the rat can burrow.

Figure 374. *Proechimys* spp. captured on a camera trap just after midnight near the Kutari River, Suriname; there are two species of terrestrial spiny rats, but identification needs closer examination of for instance degree of hairiness of tail (image © Esteban Payan, Panthera, August 2010).

Description: Large rat with a head-body length reaching 250 mm and a naked tail that is a little bit shorter than this. Sometimes, individuals are found without a tail, which breaks off easily at the first caudal (tail) vertebra when grabbed. This is considered an escape mechanism to elude predators, but it has never been observed. Fur is brownish red on the body and contrasts with a pure white belly. Hairs on the back are stiff, hence its common name.

Distribution: Found throughout the Guiana Shield in both lowland and highland regions. In Suriname, occurs in forest and savanna.

Use: This species has been used in epilepsy research because of its resistance to the neurological disease.

Similar species: Guianan Terrestrial Spiny Rats are essentially indistinguishable from Cuvier's Terrestrial Spiny Rat (*Proechimys cuvieri*), which is found in the same areas. There are subtle differences between these two species, such as a less hairy tail in the former species, but this cannot be observed without capturing the animals.

Hylaeamys megacephalus
Broad-headed Rice Rat
Alata (general local term for rat)
Rodentia, Cricetidae

The Broad-headed Rice Rat (Figure 375) may be one of the more common species of small rodents found in Suriname. For example, during a Conservation International Rapid Assessment Program survey of the rainforest region around the Trio village of Kwamalasamutu in southern Suriname, *Hylaeamys megacephalus* was more than twice as abundant as the next species of rat (Lim & Joemratie 2011). However, it does not seem to occur in the savannas in the south.

It is nocturnal, primarily terrestrial, and assumed to be herbivorous or omnivorous, but no studies have been done on feeding behavior. As with other rats and mice, this species is potential prey to any vertebrate predator, including documented evidence of owls preying on them in Brazil (Rocha *et al.* 2011).

This species inhabits houses in indigenous villages in the remote interior of French Guiana (Catzeflis 2012) and would be expected to have similar behavior in Suriname. Because of relative ease of accessibility around human settlements, they consume food such as cassava.

As with most species of small mammals, there has been very little direct research on reproductive activity, but some general information can be gleaned from museum specimens. For *Hylaeamys megacephalus*, no females were pregnant in July or August, but data is available for only four other months of the year.

Figure 375. *Hylaeamys* spp. near Werehpai; rats in this genus are difficult to identify to species by only external features and typically need examination of internal anatomy, such as teeth, or analysis of DNA data (image © Burton Lim, September 2010).

Description: Small rat with a head-body length of about 120 mm and a tail slightly shorter than this. Fur is brownish gray with belly hairs paler than the back.

Distribution: Found primarily throughout most of the Amazonian rainforest region. In Suriname, it occurs in both the lowlands and highlands (animals captured at Brownsberg Nature Park, along the Sipaliwini River, and Tafelberg, Bakhuis and Kasikasima mountains).

Use: (No known use).

Similar species: The Yungus Rice Rat (*Hylaeamys yunganus*) is not as common in Suriname but indistinguishable unless dental morphology can be examined or DNA variation analyzed.

Didelphis marsupialis
Common Opossum
Dagu-awari, foto-awari
Didelphimorphia, Didelphidae

The Neotropical Common Opossum (Figure 376) is an opportunistic omnivore consuming just about anything edible. Their diet includes fruits, nectar of flowers, insects and meat, either by hunting small vertebrates or scavenging carrion of abandoned larger carcasses (Reid 1997).

Didelphis marsupialis are typical marsupials in that the females have a pouch for nursing 4-9 under-developed newborns in at least two litters per year. After a short gestation period of approximately two weeks, the 1 cm long offspring crawl into the pouch and latch onto a nipple for almost two months. The young then ride on the back of the mother and within about another three months or more, they are weaned and become independent. By that time, females are reproductively active and can begin to breed in their first year.

Although *Didelphis marsupialis* do most of their foraging for food on the ground, they have some adaptations that allow them to move equally well up in the forest canopy. As also seen in the small mouse opossums, the Common Opossum has opposable big toes on the feet and a prehensile tail that helps to grab branches while climbing in trees. Similarly, they can live in burrows in the ground and in tree cavities (Husson 1978). Denning sites are changed on a regular basis after a few days. Other than activities associated with reproduction, this species is solitary.

Natural predators include smaller species of wild cats and snakes. An interesting potential co-evolutionary interaction is venom potency in pit vipers and venom resistance in species of *Didelphis* (Voss 2013). Not only are they preyed upon, but Common Opossums also are predators of venomous snakes.

Figure 376. *Didelphis marsupialis* walking in a clearing near the headquarters of Brownsberg Nature Park (image © Burton Lim, January 2016).

Description: Domestic cat-sized opossum weighing about 1 kg. Head-body length is approximately 400 mm with a white-tipped tail that is slightly longer than the body. Fur is long and scruffy with a salt-and-pepper appearance because guard hairs are black and the underfur is typically yellowish white.

Distribution: Most of tropical Central and South America. Found throughout Suriname.

Use: There are no obvious uses of this species by people in Suriname.

Similar species: *Didelphis albiventris* has a distinctive V-shaped black marking on the forehead and tail that has at least one-quarter of the base furred. It is also found in Suriname, but not as common as *Didelphis marsupialis*.

Marmosops parvidens
Delicate Slender Mouse Opossum
Moismoisi-awari (general local term for mouse opossum)
Didelphimorphia, Didelphidae

Unlike most other marsupials, such as the Common Opossum, the small mouse opossums of the Neotropics do not have pouches. *Marmosops parvidens* (Figure 377) produce about seven under-developed young that stay latched to the nipples of the mother. Little is known about reproductive behavior.

Sample size is small, but *Marmosops parvidens* is the most abundant species of mouse opossum caught in traps. Prior to 2002, only two specimens of this species were known from Suriname. However, this species and mouse opossums in general may not be as rare as previously thought, but just difficult to catch. The usual method of surveying small mammals, such as rats and opossums, are baited box-style live traps. But the use of pitfall traps at Brownsberg Nature Park captured 11 individuals, whereas box traps caught only one (Lim *et al*. 2005).

As with the Common Opossum (*Didelphis marsupialis*), the Delicate Slender Mouse Opossum has arboreal adaptations for living in the forest canopy such as an opposable big toe on the foot and a prehensile tail to help it grasp branches as it forages in the trees for fruits. But similar to its fellow marsupial, the Slender Delicate Mouse Opossum spends some time on the ground, as indicated by its successful capture in pitfall traps (Voss *et al.* 2001), foraging for other food such as insects during the evening.

Figure 377. Male *Marmosops parvidens* from the upper Palumeu River (image © Piotr Naskrecki, March 2012).

Description: Reddish brown fur on the back and extensive solid, paler color on the belly. The Delicate Slender Mouse Opossum has large ears and a blackish ring around the eyes. They have a head-body length of about 100 mm with a tail approximately 50 percent longer.

Distribution: This species is essentially a Guiana Shield endemic, occurring primarily north of the Amazon River in Brazil. Only one specimen was found south, and west of the Orinoco River and its tributaries in Venezuela. In Suriname, typically occurs in rainforest and has been captured at Brownsberg Nature Park, Bakhuis Mountains, Kutari River and Upper Palumeu River.

Use: No known use.

Similar species: Pinheiro's Slender Mouse Opossum (*Marmosops pinheiroi*) is also found in Suriname, but more brownish in color on the back and a slightly paler belly. However, the fur coloration is variable so identification can only be done reliably with examination of internal anatomy such as cranial morphology in museum specimens or DNA analysis to differentiate it from *Marmosops parvidens*.

Desmodus rotundus
Common Vampire Bat
Fremusu (general local term for bat)
Chiroptera, Phyllostomidae

This species is responsible for the inaccurate stereotype of bats as blood-feeding vampires. In fact only three of the more than 1,300 species of bats worldwide are sanguivores and are only found in Central and South America. However, even in scientific circles there was a 'propensity of early naturalists to regard all Neotropical bats as vampires' (Lewis & Wilson 1987:4). This is reflected in new scientific names such as *Vampyrum* and *Vampyrops* being described in the 19th century, but these are not blood feeders. Ironically, none of the vampire bats actually have vampire-sounding scientific names.

Common Vampire Bats (Figure 378) are not scary vicious species, but are very social animals with complex behavior. For example, individuals in a colony will regurgitate their blood meal to those who were not fortunate in feeding that night (Carter & Wilkinson 2015). Although complete weaning takes time, mothers begin regurgitating blood to their young in the second month. At about four months, the offspring accompany the mother to their prey and start to directly feed on blood.

Unlike many species of bats that can vary their diet depending on the availability of food, vampire bats are obligate blood feeders that cannot eat anything else (Busch 1988). They have adaptations or modifications, such as an elongated and highly vascularized stomach that is well supplied with blood vessels to allow them to convert the blood-meal volume to urine very quickly. During feeding, they may consume up to a tablespoon of blood from any large mammal.

Figure 378. *Desmodus rotundus* from the upper Palumeu River (image © Burton Lim, March 2012).

Description: Although the Common Vampire Bat is in the leaf-nosed family Phyllostomidae, it does not have a noticeable fleshy flap of skin on the face. Instead, there is a much-reduced ridge of skin over the nose. For bats, this species has well-developed thumbs that are modified for walking on the ground and in fact the vampire is the only bat known to be able to run (Riskin & Hermanson 2005).

Distribution: Found in tropical areas of Central and South America. It occurs throughout Suriname in rainforest and savanna.

Use: The anti-coagulant in vampire bat saliva has been isolated (Fernandez *et al.* 1998) and is being developed as the drug Draculin for the medical treatment of strokes and as a blood thinner for reducing heart attacks.

Similar species: There is one other species of vampire bat that occurs in Suriname, but *Diaemus youngi* feeds exclusively on bird blood and is not as common. It has white on the tips of the wings but this would need to be seen up close to be noticed.

Carollia perspicillata
Seba's Short-tailed Fruit Bat
Fremusu (general local term for bat)
Chiroptera, Phyllostomidae

Seba's Short-tailed Fruit Bat (Figure 379) is one of the most common and widely occurring bats in tropical Central and South America. It is also an important seed disperser of pioneering plant species, such as shrubs in the genus *Piper*, in treefall gaps and disturbed areas that have been cleared (Fleming 1988). This enables natural forest succession to continue and for climax tree species to emerge above the canopy. During times of low fruit availability, this species will vary its diet by feeding on nectar, pollen and insects.

Carollia perspicillata is an indicator species of disturbed environments (Wilson *et al.* 1996; Lim 2009). During biodiversity surveys in Suriname, Seba's Short-tailed Fruit Bat was more than twice as common as the next most abundant species at Brownsberg Nature Park (Lim *et al.* 2005) and Blanche Marie Vallen in the Bakhuis Mountains (Lim 2009), both of which are human-modified rainforest habitats. By contrast, it was a species of average abundance during two surveys in the pristine environments of southern Suriname (Lim & Joemratie 2011, Lim & Banda 2013).

Roost sites are typically tree hollows, but they adapt relatively well to moderate levels of human disturbance and will roost in manmade structures such as buildings. Females are usually pregnant during the rainy seasons, in the months of January, June, and July.

Figure 379. *Carollia perspicillata* from the Bakhuis Mountains (image © Alex Borisenko, January 2006).

Description: As alluded to by its common name, this species has a short tail of about 10 mm. It has tri-colored fur with a pale middle band separating dark basal and terminal bands on the body hairs. Paradoxically, identification is usually done by a process of elimination because it does not have characteristics that other groups of bats in the same family have, such as white stripes on the face or back, large ears, nose leaf or elongated snout.

Distribution: Common and broadly distributed across the Neotropics from Mexico to Argentina. In Suriname, Seba's Short-tailed Fruit Bat is found throughout the country in lowland and highland rainforest and savanna regions.

Use: (No known use).

Similar species: The Silky Short-tailed Fruit Bat (*Carollia brevicauda*) is also found in most parts of Suriname, but is not as common. It is slightly darker and smaller but positive identification can only be done after capture, examination and measurements.

Glossophaga soricina
Pallas's Long-tongued Bat
Fremusu (general local term for bat)
Chiroptera, Phyllostomidae

Pallas's Long-tongued Bat (Figure 380) is an important pollinator as a consequence of feeding on nectar in flowers. It has one of the highest mammalian metabolic rates and utilizes most of its body fat on a nightly basis (Voigt & Speakman 2007). In general, the eight species of primarily nectar-feeding bats in Suriname are the nocturnal ecological counterparts to the diurnally active hummingbirds. They all also have the same behavior of hovering in front of flowers while feeding on the nectar. Although *Glossophaga soricina* has anatomical adaptations such as an elongated tongue and snout for nectarivory, this species has a varied diet including fruit and insects when flowers are not in bloom.

This species naturally roosts primarily in tree hollows, but adapts relatively well to human disturbance by living in buildings (Alvarez *et al.* 1991). In other parts of its distribution, roost sites also include caves, which are not known in Suriname because of the geological history of the ancient Guiana Shield.

Although sampling is limited in Suriname, pregnant females have been documented during the rainy season in July. As with all species of tropical bats, one young is born per litter, but there are usually two litters per year. This makes bats unusual for mammals because typically small-sized species, such as rats and mice, have many offspring, whereas large-sized species, such as tapirs, most monkeys and manatees, have only one young.

Figure 380. *Glossophaga soricina* from the Sipaliwini River region (image © Burton Lim, July 2009).

Description: This species has a long snout and tongue as noted in its common name. The tongue can extend beyond the mouth by more than the length of the head and the tip of the tongue has a brush-like pad that sops up the nectar. It is a small bat measuring only about 60 mm from tip of nose to tip of short tail. Fur is primarily brownish in color.

Distribution: Widespread from Mexico to Argentina. Similarly, it occurs in lowland and highland rainforest and savanna throughout Suriname.

Use: (No known use).

Similar species: Thomas's Nectar Bat (*Lonchophylla thomasi*) is also found in most parts of Suriname and is slightly more common. The inner upper front incisor teeth are larger and the fur on the back is paler with a more extensive white base to the hairs. But these differences can only be seen if the bat is captured and in the hand for examination.

Rhynchonycteris naso
Proboscis Bat
Fremusu (general local term for bat)
Chiroptera, Emballonuridae

This species is one of the few bats that can be easily observed in the wild in Suriname. Ecotourists sightseeing on a boat trip may mistake the sudden fluttering near the river bank for butterflies, but there is a good chance it is the Proboscis Bat (Figure 381), as they are quite common along streams in the interior. Colonies roost head to toe in a straight line on a tree trunk overhanging the water and fly off when disturbed (Husson 1978). But much like kingfishers, some land not too far away to take off as the boat approaches again. The Dutch common name (riviervleermuis) captures its habitat perfectly.

Normally, the Proboscis Bat is active at night when it goes hunting for insects over the water (Plumpton & Jones 1992). They are aerial insectivores because their prey is caught while flying, as opposed to some bats that are gleaners picking off stationary insects on the ground or in vegetation.

As with many species of bats, there is little data on reproduction because their small size, nocturnal activity and flying behavior make them difficult to study. Nonetheless, three pregnant females were caught at the Bakhuis Mountains with weights ranging from 4-6 g for an average of 5 g, head-body lengths of 44-46 mm for an average of 45 mm, and head-body lengths of the embryos ranging from 10-24 mm indicating different stages of development. Four non-pregnant females were all 4 g in weight. While a small sample, this suggests that newborns are probably at least half the weight and length of the mother.

Figure 381. ***Rhynchonycteris naso*** **roosting on a branch near the bank of the Sipaliwini River (image © Burton Lim, July 2009).**

Description: This tiny bat weighs only 4 grams or about the same as a 25-cent coin and has a head-body length of less than 50 mm. It has a pointy nose, hence the common name. But the scientific names of '*rhyncho*' and '*naso*' are Greek and Latin words for nose, respectively ('Nycteris' is Greek for bat). The fur is grizzled brown with two indistinct wavy pale lines on the back. There are tufts of cream-colored hair on the forearms, which also gives it another common name of the tuft-armed bat. These markings help it blend in with the bark on tree trunks, where colonies roost closely spaced, making the bats well camouflaged.

Distribution: Widely distributed from southern Mexico to eastern Brazil. In Suriname, the Proboscis Bat is never found too far from water in forested habitats.

Use: No known use.

Similar species: Frosted Sac-winged Bat (*Saccopteryx canescens*) is similar in size, but has two distinct wavy white lines on its back, a more solid brown color, and no tufts of hair on the forearms.

Use, Threats and Conservation

Chantal Landburg, Jan Mol

The natural environment in Suriname cannot be understood without taking into account human influences over the centuries. In Amazonia, Pre-Columbian population densities of Indigenous peoples (formerly referred to as Amerindians) were much higher than they are today, and large areas of apparently undisturbed rainforest are actually not pristine (Heckenberger *et al.* 2003, Levis *et al.* 2017). Ever since people set foot in Suriname, they have changed the landscapes and associated biodiversity. This chapter looks into the ways people have used the ecosystems and biodiversity of Suriname; first Indigenous peoples, but later also the European colonizers, Africans originally brought in as slaves, and Asians originally brought in as contract laborers. Their status, lifestyles and activities have changed over the years and so has their impact. People use renewable, biological resources, and this use may be in part beneficial. But in many cases use will have a number of negative impacts on these resources, and may then be qualified as a threat. Humans also pose threats to these resources as a result of a range of development activities, especially the mining of non-renewable resources such as gold, bauxite, oil and sand. Moreover, the cumulative and synergistic effects of a wide range of activities often further increase negative impacts. Conservation and restoration efforts can reduce these impacts. They may allow us to maintain ecosystems or return them to a natural condition, thus maintaining or restoring natural ecosystem services.

Use

Indigenous people. Artifacts found in the Sipaliwini Savanna are considered the oldest evidence of human presence in Suriname (Versteeg 2003). They are similar to those found at sites in eastern Venezuela (Boomert 1980), which at one site are associated with charcoal dated at around 9000 years before present (BP) (Barse 1990). The Indigenous people associated with these artifacts are assumed to have lived in savannas and savanna-forest margins. The artifacts include projectile points, arrowheads, knives and scrapers, which indicates that the early inhabitants were hunter-gatherers, likely living in small nomadic family groups (Fontaine *et al.* 1980, Bakker *et al.* 1998, Versteeg 2003). They may have regularly burned the savannas to facilitate hunting and moving about on foot; the young grassy vegetation was likely attractive for certain game species, such as deer (Teunissen & Wildschut 1970, Krook 1979, Fontaine *et al.* 1980, Teunissen 1980a). By burning the savannas, the early inhabitants would have contributed to the conservation of savanna ecosystems and their many characteristic species (see introductory chapters in the present book). In present-day Suriname, savannas persist mainly in the proximity of Indigenous villages; in areas no longer burnt by people, savannas typically turn into forests (Teunissen 1980a). The first evidence of sedentary human communities in Suriname – as opposed to older nomadic communities – comes from a rainforest site near Kwamalasamutu in southern Suriname, the Werehpai Caves; at this site, rock art is associated with charcoal and pottery dated 4200 to 5000 BP (Versteeg 2007).

The arrival of Humans in the Americas has been identified as a factor contributing to the extinction of the Pleistocene megafauna 60,000-11,650 BP (Metcalf *et al.* 2016), which made subsequent hunting of other, smaller game species inevitable. Agriculture may have originated when Indigenous people observed that leftovers of gathered fruits and plants sprouted spontaneously (Versteeg 2003). A transition to more deliberate and organized agriculture took place in Amazonia around 5000 BP (Fontaine *et al.* 1980, Oliver 2008). This may have been the case in Suriname as well, based on the presence of pottery at Werehpai (Versteeg 2007), which is associated with agriculture (Oliver 2008). Before the discovery of the Werehpai site, the evidence pointed to slash-and-burn agriculture in Suriname from 2400 BP onward (Versteeg 2003). Today, this agricultural practice involves clearing (slashing) and burning of small forest plots (Figure 382), often along river banks, with the ashes temporarily increasing the fertility of the soil. The plots are used for three to five years, after which they are left fallow; based on present-day practices, it may be assumed that such plots were also used in the past to grow crops and other useful plants native to the Neotropical

Figure 382. Present-day slash-and-burn agricultural plot, with cassava and plantain, near Indigenous village of Powakka on the left bank of the lower Suriname River (image © Chantal Landburg).

region such as cassava (*Manihot esculenta*; *kasaba*), pineapple (*Ananas comosus*; *nanasi*), tobacco (*Nicotiana tabacum*; *tabaka*), pepper (*Capsicum annuum*; *pepre*), giant agave (*Furcraea foetida*; *ingisopo*) and giant reed (*Gynerium sagittatum*; *peiri*) (Fontaine *et al.* 1980, Bakker *et al.* 1998, Versteeg 2003). Teunissen (1978a,d) assumes that Indigenous people have introduced and dispersed useful plants in Suriname such as cacao (*Theobroma cacao; kakaw)*, the palm *Astrocaryum vulgare* (*awara*), papaya (*Carica papaya*; *papaya*), cotton (*Gossypium* spp.; *katun)*, some bamboo species (*Guadua* spp.) and cashew (*Anacardium occidentale*; *kasyu*) (Figure 383). Cacao, *awara* and bamboo can prevail in previously inhabited areas even after the forest has taken over most of the formerly cultured land (Teunissen 1978a,d).

Figure 383. Trees thought to have been cultivated or at least used by early Indigenous people in Suriname which can still be found in or near Indigenous villages today (image taken at Pierre Kondre Kumbasi along the Suriname River; © Chantal Landburg):

a. *Awara* palm (left; *Astrocaryum vulgare*) and cashew tree (right; *Anacardium occidentale*), and

b. Neotropical *Guadua* sp. bamboo.

In parts of the Young Coastal Plain that are seasonally flooded, a more permanent form of agriculture was developed around 2000 BP. Arrays of small square raised fields ('beds') were made, likely by piling up material dug up from the spaces between the beds, thus also creating a network of small ditches. These beds were likely kept fertile and dry by regularly re-digging the surrounding ditches. Remnants of raised beds can be found at Hertenrits, Buckleburg and Peruvia in northwest Suriname, and several sites in northeast Suriname (Figure 384) (Fontaine *et al*. 1980, Versteeg 2003). The present-day vegetation at these sites includes species that suggest previous occupation; the percentage of useful trees and shrubs at the sites proved to be higher than in comparable surrounding areas (Werkhoven & Versteeg 1980).

Pre-Columbian coastal communities in northeast Suriname built their villages on natural sandy ridges (cheniers), while those in northwest Suriname, where cheniers are not common, developed their communities on artificially raised mounds near agricultural beds. These mounds are easily

Figure 384. A complex of raised fields near Galibi in northeast Suriname (image © Frans Bubberman).

spotted in the landscape: they are raised a few meters above the surrounding swamp level and have tall, mixed forest at the core, as opposed to herbaceous swamp or mangrove forest (Figure 385) (Fontaine *et al.* 1980, Rostain 2010, Versteeg 2003).

Species hunted by present-day Indigenous peoples – such as deer, birds, monkeys, sloths, tapirs, peccaries, armadillos, anteaters, iguanas, agoutis and other rodents (Heemskerk & Delvoye 2007) – may have been targeted by the early hunter-gatherers as well (Bakker *et al.* 1998, Versteeg 2003). Similar to present-day practices, hunting trips were probably also used to gather food, medicine, tools and other materials. Gathered foods may have consisted of berries, honey, insect larvae, caterpillars, ants, snails, frogs, and eggs of sea turtles and iguana (Bakker *et al.* 1998). Actual records of the diet of Pre-Columbian inhabitants come from animal bones found at the Peruvia site in 'terra preta', a black soil layer which indicates accumulation of organic matter as a result of many years of refuse dumping by people. Remnants found at this archaeological site showed that early coastal inhabitants ate mainly deer and caiman, in addition to catfishes, manatees, birds and crabs (Versteeg 2003). Like present-day Indigenous people, Pre-Columbian inhabitants likely gathered fruits and honey for food, palm

Figure 385. The Wageningen-1 mound in northwest Suriname (image © Aad Versteeg).

leaves for roofing thatch and wood for construction of houses, canoes, paddles and furniture. Rubber and non-cooking oils, plants and animals were likely collected for medicine and body care, pesticides and repellents, clothing, adornments and musical instruments, weapons, twining rope, fishing gear, fishing poison, and initiation rituals (Versteeg 2003, Plotkin 1993, Heemskerk & Delvoye 2007, van Andel & Ruysschaert 2011).

The arrival of Europeans in South America, first as explorers and traders, later as invaders and settlers (see below and General Introduction / History chapter), had a dramatic impact on the Indigenous people. Many moved upriver as their territories in the coastal area were increasingly occupied, first by Europeans and later on also by Africans who fled from the slave plantations and settled in the forest (Boven 2006, Heemskerk & Delvoye 2007, Heemskerk *et al.* 2007). The lifestyle of many Indigenous peoples changed following contact with Western culture (Boven 2006, Heemskerk & Delvoye 2007, Heemskerk *et al.* 2007). This resulted in a mix of traditional and Western customs (Bakker *et al.* 1998). The most common hunting weapon nowadays is the shotgun, while nets and nylon lines are used for fishing. Hunting and fishing is often no longer only for subsistence, but also for commercial purposes. Nowadays, bush meat, fish and also trapped live animals are sold to local middlemen, as well as to clients in Suriname's capital and abroad. Brazil nuts (*Bertholletia excelsa*; *inginoto*) and traditional handicrafts are also sold.

In the 1960s, a USA-based missionary group (World Team 2016) convinced Indigenous people in the remote south of Suriname to leave behind their semi-nomadic lifestyle and live in permanent villages to facilitate baptism, healthcare and education. This resulted in the loss of traditional knowledge related to the use of forest products for spiritual and medicinal purposes. Although western products have become an important part of their lives, Indigenous people often still use traditional non-timber forest products (NTFPs) (Plotkin 1993, Heemskerk & Delvoye 2007, Heemskerk *et al.* 2007, Hoffman 2009, van Andel & Ruysschaert 2011).

Development of a European colony. Europeans first arrived in the Guianas in the 16th century to trade textiles and metal products for tobacco, cacao, indigo, cotton, cassava, fish, and what at the time was known as specklewood (*Brosimum guianensis*; *letterhout*). In the 17th century, trading posts were set up in Suriname, followed by larger settlements along the Suriname River where colonists logged the forest and started to grow their own tobacco (*Nicotiana tabacum*; *tabaka*) and sugarcane (*Saccharum officinarum*; *ken*), on a small scale (Bakker *et al.* 1998, Fontaine *et al.* 1980).

In 1651, the English governor of Barbados sent Francis Willoughby to Suriname, accompanied by 300 men (Parker 2015; see also General Introduction / History chapter). They established the first plantations along the Suriname and Commewijne rivers in what was then known as Willoughbyland By 1667, when Suriname was taken by the Dutch, the number of plantations had already increased to 178. By 1785 there were 571 plantations: 432 at which export crops were cultivated, and 139 at which wood was cut or crops were planted for the local (Surinamese) market (van Lier 1971). By that time, Suriname had achieved great prosperity by producing sugar, cacao and cotton for export. The dark side of this development was the labor by enslaved people and the violent repression of escape from enslavement (Stedman 1806; de Kom 1934). Plantations expanded mainly in the low-lying coastal areas with their fertile soils drained by canals and sluices. In the 19th century, plantations were also established in the coastal Saramacca, Coronie and Nickerie districts (Bakker *et al.* 1998, Fontaine *et al.* 1980). The substantial increase of area under plantation over time greatly changed ecosystems in the Coastal Plain: large areas were deforested and drained to grow crops. A detailed plantation map shows the extent of the affected area by the early 18th century (Figure 386).

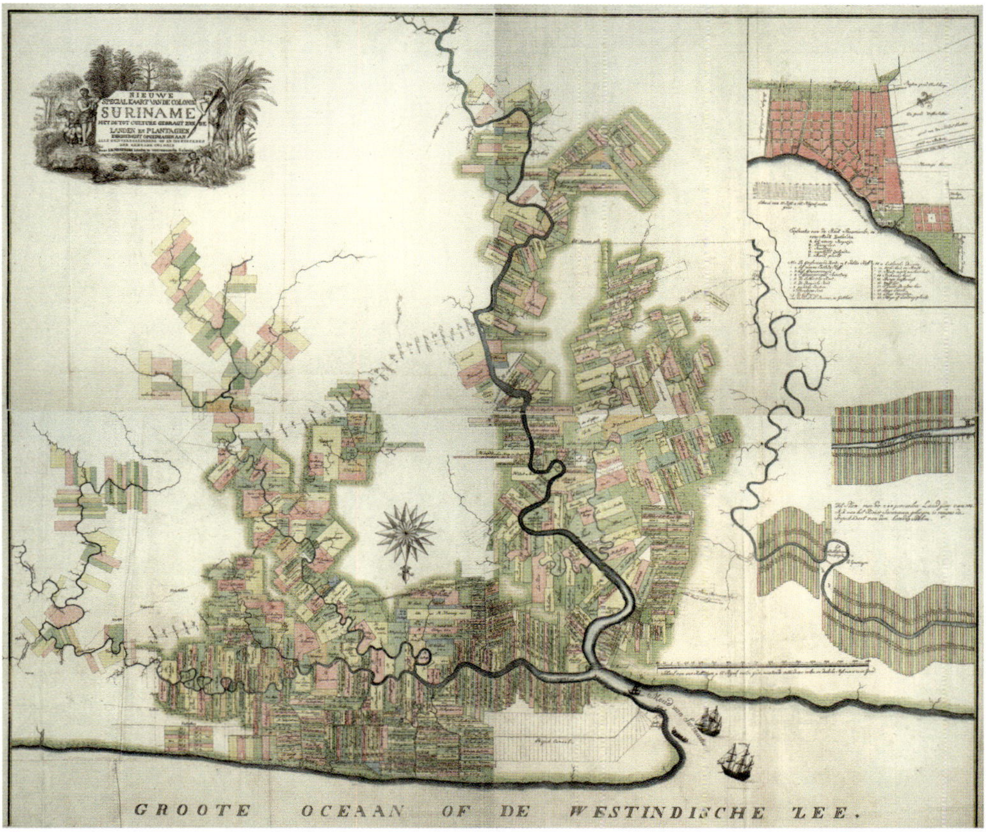

Figure 386. Map by J.H. Moseberg of the plantations in Suriname in 1801 ("Nieuwe speciaalkaart van de kolonie Suriname met de tot culture gebragt zyn de landen en plantagien"; image © Universiteit van Amsterdam).

Wood was an important building material on the plantations and in the capital Paramaribo. It was harvested from forested areas surrounding cash-crop plantations or from so-called wood plantations, which were in fact simply concessions to harvest logs from natural forest. During the early 20th century, efforts were made to commercialize timber and other forest products, including black mangrove bark (*Avicennia germinans*; *parwa*) as a tanning agent, *ceder* (*Cedrela odorata*) and crabwood (*Carapa* spp.; *krapa*) for cellulose production, rosewood oil (*Aniba rosaeodora*; *rowsudu*) for perfumes, and chicle for chewing gum (probably from the *Achras zapota* tree; Ostendorf 1962). However, most of these initiatives proved unsuccessful (van Traa 1946). Export of *kwasibita* (*Quassia amara*) had considerable success though; this plant was named after the formerly enslaved African and traditional healer Quassi (Dragtenstein 2004), and was used as a treatment for malaria fever (van Andel and Ruysschaert 2011). The harvesting of *balata*, a natural rubber derived from the *boletri* tree (*Manilkara bidentata*), was also successful. A firm from British Guiana (present-day Guyana) started the extraction and export of this product – as much as 75,000 kg per year – in 1890. By 1903, 621,000 ha of harvesting concessions for *balata* were granted in west and central Suriname, along the Corantijn, Nickerie, Marataka, Coppename, and Coesewijne rivers, where the highest densities of *boletri* were found (Ostendorf 1962, Bakker *et al.* 1998).

Tribal Maroon communities. Indigenous people were the first enslaved workers on the plantations. Later, during the 17th century, the Europeans brought Africans to the country because they better survived the tropical diseases and the harsh slave labor (Fontaine *et al.* 1980). In 1712, the French admiral Cassard attacked Suriname and plundered both Paramaribo and several plantations. Many people, both Europeans and Africans, fled from the plantations to escape. The European plantation masters hid their African slaves in the forest to avoid high ransom demands by the French (wealth was assessed based on number of slaves). This was an ideal opportunity for slaves to flee, and seems to have resulted in one of the largest events of escape from slavery in Suriname. The Africans who escaped from enslavement, as well as their descendants, are called Maroons (*marrons*) (Fontaine *et al.* 1980). Successive waves of organized and opportunistic escapes of African slaves resulted in the establishment of Maroon villages along the main rivers of central and east Suriname.

Maroons practiced slash-and-burn agriculture, also known as swidden agriculture, as well as hunting and fishing for their subsistence, much like Indigenous peoples did. Crops they planted included cassava, peppers, pineapple, sugarcane, tobacco, coconut (*Cocos nucifera*; *kronto*), corn (*Zea mays*; *karu*), okra (*Abelmoschus esculentus*; *okro*), peanuts (*Arachis hypogaea; pinda*), plantain (*Musa* spp.; *bana*), rice (*Oryza sativa; alesi*), sweet potato (*Ipomoea batatas; switipatata*), and watermelon (*Citrullus lanatus*; *watramun*) (Price 1991). Today, these crops are still grown by Maroons who live more or less traditional lifestyles.

Maroons used to catch fish with bow-and-arrow, nets and poisonous plants such as *Lonchocarpus* spp. lianas (*neku*) (Geijskes 1954). The most favored fishes included the wolf fishes *anyumara* (*Hoplias aimara*) and *pataka* (*Hoplias malabaricus)*, both of which are still popular for consumption today. Price (1991) provides a long list of game species hunted by Maroons. The number of game animals near Maroon settlements tends to dwindle over time as a result of overharvesting since Maroons do not have a nomadic lifestyle. Animals are also used for non-food purposes, such as deerskin for drumheads and hunting sacks, tails of stingrays for medicine, monkey tails as whisk brooms, opossum bones as repellents for cockroaches, and feathers to clean out earwax (Bakker *et al.* 1998, Price 1991).

The enslaved Africans brought a diversity of useful species with them from Africa to cultivate in their gardens. These species include the herb klip dagga (*Leonotis nepetifolia*), the spider plant vegetable (*Cleome gynandra*), the Senegal date palm (*Phoenix reclinata*), pigeon peas (*Cajanus cajan; loangopesi*) and sesame (*Sesamum indicum; abongra*). Many of these species are no longer cultivated, but some still grow locally in the wild, such as the Senegal date palm (van Andel *et al.* 2012a).

Nowadays, Maroons in rural and urban areas such as Paramaribo, as well as those who later migrated to the Netherlands, still use a wealth of plants for both subsistence and commercial purposes (van Andel & Havinga 2008, van Andel & Westers 2010). A total of 245 species of medicinal and ritual plants were recorded at the markets of Paramaribo in 2005; the total annual value of these plants on domestic and export markets was estimated at over US$ 1.5 million (van Andel *et al.* 2007).

From the abandonment of plantations to the present. When slavery was abolished in Suriname in 1863, plantation agriculture was already in decline, but for many plantations this was the final straw. Within 50 years the number of plantations in use declined from 216 (in 1863) to 79 (in 1913) (Bakker *et al.* 1998). At abandoned plantations, forest slowly took over again and a secondary forest developed with relicts of plantation crops and garden plants (Heilbron 2012)

In an effort to support Suriname's agricultural economy after the decline of the plantation system, the government lured in immigrants from China (since 1858), India (since 1873) and Indonesia (since 1890) (Fontaine *et al.* 1980). These immigrants were contracted to work for minimal wages on the plantations; they retained the right to return to their home country after expiration of the contract. The government made many efforts to convince these immigrants to remain in Suriname, such as giving out small plots of land. Those who did stay, settled down to become self-supporting small farmers. Over time these small farmers became important producers of agricultural crops, surpassing the role of the larger plantations in providing food for the local market (van Traa 1946, Fontaine *et al.* 1980). Efforts to retain immigrants included government-driven projects to develop polders, which is land that used to be swampy or marshy, cleared of natural vegetation and drained by means of ditches and sluices (additional ditches for water supply or storage are often also present). Polders were created on a large scale in the Nickerie District: for small-scale dryland farming at Nani and Henar, and for wet rice cultivation at Prins Bernard and Wageningen.

The National Planning Atlas (National Planning Office Suriname 1988) provides an overview of the different land uses in the 20th century (summarized in Table 14), such as shifting cultivation (slash-and-burn agriculture), fisheries (in estuaries, lagoons, rivers, swamps, creeks, and the ocean), forestry (timber concessions, wood-cutting licenses, forest plantations), nature conservation (reserves and parks), and recreation. This overview is still relevant today, although the location and scale of activities has changed somewhat. Current sectors that involve the use of ecosystems and biodiversity include extractive sectors such as fisheries, forestry, hunting, and collecting of wild flora and fauna; non-extractive sectors include tourism and research (Algemeen Bureau Statistiek 2015). Industrial fisheries are largely restricted to the marine area, while artisanal fisheries extend from inland waters – rivers, creeks and swamps – to estuaries and the near-shore sea. Forestry activities take place in the 'forestry belt': a belt of forested land south of the coastal swamps and into the northern part of the Interior (see Werger 2011, for a historic map; for updated map, see **sbbsur.com/wp-content/uploads/2017/02/Bosbouwlegger_A0_20170201_website.pdf**). At present, multiple types of forestry concessions exist, and extraction methods vary accordingly (see Werger 2011). Through the development of infrastructure during the 20th and 21st centuries, access to hunting grounds has become easier, as well as the collection of plants and animals for the wildlife trade (see Threats section below). Plant collection is often intended for medicinal use of plant parts. As for tourism, there are increasing numbers of mainly small nature tourism resorts popping up along accessible creeks and rivers throughout the country. At the coast, natural sites such as Braamspunt, Galibi and Bigi Pan have become the focus of tours that show visitors the native dolphins, sea turtles and coastal birds. Such tourism represents non-consumptive use of natural ecosystems and wild species.

Sector	Sub-sector	Resources used	Areas involved
Land-use and concessions	Shifting cultivation	Natural forest	In vicinity of villages
	Timber concessions	Native trees in natural forest	Forest belt: a strip of forest between coastal swamps and hilly Interior (see concession map)
	Wood-cutting licenses	Native trees in natural forest	Forests surrounding tribal communities
	Forest estates	Native & introduced trees in plantations	22,000 ha: Zanderij, Mijnzorgweg, Palissadeweg, Blakawatra, Coese-wijne, Kamp 8, Perica (NIMOS 2003)
	Protected areas	Natural areas	See protected areas map
	Recreation resorts	Rivers, creeks, forests	Young Coastal Plain: river estuaries, sandy beaches & brackish ponds; rest of the country: creeks & rivers
Fisheries	Coastal	Coastal fish, mainly Sciaenidae (drums, croakers), ariid catfishes	Mudflats along the coast: Potosi, Coronie, Hermina, Wia Wia banks
	Estuarine	Shrimp, ariid catfishes, juvenile sciaenids (*tri*)	River estuaries: Corantijn, Coppename, Suriname rivers, and coast north of Matapica canal
	Riverine	Downstream of most seaward rapids: *kubi*, goliath catfishes, *kwasimama*; upstream: such as *anyumara, tukunari*	Entire river stretch
	In ponds in mangroves	Snook, mullet, *tilapia*, juveniles of Southern Brown Shrimp (*Penaeus subtilis*)	Brackish lagoons along the coast: Bigi Pan, Boskamp, north Commewijne
	In swamps and creeks	*Kwikwi, pataka, walapa* and *krobia*	Freshwater swamps and creeks

Table 14. Production sectors involving the use of biodiversity (adapted from National Planning Office Suriname 1988).

Threats

HIPPCO. People are increasingly impacting the natural ecosystems of Planet Earth. The current era is even referred to as the geological Anthropocene Epoch (began about 1950) which literally means the new age of man (Waters *et al.* 2016). Habitat loss, Invasive (non-native) species, human Population growth – associated with increasing consumption, the ultimate cause of many environmental problems – Pollution, Climate change and Overexploitation (HIPPCO factors) are identified as the most important direct causes of extinction and threats to species and ecosystems around the world (Helfman 2007, Miller & Spoolman 2016, WWF 2016a). Over the years, multiple authors have attempted to list and describe the many threats to ecosystems and biodiversity in Suriname (Teunissen 1978a,b,c,d; 1980; 1987b,c,d; Ouboter 2000, Griffith *et al.* 2016).

Table 15 lists threats to Suriname's ecosystems and biodiversity; much depends on location and scale. In the Coastal Plain, where most people in Suriname live, residential and agricultural developments have a high impact. Further inland, the ecosystems of the Zanderij Belt suffer impacts from forestry and mining. In the hilly Interior, mining and infrastructure, such as new roads and bridges, contribute greatly to loss of habitat and species. An important consideration when evaluating these different threats is that multiple pressures have a much larger cumulative effect than the sum of their separate parts. Indirect and long-term large-scale effects are also often not taken into account (see for example Angermeier *et al.* 2004). Humans will ultimately suffer because 'healthy' natural ecosystems and wild species provide a variety of renewable goods and services that people depend on, such as nutrient cycling, food, drinking water, medicines, and recreation. Inspiration for art and innovation is also provided; see for example the artwork on the cover of this book, and the electric eels which inspired the design of first electric battery by the inventor Volta. These goods and services are degraded or lost when native biodiversity is affected and natural ecosystem function impaired (see for example WWF 2016a). A single damaging activity by some people, such as artisanal 'small-scale' gold miners, may imperil ecosystem services of critical importance to many others. It should be noted that threats to biodiversity and ecosystems should be viewed within a larger cost-benefit context, the costs often being loss of natural environment and traditional social structures, the benefits are usually income and 'development'. This larger context and the issues it raises are not the focus of the present chapter.

To better understand the main threats to the ecosystems and biodiversity of Suriname, it helps to look at different economic sectors and different types of impact (see below). In the Guianas, the mining industry at present poses the greatest threats; it mainly focusses on bauxite, gold and oil (WWF 2012). Rahm *et al.* (2014) identified mining as the most important driver of deforestation in Suriname, reporting that between 2008 and 2014, more than 92,000 ha of forest in the Guiana Shield was newly impacted by gold mining and 9000 km of waterways were in direct contact with mine sites. Almost 5000 km of these impacted waterways were in Suriname, where artisanal gold mining is pervasive in the eastern part of the country where the geological Greenstone Belt is located (Figure 387) (Ouboter *et al.* 2012).

Bauxite mining. While gold mining has a longer history, the bauxite mining sector was the first to scale up considerably in Suriname. Bauxite mining started small in 1916 in and around Moengo, in the northeast of the country, and later expanded to areas located between the district of Paramaribo and the Zanderij international airport. Recently, expansion has been proposed to the Bakhuis and Nassau mountains. As this sector expanded, environmental impacts (see below) became more significant, to the extent that the proposed mining in the Nassau Mountains would threaten several narrowly endemic species with extinction (Alonso & Mol 2005). At first unprocessed bauxite was exported, but in the 1940s a refinery was built to produce alumina, a precursor product in the process of making aluminum. In the 1950s it was decided to expand the refinery to include a smelter, which allows for actual aluminum production on site. A smelter requires massive amounts of energy, and to supply it, a dam was constructed in the Suriname River at Afobaka. The dam resulted in the creation of the 1560 km^2 Brokopondo Reservoir (Fawcett 1981; see also chapter on Climate and Hydrology).

Specific threat / Sector	Habitat loss	Invasive species	Population growth	Pollution	Climate change	Overexploitation	Affected ecosystems/ areas	Affected species
Climate change & sea level rise	2	1	1	0	3	1	Mangroves, brackish lagoons, river estuaries, freshwater stretches of rivers, freshwater swamps	Fish, birds, invertebrates, sea turtles
Oil exploration and exploitation	1	?	0	1	1	2	Shoreline mud banks, sandy beaches, coastal waters, freshwater swamps	Marine and estuarine mammals, fish, birds, and sea turtles, mangroves
Large-scale mining	3	1	1	3	1	2		Landscape-level biodiversity
Sand mining	2	?	0	1	0	2	Cheniers, freshwater swamps behind cheniers, beaches, river-bed, savannas	Sea turtles, flora and fauna in freshwater swamps, landscape-level biodiversity (both cheniers and savannas)
Small-scale goldmining	3	?	1	3	1	2	Forests, freshwater ecosystems, shoreline mud-banks, sandy beaches	Landscape-level biodiversity, aquatic plants (phytoplankton & vascular plants), fish, food web
Urban development	3	3	3	3	2	3	Mangrove forests, freshwater swamps, rivers and creeks	General change in biodiversity and species distribution
Infrastructure: roads, bridges	3	2	3	2	1	2	All	All
Infrastructure: hydro dams, canals	3	?	0	1	1	0	Suriname River, Poketi, Palumeu, Nani swamp, Coastal Plain (impoldering)	
Waste water	2	?	0	3	0	0	Rivers and creeks, estuaries, near-shore sea	Aquatic flora and fauna and their predators
Water export	2	2	0	0	?	2	Rivers and creeks, estuaries (possibly near-shore sea)	Freshwater, riverbank and estuarine flora and fauna
Industrial fishing	2	0	0	1	1	2	Coastal, river estuaries	Targeted fish and bycatch, Guiana Dolphin, rays, sharks, sea turtles, bottom fauna
Artisanal fishing	0	0	0	0	0	2	Freshwater swamps, rivers and creeks	
Agriculture	3	2	1	2	1	2	Young and Old Coastal Plain, savannas, freshwater swamps (for irrigation)	Landscape-level biodiversity (terrestrial and aquatic)
Slash-and-burn agriculture	2	?	0	0	1	1	Forests, especially along river banks	Landscape-level biodiversity
Forestry	1	0	1	1	1	1	Forests	
Hunting, collecting, and wildlife trade	0	0	0	0	0	3	Cheniers, forests, rivers and creeks	Sea turtles, game animals and cage species
Nature tourism	1	?	1	2	0	1	General habitat degradation at tourism sites	Increased pressure on wildlife
Absence of (intervention by) people	3	0	0	0	0	0	Savannas	Landscape-level biodiversity
Fire	2	?	0	0	1	0	Swamps and creeks, e.g. at Perica	Landscape-level biodiversity

Table 15. Threats to biodiversity and ecosystems in Suriname.

Each threat / sector gets scores that reflect the relative magnitude of its impact as to each of the six HIPPCO factors (colored columns); ecosystems, areas and species that are significantly impacted are listed. Impact scores: ? = unknown (white), 0 = none / insignificant (light blue), 1 = slight (yellow), 2 = medium (orange), 3 = strong (red).

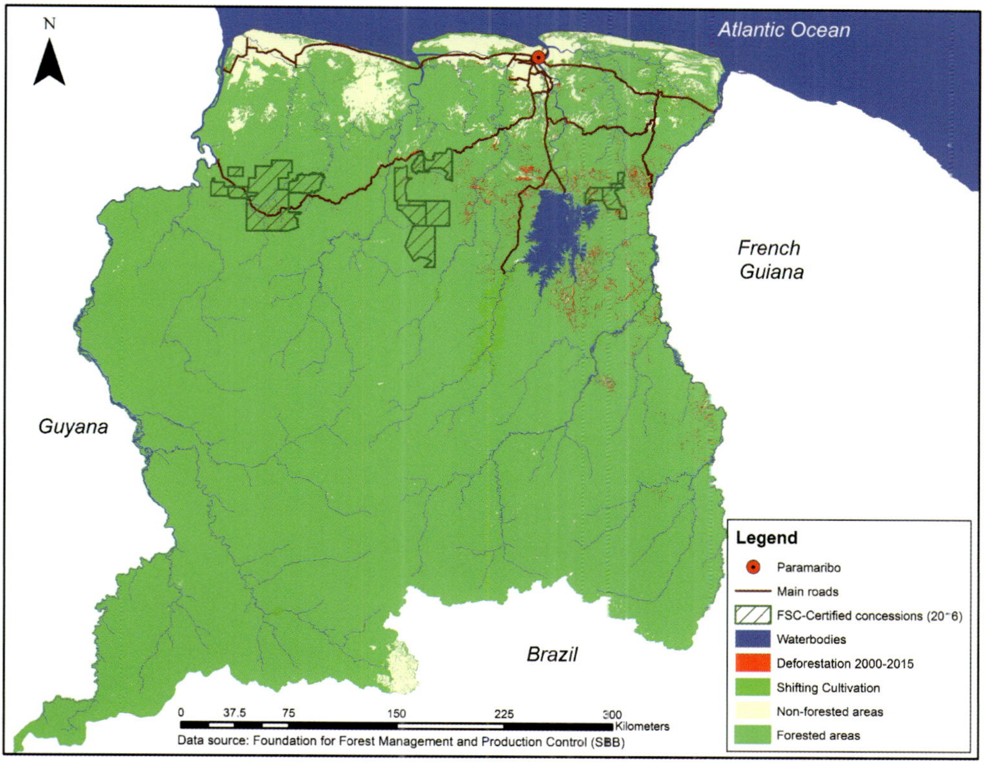

Figure 387. Areas in Suriname recently deforested, mostly as a result of gold mining, and FSC certified logging concessions (image © Foundation for Forest Management and Production Control, see also www.gonini.org/portal)

The bauxite industry has lasted for 100 years in Suriname, leaving behind two areas that together amount to an estimated 0.1 percent of Suriname's land surface with destroyed habitats, open pits, abandoned infrastructure, and buried chemical waste; limited rehabilitation efforts are ongoing. This does not measure up to the scale of the Brokopondo Reservoir, whose creation has drowned creeks, rapids, forests and other habitats over an area of about 1 percent of Suriname's land surface. While 172 fish species used to live in the area, only 41 now survive in the reservoir (Mol *et al.* 2007a). The Brokopondo Dam blocks fish and crustacean migrations, and has eliminated the food fish *Prochilodus rubrotaeniatus* (*kwimata*) from the river downstream of the dam (Mol *et al.* 2007a). Other downstream effects include low oxygen levels in the Suriname River water just below the dam, and flow regulation reducing seasonal inundation of the floodplain, which negatively impacts fish reproduction and growth (Panday-Verheuvel & Buys 1976, Panday-Verheuvel 1982, 1983). Conditions in the reservoir also seem to favor the accumulation of mercury in fishes (Mol *et al.* 2001; Ouboter *et al.* 2012) and the emission of carbon dioxide and methane, thus contributing to climate change (Abril *et al.* 2005).

Gold mining. Gold mining inflicts wounds on the lowlands in the east of the Interior, in the area around Brokopondo Reservoir and east and southeast of the reservoir; an estimated 3 percent of Suriname's land surface is heavily affected, another 10 percent to a much lesser extent. Historically, there was a first gold mining boom in Suriname in late 19th – early 20th century with a limited impact on the landscape. The second, current gold rush in Suriname took off in the late 1980s

(De Vletter & Hakstege 1998) and resulted in more extensive impacts. Currently, the gold-mining industry in Suriname includes both smaller-scale artisanal mining and larger-scale industrial mining.

The artisanal mining of gold involves excavation and hydraulicking, the loosening the gold-containing ore with high-pressure water sprays (Arets *et al.* 2006, Rahm *et al.* 2014). This is in fact artificial erosion at a very high rate. Artisanal gold mining may also involve suction dredging of riverbeds. In both cases the resulting slurry is pumped into so-called sluice boxes where gold is separated and amalgamated with mercury. The amalgam – a liquid mix of gold and mercury – is heated to vaporize the mercury, leaving behind the gold (Duijves & Heemskerk 2014, Rahm *et al.* 2014). Wastewater is usually discharged directly into the stream, and no form of rehabilitation takes place (Hays & Vieira 2005, Rahm *et al.* 2014).

The environmental impacts of artisanal gold mining include the loss of riparian forest (Peterson & Heemskerk 2001; Jan Mol, personal observation), flash floods, and intense soil erosion which results in streams with water the color of coffee with milk due to the extremely high sediment load (Mol & Ouboter 2004, Cremers *et al.* 2013). This has negative impacts on life in stream ecosystems (see for example Mol & Ouboter 2004). The loss of riparian forest impacts both aquatic and terrestrial biota (Davis *et al.* 1996). The original aquatic habitat is often completely lost by transforming a cool, clear creek underneath the forest canopy into a series of muddy, sunlit pools. Soils, water and air are polluted with mercury. Sediment and mercury are transported downstream of the mine and thus impact downstream waterways. Brownsberg Nature Park is an example of a formally protected area where these impacts are clearly visible and damage the park (Arets *et al.* 2006, White 2012). Suction dredging in rivers negatively affects the aquatic fauna and their spawning grounds at the dredging site; it also results in pollution of the water with sediment and mercury, and even the physical shape of the river is changed (Harvey & Lisle 1998, Sandoval *et al.* 2006). The gold miners negatively impact wildlife through hunting; they themselves often suffer from malaria and sexually transmitted diseases, and disrupt the social fabric of nearby Indigenous and Maroon communities (Peterson & Heemskerk 2001). Peterson and Heemskerk (2001) found regeneration of forests on old mining sites extremely slow. Typically, these mined-out areas are long stretches of bare soil, sometimes overgrown with grass and vines, and old mine pits with stagnant water. The environment is sunlit, hot and dry – nothing like the conditions at old-growth forest that used to cover the land – and this slows down initial forest regeneration. Mercury vapor is also released into the environment in the capital Paramaribo near shops were gold is refined (Wip *et al.* 2013). The online storyline 'The Amazon Gold Rush: Gold Mining in Suriname' (**amazonteam.org/maps/suriname-gold**) provides a visualization of the gold mining and its environmental impacts.

The mercury that finds its way into the environment makes a long journey through the food web. Bacteria convert inorganic mercury to highly toxic methyl-mercury, a form which can be incorporated in organisms and is 'biologically magnified' up the food chain (Morel *et al.* 1998), meaning that it ultimately leads to higher concentration in such organisms as predatory fish (Mol *et al.* 2001, Ouboter *et al.* 2012). Ouboter *et al.* (2012) hypothesized that mercury disperses not only in water, but also in the air, and is carried by the wind to areas without gold mining. When contaminated fish are staple food, the mercury accumulates in people, and may cause brain damage, especially in children and fetuses (Hays & Vieira 2005, Miller & Spoolman 2008). The mercury gets incorporated in growing hair: Mohan *et al.* (2005) found elevated levels of mercury in hair samples of people in both urban and Interior areas in Suriname. Fréry *et al.* (2001) found that the weekly mercury intake from consumption of contaminated fish by the Wayana indigenous people living along the Upper Marowijne River was ten times the weekly level set by the United States Environmental Protection Agency.

Threats from larger-scale industrial gold mining are much better contained, in the sense that these large companies, typically multinationals, do not use mercury, and have environmental management plans to reduce their impacts; their activities are also restricted to well-defined concessions. However, local habitat destruction is an inevitable result of industrial surface mining: large, deep pits are excavated that are not backfilled, and 'waste rock' is piled up and shaped into artificial hills that cover the original landscape. The cyanide used for industrial-scale gold recovery, and the wastewater produced in the refining process, can pose threats to animals coming in contact with wastewater ponds and to creeks and rivers in and downstream of the concession, for example in the event of leakages (Ramjai 2001, Eisler & Wiemeyer 2004). However, the toxic cyanide is broken down quickly into less toxic N-compounds (leading to eutrophication) and these are diluted with rain water and taken up by micro-organisms (Jan Mol, personal observations).

Oil. Oil exploration and exploitation is increasing in the Coastal Plain and in the country's exclusive economic zone (EEZ), which extends 200 nautical miles (370 km) from the coastline (see **opportunities.staatsolie.com**). The US Geological Survey (2012) estimates that some 49 million m^3 of oil is present in the Guyana-Suriname basin. Since the 1980s the state oil company Staatsolie has developed onshore oil production fields in the Saramacca District, west of Paramaribo, and is constantly exploring new areas in other coastal districts. During exploration, coastal ecosystems suffer some damage as a result of the creation of transportation trails and construction of drilling pads. These effects are generally very localized and not likely to result in significant impacts. However, old exploitation methods in the swamps in the Coastal Plain included impoldering (the creation of polders), which disrupted connections between wetlands (Figure 388.a). Such hydrological disturbance can cause wetland loss, vegetation changes and mangrove die-off (Erftemeijer & Teunissen 2009) and can affect fish populations. In the recently developed 'wet operation' exploitation system there is no need for impoldering, thus reducing damage to the swamp ecosystem (Figure 388.b; Jan Mol, personal observation). Offshore oil exploration and exploitation creates another set of environmental problems. The seismic exploration ships produce intense sound blasts which disturb marine mammals and fishes; these animals are also impacted by drilling and piling (McCauley *et al.* 2000). Oil spills are a significant risk, and if they occur will affect both the sea and coastal lands (Smith *et al.* 2011).

Sand mining. River beds, savanna areas in the Zanderij Belt, cheniers and coastal beaches are mined for sand. Dredging of rivers causes changes in the hydrological regime and destroys aquatic habitats. Mining in savanna areas ravages these unique habitats, such as near the International Airport at Zanderij (Chantal Landburg, personal observation). Sand mining of coastal beaches threatens their role in the protection of the coast from erosion, and destroys nesting habitat of marine turtles (Anthony 2016).

Infrastructure. Road construction causes direct localized but permanent damage to the environment in Suriname, and often leads to spin-off activities that cause additional damage. Roads

Figure 388. Oil exploration and exploitation by Staatsolie in wetlands in Suriname (images © Dirk Noordam):
a. impoldering of coastal swamps, and b. newly developed 'wet operations'.

directly result in animal mortality (roadkill), which causes local populations to decline (Benítez-López *et al.* 2010); they also facilitate the activities of hunters and plant and animal collectors, with the same effect (Laurance *et al.* 2009; De Dijn, personal observation). In addition, they lead to altered, polluted roadside habitats and habitat fragmentation, meaning that roads will cut previously continuous habitat in smaller, disconnected pieces that are often less suitable for wild species survival (Laurance *et al.* 2009, Hill 2016). Roads, much like dams, also adversely affect wetland and stream ecosystems, for example by disrupting (longitudinal) connectivity and impairing movement of aquatic biota (Angermeier *et al.* 2004, Januchowsky-Hartley *et al.* 2013). The Interior of Suriname has few roads, but the road system is expanding. If roads projected as part of the regional IIRSA infrastructure project (Figure 389) are indeed constructed, they are expected to lead to substantial land-use conversion, and habitat and biomass loss (van Dijck 2013).

An example of strong habitat alteration due to road construction is provided by the coastal road in the west of Suriname which cuts through a wetland area. Because initially culverts were not put in place, the road functioned as a dam that prevented the exchange and mixing of fresh and salt water between the southern and northern part of the swamp. The area north of the road suffered from salinization resulting in the die-off of the original vegetation and subsequent development of large areas of open water. Over time, a different vegetation developed on both sides of the road. The construction of this road is also thought to have caused a decrease in natural productivity and biodiversity (van der Steege 1986, Teunissen 1987b). When in the 1990s the pressure of accumulated rainwater north of the road caused it to break, culverts were finally constructed under the road and the free exchange of water is now again possible (Erftemeijer & Teunissen 2009). However, culverts can impair fish movement (Benton *et al.* 2008, Mariano *et al.* 2012).

Other infrastructure such as dams can cause several negative impacts as well (see section on bauxite mining above). In addition to the existing Brokopondo Dam, plans exist to increase the capacity of the Afobaka energy generation facility by diverting the Tapanahoni River via Jai Creek to the Brokopondo Reservoir(see Figure 387) (Boksteen 2009). This would aggravate existing environmental effects and add new ones, such as low water levels in the Marowijne River and introductions of endemic fishes of the Marowijne River to the Suriname River and vice versa.

Agriculture. In Suriname, agricultural activities are limited to areas that are inhabited and accessible by road or river. Intensive permanent agriculture is concentrated in parts of the Coastal

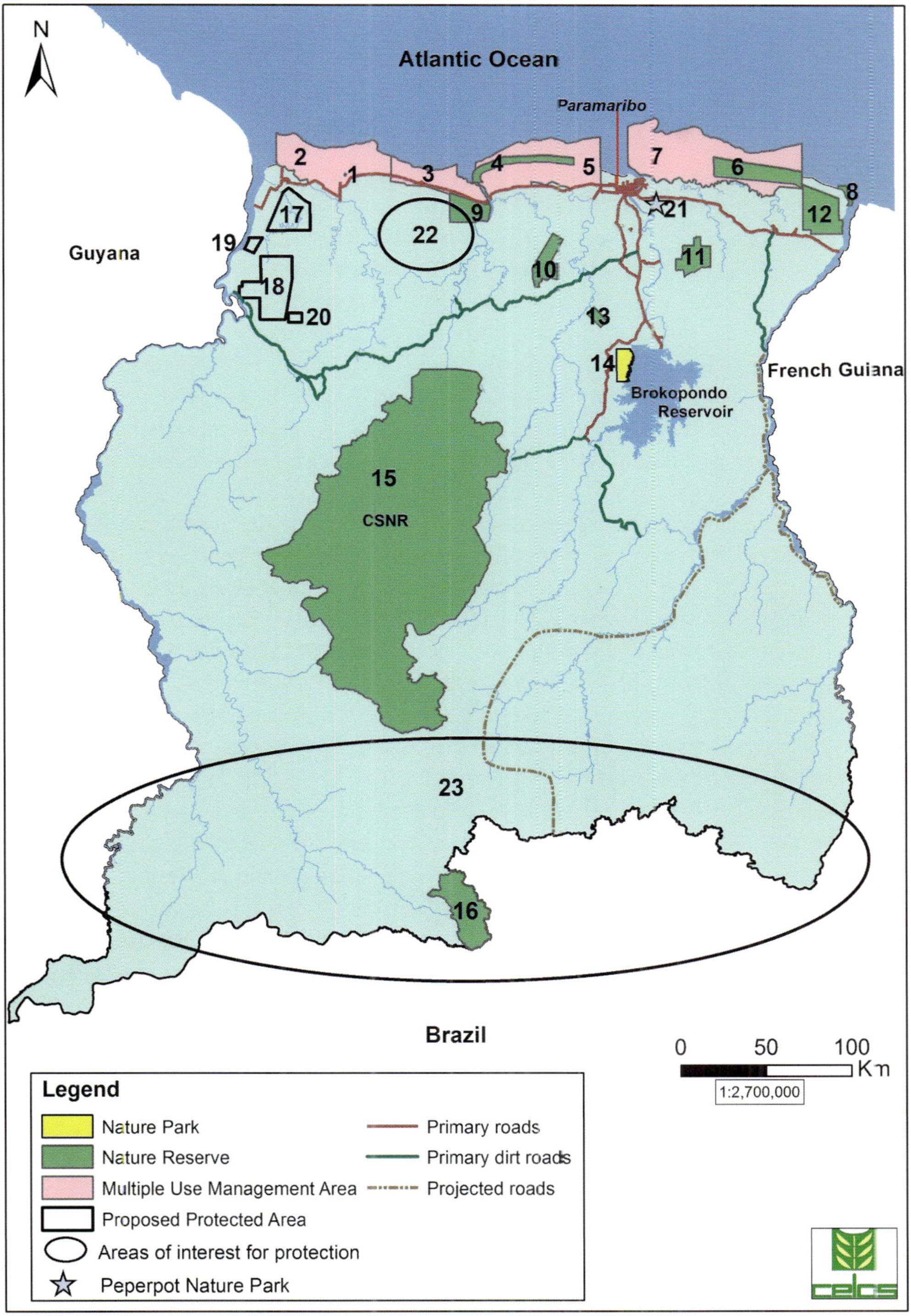

Figure 389. Locations of formal, informal and proposed protected areas, as well as primary and projected roads in Suriname (numbers refer to those in table 16).
* Central Suriname Nature Reserve

Plain (on an estimated 1 to 2 percent of Suriname's land surface), extensive slash-and-burn agriculture in parts of Interior (another 1 to 2 percent), such as the area along the Upper Suriname River south of Brokopondo Reservoir (see Figure 387). Agriculture, especially industrial agriculture, is the leading source of pollution in many countries. Pesticides, fertilizers and other agrochemicals are introduced into the environment either by aerial spraying or leaching of contaminated water (WWF 2016a, b). In recent years, Suriname has imported over 33 million kg of fertilizer and more than a million liters of pesticides annually (Algemeen Bureau voor de Statistiek 2014). Great quantities of these chemicals are used in rice farming in the Nickerie and Coronie districts, altering the water quality of, for instance, the Bigi Pan area (Van der Steege 1990). Vermeer *et al.* (1974) found that spraying of pesticides against snails in the Nickerie District resulted in substantial fish and bird mortality. The situation regarding the use of pesticides in Suriname seems to have improved over the last decades (Jan Mol, personal observation). Applying large quantities of fertilizers may cause eutrophication of surface waters – an overabundance of nutrients which can cause harmful algal blooms, and a resulting loss of oxygen in the aquatic environment (Chorus & Bartram 1999). Even with increasing pesticide use, there is still no treatment of polluted agricultural runoff (Mahabali & Spanoghe 2013).

Forestry. In Brazil, logging results directly in high levels of forest canopy damage and indirectly in widespread deforestation. High levels of canopy damage and deforestation typically occurred within 25 km of the main access roads (Asner *et al.* 2006). Deforestation contributes to carbon emissions and a loss of carbon sinks, contributing to climate change (Fonseca *et al.* 2007). While widespread forest canopy damage along access roads is common in Suriname, deforestation is not (De Dijn, personal observation). In Suriname forestry is currently limited to the forestry belt and is strictly managed by the Foundation for Forest Management and Production Control (SBB). SBB was created in the 1990s, in a tumultuous period when several Asian companies were vying – ultimately unsuccessfully – for huge logging concessions in Suriname (Sizer & Rice 1995). Illegal logging and timber overexploitation is an issue in parts of Suriname; examples are the illegal harvest of mangrove trees by fishermen for poles and fuelwood, and the overharvesting of *watrakan* (*Cordia* spp.) and *Cedrela odorata* in the coastal area (Playfair 2007).

Fisheries. Fisheries contribute greatly to global food security and income, but can damage aquatic ecosystems in the absence of sustainable management. Overfishing has become an increasing problem worldwide with fish stocks declining by over 50 percent between 1970 and 2010 (WWF 2015). In Suriname, overfishing has depleted Southern Brown Shrimp (*Penaeus subtilis*) populations (Chin-A-Lin & Yspol 2000); their populations have failed to recover (Charlier 2001, Miglino *et al.* 2005). The destruction of breeding grounds by dragging bottom-trawl nets over the sea floor in shallow waters may have also contributed to the depletion of shrimp (Hornby *et al.* 2015). Stocks of the large catfish *yaribaka* (*Sciades parkeri*) may also be depleted, as evident from high prices and rarity of large specimens at local markets. Bycatch – the landing of species that are not the principal targets of the fishermen – is estimated to account for as much as 80 percent of total catches in Surinamese waters (Hornby *et al.* 2015, Kelleher 2005). The bycatch consists of unwanted species, juveniles of commercial species, marine turtles (WWF 2011) and rays (Willems *et al.* 2016c), but also some species that are then sold opportunistically (Clucas 1997).

Hunting, collecting and wildlife trade. Worldwide, many people depend on wildlife hunting and trade, whether for food, medicines, rituals or cash income. Properly managed and regulated systems of hunting and trade in wildlife specimens can potentially result in sustainable harvest and contribute to conservation. It can do this by engaging hunters and collectors and protecting the habitat of the hunted species. However, a large proportion of hunting and collecting activities is unsustainable and often illegal (Roe 2008). Moreover, wildlife trafficking is estimated to be the fourth largest illegal international trade in the world, behind trafficking in drugs, counterfeit goods and humans (McLellan & Allan 2015).

In Suriname, hunting and the pet trade are increasingly threatening targeted species (Griffith *et al.* 2016). Subsistence hunting for food has historically been important for Indigenous and Maroon communities (Bakker *et al.* 1998, Fontaine *et al.* 1980, Versteeg 2003, Price 2011). However, the impacts are increasing because of the growth of these populations (Heemskerk & Delvoye 2007, Heemskerk *et al.* 2007, Ouboter 2000), the use of shotguns and motorized boats, and the bush meat trade. Ouboter *et al.* (2010) found that mammals and reptiles are most commonly sold as bush meat in local markets, while birds are often sold as pets. Sports hunting is also becoming more popular in areas easily accessible by roads, rivers or creeks (Ouboter 2000). As many as 1500 to 2000 hunting permits were granted annually to sport hunters between 2009 and 2013 (Algemeen Bureau Statistiek 2014).

The pet trade focusses mainly on amphibians, reptiles, birds and monkeys (Ouboter 2000). However, some of Suriname's endemic freshwater fishes are also caught for a specialized segment of this trade, the aquarium trade. An example is the beautiful river stingray of the Corantijn River *Potamotrygon boesemani* (see Freshwater Fishes chapter). The local market is supplied with parrots, song birds, monkeys and sloths (Chantal Landburg, personal observation), foreign markets with a much larger variety of animals. In the period 2009-2013, US$ 900,000 worth of animals was exported annually, with the highest revenues coming from selling parrots and snakes (Algemeen Bureau Statistiek 2014, Kerman 2010). Animal traders in Paramaribo often buy animals from indigenous animal trappers for whom this is an important source of income; however, the price paid for the animals to the trappers is very low (Heemskerk & Delvoye 2007). A worrying recent development is the illegal hunting of Jaguars (*Panthera onca*) and trade in their meat and body parts on both national and international markets (Kerman 2010).

There are several examples of how hunting and pet trade impact the population of targeted animals. Caimans are absent within a 60 km radius of the Indigenous village of Kwamalasamutu (Ouboter 1989). The *twatwa* (*Oryzobourus crassirostris*), a beloved song bird from the savannas, has virtually disappeared from the wild in Suriname due to over-collecting (Ouboter *et al.* 2010, Griffith *et al.* 2016). The blue 'azureus' and other unique populations of the poison dart frog *Dendrobates tinctorius*, are collected for the pet trade and extremely vulnerable because of their restricted distribution in the Interior. Marine turtles make their nests on the oceanic beaches of Suriname; although it is forbidden to hunt the turtles or harvest their eggs, poachers continue to collect their eggs. In the absence of patrolling by game wardens all eggs may be poached on certain beaches (Hilterman & Goverse 2006). The Eilanti beach near the mouth of the Marowijne River was an important site of mass nesting (so-called arribadas) of the Olive Ridley turtle (*Lepidochelys olivacea*); these arribadas no longer occur, most likely due to the massive poaching of the highly prized eggs of this species over many decades (Goverse 2003).

Invasive species. With people and goods constantly moving across the globe, wildlife has been transported along, purposefully or inadvertently transplanting species to new environments (for a review, see Lockwood *et al.* 2013). Species introductions in Suriname include, among others, ruderal herbs which such as weeds and escaped ornamentals and useful plants (Ostendorf 1962), some worms, ants and other invertebrate animals, many of which are pests (van Dinther 1960; Bart De Dijn, personal observations), fish (Lijding 1958, Mol & van der Lugt 1996), gecko's (Hoogmoed 1981) and the mongoose (*Mangusta auropunctata*; Husson 1978). A few examples are discussed in more detail below.

Marine and estuarine fishes are often introduced when large ships discharge ballast water they took in across the ocean. Manmade connections can also lead to introduction, such as canals connecting different river systems and allowing aquatic species to move into non-native river systems. An example is the connection of the Suriname and Saramacca rivers by the Saramacca canal; similarly, if the so-called Tapajay project (Boksteen 2009) is implemented, a connection will be created between the Suriname River and the Tapanahoni River (part of the Marowijne River system). Non-native species entering a new area often fail to survive or reproduce, but if they do, they may become invasive (Lockwood *et al.* 2013).

The various effects of invasive fish species include predation on native species (often on eggs and young), competition with native species for food and habitat, and transmission of diseases and parasites. Examples of invasive fish species in Suriname are the tilapia *Oreochromis mossambicus*, a food fish imported for aquaculture (Lijding 1958), and the Amazon Sailfin Catfish *(Pterygoplichthys disjunctivus)*, an ornamental fish imported by aquarium hobbyists (Jan Mol, personal observation). The tilapia species now dominates the fish fauna of coastal drainage canals, for instance in Paramaribo, and brackish lagoons, for instance Bigi Pan (Mol & van der Lugt 1996). The Amazon Sailfin Catfish seems to outcompete local *Hypostomus* spp. (*warawara*), some of which are endemic to Suriname; it burrows in the banks of drainage canals and natural streams, in doing so destabilizing and eroding the banks (Nico *et al.* 2009; Jan Mol, personal observation). Of the many different reptiles introduced in Suriname, the gecko *Hemidactylus mabouia* (*kamrawenke*) has successfully expanded its range and can be found in almost every house, especially in the Coastal Zone (Hoogmoed 1981; Chantal Landburg, personal observation).

Climate change. Climate change is increasingly threatening the ecosystems of our planet (see for example McGranahan *et al.* 2007), including those in Suriname. Climate models predict an increase in average atmospheric temperature, reduced average annual rainfall, increased sea surface temperatures, and sea-level rise in Suriname (CARIBSAVE 2012, Conrads 2015; see also Climate and Hydrology chapter). A combination of reduced rainfall and sea-level rise can cause salinization and vegetation die-off in ecosystems such as marshes and a retreating coast (loss of land and Coastal Plain ecosystems). Mol *et al.* (2000) found that the El Niño Southern Oscillation (ENSO), which in Suriname is associated with reduced rainfall, resulted in the drying up of the Bigi Pan lagoon and subsequent die-off of ariid catfish, followed by snook (*Centropomus* spp.), tilapia (*Oreochromis mossambicus*), mullet (*Mugil* spp.) and tarpon (*Megalops atlanticus*) in 1997-1998. Although ENSO is not the same as climate change, the shared effect of reduced rainfall provides insight into some possible impacts of climate change.

Transfer of knowledge. A subtler yet important threat to ecosystems and wildlife species is the limited transfer of both traditional and scientific knowledge in Suriname. This concerns knowledge about the structure and functioning of ecosystems, how and where the species live, and how both ecosystems and species can be used, managed and maintained. In general, traditional Indigenous and Maroon knowledge is not sufficiently transferred to younger generations (Plotkin 1993, Heemskerk & Delvoye 2007, Heemskerk *et al*. 2007, Hoffman 2009). Organizations such as ACT are trying to remedy this (**www.ethnobiobase.act-suriname.org**). The fading role of Indigenous people in the conservation of savannas through burning practices is also taking its toll. Savannas that are no longer burned are progressively invaded by surrounding forests (Teunissen 1980a) and scientific knowledge of managed burning practices to maintain savanna is lacking. Suriname has a rich research history and, in recent times, has been attracting more international researchers. Yet the historical and recent science-based knowledge appears to be poorly transferred across generations and from expatriates to home-grown academics (Bart De Dijn, personal communication; Jan Mol, personal observation).

Conservation

Landscapes and ecosystems. Conservation can be focused on multiple levels: landscapes, ecosystems, species, and genetic resources. Conservation of larger natural landscapes and ecosystems, setting them aside with minimal interference by humans, protects multiple species or even ecosystems. This also protects natural ecological and evolutionary processes (Miller & Spoolman 2016); for example, conserving Suriname's estuarine ecosystems protects the connection between the sea and upriver freshwater reaches of major rivers, thus allowing certain marine and freshwater fishes to spawn in the estuary or periodically return there to feed. Intact ecosystems provide valuable environmental and socio-economic services to people. These services are easily lost as a result of disturbance and degradation, and are very hard to regain through often costly rehabilitation and restoration (Holmlund & Hammer 1999). Economic benefits from ecosystem services should not be underestimated: many people's daily lives, food and income depend on them. For example, over 70 percent of total fisheries landings in Suriname originate from artisanal fisheries in the estuaries of the Suriname and Commewijne rivers, totaling over 22,000 tons of fish in 2011; this industry not only provides an income to people – including some of the poorest and most deprived – but also cheap food for most inhabitants of Suriname (CARICOM Fisheries Unit Belize 2000, Ministerie van Landbouw, Veeteelt en Visserij 2013). These important estuarine fishing grounds are also increasingly popular with tourists, thus adding to their value. Dolphin and sea turtle tours provide income not only for tour operators, but also for boatmen, tour guides, restaurants, and even a dolphin research program of the local NGO Green Heritage Fund Suriname (Chantal Landburg, personal observation). Keeping ecosystems intact is therefore important from both an environmental and economic point of view.

Inaccessibility often helps to protect ecosystems. Once an area becomes accessible, especially by road, a chain reaction of harmful activities typically follows, causing fragmentation, degradation and destruction of ecosystems (Laurance *et al.* 2009, Miller & Spoolman 2016). Since most of Suriname's infrastructure is concentrated in the Coastal Plain (see figure 389), the inaccessible Interior is to some extent protected by default.

#	Name	Area in km²
Formal PAs in brackish-freshwater part Coastal Zone		
1	Hertenrits NR	1
2	Bigi Pan MUMA	679
3	North Coronie MUMA	272
4	Coppename-monding NR	120
5	North Saramacca MUMA	884
6	Wia-Wia NR	360
7	North Commewijne-Marowijne MUMA	615
8	Galibi NR	40
Formal PAs in strictly freshwater part Coastal Zone		
9	Peruvia NR	310
10	Boven Coesewijne NR	270
11	Copi NR	280
12	Wanekreek NR	450
Formal PAs in Interior (Precambrian basement)		
13	Brinckheuvel NR	60
14	Brownsberg NP	120
15	Centraal Suriname NR	15920
16	Sipaliwini NR	1000
Previously proposed protected areas (not effective)		
17	Nani PP	540
18	Kaburi kreek PP	680
19	Mac Clemen PP	60
20	Snake Creek PP	40
Private protected areas (informal)		
21	Peperpot NP	26
Areas recently proposed for protection (not effective)		
22	Coronie zwamp PP	Unknown
23	South Suriname Conservation Corridor	Unknown

Table 16. Formally recognized, informal and proposed protected areas (PAs) in Suriname.
NR = nature reserve, MUMA = multiple-use management area, NP = nature park, PP = proposed PA.

Suriname has a system of legally protected areas, which was designed from the 1950s onward to include as much of the diversity of the country's natural ecosystems as realistically possible (Teunissen 2003). However, Suriname's system of protected areas primarily targets terrestrial ecosystems. Aquatic (stream) ecosystems need protection based on their catchment (Sheldon 1988; see below CSNR) and especially headwaters (Saunders *et al.* 2002) taking into account connectivity. At present, the system includes 11 nature reserves, 1 nature park, and 4 multiple use management areas. Four additional areas were proposed for protection, but still have no formal status (Table 16; Figure 389) (Ouboter 2002, Teunissen 2003). The system of protected areas is based on the Nature Preservation Act ('Natuurbeschermingswet'; Government Bulletin 1952, No.26), which is the basis for the establishment of nature reserves. The Decree Issuance of State-Owned Lands ('Decreet Uitgifte Domeingrond'; Government Bulletin 1982, No.11) has been used to establish the Brownsberg Nature Park and the multiple use management areas. The Central Suriname Nature Reserve (CSNR), a consolidation and expansion of three previously existing nature reserves (Ouboter 2002), is not only protected by national law, but has also been registered as a UNESCO World Heritage site since 2000 (WHC Nomination Document 1017; Ministry of Natural Resources 1999). This listing was approved because of CSNR's unique potential to protect ecosystems and species, many of which are globally threatened or endemic (unique) to the Guiana Shield, Suriname or even just the Coppename River. Its large size protects most, but unfortunately not all, of the Coppename River watershed and covers a wide range of lowland and upland landscapes and ecosystems, meaning that it also incorporates a high floral and faunal diversity. In 2009, a small (820 ha), privately owned nature park was set up to protect part of the abandoned historical Peperpot plantation (see **www.plantagepeperpot.nl**; Ramcharan 2011). Conservation International and several partners, including WWF, are currently lobbying for some form of protection for the large Coronie Swamp and the South Suriname Conservation Corridor (Ramirez-Gomez *et al.* 2016). There are currently no marine protected areas in Suriname; in January 2017, WWF and partners launched an initiative to protect 10 percent of Suriname's marine territory by 2020. In recent years, progress in establishing new protected areas has been slow, and additional work needs to be done in Suriname to meet the 2020 objective (under the Convention of Biological Diversity; CBD 2017) of assigning 17 percent of total land surface to protected areas and ensuring effective protection. Land use and ownership claims by local communities (see for example IACHR 2015) and uncertainties as to the socio-economic benefits of protected areas have tended to slow things down.

In areas such as the Central Suriname Nature Reserve (CSNR), pressures on the environment are generally low due to poor accessibility, lack of evidence of the presence of highly valuable mineral resources, and few people living in the area. The government can effectively ensure the conservation of these areas by not allowing the building of access infrastructure, and not granting concessions in the area and its immediate surroundings. In areas under high pressure, typically more accessible areas with either a larger resident population or some valuable commodity, conservation requires a more active approach. An example would be the Brownsberg Nature Park which is under severe pressure due to illegal gold mining (see for example Arets *et al.* 2006, De Dijn *et al.* 2005). Comprehensive management plans are important to guide conservation action in protected areas. So far, the majority of protected areas in Suriname have a management plan, but implementation remains a challenge. Inadequate management is mainly due to limited capacity, budget, political will and community involvement (Symbiont Consulting 2007, Hendrison 2009). Innovative approaches, including more transparent and participatory management systems (Thomas & Middleton 2003), may result in better conservation outcomes. In particular, consulting with local communities and involving them in the management of protected areas can support self-governance and sustainable livelihoods, while conserving their environment (Ramirez-Gomez *et al.* 2016). Engaging with the private sector may also create new possibilities, not only to increase commitments to environmentally sound business practices, but also to contribute to sustainable financing of protected areas (see for example Hardner *et al.* 2015, on corporate offset and compensation for environmental impacts). Business models can be based on best practices examples (such as Sutton 1999, Stronza & Durham 2008, Wei 2014).

Protected areas can easily become less disturbed natural islands in a sea of disturbed or manmade ecosystems (Miller & Spoolman 2016). Isolation from similar natural areas reduces or stops wild species gene flow into the protected area and makes their populations vulnerable to inbreeding and extinction (Reece *et al.* 2009). Wild animals may be harmed if they venture outside the protected area; if this happens on a regular basis, populations inside the protected area may ultimately crash. In French Guiana, for example, a marine protected area has proved to be insufficient to protect the IUCN red listed Atlantic Goliath Grouper (*Epinephelus itajara; granmorgu*) because this fish migrates outside of the protected area (Artero *et al.* 2015). It is therefore important to maintain or create connections between similar protected areas, and establish buffer zones around these areas where human activities are limited (Reece *et al.* 2009, Miller & Spoolman 2016).

Species. Species extinction is a worldwide problem, and conservation efforts are often required that specifically target species that are endangered or endemic, especially when endemic to a small area (see for example Miller & Spoolman 2016). Suriname is home to many species that are endemic to the Guiana Shield, to the country or even to particular parts of the country (see examples in for instance the Freshwater Fishes chapter). Several species are listed as endangered by the International Union for the Conservation of Nature (IUCN), the Convention on International Trade in Endangered Species of Wild Fauna and Flora (CITES), or both (see below). In Suriname, the Game Act ('Jachtwet'; Government Bulletin 1954, No.25) offers protection to animal species, along with the Fish Protection Act ('Visstandbeschermingswet'; Government Bulletin 1961, No.44; Rondeel 1965). The Forest Management Act ('Wet Bosbeheer'; Government Bulletin 1992, No. 80) regulates the harvest and management of timber and non-timber plant species, and related older laws identify a number of protected tree species. Furthermore, multiple international treaties ratified by Suriname, notably CITES, also provide protection. However, laws and regulations to protect endangered species need better enforcement. In addition, it is important to actively involve local people in conservation efforts, reducing potential conflicts and empowering them to conserve their natural resources (Sobrevila 2008). This involvement of local people in decision-making is slowly materializing in Suriname (see for example Ramirez-Gomez *et al.* 2016), but has yet to become more structured and embedded in national legislation.

The Nature Conservation Division (NCD) of the Ministry of Spatial Planning, Land and Forest Management (Min. RGB) is the designated protected area and wildlife management authority of Suriname. It enforces the Game Act and the Fish Protection Act and supports the implementation of the Forest Management Law as to non-timber plant species. The NCD employs game wardens who exercise control in the field. The NCD is also the national CITES authority of Suriname, and applies collecting and export restrictions (see **www.cites.org** for more information).

The IUCN Red List (**www.iucnredlist.org**) assesses the global conservation status of wild species and certain populations of these species. It highlights those that are threatened with extinction, and promotes their protection. Species are assigned a status from least concern through near threatened, vulnerable, endangered and critically endangered to extinct in the wild and extinct (some are classified as data deficient). Many governments, NGOs and scientific institutions use this listing as a reference for conservation decision-making. IUCN (2012) list species as either:

- Extinct (EX) when there is no reasonable doubt that the last individual has died,
- Extinct in the Wild (EW) when it is known only to survive in cultivation, in captivity or as a naturalized population (or populations) well outside the past range,
- Critically Endangered (CR) when it is considered to be facing an extremely high risk of extinction in the wild,

- Endangered (EN) when considered to be facing a very high risk of extinction in the wild,
- Vulnerable (VU) when it is considered to be facing a high risk of extinction in the wild,
- Near Threatened (NT) when it has been evaluated against the criteria but does not qualify for Critically Endangered, Endangered or Vulnerable now, but is close to qualifying for or is likely to qualify for a threatened category in the near future,
- Least Concern (LC) when it has been evaluated against the criteria and does not qualify for Critically Endangered, Endangered, Vulnerable or Near Threatened: widespread and abundant taxa are included in this category, and
- Data Deficient (DD) when there is inadequate information to make a direct, or indirect, assessment of its risk of extinction based on its distribution and/or population status.

Genetic resources. The Convention on Biological Diversity recognizes the need for conservation of genetic resources, meaning the variety of genes and genetic information contained in organisms (**www.cbd.int**; CBD 2017). Suriname has developed a National Biodiversity Strategy and Action Plan to regulate and monitor access to genetic resources and associated traditional knowledge, and to ensure equitable benefit-sharing (Ministry of Labour, Technological Development, and Environment 2006, 2013b). The conservation, sustainable use and development strategy of plant genetic resources has been formulated in accordance with the International Treaty on Plant Genetic Resources for Food and Agriculture (ITPGRFA) (Milton 2012). In line with these plans, the Center for Agricultural Research in Suriname (CELOS) implements a project in cooperation with indigenous and Maroon communities for the conservation of agro-biodiversity (see Figure 390; Small Grants Programme 2016).

Figure 390. Plantain, tubers (cassava and sweet potato) and seeds (pumpkin and cucumber) of plant species conserved in the context of CELOS project (image © Chantal Landburg).

Sustainable use. Strict conservation in protected areas and targeting specific endangered and endemic species has a limited impact. Regulating the sustainable use of biological resources outside protected areas can help to conserve a wider range of ecosystems and species. In Suriname this is promoted by means of sustainable forest management and certification, and fisheries certification.

In 1978, a cooperation agreement was signed between the Anton de Kom University of Suriname and Wageningen University in the Netherlands to develop a sustainable forest management system. Experimental stations were set up at Mapane and Kabo in the district of Para, and a silvicultural system was developed: the CELOS Management System (CMS; review in Werger *et al.* 2011). The CMS was never applied at a commercial scale in Suriname and remains in an experimental stage. The rules and regulations set by the forest management authority SBB aim to ensure forestry is sustainable and require concessionaires to produce a management plan for sustainable harvest (see **sbbsur.com/bosbeheer/richtlijnen-houtkap**), but at present do not adequately take collateral damage control into account (Bart De Dijn, personal observation) . Recently, several logging companies in Suriname have started applying reduced impact logging. This involves carrying out pre-harvest inventories of timber trees, careful planning of roads, skid trails and actual logging using low-impact techniques, and conducting a post-harvest assessment (see **www.itto.int/feature15**). Some logging companies have chosen to seek certification of their operations by the Forest Stewardship Council (FSC). This certification system strives to ensure environmental, social, and economic benefits based on agreed principles and criteria and carefully developed guidelines (see **ic.fsc.org/en/certification**). At present, over 400,000 ha of production forests in Suriname are FSC certified (Figure 387; see **info.fsc.org/certificate.php**; van Benthem & van Dijk 2015). The commercial harvest of NTFPs in Suriname can also be sustainable (van Andel & Havinga 2008) and NTFP trade from well-managed forests may be both sustainable and profitable (van Andel *et al.* 2003, Sizer and Rice 1995).

In the fisheries sector, two independently operating companies harvesting Atlantic Seabob Shrimp (*Xiphopenaeus kroyeri*) in Suriname received Marine Stewardship Council (MSC) certification (see **www.seabob.sr/msc-certification**) in November 2011. MSC certification provides assurance that a fishery is well managed and that harvest levels and impacts on marine ecosystems are sustainable. The companies worked with government and conservation NGOs to develop management and research plans to assess the fishery's impacts and adjust fishing effort if necessary to avoid the depletion of stocks (Willems 2016). The shrimping fleet (Figure 391) is tied to a code of practice to ensure that operations at sea are in line with sustainable harvest and reduced impact (Southall *et al.* 2011; for the most recent certification renewal, see **https://fisheries.msc.org/en/fisheries/suriname-atlantic-seabob-shrimp**).

Environmental impact assessments. Another tool supporting measures to achieve sustainability and conservation is environmental impact assessment (EIA). EIAs help to assess the future consequences of a proposed action or project, for instance the development of a new mine or construction of a new road, and seek ways to mitigate any negative impacts on the environment (see **www.iaia.org**). It is important that an EIA does not focus on direct and local, short-term impacts only, but also includes indirect, long-term, cumulative, large scale impacts (Angermeier *et al.* 2004). EIAs are not mandatory by law in Suriname, yet ministries often request advice from NIMOS – the national environmental protection agency – on specific projects before granting a permit. NIMOS can assess, based on their own guidelines (NIMOS 2005), if and to what extent the proposed projects need to be subject to an EIA, and NIMOS advises the ministry based on its outcome. At present, multinational companies operating in Suriname often implement EIA procedures as part of their own environmental and corporate social responsibility policies.

A look ahead. Suriname is endowed with a rich environmental resource base and a relatively small population, which is ideal to achieve sustainable freshwater, food and climate security for current and future generations. The globally important landscapes and seascapes of the country are generally in good condition. A strong environmental governance system is key to an economy based on sustainable natural resource extraction. Innovative ways should be explored for more transparent and participative natural resource management systems with delegated responsibilities ('ecosystem management'; Grumbine 1994, Willems 2016; but see Simberloff 1998, Lindenmayer *et al.* 2007). Local communities should be closely involved and empowered. Engaging with the private sector will also create new possibilities. Although structured payment for ecosystem services (PES) can offer a range of opportunities to cope with the environmental challenges in the region, it is still a rather new concept and relevant national policies are generally lacking. Nature-based tourism (Olsder 2004), REDD+ and other low carbon development strategies provide opportunities for PES in the near future.

The government of Suriname, traditional and scientific communities, NGOs and other stakeholders should grasp the opportunity to build on a broad existing knowledge base and work progressively towards sound management of Suriname's rich natural ecosystems. The current book is an important step toward providing essential information on Suriname's natural history and ecology to the general public, and will hopefully contribute to informed decision-making for the sustainable development of our beloved Suriname.

Glossary

Acrodont teeth – teeth attached on top of the jawbone without sockets

Aestivate – state of animal rest in response to high temperatures or dryness

aff. – refers to a probably undescribed species that is closely related to another one that has been described; so *Hura* aff. *crepitans* would be a tree related to *Hura crepitants*, but distinct from *Hura crepitants*

Amplexus – mating embrace of frogs in which the male is on the back of the female to fertilize the eggs

Anomaly (as to climate) – climatological conditions that diverge significantly from the norm (long term average)

Anthropophilic – preferring the vicinity of human beings or the resulting habitat changes

Atria – upper chambers of the heart that receive the blood from the veins and pump it to the ventricles (larger, lower chambers of the heart)

Biomass – the total mass of organisms; in studies of plants or the vegetation typically determined destructively, by collecting all plant material and drying it in a stove, then weighing it

cf. – refers to an organism whose identify has not been confirmed, but looks similar to a described species; so bees referred to as *Melipona* cf. *fulva* look very similar to the stingless bee *Melipona fulva*

Cloaca – the opening in the body of animals, such as amphibians, reptiles and birds, which serves both to get rid of waste products (digestive system, kidneys), and to release sperm or eggs, or to copulate

Cryptic – hidden or difficult to discern; for more context: see introduction of Amphibians and Reptiles chapter

Desmids – a type of green algae

Detritus – decaying organic matter of mostly plant origin

Dewlap – flap of skin under the throat of some lizards

Diatoms – algae, mostly unicellular (individual consists of just a single cell), that form a shell

Digital lamellae – microscopic hairs under the toes of some lizards

Dorsal – the upper parts of the body

Dorso-lateral – positioned between the upper parts of the body and the flanks

Dorso-ventrally flattened – flattened in such a way that the upper and lower parts of the body are very wide and the flanks are small in height

Duricrust (as to geology / soils) – hardened layer in the soil, which can be several meters thick, that contains a high concentration of iron and/or aluminum-rich minerals

Ecosystem – an (entire) biological community – the assemblage of (all) species populations (plant, animal, other) which occur together in space and time – and its (non-living) physical environment

Ectothermy – optimal body temperature is reached through external sources, mostly sunlight, not through internal physiological processes

Endemic (as to species) – naturally occurring in a specific area only (as opposed to cosmopolitan); restricted range endemic (RRE) is a term used for species that occur in particularly small areas only

Epiphyte – plant that is not a parasite, but does grow on part of another plant, typically on a woody plant part, such as on a tree stem or brach (see also Box 7 in Plants chapter)

Estuary – part of a river and nearby flooded land whose water is in contact with the sea, and where the water level goes up and down with the tides; this is typically an area with brackish water, but the salinity of the water is often quite variable (daily and/or seasonally) (for more context see Ecosystems chapter)

Evapotranspiration – evaporation of water from water bodies, as well as ground and plant surfaces (evaporation), as well as water vapor released by plants (transpiration) via the leaf stomata (tiny openings in the leaf surface)

Genus – see Box 4 in first chapter

Gondwana – the continent that was unbroken till roughly 200 million years ago, and gave rise – after its gradual breakup over a period of more than 100 million years – to present-day South America, Africa, Antarctica, India and Australia

Greenhouse gases (as to climate) – gases that contribute to the warming of the atmosphere, such as carbon dioxide and methane

Habitat – the location where on organisms occurs, often characterized in broad ecological terms, such as climate / elevation, and vegetation type or type of water body / water quality

Heliotherm – organism reaching its optimal body temperature through basking in the sun

Hemotoxic – having toxins that damage or destroy tissues, including blood cells

Holocene – the most recent period (till present) of about 10,000 years with a relatively warm global climate

Hydrophyte – aquatic plant, adapted to open water or flooding of the soil and (most of) its root system

Hypersaline – salinity of a water body that is higher than that of the sea water of the open ocean (see Ecosystems chapter for more context)

Invasive (species) – species that are not native to a certain area, yet easily disperse and establish themselves in new areas, while doing so displacing the local native species; invasive species are typically transported over long distances – knowingly or inadvertently - by people, such as via planes and ships

Lateral – positioned at the flanks

Laterite / lateritic – strongly weathered rock that is reduced to mostly clay and related minerals rich in iron and aluminum, typically with a reddish color due to the oxidation of iron; see also Geology, Landscape and Soils chapter

Lineage – see Box 4 in first chapter

Litter (forest / leaf litter) – fallen (dead) leaves, twigs and other woody and non-woody plant parts that remain on the soil in relatively intact form, at least temporarily (not yet broken down to humus)

Macro-invertebrates – larger invertebrate animals, such as insects, clearly visible to the unaided human eye

Macrophytes – larger plants, clearly visible to the unaided human eye

Mesic – 'favorable' conditions on land in terms of water availability - not too dry, not too wet – which are suitable for a wide variety of organisms

Metamorphic / prefix 'meta' (as to geology) – rock that has (partly) melted and re-crystallized after its original formation due to exposure to great pressure and/or heat, typically deep in the earth's crust

Mimicry – similarity to another species or part of it for protection against predators or predation on another species

Opisthoglyph – colubrid snakes with enlarged, grooved teeth in the posterior maxilla, connected to the Duvernoy's gland (gland located behind the eye in colubrid snakes producing a secretion that helps with swallowing and digestion)

Oviparous – reproduction by laying eggs that hatch outside the body of the mother

Ovoviviparous – reproduction by retaining eggs in the ovary till the time of hatching; the embryos are nourished by the egg yolk, not by the mother

Paratoid gland (in amphibians) – gland above the ear drum, behind the eye excreting a toxic, alkaloid substance

Pelvic bone – the pelvic bone is in most animals composed of several bones and forms a connection between the backbone and the hindlegs

Phoretic – using a larger animal as a 'ride' to disperse, typically to reach a food source

Pleistocene – the most recent period of about 2 million years, minus the Holocene, characterized by dramatic cyclical changes in the global climate, from cold periods during which massive, extensive ice caps form, to warmer periods during which these caps retreat

Pleurodont teeth – teeth attached to the inner side of the jawbone without sockets

Poikilotherm – optimal body temperature is reached through external sources, mostly sunlight, not through internal physiological processes

Precambrian – geological period which lasted till some 540 million years ago, when there was no life – except maybe bacterial – on land, and only "primitive" life in the oceans

Rotifers – a group of small 'primitive' multicellular animals (individual consist of many cells that form tissues) that live in water as plankton or attached to substrates

Salinization – increase in (dissolved) salt content in a water body, towards levels of sea water or beyond (see also: hypersaline)

Salt gland (in reptiles) – gland excreting excess salt; located in the area of the eye

Sarcodina – amoebas and similar unicellular animals (individual consists of just one cell), most living in water

Scutes (in reptiles) - resemble scales but have a bony component in the skin and are often larger; they are common in crocodilians and turtles

Sexual selection – one of the partners, usually the female, has a sexual preference for a partner with specific characteristics, leading to the enhancement of this characteristic, such as brightly colored plumage in male birds

Snout-vent length – the measurement of (part of) the length of an animal from the tip of the snout to the cloaca or anus

sp. – refers to a species whose exact identity is unknown; so *Rattus* sp. would be a rat of the genus *Rattus*, but it's not known which species exactly

Species – see Box 4 in first chapter

spp. – refers to multiple species; so *Citrus* spp. would refer to several species of citrus trees (not just a single species)

Subspecies – see Box 4 in first chapter

Taxon / taxa – see Box 4 in first chapter

Thecodont teeth – teeth of the jawbones located in sockets

Thermoregulation - in ectotherm or poikilotherm organisms, the behaviour to reach an optimal body temperature, for instance by changing position to warmer or cooler locations

Ventral – the underside of the body

Ventricles – the large lower chambers of the heart that receive the blood from the atria and pump it to the lungs and other parts of the body

Vestigial limbs – limbs that have become very small and lost their function

Viviparous – reproduction by giving birth to living young; embryos are nourished through a special organ in the uterus of the mother

Vomeronasal organ (in reptiles) – organ for chemoreception (taste / smell perception) located in the roof of the mouth

Webbed – having skin between the toes and/or fingers; this skin can extend to the tip of the digits or only partly so

Xeric – dry from a physiological perspective, meaning that there is a shortage of available water on land for organisms

Xerophyte – plant adapted to a shortage of available water, such as by having small, hard leaves that lose less water, or by having tissue (in leaves, stem or root) that can store water

Information sources / references

Abee, C. R. 1989. The squirrel monkey in biomedical research. Journal of the Institute for Laboratory Animal research (ILAR) 31:11-20.

Abell, R., M. L. Thieme, C. Revenga, M. Bryer, M. Kottelat, N. Bogutskaya, B. Coad, N. Mandrak, S. C. Balderas, W. Bussing, M. L. J. Stiassny, P. Skelton, G. R. Allen, P. Unmack, A. Naseka, R. Ng, N. Sindorf, J. Robertson, E. Armijo, J. V. Higgins, T. J. Heibel, E. Wikramanayake, D. Olson, H. L. Lopez, R. E. Reis, J. G. Lundberg, M. H. Sabaj Perez, and P. Petry. 2008. Freshwater ecoregions of the world: a new map of biogeographic units for freshwater biodiversity conservation. BioScience 58: 403-414.

Abril, G., F. Guerin, S. Richard, R. Delmas, C. Galy-Lacaux, P. Gosse, A. Tremblay, L. Varfalvy, M. A. Dos Santos, and B. Matvienko. 2005. Carbon dioxide and methane emissions and the carbon budget of a 10-year old tropical reservoir (Petit Saut, French Guiana). Global Biogeochemical Cycles 19, GB4007. **doi: 10.1029/2005GB002457**

Acevedo, M., K. McGregor, R. Andressen, R. Hirma, and M. Ramirez. 1999. Relations of climate variability in Venezuela to tropical Pacific SST anomalies. *In* American Meteorological Society (Ed.). 10th Symposium on Global Change Studies, pp. 81-84. American Meteorological Society, Dallas, TX, USA.

Achee N. L., J. P. Grieco, P. Masuoka, R. G. Andre, D. R. Roberts, J. Thomas, I. Briceno, R. King, and E. Rejmankova. 2006. Use of remote sensing and geographic information systems to predict locations of *Anopheles darlingi*–positive breeding sites within the Sibun River in Belize, Central America. Medical Entomology 43: 382-392

Ackerman, J. D., M. Mesler, L. Karen, and A. M. Montalvo. 1982. Food-foraging behavior of male Euglossini (Hymenoptera: Apidae): vagabonds or trapliners? Biotropica 14: 241-248.

Adams, M., T. A. Raadik, C. P. Burridge, and A. Georges. 2014. Global biodiversity assessment and hyper-cryptic species complexes: more than one species of elephant in the room? Systematic Biology 63: 518-533.

Adams, P. J. 1999. *Choloepus didactylus*. Mammalian Species 621: 1-8.

Adis, J. 1993. Überlebensstrategien terrestrischer Invertebraten in Überschwemmungswäldern Zentralamazoniens. Verhandlungen des Naturwissenschaftlichen Vereins Hamburg (Neue Folge) 33: 21–114.

Adis, J., and M. S. Harvey. 2000. How many Arachnida and Myriapoda are there world-wide and in Amazonia? Studies on Neotropical Fauna and Environment, 35:139-141.

Afonso R., P. G. Lemes, R. A. Sarmento, P. J. B. Leite, M. P. Neto, and N. dos Anjos. 2014. First report of giant grasshopper *Tropidacris collaris* (Orthoptera: Acridoidea: Romaleidae) attacking plantations of *Acacia mangium* (Fabaceae) in Brazil. Journal of the Kansas Entomological Society 87:102-105.

Agosti, D., J. D. Majer, L. E. Alonso, and T. R. Schultz (Eds.). 2000. Ants: Standard Methods for Measuring and Monitoring Biodiversity. Smithsonian Institution Press, Washington, DC, USA, and London, England, UK.

Aguiar, N. O., and P. F. Bührnheim. 1998. Phoretic pseudoscorpions associated with flying insects in Brazilian Amazonia. Journal of Arachnology 26: 452–459.

Aiello, B. R., T. A. Stewart, and M. E. Hale. 2016. Mechanosensation in an adipose fin. Proceedings of the Royal Society B 283: 20152794. **dx.doi.org/10.1098/rspb.2015.2794**

Aleixo, A., and D. de Fátima Rossetti. 2007. Avian gene trees, landscape evolution, and geology: Towards a modern synthesis of Amazonian historical biogeography? Journal of Ornithology 148: 443–453.

Algemeen Bureau Statistiek. 2014. Environment statistics. Suriname in Cijfers no. 309 – 2014 / 10. Algemeen Bureau voor de Statistiek, Paramaribo, Suriname.

Algemeen Bureau Statistiek. 2015. Demografische data 2009-2014. Algemeen Bureau voor de Statistiek, Paramaribo, Suriname.

Allaby, M. 2009. A Dictionary of Zoology (3rd edition). Oxford University Press, UK.

Allwood, A. J., A. Chinajariyawong, S. Kritsaneepaiboon, R. A. I. Drew, E. L. Hamacek, D. L. Hankock, C. Hengsawad, J. C. Jipanin, M. Jirasurat, C. Kong Krong, C. T. S. Leong, and S. Vijaysegaran. 1999. Host plant records for fruit flies (Diptera: Tephritidae) in Southeast Asia. Raffles Bulletin of Zoology, supplement 7: 1-92.

Almeida, A. V., and C. A. G. Câmara. 2008. Distribution of grasshoppers (Othoptera: Acridoidea) in the Tapacurá ecological station (São Lourenço da Mata, PE / Brazil). Brazilian Journal of Biology 68: 21-24.

Almeida-Santos, S. M., and M. G. da Salomao. 2002. Reproduction in Neotropical pitvipers, with emphasis on species of the genus *Bothrops*. *In* G. W. Schuett, M. Höggren, M. E. Douglas, and H. W. Greene (Eds.). Biology of the vipers, pp. 445-462. Eagle Mountain Publications, Eagle Mountain, Utah, USA.

Alonso, L. E., and J. H. Mol (Eds.). 2005. A rapid biological assessment of the Lely and Nassau plateaus, Suriname (with additional information on the Brownsberg Plateau). RAP Bulletin of Biological Assessment 43, Conservation International, Arlington, VA, USA.

Alonso, L. E., and T. H. Larsen (Eds.). 2013. A rapid biological assessment of the Upper Palumeu River watershed (Grensgebergte and Kasikasima) of southeastern Suriname. RAP Bulletin of Biological Assessment 67, Conservation International, Arlington, VA, USA.

Altrichter, M., C. Drews, E. Carrillo, and J. C. Sáenz. 2001. Sex ratio and breeding of white-lipped peccaries *Tayassu pecari* (Artiodactyla: Tayassuidae) in a Costa Rican rain forest. Revista de Biología Tropical 49: 383-389.

Altrichter, M., C. Drews, J. C. Sáenz, and E. Carrillo. 2002. Presupuesto de Tiempo del Chancho Cariblanco (Tayassu pecari) en un Bosque Hu medo de Costa Rica. Biotropica 34: 126-143.

Alvarez, J., M. R. Willig, J. K. Jones, Jr., and W. D. Webster. 1991. *Glossophaga soricina*. Mammalian Species 379: 1-7.

Amatali, M. A. 1993. Climate and surface water hydrology. *In* P. E. Ouboter (Ed.). The Freshwater Ecosystems of Suriname, pp. 29–51. Kluwer Academic Publishers, Dordrecht, The Netherlands.

Amazon Conservation Team. 2004. The importance of Brazil nut (*Bertholletia exelsa*) harvesting and their uses in Suriname. Unit for Sustainable Development and Environment, Organization of American States, Paramaribo, Suriname.

Ambrizzi, T., E. B. de Souza, and R. S. Pulwarty. 2005. The Hadley and Walker regional circulations and associated ENSO impacts on South American seasonal rainfall. *In* H. F. Diaz, and R. S. Bradley (Eds.). The Hadley Circulation: present, past and future, pp. 203-231. Kluwer Academic Publishers, Dordrecht, The Netherlands.

Amici, F., Call, J., and F. Aureli. 2009. Variation in withholding of information in three monkey species. Proceedings of the Royal Society B 276: 3311-3318.

Amusant, N., C. Moretti, B. Richard, E. Prost, J. M. Nuzillard, and M. F. Thévenon. 2007. Chemical compounds from *Eperua falcata* and *Eperua grandiflora* heartwood and their biological activities against wood destroying fungus (*Coriolus versicolor*). Holz als Roh- und Werkstoff 65: 23–28.

Anderson, A. B. 1981. White-Sand Vegetation of Brazilian Amazonia. Biotropica 13: 145-151.

Anderson, C. 1997. Monograph of Stigmaphyllon (Malpighiaceae). Systematic Botany Monographs 5[illegible]: 1-313.

Angermeier, P. L., A. P. Wheeler, and A. E. Rosenberger. 2004. A conceptual framework for assessing impacts of roads on aquatic biota. Fisheries 29: 19-29.

Anhuf, D., M. P. Ledru, H. Behling, F. W. Da Cruz, R. C. Cordeiro, T. van der Hammen, I. Karmann, J. A. Marengo, P. E. De Oliveira, L. Pessenda, A. Siffedine, A. L. Albuquerque, and P. L. Da Silva Dias. 2006. Paleo-environmental change in Amazonian and African rainforest during the LGM. Palaeogeography, Palaeoclimatology, Palaeoecology 239: 510–527.

Annan, K., R. A. Dickson, I. K. Amponsah, J. Jato, and I. K. Nooni. 2013. Pharmacognostic evaluation and physicochemical analysis of *Paullinia pinnata* L. (Sapindaceae). Journal of Pharmacognosy and Phytochemistry 2: 203-208.

Anthony, E. J. 2016. Impacts of sand mining on beaches in Suriname. WWF-Guianas, Paramaribo, Suriname.

Araujo, Y., and P. Beserra. 2007. Diversidad de invertebrados consumidos por las etnias yanomami del alto Orinoco, Venezuela. Interciencia: Revista de ciencia y tecnología de América 32: 318-323.

Ardila-Camacho, A. 2014. A new species of *Corydalus* Latreille, 1802 (Megaloptera, Corydalidae) and first record of *C. clavijoi* Contreras-Ramos, 2002 and *C. nubilus* Erichson, 1848 from Colombia. Zootaxa 38[illegible]1: 107-118.

Arets, E. J. M. M., P. J. van der Meer, N. W. van den Brink, K. Tjon, V. P. Atmopawiro, and P. E. Ouboter. 2005. Assessment of the impacts of gold-mining on soil and vegetation in Brownsberg Nature Park, Suriname. Alterra-Rapport 1359, Alterra, Wageningen, The Netherlands.

Armbruster, W. S. 1982. Seed production and dispersal in *Dalechampia* (Euphorbiaceae): divergent patterns and ecological consequences. American Journal of Botany 69: 1429-1440.

Armbruster, W. S. 1984. The role of resin in angiosperm pollination: ecological and chemical considerations. American Journal of Botany 71: 1149-1160.

Armbruster, W. S. 1993. Evolution of plant pollination systems: hypotheses and tests with the Neotropical vine *Dalechampia*. Evolution 47: 1480–1505.

Armbruster, W. S., and A. Herzig. 1984. Partitioning and sharing of pollinators by four sympatric species of *Dalechampia* (Euphorbiaceae) in Panama. Annals of the Missouri Botanical Garden 71: 1-16.

Armbruster, W. S., and G. L. Webster. 1982. Divergent pollination systems in sympatric species of South American *Dalechampia* (Euphorbiaceae). The American Midland Naturalist 108: 325-337

Arrington, D. A., and K. O. Winemiller. 2003. Diel changeover in sandbank fish assemblages in a Neotropical floodplain river. Journal of Fish Biology 63: 442–459.

Arrington, D. A., K. O. Winemiller, W. F. Loftus, and S. Akin. 2002. How often do fishes run on empty? Ecology 83: 2145–2151.

Artero, C., D. J. Murie, C. C. Koenig, R. Berzins, C. Bouchon, and L. Lampert. 2015. Age, growth, and mortality of the Atlantic goliath grouper *Epinephelus itajara* in French Guiana. Endangered Species Research 28: 275-287.

Artigas, L. F., P. Vendeville, M. Leopold, D. Guiral, and J. F. Ternon. 2003. Marine biodiversity in French Guiana: estuarine, coastal and shelf ecosystems under the influence of Amazonian waters. Gayana 67: 302-326.

Asner, G. P., E. N. Broadbent, P. J. C. Oliveira, M. Keller, D. E. Knapp, and J. N. M. Silva. 2006. Condition and fate of logged forests in the Brazilian Amazon. Proceedings of the National Academy of Sciences 103: 12947–12950.

Assunção, M. I. S., and H. O. Schwassmann. 1995. Reproduction and larval development of *Electrophorus electricus* on Marajó Island (Pará, Brazil). Ichthyological Explorations Freshwaters 6: 175-184.

Aublet, J. B. C. F. 1775. Histoire des Plantes de la Guiane Françoise. Didot, Paris, France

Augustinus, P. G. E. F. 1978. The changing shoreline of Surinam (South America). PhD dissertation / Publication of the Foundation for Scientific Research in Surinam and the Netherlands Antilles, no. 95, Utrecht University, Utrecht, The Netherlands.

Augustinus, P. G. E. F. 2004. The influence of the trade winds on the coastal development of the Guianas at various scale levels: a synthesis. Marine Geology 208: 145-151.

Augustinus, P. G. E. F., L. Hazelhoff, and A. Kroon. 1989. The chenier coast of Suriname: modern and geological development. Marine Geology 90: 269-281.

Austin, D. F. 2007. Water spinach (*Ipomoea aquatica*, Convolvulaceae): A food gone wild. Ethnobotany Research & Applications 5: 123-146

Autran, E., I. Neves, C. Dasilva, G. Santos, C. Camara, and D. Navarro. 2009. Chemical composition, oviposition deterrent and larvicidal activities against *Aedes aegypti* of essential oils from *Piper marginatum* Jacq. (Piperaceae). Bioresource Technology 100: 2284–2288. **doi:10.1016/j.biortech.2008.10.055**

Avery, R. A. 1979. Lizards – A study in thermoregulation. Edward Arnold, London, England, UK.

Avila-Pires, T. C. S. 1995. Lizards of Brazilian Amazonia. Zoologische Verhandelingen. 299: 1-706.

Avila-Pires, T. C. S. 2005. Reptiles. *In* T. Hollowell, and R.P. Reynolds (Eds.). Checklist of the terrestrial Vertebrates of the Guiana Shield. Bulletin of the Biological Society of Washington 13: 25-40.

Aymard, G. C. 1998. Four new species of *Doliocarpus* (Dilleniaceae) from the Brazilian Amazonian Region. Kew Bulletin 53: 713-721.

Ayoka, A. O., R. O. Akomolafe, E. O. Iwalewa, M. A. Akanmu, and O. E. Ukponmwan. 2006. Sedative, antiepileptic and antipsychotic effects of *Spondias mombin* L. (Anacardiaceae) in mice and rats. Journal of Ethnopharmacology 103: 166-175.

Bakker, E., L. Dalhuisen, R. Donk, M. Hassankhan, and F. Steegh. 1998. Geschiedenis van Suriname. Van stam tot staat. Uitgeversmaatschappij Walburg Pers, Zutphen, The Netherlands.

Balée, W. 1989. The culture of Amazon forests. Advances in Economic Botany 7: 1-21.

Balian, E. V., H. Segers, C. Lévèque, and K. Martens. 2008. The freshwater animal diversity assessment: an overview of the results. Hydrobiologia 595: 627-637.

Balon, E. K. 1975. Reproductive guilds of fishes: a proposal and definition. Journal Fisheries Research Board of Canada 32: 821-864.

Bancroft, E. 1769. An essay on the natural history of Guiana, in South America: containing a description of many curious productions in the animal and vegetable systems of that country. Together with an account of the religion, manners, and customs of several tribes of its Indian inhabitants. Interspersed with a variety of literary and medical observations. In several letters. T. Becket and P. A. De Hondt, The Strand [London, England, UK].

Bandeira, A. G., A. Vasconcellos, M. P. Silva, and R. Constantino. 2003. Effects of habitat disturbance on the termite fauna in a highland humid forest in the Caatinga domain, Brazil. Sociobiology 42: 117-128.

Banki, O. S. 2010. Does Neutral Theory explain community composition in the Guiana Shield forests? PhD dissertation, Institute of Environmental Biology, Utrecht University, Utrecht, The Netherlands.

Bánki, O. S., H. ter Steege, M. J. Jansen-Jacobs, and U. P. Raghoenandan. 2003. Plant diversity of the Nassau Mountains, Suriname. Report of the 2003 Expedition. University of Utrecht, Utrecht, the Netherlands.

Banki, O.S., and C.R. Bhikhi. 2011. Plant diversity and composition of the forest in the surroundings of Kwamalasamutu. *In* B. J. O'Shea, L. E. Alonso, and T. H. Larsen (Eds.). A rapid biological assessment of the Kwamalasamutu region, southwestern Suriname, pp. 43-55. RAP Bulletin of Biological Assessment 63, Conservation International, Arlington, VA, USA.

Banki, O.S., and C.R. Bhikhi. 2013. Floristic assessment of the Upper Palumeu River, the Grensgebergte, and the Kasikasima areas. *In* L. E. Alonso, and T. H. Larsen (Eds.). A rapid biological assessment of the upper Palumeu River watershed (Grensgebergte and Kasikasima) of southeastern Suriname, pp. 57-78. RAP Bulletin of Biological Assessment 67, Conservation International, Arlington, VA, USA.

Barbarino Duque, A., and K. O. Winemiller. 2003. Diet segregation among large catfishes of the Apure and Arauca rivers, Venezuela. Journal of Fish Biology 63: 410-427.

Barbosa, A. P., C. S. Nascimento, and J. W. Morais. 2007. Estudos de propriedades antitermíticas de extratos brutos de madeira e casca de espécies florestais da Amazônia Central, Brasil. Acta Amazonica 37: 213-218.

Barceloux, D. G. 2008. Medical toxicology of natural substances: Foods, fungi, medicinal herbs, plants, and venomous animals. Wiley Publications. New York, NY, USA.

Barker, F. K., K. J. Burns, J. Klicka, S. M. Lanyon, and I. J. Lovette. 2013. Going to extremes: contrasting rates of diversification in a recent radiation of New World passerine birds. Systematic Biology 62: 298-320.

Barré, N., J. Theuerkauf, L. Verfaille, P. Primot, and M. Saoumoe. 2010. Exponential population increase in the endangered Ouvéa Parakeet (*Eunymphicus uvaeensis*) after community-based protection from nest poaching. Journal of Ornithology 151: 695-701.

Barreto, G. R., and R. D. Quintana. 2010. Foraging strategies and feeding habits of capybaras. *In* J. R. Moreira (Ed.). Capybara: biology, use and conservation of an exceptional neotropical species, pp. 83-96. Springer, New York, NY, USA.

Barse, W. P. 1990. Preceramic occupations in the Orinoco River valley. Science 250: 1388-1390.

Barthe, S., G. Binelli, B. Hérault, C. Scotti-Saintagne, D. Sabatier, and I. Scotti. 2016. Tropical rainforests that persisted: inferences from the Quaternary demographic history of eight tree species in the Guiana shield. Molecular Ecology 26: 1161–1174.

Barthem, R. B., and M. Goulding. 1997. The catfish connection: ecology, migration, and conservation of Amazon predators. Columbia University Press, New York, USA.

Barthem, R. B., M. C. L. B. Ribeiro, and M. Petrere. 1991. Life strategies of some long-distance migratory catfish in relation to hydroelectric dams in the Amazon basin. Biological Conservation 55: 339-345.

Basset, Y., L. Cizek, P. Cuénoud, R. K. Didham, F. Guilhaumon, O. Missa, V. Novotny, F. Ødegaard, T. Roslin, J. Schmidl, A. K. Tishechkin, N. N. Winchester, D. W. Roubik, H.-P. Aberlenc, J. Bail, H. Barrios, J. R. Bridle, G. Castaño-Meneses, B. Corbara, G. Curletti, W. D. da Rocha, D. De Bakker, J. H. C. Delabie, A. Dejean, L. L. Fagan, A. Floren, R. L. Kitching, E. Medianero, S. E. Miller, E. G. de Oliveira, J. Orivel, M. Pollet, M. Rapp, S. P. Ribeiro, Y. Roisin, J. B. Schmidt, L. Sørensen, and M. Leponce. 2012. Arthropod Diversity in a Tropical Forest. Science 338: 1481-1484.

Baumann, T. W., B. H. Schulthess, and K. Hänni. 1995. Guaraná (*Paullinia cupana*) rewards seed dispersers without intoxicating them by caffeine. Phytochemistry 39: 1063–1070.

Bawa, K. S., S. H. Bullock, D. R. Perry, R. E. Coville, and M. H. Grayum. 1985. Reproductive biology of tropical lowland rain forest trees. II. Pollination systems. American Journal of Botany 72: 346-356.

BDGS. 2016. BDGS Database. Biological Diversity of the Guiana Shield Program, Department of Botany, US. National Museum of Natural History, Smithsonian Institution, Washington, DC, USA. botany.si.edu/bdg/specimenquery/query.cfm [database queried Nov. 2016]

Beard, J.S. 1944. Climax vegetation in tropical America. Ecology 25: 127-158.

Beard, J.S. 1953. The savanna vegetation of northern tropical America. Ecological Monographs 23: 149-215.

Beard, J.S. 1955. The classification of tropical American vegetation-types. Ecology 36: 89-100.

Beati L., S. Nava, E. J. Burkman, D. M. Barros-Battesti, M. B. Labruna, A. A. Guglielmone, A. G. Cáceres, C. M. Guzmán-Cornejo, R. León, L. A. Durden, and J. L. H. Faccini. 2013. *Amblyomma cajennense* (Fabricius, 1787) (Acari: Ixodidae), the Cayenne tick: phylogeography and evidence for allopatric speciation. Biomed Central Evolutionary Biology 13: 267.

Beccaloni, G. 2014. Cockroach Species File Online, Version 5.0. **Cockroach.SpeciesFile.org** [accessed Jan. 2016]

Beck, D. L., J. Zavala, E. O. Montalvo, and F. G. Quintana. 2011. Meteorological indicators for *Amblyomma cajennense* and population dynamics in the Tamaulipan Biotic Province in Texas. Journal of Vector Ecology 36: 135-146.

Becker, G. 1961. Beobachtungen und Versuche über den Beginn der Kolonie-entwicklung von *Nasutitermes ephratae* Holmgren (Isoptera). Zeitschrift für Angewandte Entomologie 49: 78-93.

Becker, M, and J. Grazia. 1971. Sobre o gênero *Rhyncholepta* Bergroth, 1911, com descrição de uma nova espécie (Hemiptera, Pentatomidae, Pentatomini). Revista Brasileira de Biologia 31: 389-399.

Beebe, W. 1945. Vertebrate fauna of a tropical dry season mud hole. Zoologica 30: 81-90.

Begon, M., C. R. Townsend, and J. L. Harper. 2006. Ecology: from individuals to ecosystems (4th ed.). Blackwell Publishing, Ltd., USA.

Beier, M. 1948. Phoresie und Phagophilie bei Pseudoscorpionen. Österreichische Zoologische Zeitschrift 1: 441–497.

Beisiegel, B. M., and G. L. Zuercher. 2005. *Speothos venaticus*. Mammalian Species 783: 1-6.

Beissinger, S. R. 1990. Alternative foods of a diet specialist, the Snail Kite. Auk 107: 327-333.

Benítez-López, A., R. Alkemade, and P. A. Verweij. 2009. The impacts of roads and other infrastructure on mammal and bird populations: a meta-analysis. Biological Conservation 143: 1307–1316.

Bennett, F. D. 1972. Observations on *Exaerete* spp. and their hosts *Eulaema terminata* and *Euplusia surinamensis* (Hymen., Apidae, Euglossinae) in Trinidad. Journal New York Entomological Society 80: 118-124.

Benton, P. D., W. E. Ensign, and B. J. Freeman. 2008. The effect of road crossing on fish movements in small Etowah Basin streams. Southeastern Naturalist 7: 301-310.

Benzaquem, D. C., C. Oliveira, J. S. Batista, J. Zuanon, and J. I. R. Porto. 2015. DNA barcoding in pencilfishes (Lebiasinidae: *Nannostomus*) reveals cryptic diversity across the Brazilian Amazon. PLoS ONE 10: e0112217. **doi: 10.137/journal.pone.0112217**

Béreau, M., M. Gazel, and J. Garbaye. 1997. Les symbioses mycorrhiziennes des arbres de la forêt tropicale humide de Guyane française. Canadian Journal Botany 75: 711-716.

Berenger J. M, D. Pluot-Sogwalti, F. Pages, D. Blanchet, and C. Aznar. 2009. The Triatominae species of French Guiana (Heteroptera: Reduviidae). Memorias do Instituto Oswaldo Cruz 104: 1111-1116. **doi: 10.1590/S0074-02762009000800007**

Berg, C. C., and P. Franco. 2005. *Cecropia*. Flora Neotropica Monographs 94: 1-230.

Bergmann, F. B., H. L. C. Amaral, D. P. Pinto, C. C. Chivittz, and A. M. Tozetti. 2013. Foraging activity of the Snail Kite, *Rostrhamus sociabilis* (Aves: Accipitridae) in wetlands of southern Brazil. Brazilian Journal of Biology 73: 245-252.

Berrenstein, H. J. 2012. Stress factors and ecological conditions of the mangrove ecosystems along the coast of Suriname, with special emphasis on the district of Coronie. Faculty of Technological Sciences, Anton de Kom University of Suriname, Paramaribo, Suriname.

Bertani R., C. Sayuri Fukushima, R. Martins. 2007. Sociable widow spiders? Evidence of subsociality in *Latrodectus* Walckenaer, 1805 (Araneae, Theridiidae). Journal of Ethology 26: 299-302. **doi:10.1007/s10164-007-0082-8.**

Bhikhi, C., P. Maas, J. Koek-Noorman, and T. van Andel. 2016. Timber trees of Suriname. LM Publishers, Volendam, The Netherlands.

Bickford, D., D. J. Lohman, N. S. Sodhi, P. K. L. Ng, R. Meier, K. Winker, K. K. Ingram, and I. Das. 2007. Cryptic species as a window on diversity and conservation. Trends in Ecology and Evolution 22: 148-155.

Bisset, G. B. 2002. War and hunting poisons of the New World. Part I. Notes on the early history of curare. Journal of Pharmacology 36: 1-26.

Blake, J. G., and B. A. Loiselle. 2009. Species composition of Neotropical understory bird communities: local versus regional perspectives based on capture data. Biotropica 41: 85-94.

Blassioli-Moraes, M. C., D. M. Magalhaes., A. Čokl, R. A. Laumann, J. P. Da Silva, C.C. Silva, and M. Borges. 2014. Vibrational communication and mating behaviour of *Dichelops melacanthus* (Hemiptera: Pentatomidae) recorded from loudspeaker membranes and plants. Physiological entomology 39: 1–11.

Bleckmann, H. 1993. Role of the lateral line in fish behaviour. *In* T.J. Pitcher (Ed.). Behaviour of teleost fishes, pp 201-246. Chapman and Hall, London, UK.

Blendinger, P. G., J. G. Blake, and B. A. Loiselle. 2011. Connecting fruit production to seedling establishment in two co-occurring *Miconia* species: consequences of seed dispersal by birds in upper Amazonia. Oecologia 167: 61-73.

Bodmer, R. E. 1990. Responses of ungulates to seasonal inundations in the Amazon floodplain. Journal of Tropical Ecology 6: 191-201.

Bodmer, R. E. 1991. Strategies of seed dispersal and seed predation in Amazonian ungulates. Biotropica 23: 255–261.

Boinski, S. 1987. Mating patterns in squirrel monkeys (*Saimiri oerstedi*): implications for seasonal sexual dimorphism. Behavioral Ecology and Sociobiology 21: 13-21.

Boinski, S. 2002. De apen van Suriname / The monkeys of Suriname. Stinasu, Paramaribo, Suriname.

Boinski, S., and S. J. Cropp. 1999. Disparate data sets resolve squirrel monkey (*Saimiri*) taxonomy: implications for behavioral ecology and biomedical usage. International Journal of Primatology 20: 237-256.

Boisselier-Dubayle, M., R. Leblois, S. Samadi, J. Lambourdière, and C. Sarthou. 2010. Genetic structure of the xerophilous bromeliad *Pitcairnia geyskesii* on inselbergs in French Guiana - a test of the forest refuge hypothesis. Ecography 33: 175–184.

Boksteen, L. W. 2009. Strategische analyse en participatief actieplan voor Zuid Oost Suriname. Deelstudie impact vergroting beschikbare hoeveelheid water in het bestaande Brokopondo stuwmeer. L. W. Boksteen, Paramaribo, Suriname.

Bondar, G. 1926. Ainda o gigantesco besouro *Acrocinus longimanus*. Chácaras e Quintais 33: 234-235.

Bone, Q., and R. H. Moore. 2008. Biology of fishes (3rd ed.). Taylor and Francis Group, New York, USA.

Bongers, F., P. Charles-Dominique, P. M. Forget, and M. Théry (Eds.). 2013. Nouragues: Dynamics and plant-animal interactions in a neotropical rainforest. Kluwer Academic Publishers, Dordrecht, Netherlands.

Bonne, C., and J. Bonne-Wepster. 1925. Mosquitos of Suriname; a study on neotropical mosquitos. Mededelingen Vereniging Koloniaal Instituut 13, Amsterdam, The Netherlands.

Boomert, A. 1980. The Sipaliwini archeological complex of Surinam, a summary. Nieuwe West-Indische Gids 54: 94-108.

Bordenave, B. G., J.-J. De Granville, and K. Steyn. 2011. Quantitative botanical diversity descriptors to set conservation priorities in Bakhuis Mountains rainforest, Suriname. Botanical Journal of the Linnean Society 167: 94-130.

Borges, M., P. C. Jepson, and P. E. Howse. 1987. Long-range mate location and close-range courtship behaviour of the green stink bug, *Nezara viridula* and its mediation by sex pheromones. Entomologia Experimentalis et Applicata 44: 205–212.

Born, M., F. Bongers, E. H. Poelman, and F. J. Sterck. 2010. Dry-season retreat and dietary shift of the dart-poison frog *Dendrobates tinctorius* (Anura: Dendrobatidae). Phyllomedusa 9: 37-52.

Bosma, W., S. B. Kroonenberg, K. Maas, and E. W. F. De Roever. 1983. Igneous and metamorphic complexes of the Guiana Shield in Suriname. Geologie en Mijnbouw 62: 241-254.

Bouchet, P. 2006. The magnitude of marine biodiversity. In C. M. Duarte (Ed.). The exploration of marine biodiversity: scientific and technological challenges, pp. 33-64. Fundación BBVA, Bilbao, Spain.

Boujard, T., and R. Rojas-Beltrán. 1988. Zonation longitudinale du peuplement ichtyque du fleuve Sinnamary (Guyane fran aise). Revue Hydrobiologie Tropicale 21: 47-61.

Boujard, T., D. Sabatier, R. Rojas-Beltran, M. F. Prevost, and J. F. Renno. 1990a. Food habits of three allochthonous feeding characoids in French Guiana. Revue d'Ecologie (La Terre et la Vie) 45: 247-258.

Boujard, T., P. Keith, and P. Luquet. 1990b. Diel cycle in *Hoplosternum littorale* evidence for synchronization of locomotor, air breathing and feeding activity by aracadian alternation of light and dark. Journal of Fish Biology 36: 133-140.

Boerboom, J. H. A. 1974. Succession studies in the humid tropical lowlands of Surinam. Proceedings 1st International Congress Ecology, Junk, The Hague, pp. 343-347.

Boven, K. 2006. Overleven in een grensgebied. Bronnen voor de studie van Suriname, No. 26, IBS/Rozenberg Publishers, Reeuwijk, the Netherlands.

Bowen, S. H. 1983. Detritivory in Neotropical fish communities. Environmental Biology of Fishes 9: 137-144.

Boyle, K. S., S. Riepe, G. Bolen, and E. Parmentier. 2015. Variation in swim bladder drumming sounds from three doradid catfish species with similar sonic morphologies. Journal of Experimental Biology 218: 2881-2891.

Bradbury, J. W., and S. L. Vehrencamp. 1977. Social organization and foraging in emballonurid bats: III. Mating systems. Behavioral Ecology and Sociobiology 2: 1–17.

Branan, W. V., and R. L. Marchinton. 1987. Reproductive ecology of white-tailed and red brocket deer in Suriname. *In* C. Wemmer (Ed.). Biology and management of the Cervidae, pp. 344-351. Smithsonian Institution Press, Washington, DC.

Branan, W. V., M. C. M. Werkhoven, and R. L. Marchinton. 1985. Food habits of brocket and white-tailed deer in Suriname. Journal of Wildlife Management, 49: 972-976.

Brandão, M. G. L., T. S. M. Grandi, E. M. M. Rocha, D. R. Sawyer, and A. U. Krettli. 1992. Survey of medicinal plants used as antimalarials in the Amazon. Journal of Ethnopharmacology 36: 175-182.

Breault, J. L. 1991. Candirú: Amazonian parasitic catfish. Journal of Wilderness Medicine 2: 304-312.

Breder, C. M., and D. E. Rosen. 1966. Modes of reproduction in fishes. T.F.H. Publications, Jersey City, NJ, USA.

Brenner, M., and U. Krumme. 2007. Tidal migration and patterns in feeding of the four-eyed fish *Anableps anableps* L. in a north Brazilian mangrove. Journal of Fish Biology 70: 406-427.

Briceño, H., C. Schubert, and J. Paolini. 1990. Table-mountain geology and surficial geochemistry: Chimantá Massif, Venezuelan Guayana Shield. Journal of South American Earth Science 3: 179–194.

Brickham, J. W., P. K. Tucker, and J. M. Legler. 1985. Diploid-Triploid Mosaicism: An Unusual Phenomenon in Side-Necked Turtles (*Platemys platycephala*). Science 29, 227(4694): 1590-1593.

Briggs, J. C. 2003. Fishes and birds: Gondwana life rafts reconsidered. Systematic Biology 52: 548–553.

Brightsmith, D. J. 2004. Nest sites of termitarium nesting birds in SE Peru. Ornitología Neotropical 15: 319-330.

Brightsmith, D. J. 2005a. Competition, predation and nest niche shifts among tropical cavity nesters: ecological evidence. Journal of Avian Biology 36: 74-83.

Brightsmith, D. J. 2005b. Competition, predation and nest niche shifts among tropical cavity nesters: phylogeny and natural history evolution of parrots (Psittaciformes) and trogons (Trogoniformes). Journal of Avian Biology 36: 64-73.

Brightsmith, D., and A. Bravo. 2006. Ecology and management of nesting blue-and-yellow macaws (*Ara ararauna*) in *Mauritia* palm swamps. Biodiversity and Conservation 15: 4271-4287.

Brinkman, R., and L. J. Pons. 1968. A pedo-geomorphological classification and map of the Holocene sediments in the Coastal Plain of the three Guianas. Soil Survey Papers 4, Netherlands Soil Survey Institute, Wageningen. The Netherlands.

Brosse, S., J. I. Montoya-Burgos, G. Grenouillet, and N. Surugue. 2013. Determinants of fish assemblage structure in Mount Itoupé mountain streams (French Guiana). Annales de Limnologie – International Journal of Limnology 49: 43-49.

Brosset, A. 1997. Aggressive mimicry by the characid fish *Eythrinus eythrinus*. Ethology 103: 926-934.

Brouard, O., R. Cereghino, B. Corbara, C. Leroy, L. Pelozuelo, A. Dejean, and J. F. Carrias. 2012. Understorey environments influence functional diversity in tank-bromeliad ecosystems. Freshwater Biology 57: 815-823.

Brousseau, L. 2013. Diversity and evolution in tropical rainforest trees: example of *Eperua falcata* in French Guiana. PhD Dissertation, Université de Lorraine, Nancy, France.

Brower, L. 1984. Chemical defense in butterflies. *In* R. I. Vane-Wright, and P. R. Ackery (Eds.). The Biology of Butterflies, pp. 109-134. Academic Press, London, England, UK.

Bubberman, F. C. 1973. De bosbranden van 1964 in Suriname. Nieuwe West-Indische Gids 49: 163-173.

Buchmann, S. L. 1983. Buzz pollination in angiosperms. *In* C. E. Jones and J. R. Little (Eds.). Handbook of Experimental Pollination Biology, pp. 73-113. Van Nostrand Reinhold, New York, NY, USA.

Buchmann, S. L. 1987. Ecology of oil flowers and their bees. Annual Review of Ecology and Systematics 18: 343-369.

Buitelaar, R., A. Kambon, M. Hendrickson, and E. Blommestein. 2007. Suriname: the impact of the May 2006 floods on sustainable livelihoods. United Nations, Santiago, Chile.

Bunkley-Williams L., and E. Williams. 1998. Ability of Pederson Cleaner Shrimp to remove juveniles of the parasitic cymothoid Isopod, *Anilocra haemuli*, from the host. Crustaceana 71: 862–869.

Burgess, W. E. 1989. An atlas of freshwater and marine catfishes. T.F.H. Publications, Neptune City, NJ. USA.

Burns, K. J., A. J. Schultz, P. O. Title, N. A. Mason, F. K. Barker, J. Klicka, S. M. Lanyon, and I. J. Lovette. 2014. Phylogenetics and diversification of tanagers (Passeriformes: Thraupidae), the largest radiation of Neotropical songbirds. Molecular Phylogenetics and Evolution 75: 41–77.

Busch, C. 1988. Consumption of blook, renal function and utilization of free water by the vampire bat, *Desmodus rotundus*. Comparative Biochemistry and Physiology 90: 141-146.

Bush, M. B. 1994. Amazonian speciation – a necessarily complex mode. Journal of Biogeography 21: 5–17.

Cadée, G. C. 1975. Primary production off the Guyana coast. Netherlands Journal of Sea Research 9: 128-143.

Caldwell, J. P. 1996. The evolution of myrmecophagy and its correlates in poison frogs (Family Dendrobatidae). Journal of Zoology, London 240: 75-101.

Calouro, A. M. 2000. Attempted predation on Brazilian rabbit (*Sylvilagus brasiliensis* – Lagomorpha Leporidae) by tayra (*Eira Barbara* – Carnivora: Procyonidae). Revista de Biología Tropical 48: 267-268.

Camargo J. M. F., and S. R. M. Pedro. 2013. Meliponini Lepeletier, 1836. *In* J. S. Moure, D. Urban, and G. A. R. Melo (Eds.). Catalogue of Bees (Hymenoptera, Apoidea) in the Neotropical Region - online version. **www.moure.cria.org.br/catalogue** [accessed Oct. 2015]

Cameron, E. (Ed.). 2001. Early Modern Europe, an Oxford history. Oxford University Press, Oxford, England, UK.

Campbell, J. A., and W. W. Lamar. 1989. The Venomous Reptiles of Latin America. Comstock Publishing Associates, Ithaca, NY, USA, and London, England, UK.

Campbell, J. A., and W. W. Lamar. 2004. The Venomous Reptiles of the Western Hemisphere. Comstock Publishing Associates, Ithaca, NY, USA and London, England, UK.

Campos e Silva, J. V., and F. De Almeida Meirelles. 2015. A small homage to Maria Sibylla Merian, and new records of spiders (araneae: Theraphosidae) preying on birds. Revista Brasileira de Ornitologia 24: 30-33.

Campos, Z., and W. E. Magnusson. 2013. Thermal relations of dwarf caiman, *Paleosuchus palpebrosus*, in a hillside stream: Evidence for an unusual thermal niche among crocodilians. Journal of Thermal Biology 38: 20–23.

Campos, Z., T. Sanaiotti, and W. E. Magnusson. 2010. Maximum size of dwarf caiman, *Paleosuchus palpebrosus* (Cuvier, 1807), in the Amazon and habitats surrounding the Pantanal, Brazil. Amphibia-Reptilia 31: 439-442.

Campos, Z., T. Sanaiotti, F. Muniz, I. Farias, and W. E. Magnusson. 2012. Parental care in the dwarf caiman, *Paleosuchus palpebrosus* Cuvier, 1807 (Reptilia: Crocodilia: Alligatoridae). Journal of Natural History 46(47-48): 2979-2984.

Caramaschi, U. 2010. Notes on the taxonomic status of *Elachistocleis ovalis* (Schneider, 1799) and description of five new species of *Elachistocleis* Parker, 1927 (Amphibia, Anura, Microhylidae). Boletim do Museu Nacional. Nova Serie, Zoologia, Rio de Janeiro 527: 1–30.

Carbonell, C. S. 1986. Revision of the Neotropical Genus *Tropidacris* (Orthoptera, Acridoidea, Romaleidae, Romaleinae). Proceedings of the Academy of Natural Sciences of Philadelphia 138: 366-402.

Cardoso, Y. P., and J. I. Montoya-Burgos. 2009. Unexpected diversity in the catfish *Pseudancistrus brevispinis* reveals dispersal routes in a Neotropical center of endemism: the Guyanas Region. Molecular Ecology 18: 947–964.

Caribsave. 2012. CARIBSAVE climate change risk profile for Suriname: summary document. Caribsave, Barbados.

CARICOM Fisheries Unit Belize. 2000. Report of the multidisciplinary survey of the fisheries in Suriname. Project No. 7: ACP: RPR: 385. CARICOM Fisheries Unit, Belize City, Belize.

Carolsfeld, J., B. Harvey, C. Ross, and A. Baer (Eds.). 2003. Migratory fishes of South America: biology, fisheries and conservation Status. International Development Research Centre and the World Bank, Ottawa, Canada.

Carrasco, H. J., A. Torrellas, C. García, M. Segovia, and M. D. Feliciangeli. 2005. Risk of *Trypanosoma cruzi* (Kinetoplastida: Trypanosomatidae) transmission by *Panstrongylus geniculatus* (Hemiptera: Reduviidae) in Caracas (Metropolitan District) and neighboring States. Venezuela International Journal of Parasitology 35: 1379-84.

Carrera, M. 1980. Entomologia para você. Nobel, Sao Paulo, Brazil.

Carter, G. G., and G. S. Wilkinson. 2015. Social benefits of non-kin food sharing by female vampire bats. Proceedings of the Royal Society Biological Sciences 282: 20152524. **dx.doi.org/10.1098/rspb.2015.2524**

Carter, G. S., and L. C. Beadle. 1931. The fauna of the swamps of the Paraguayan Chaco in relation to its environment. II. Respiratory adaptations in the fishes. Zoological Journal of the Linnean Society 37: 327-368.

Carter, T. S., and C. D. Encarnaçao.1983. Characteristics and use of burrows by four species of armadillos in Brazil. Journal of Mammalogy 64: 103-108.

Carthew, S. M., and R. L. Goldingay. 1997. Non-flying mammals as pollinators. Trends in Ecology and Evolution 12: 104-108.

Carvalho, F. M. 1980. Alimentação do mapará (*Hypophthalmus edentatus* Spix, 1829) do lago Castanho, Amazonas (Siluriformes, Hypophthalmidae). Acta Amazonica 10: 545-555.

Carvalho, L. N., J. Zuanon, and I. Sazima. 2007. Natural history of Amazon fishes. *In* K. Del Claro, P. S. Oliveira, V. Rico-Gray, A. Ramirez, A. A. A. Barbosa, A. Bonet, F. R. Scarano, F. L. Consoli, F. J. M. Garzon, J. N. Nakajima, J. A. Costello, M. V. Sampaio, M. Quesada, M. R. Morris, M. P. Rios, N. Ramirez, O. M. Junior, R. H. F. Macedo, R. J. Marquis, R. P. Martins, S. C. Todrigues, and U. Luttje (Eds.) Encyclopedia of life support systems (EOLSS), pp 1-32. UNESCO, Eolss publishers, Oxord, England, UK. **www.eolss.net**

Carvalho, L. N., L. Fidelis, R. Arruda, A. Galuch, and J. Zuanon. 2013. Second floor, please: the fish fauna of floating litter banks in Amazonian streams and rivers. Neotropical Ichthyology 11: 85-94.

Carvalho, L. N., R. Arruda, and J. Zuanon. 2003. Record of cleaning behavior by *Platydoras costatus* (Siluriformes: Doradidae) in the Amazon Basin, Brazil. Neotropical Icthyology 1:137-139.

Castner, J. L. 2000. Amazon insects, a photo guide. Feline Press, Gainesville, FL, USA.

Catania, K. 2014. The shocking predatory strike of the electric eel. Science 346: 1231-1234.

Catzeflis, C. 2012. A survey of small non-volant mammals inhabiting Wayampi Amerindian houses in French Guiana. Mammalia 76: 327–330.

CBD. 2017. Convention on Biological Diversity. United Nations, Washington, DC, USA. **www.cbd.int/convention/**

Chagas R. A., M. Barros, R. Salimos, W. Santos, and M. Herrman. 2015. Ocorrência de *Braga patagonica* (Isopoda, Cymothoidae) parasitando *Peprilus paru* (Osteichthyes: Stromateidae) em águas costeiras tropicais de São João de Pirabas, Pará, Norte do Brasil. Acta of Fisheries and Aquatic Resources 3: 1-9.

Chagas, M. C. M., M. A. B. Moreira, and M. F. P. Barreto. 1995. Biological aspects of the *Schistocerca pallens*, *Stiphra robusta* and *Tropidacris collaris* grasshopper species at Rio Grande do Norte State, Brazil. Acta Horticulturae 370: 83-88.

Chambers, J. Q., N. Higuchi, and J. P. Schimel. 1998. Ancient trees in Amazonia. Nature 391: 135-136.

Chapman, C. A., L. J. Chapman, and K. Vulinec. 2003. Fragmentation and alteration of seed dispersal processes: an initial evaluation of dung beetles, seed fate, and seedling diversity. Biotropica 35: 382-393.

Chapman, L. J. 2015. Low-oxygen lifestyles. *In* R. Riesch, M. Tobler, and M. Plath (Eds.). Extremophile fishes: ecology, evolution, and physiology of teleosts in extreme environments, pp 9-33. Springer, Cham, Heidelberg, New York, Dordrecht, and London.

Charles-Dominique, P. 1991. Feeding strategies and activity budget of the frugivorous bat *Carollia perspicillata* (Phyllostomidae) in French Guiana. Journal of Tropical Ecology 7: 243–256.

Charles-Dominique, P., J. Chave, M. Dubois, D. Granville, B. Riera, and C. Vezzoli. 2003. Colonization front of the understorey palm *Astrocaryum sciophilum* in a pristine rain forest of French Guiana. Global Ecology and Biogeography 12(3): 237-248.

Charles-Dominique, P., M. Atramentowicz, M. Charles-Dominique, H. Gerard, A. Hladik, C. M. Hladik, and M. F. Prevost. 1981. Les Mammifères frugivores arboricoles nocturnes d'une forêt Guyanaise: Inter-Relations Plantes-Animaux. Revue d'Ecologie (Terre et Vie) 35: 341-435.

Charlier, P. 2001. Report of a national workshop on fisheries management in Suriname 21-22 June 2000. Ministerie van Landbouw, Veeteelt en Visserij, Paramaribo, Suriname.

Charvet-Almeida, P., M. L. G. De Araújo, and M. P. De Almeida. 2005. Reproductive aspects of freshwater stingrays (Chondrichthyes: Potamotrygonidae) in the Brazilian Amazon Basin. Journal Northwest. Atlantic. Fisheries. Science 35: 165-171.

Chave, J. 2004. Neutral theory and community ecology. Ecology Letters 7: 241-253.

Chaves-Campos, J. 2004. Elevational movements of large frugivorous birds and temporal variation in abundance of fruits along an elevational gradient. Ornitología Neotropical 15: 433-445.

Chaves-Campos, J. 2005. Bare-necked Umbrellabird (*Cephalopterus glabricollis*) foraging at an unusually large assemblage of army ant-following birds. Wilson Bulletin 117: 418-420.

Chen, T., R. Gagliardo, B. Walker, M. Zhou, and C. Shaw. 2005. Partial structure of the phylloxin gene from the giant monkey frog, *Phyllomedusa bicolor:* Parallel cloning of precursor cDNA and genomic DNA from lyophilized skin secretion. Peptides 26: 2624-2628.

Cherrett, J. M. 1968. The foraging behavior of *Atta cephalotes* L. (Hymenoptera, Formicidae) I. Foraging pattern and plant species in tropical rain forest. Journal of Animal Ecology 37: 387-403.

Cherrett, J. M., and R. J. Quinlan. 1979. The role of fungus in the diet of the leaf-cutting ant *Atta cephalotes* (L.). Ecological Entomology 4: 151-160.

Chesser, R. T. 1997. Patterns of seasonal and geographical distribution of austral migrant flycatchers (Tyrannidae) in Bolivia. Ornithological Monographs 48: 171-204.

Chinajariyawong, A., A. R. Clarke, and M. Jirasurat. 2000. Survey of opiine parasitoids of fruit flies (Diptera: Tephritidae) in Thailand and Malaysia. The Raffles Bulletin of Zoology 48: 71-101.

Chin-A-Lin, T., and M. Yspol. 2000. Suriname, groundfish and shrimp fisheries. *In* Fourth workshop on the assessment and management of shrimp and groundfish fisheries on the Brazil-Guianas shelf. FAO Fisheries Report No. 651, pp. 94 104. Food and Agriculture Organization of the United Nations (FAO), Rome, Italy.

Chorus I., and J. Bartram. 1999. Toxic cyanobacteria in water: a guide to their public health consequences, monitoring and management. E & FN Spon, An imprint of Routledge, London, England, UK.

Christianini, A. V., and P. S. Oliveira. 2010. Birds and ants provide complementary seed dispersal in a neotropical savanna. Journal of Ecology 98: 573-582.

Cisneros-Heredia, D. F., and R. W. McDiarmid. 2007. Revision of the characters of Centrolenidae (Amphibia: Anura: Athesphatanura), with comments on its taxonomy and the description of new taxa of glassfrogs. Zootaxa 1572: 1-82.

CITES. 2015. Convention on International Trade in Endangered Species of Wild Fauna and Flora – Appendices I, II and III. CITES, Geneva, Switzerland. **www.cites.org** [accessed Nov. 2015]

Clark, P., A. Dyke, J. D. Shakun, A. E. Carlson, J. Wohlfarth, J. Mitrovica, S. Wostetler, and A. McCabe. 2009. The Last Glacial Maximum. Science 325: 710-714.

Clausing, G., and S. Renners. 2001. Molecular phylogenetics of Melastomataceae and Memecylaceae: implications for character evolution. American Journal of Botany 88: 486–498.

Clavier, S., A. Contreras-Ramos, and L. Guillemet. 2010. First report of *Protosialis* Weele, 1909 in French Guiana (Megaloptera, Sialidae). Ephemera 11: 133-138.

Clay, J., and C. R. Clement. 1993. Selected species and strategies to enhance income generation from Amazonian forests. FAO Working Paper FO: Misc/93/6. **www.fao.org/docrep/v0784e/v0784e0b.htm**

Clements, F. E. 1916. Plant succession: an analysis of the development of vegetation. 242. Carnegie Institution of Washington, Washington, DC, USA

Cloudsley-Thomson, J., and C. Constantinou. 1985. Diurnal rhythm of activity in the arboreal tarantula *Avicularia avicularia* (L.) (Mygalomorphae: Theraphosidae). Journal of Interdisciplinary Cycle Research 16: 1[illegible]3-116.

Clucas, I. 1997. A study of the options for utilization of bycatch and discards from marine capture fisheries. FAO Fisheries Circular No. 928 FIIU/C928. Food and Agriculture Organization of the United Nations (FAO), Rome, Italy. **www.fao.org/docrep/W6602E/w6602E09.htm**

Coates, C. W., and R. T. Cox. 1945. A comparison of length and voltage in the electric eel, *Electrophorus electricus* (Linnaeus). Zoologica New York 30: 89-93.

Cockerell, T. D. A. 1925. A fossil fish of the family Callichthyidae. Science 62: 397-398.

Coelho de Souza, F., K. G. Dexter, O. L. Phillips, R. J. W. Brienen, J. Chave, R. D. Galbraith, G. L. Gonzalez, A. M. Mendoza, R. T. Pennington, L. Poorter, M. Alexiades, E. Álvarez-Dávila, A. Andrade, L. E. O. C. Aragão, A. Araujo-Murakami, E. J. M. M. Arets, G. A. Aymard C, C. Baraloto, J. G. Barroso, D. Bonal, R. G. A. Boot, J. L. C. Camargo, J. A. Comiskey, F. Cornejo Valverde, P. B. de Camargo, A. Di Fiore, F. Elias, T. L. Erwin, T. R. Feldpausch, L. Ferreira, N. M. Fyllas, E. Gloor, B. Herault, R. Herrera, N. Higuchi, E. N. Honorio Coronado, T. J. Killeen, W. F. Laurance S. Laurance, J. Lloyd, T. E. Lovejoy, Y. Malhi, L. Maracahipes, B. S. Marimon, B. H. Marimon-Junior, C. Mendoza, P. Morandi, D. A. Neill, P. Núñez Vargas, E. A. Oliveira, E. Lenza, W. A. Palacios, M. C. Peñuela-Mora, J. J. Pipoly III, N. C. A. Pitman, A. Prieto, C. A. Quesada, H. Ramirez-Angulo, A. Rudas, K. Ruokolainen, R. P. Salomão, M. Silveira, J. Stropp, H. ter Steege, R. Thomas-Caesar, P. van der Hout, G. M. F. van der Heijden, P. J. van der Meer, R. V. Vasquez, S. A. Vieira, E. Vilanova, V. A. Vos, O. Wang, K. R. Young, R. J. Zagt, and Timothy R. Baker. 2016. Evolutionary heritage influences Amazon tree ecology. Proceedings of the Royal Society B: Biological Sciences. 283: 20161587.

Cohen, A., and J. J. van der Eyk. 1953. Klassificatie en ontstaan van savannen in Suriname. Geologie en Mijnbouw, new series 15: 202-214.

Cohen, M. A., W. R. C. Beaumont, and N. C. Thorp. 1999. Movement and activity patterns of the black piranha. Environmental Biology of Fishes 54: 45-52.

Coley P. D. 1986. Costs and benefits of defense by tannins in a neotropical tree. Oecologia 70: 238-241.

Colinvaux, P. A., P. E. De Oliveira, and M. B. Bush. 2000. Amazonian and neotropical plant communities on glacial time-scales: The failure of the aridity and refuge hypotheses. Quaternary Science Reviews 19: 141–169.

Collar, N. J. 2001. Family Trogonidae (Trogons). *In* J. Del Hoyo, A. Elliot, and J. Sargatal (Eds.). Handbook of Birds of the World, Vol. 6, pp. 80-129. Lynx Edicions, Barcelona, Spain.

Collar, N. J., and D. C. Wege. 1995. The distribution and conservation status of the Bearded Tachuri (*Polystictus pectoralis*). Bird Conservation International 5: 367-390.

Collins, C. T. 2015. Food habits and resource partitioning in a guild of Neotropical swifts. Wilson Journal of Ornithology 127: 239-248.

Collins, R. A., R. Britz, and L. Reuber. 2015. Phylogenetic systematics of leaffishes (Teleostei: Polycentridae, Nandidae). Journal of Zoological Systematics and Evolutionary Research 53: 259—272.

Compton, S. G., J. T. Wiebes, and C. C. Berg. 1996. The biology of fig trees and their associated animals. Journal of Biogeography 23: 405–407.

Conrads, L. A. 2015. Surinaamse kust onder water! Onvermijdelijk? Meteorologica 2: 4-5.

Constantino, R. 2002. The pest termites of South America: taxonomy, distribution and status. Journal of Applied Entomology 126: 355-365.

Contreras-Ramos, A. 2007. Recent accounts on the systematic and biogeography of Neotropical Megaloptera (Corydalidae, Sialidae). Annali del Museo Civico di Storia Naturale di Ferrara 8: 67-72.

Contreras-Ramos, A. 2011. Phylogenetic review of dobsonflies of the subfamily Corydalinae and the genus *Corydalus* Latreille (Megaloptera: Corydalidae). Zootaxa 2862: 1–38.

Contreras-Ramos, A., and K. von der Dunk. 2010. A new species of *Corydalus* Latreille from Venezuela (Megaloptera, Corydalidae). ZooKeys 67: 11–19.

Cook, J. M., and J.-Y. Rasplus 2003. Mutualists with attitude: coevolving fig wasps and figs. Trends in Ecology and Evolution 18: 241-248.

Cooper, S. 2015. Fork-tailed Palm-Swift (*Tachornis squamata*), Neotropical Birds Online (T. S. Schulenberg, Ed.). Cornell Lab of Ornithology. neotropical.birds.cornell.edu/portal/species/overview?p_p_spp=229211

Costa, A. A., and E. M. Varanda. 2002. Building of leaf shelters by *Stenoma scitiorella* Walker (Lepidoptera: Elachistidae): Manipulation of host plant quality? Neotropical Entomology 34: 537-540.

Costello, M. J., R. M. May, and N. E. Stork. 2013 Can we name Earth's species before they go extinct. Science 339: 413-416.

Courtois, E. A., C. E. T. Paine, P. A. Blandinieres, D. Stien, J. M. Bessiere, and E. Houel. 2009. Diversity of the volatile organic compounds emitted by 55 species of tropical trees: a survey in French Guiana. Journal of Chemical Ecology 35: 1349–1362.

Covain, R., S. Fisch-Muller, J. Montoya-Burgos, J. Mol, P. Y. Le Bail, and S. Dray. 2012. The Harttiini (Siluriformes, Loricariidae) from the Guianas: a multitable approach to assess their diversity, evolution, and distribution. Cybium 36: 115-161.

Cowan, R.S. 1975. A monograph of the genus *Eperua* (Leguminosae: Caesalpinioideae). Smithsonian Institution Contributions to Botany, no. 28, Smithsonian Institution, Washington, DC, USA.

Crabbe, S. 2012. Results of forest carbon assessment and monintoring project Suriname, period 2010-2011. Stichting Bosbeheer en Bostoezicht, Paramaribo, Suriname.

Crabtree, R. E., E. C. Cyr, D. C. Chaverri, W. O. McLarney, and J. M. Dean. 1997. Reproduction of tarpon, *Megalops atlanticus,* from Florida and Costa Rican waters and notes on their age and growth. Bulletin of Marine Science 61: 271–285.

Crabtree, R. E., E. C. Cyr, R. E. Bishop, L. M. Falkenstein, and J. M. Dean. 1992. Age and growth of tarpon, *Megalops atlanticus*, larvae in the eastern Gulf of Mexico, with notes on relative abundance and probable spawning areas. Environmental Biology of Fishes 35: 361-370.

Cracraft, J. 1985. Historical biogeography and patterns of differentiation within the South American avifauna: areas of endemism. Ornithological Monographs 36: 49-84.

Craig, T. P., L. M. Araújo, J. K. Itami, and G. W. Fernandes. 1991. Development of the insect community centered on a leaf-bud gall formed by a weevil (Coleoptera: Curculionidae) on *Xylopia aromatica* (Annonaceae). Revista Brasileira de Entomologia 35: 311-317.

Crampton, W. G. R. 2008. Diversity and adaptation in deep channel Neotropical electric fishes. *In* P. Sebert, D. W. Onyango and K. G. Kapoor (Eds.), Fish life in special environments, pp 283-339. Science Publishers, Enfield, CT, USA.

Crampton, W. G. R., and C. D. Hopkins. 2005. Nesting and parental care in the weakly electric fish *Gymnotus* (Gymnotiformes: Gymnotidae) with descriptions of larval and adult electric organ discharges of two species. Copeia 2005: 48-60.

Crayn, D. M., K. Winter, and J. A. C. Smith. 2004. Multiple origins of crassulacean acid metabolism and the epiphytic habit in the Neotropical family Bromeliaceae. Proceedings of the National Academy of Sciences of the United States of America 101: 3703-3708.

Cremers, L., J. Kolen, and M. De Theije (Eds.). 2013. Small-scale gold mining in the Amazon: the cases of Bolivia, Brazil, Peru, and Suriname. Cuadernos de CEDLA no. 26, Amsterdam, the Netherlands.

Cron, G. V., C. Pirone, M. Bartlett, W. J. Kress, and C. Specht. 2012. Phylogenetic relationships and evolution in the Strelitziaceae (Zingiberales).Systematic Botany 37: 606-619.

Crul, R., and L. Reyrink. 1980. Ecologische studie Kabalebo project. Hydrobiologisch onderzoek in het stroomgebied van de Corantijn en Kabalebo rivier in verband met de aanleg van het Devisvallen stuwmeer. Hydraulic Research Division (Waterloopkundige Afdeling), Ministry of Public Works, Paramaribo, Suriname.

Cruz, F. W. J., S. J. Burns, I. Karmann, W. D. Sharp, M. Vuille, A. O. Cardoso, J. A. Ferrari, P. L. S. Dias, and O. J. Viana. 2005. Insolation-driven changes in atmospheric circulation over the past 116,000 years in subtropical Brazil. Nature 434: 63–66.

Cuatrecasas, J. 1964. Cacao and its allies. A taxonomic revision of the genus *Theobroma*. Contributions of the U.S. National Herbarium 35: 379-605.

Curry, J. A., P. J. Webster, and G.J. Holland. 2006. Mixing Politics and Science in Testing the Hypothesis That Greenhouse Warming Is Causing a Global Increase in Hurricane Intensity. Bulletin of the American Meteorological Society. BAMS 87: 1025-1037.

D'Orgeix, C. A., I. V. Monagan, K. Robinson, P. Scott, and R. Jairam. 2015. *Ameerega trivittata* (Three-striped Poison-dart Frog). Male size and reproductive success. Herpetological Review 46: 72-73.

da Cunha, R. G. T., and R. W. Byrne. 2006. Roars of black howler monkeys (*Alouatta caraya*): evidence for a function in inter-group spacing. Behaviour 143: 1169-1199.

Da Silva, P. 2014. Avifaunal diversity in a mangrove reserve in Guyana, South America. International Journal of Science, Environment and Technology 3: 23-32.

Daly, D. C. 1987. A taxonomic revision of *Protium* (Burseraceae) in eastern Amazonia and the Guianas. PhD Dissertation, City University of New York, NY, USA.

Daly, D. C., and P. V. A. Fine. 2011. A new Amazonian section of *Protium* (Burseraceae) including both edaphic specialist and generalist taxa. Studies in Neotropical Burseraceae XVI. Systematic Botany 36: 939-949.

Daly, J. W., J. Caceres, R. W. Moni, F. Gusovsky, M. Moos Jr., K. B. Seamon, K. Milton, and C.W. Myers. 1992. Frog secretions and hunting magic in the upper Amazon: identification of a peptide that interacts with an adenosine receptor. Proceedings of the National Academy of Sciences 89: 10960-10963.

Damuth, J. E., and N. Kumar. 1975. Amazon Cone: morphology, sediments, age and growth pattern. Geological Society of America Bulletin 86: 863-878

DAOUST, C., G. VOICU, H. BRISSON, AND M. GAUTHIER. 2011. Geological setting of the Paleoproterozoic Rosebel gold district, Guiana Shield, Suriname. Journal of South American Earth Sciences 32: 222-245.

DARST, C. R., AND D. C. CANNATELLA. 2004. Novel relationships among hyloid frogs inferred from 12S and 16S mitochondrial DNA sequences. Molecular Phylogenetics and Evolution 31: 462-475.

DAVENNE, C., AND C. FLEURENT. 2011. Cabinets of wonder: a passion for collecting. Abrams, New York, NY, USA.

DAVIDSON, D. W. 1988. Ecological studies of neotropical ant gardens. Ecology 69: 1138-1152.

DAVIS, M. M., W. A. MITCHELL, J. S. WAKELEY, J. C. FISCHENICH, AND M. M. CRAFT. 1996. Environmental value of riparian vegetation. Technical Report EL-96-16, US Army Corps of Engineers, Vicksburg, MS, USA.

DAVIS, T. A. W., AND P. W. RICHARDS 1933. The vegetation of Moraballi Creek, British Guiana: an ecological study of limited area of tropica rain forest. Part I. Journal of Ecology 21: 350-384.

DAWS, M. I., D. F. R. P. BURSLEM, L. M. CRABTREE, P. KIRKMAN, C. E. MULLINS, AND J. W. DALLING. 2002. Differences in seed germination responses may promote coexistence of four sympatric *Piper* species. Functional Ecology 16: 258-267.

DBK. 1977. Overzichtsbodemkaart van Noord-Suriname ten noorden van de 5e breedtegraad, schaal 1:500.000. Dienst Bodemkartering (DBK), Paramaribo, Suriname. [soil map N part Suriname]

DE BOER, M. W. H. 1972. Landforms and soils in eastern Surinam (South America). PhD dissertation. Pudoc, Wageningen, The Netherlands.

DE BRY T. 1599. Americae pars VIII continens primo, descriptionem trium itinerum. Francofurti [Frankfurt am Main, Germany].

DE CASTRO, A. L., AND J. DE LOYOLA E SILVA. 1985. Manual de identificaçao de invertebrados límnicos do Brasil: isopoda. Conselho Nacional de Desenvolvimento Científico e Tecnológico (CNPq). Ministério da Ciência, Tecnologia, Inovações e Comunicações, Brasilia, Brazil.

DE DIJN, B. P. E. 2005. Flower-visiting insects in Guianan forests: pollinators, thieves, lovers and their foes. *In* D. S. Hammond (Ed.), Tropical forests of the Guiana Shield: Ancient forests in a modern world, pp. 321-342. CABI Publishing, Cambridge, MA, USA.

DE DIJN, B. P. E. 2012. Baseline study for the proposed bauxite mine at Nassau, Suriname (South America): Nassau concession terrestrial vegetation, final, edited report. Environmental Services & Support NV, Paramaribo, Suriname.

DE DIJN, B. P. E., I. E. MOLGO, M. A. NORCONK, L. T. GREGORY, B. O'SHEA, C. MARTY, M. LUGER, M. RINGLER, S. CROTHERS IV, B. NOONAN, K. FITZGERALD, S. MITRO, A. VREEDZAAM, AND D. SATYAWAN. 2007. The Biodiversity of Brownsberg. *In* L. E. Alonso, and J. H. Mol (Eds.). A rapid biologica assessment of the Lely and Nassau plateaus, Suriname (with additional information on the Brownsberg Plateau), pp. 135-155. RAP Bulletin of Biological Assessment 43, Conservation International, Arlington, VA, USA.

DE DIJN, B. P. E., J. BEHARI-RAMDAS, AND F. VAN TROON. 2005. Tafelberg expedition NZCS-BES March-April 2005: botany activities report, second draft. Bart De Dyn environmental consultancy, Paramaribo, Suriname.

DE FORESTA, H., AND J.-P. LESCURE. 1989. Les formations secondaires. Bois et Forêts des Tropiques 219: 61-68.

DE GRANVILLE, J.-J. 1986. Les formations végétales de la bande cotière de Guyane Francaise. *In* Anonymus (Ed.). Le littoral Guyanais, fragilité de l'environnement, pp. 47-63. Nature Guyanaise, Sepanguy, Cayenne, French Guiana, and Sepanrit, Talence, France.

DE GRANVILLE, J.-J. 1988. Phytogeographical characteristics of the Guianan forests. Taxon : 578-594.

DE GRANVILLE, J.-J. 1991. Remarks on the montane flora and vegetation types of the Guianas. Wildenowia 21: 201-213.

DE GRANVILLE, J.-J. 1994. Les formations végétales primaires de la zone intérieure de Guyane. *In* Anonymus (Ed.). Forêt guyanaise, gestion de l'écosysteme forestier et aménagement de l'espace regional, pp. 21-40. Sepanguy, Cayenne, French Guiana.

DE GRANVILLE, J.-J. 1999. Palms of French Guiana: diversity, distribution, ecology and uses. Acta Botanica Venezuelica 22: 109-125.

DE KOM, A. 1934. Wij slaven van Suriname. Uitgeverij Contact, Amsterdam, The Netherlands.

DE MORAES, P. L. R., J. DOBREFF, AND L. G. REINHAMMAR. 2014. The plants by Daniel Rolander (c. 1723-1793) in Diarium Surinamicum (1754-1756) and herbaria. Phytotaxa 165: 1-101.

DE MORAIS, J. W., J. ADIS, V. MAHNERT, AND E. BERTHI-FILHO. 1997. Abundance and phenology of Pseudoscorpiones (Arachnida) from a mixedwater inundation forest in Central Amazonia, Brazil. Revue Suisse de Zoologie 104: 475–483.

DE QUEIROZ, K. 2007. Species concepts and species delimitation. Systematic Biology 56: 879-886.

DE QUEIROZ, K., AND J. GAUTHIER. 2007. Phylogeny as a central principle in taxonomy: phylogenetic definitions of taxon names. Systematic Zoology 39: 307-322.

DE ROEVER, E. W. F., J. M. LAFON, C. DELOR, P. ROSSI, A. COCHERIE, C. GUERROT, AND A. POTREL. 2003a. The Bakhuis Ultra-high temperature granulite belt: I Petrological and geochronological evidence for a counterclockwise P-T path at 2 07-2.05 Ga. Géologie de la France 2,3,4: 175-205.

DE ROEVER, E. W. F., S. B. KROONENBERG, C. DELOR, AND D. PHILLIPS. 2003b. The Käyser dolerite, a Mesoproterozoic alkaline dyke suite from Suriname. Géologie de la France 2,3,4: 161-174.

DE SOUSA E SILVA JÚNIOR, J., J. A. OLIVEIRA, P. A. DIAS, AND T. GOMES DE OLIVEIRA. 2005. Update on the geographical distribution and habitat of the tapiti (*Sylvilagus brasiliensis*: Lagomorpha, Leporidae) in the Brazilian Amazon. Mammalia 69: 245-250.

DE THOISY, B., T. SPIEGELBERGER, S. ROUSSEAU, G. TALVY, I. VOGEL AND J. C. VIÉ. 2003. Distribution, habitat, and conservation status of the West Indian manatee *Trichechus manatus* in French Guiana. Oryx 37: 431-436.

DE VLETTER, D. R. 1984. Synthesis of the Precambrian of Suriname and review of some outstanding problems. Mededelingen Geologisch Mijnbouwkundige Dienst van Suriname 27: 11-30.

DE VLETTER, D. R., AND A. L. HAKSTEGE. 1998. The search for gold in Suriname. *In*: Th. E. Wong, D. R. De Vletter, L. Krook, J. I. S. Zonneveld, and A. J. van Loon (Eds.). The history of earth sciences in Suriname, pp. 311-349. Netherlands Institute of Applied Geoscience TNO, and Royal Netherlands Academy of Arts and Sciences, Amsterdam, The Netherlands.

De Vletter, D. R., G. J. J. Aleva, and S. B. Kroonenberg. 1998. Research into the Precambrian of Suriname. *In* T. E. Wong, D. R. de Vletter, L. Krook, J. I. S. Zonneveld, and A. J. van Loon (Eds.). The history of earth sciences in Suriname, pp. 15-64. Netherlands Institute of Applied Geoscience TNO, and Royal Netherlands Academy of Arts and Sciences, Amsterdam, The Netherlands.

Deckart, K., H. Bertrand, and J.-P. Liégeois. 2005. Geochemistry and Sr, Nd, Pb isotopic composition of the Central Atlantic Magmatic Province (CAMP) in Guyana and Guinea. Lithos 82: 289– 314.

DeFilipps, R. A., S. L. Maina, and J. Crepin. 2004 Medicinal plants of the Guianas (Guyana, Surinam, French Guiana). Department of Botany, National Museum of Natural History, Smithsonian Institution, Washington, DC, USA.

Dei Tos, C., L. C. Gomes, A. M. Ambrósio, and E. Goulart. 2010. An overview of freshwater fish aging in South America: the science, biases and future directions. Acta Scientiarum. Biological Sciences Maringá 32: 323-333.

Dejean, A., A. Quilichini, J. H. Delabie, J. Orivel, B. Corbara, and M. Gibernau. 2004. Influence of its associated ant species on the life history of the myrmecophyte *Cordia nodosa* in French Guiana. Journal of Tropical Ecology 20: 701-704.

Delariva, R. L., and A. A. Agostinho.2001. Relationship between morphology and diets of six Neotropical loricariids. Journal of Fish Biology 58: 832-847.

Delaval, M., M. Henry, and P. Charles-Dominique. 2005. Interspecific competition and niche partitioning: example of a Neotropical rainforest bat community. Revue d'Ecologie (Terre et Vie) 60: 149-166.

Delor, C., D. Lahondère, E. Egal, J. Lafon, C. Truffert, and H. Théveniaut. 2003a. Transamazonian crustal growth and reworking as revealed by the 1:500,000-scale geological map of French Guiana (2nd edition). Géologie de la France 2,3,4: 5-57.

Delor, C., E.W.F. de Roever, J.M. Lafon, D. Lahondère, Ph. Rossi, A. Cocherie, C. Guerrot, and A. Potrel. 2003b. The Bakhuis ultra-high temperature granulite belt (Suriname): II implications for late Trans-Amazonian crustal stretching in a revised Guiana Shield framework. Géologie de la France 2,3,4: 207-230.

Desbiez, A. L. J., and D. Kluyber. 2013. The role of giant armadillos (*Priodontes maximus*) as physical ecosystem engineers. Biotropica 45: 537–540.

Deutsch, L. A. 1983. An encounter between bush dog (*Speothos venaticus*) and paca (*Agouti paca*). Journal of Mammalogy 64: 532-533.

DeVries, P. J. 1987. The butterflies of Costa Rica and their natural history, Volume I: Papilionidae, Pieridae, Nymphalidae. Princeton University Press, Princeton, NJ, USA.

Di Beneditto, A. P. M., and R. M. A. Ramos. 2004. Biology of the marine tucuxi dolphin (*Sotalia fluviatilis*) in south-eastern Brazil. Journal of the Marine Biological Association of the United Kingdom 84: 1245-1250.

Di Fiore, A., Link, A., and C. J. Campbell. 2011. The atelines. *In* C. J. Campbell, A. Fuentes, K. C. MacKinnon, S. K. Bearder and R. M. Stumpf (Eds.). Primates in Perspective, pp. 155-188. Oxford University Press, Oxford U.K.

Dick, A. J., and R. M. Mitchell. 1979. Un comportement antipredateur du Gobe-mouche royal. Oiseau 49: 155-157.

Dick, C. W., E. Bermingham, M. R. Lemes, and R. Gribel. 2007. Extreme long-distance dispersal of the lowland tropical rainforest tree *Ceiba pentandra* L. (Malvaceae) in Africa and the Neotropics. Molecular Ecology 16: 3039–3049.

Diego, D., and M. D. M. Castro. 2008. Articulated anastomosing laticifers in species of Asclepiadeae (Asclepiadoideae, Apocynaceae) and their ecological significance. Revista Brasilera Botanica 31: 699-711. **dx.doi.org/10.1590/S0100-84042008000400015**

Digby, L.J., Ferrari, S.F., and W. Saltzman. 2011. Callitrichines: The role of competition in cooperatively breeding species. *In* C. J. Campbell, A. Fuentes, K. C. MacKinnon, S. K. Bearder and R. M. Stumpf (Eds.). Primates in Perspective, pp. 91-107. Oxford University Press, Oxford, UK.

Dill-McFarland, K. A., P. J. Weimer, J. N. Pauli, M. Z. Peery, and G. Suen. 2016. Diet specialization selects for an unusual and simplified gut microbiota in two- and three-toed sloths. Environmental Microbiology 18: 1391-1402.

Dixon, J. R., J. A. Wiest Jr., and J. M. Cei. 1993. Revision of the Neotropical snake genus *Chironius* Fitzinger (Serpentes, Colubridae). Museo Regionale di Scienze Naturali, Torino. MonografieXIII.

Dodson, C. H. 1962. Pollination and variation in the Subtribe Catasetinae (Orchidaceae). Annals of the Missouri Botanical Garden 49: 35-56.

Dodson, C. H. 1966. Ethology of some bees of the tribe Euglossini (Hymenoptera: Apinae). Journal Kansas Entomological Society 39: 607-629.

Dodson, C. H. 1978. The Catasetums (Orchidaceae) of Tapakuma, Guyana. Selbyana 2: 159-168.

Doerga, S. 2013. Macrophytes and water quality in running water, stagnant open water and wetlands in northern Suriname. MSc thesis, Institute for Graduate Studies and Research, Anton de Kom University Suriname, Paramaribo, Suriname.

Doody, J. S., B. Green, D. Rhind, C.M. Castellano, R. Sims, and T. Robinson. 2009. Population-level declines in Australian predators caused by an invasive species. Animal Conservation 12: 46-53.

Dost, H. (Ed.). 1973. Proceedings of the international symposium on Acid Sulphate Soils, Wageningen, 1973. ILRI publication 18, ILRI, Wageningen, The Netherlands.

Doyle, T. K., J. D. R. Houghton, P. F. O'Suilleabhain, V. J. Hobson, F. Marnell, J. Davenport, and G. C. Hays. 2008. Leatherback turtles satellite-tagged in European waters. Endangered Species Research 4: 23-31.

Dragtenstein, F. 2004. 'Trouw aan de blanken' Quassi van Nieuw Timotibo, twist en strijd in de 18de eeuw in Suriname. KIT Publishers, Amsterdam, The Netherlands.

Dressler, R. L. 1979. *Eulaema bombiformis*, *E. meriana*, and Müllerian mimicry in related species (Hymenoptera: Apidae). Biotropica 11: 144-151.

Dressler, R. L. 1982. Biology of the orchid bees (Euglossini). Annual Reviews Ecology & Systematics 13: 373-394.

Dubost, G., and O. Henry. 2006. Comparison of diets of the acouchy, agouti and paca, the three largest terrestrial rodents of French Guianan forests. Journal of Tropical Ecology 22: 641-651.

Dubs, B. 1992. Birds of southwestern Brazil: catalogue and guide to the birds of the Pantanal of Mato Grosso and its border areas. Schellenberg Druck AG, Pfäffikon, Switzerland.

Duellman, W. E. 1978. The biology of an Equatorial Herpetofauna in Amazonian Ecuador. Miscellaneous. Publications of the Univiversity of the Kansas Museum of Natural History 65: 1-352.

Duellman, W. E., and L. Trueb. 1994. Biology of Amphibians. Johns Hopkins University Press, Baltimore, MD, USA.

Duellman, W. E., and M. Lizana. 1994. Biology of a Sit-and-Wait Predator, the Leptodactylid Frog *Ceratophrys cornuta*. Herpetologica 50: 51-64.

Duijves, C., and M. Heemskerk. 2014. Gold miners' knowledge, attitudes & practices with regard to mercury: a study in four small-scale gold mining regions in Suriname. The GOMIAM research network & WWF-Guianas, Paramaribo, Suriname.

Duplaix, N. 1978. Observations on the ecology and behaviour of the giant river otter *Pteronura brasiliensis* in Suriname. Revue d'Ecologie (Terre et Vie) 34: 495-620.

Duplaix, N. 1980. Observations on the ecology and behavior of the giant river otter *Pteronura brasiliensis* in Suriname. Revue d'Ecologie (Terre et Vie) 34: 495–620.

Duplaix, N. 2001. Evaluation of the animal and plant trade in the Guianas: preliminary findings. WWF Guianas, Paramaribo, Suriname.

Duplaix, N. 2002. Giant otter final report. WWF-Guianas Rapid River Bio-assessments and Giant Otter Conservaton Project FG-40 FY2002. WWF Guianas, Paramaribo, Suriname.

Duplaix, N., M. Lingaard, and C. Sakimin. 2001. A survey of Kaburi Creek, west Suriname and its conservation implications. The Oceanic Society, San Francisco, CA, USA.

Duponchelle, F., M. Pouilly, C. Pecheyran, M. Hauser, J. F. Renno, J. Panfili, A. M. Darnaude, A. Garcia-Vasquez, F. Carvajal-Vallejos, C. Garcia-Davila, C. Doria , S. Berail, A. Donard, F. Sondag, R. V. Santos, J. Nunez, D. Point, M. Labonne, and E. Baras. 2016 Trans-Amazonian natal homing of giant catfishes Journal of Applied Ecology 53: 1611-1520.

Duputié, A., M. Delêtre, J.-J. De Granville, and D. Mckey. 2009. Population genetics of *Manihot esculenta* ssp. *flabellifolia* gives insight into past distribution of xeric vegetation in a postulated forest refugium area in northern Amazonia. Molecular Ecology 18: 2897–2907.

Dussourd, C. H. 1962. Pollination and variation in the subtribe Catasetinae (Orchidaceae). Annals of the Missouri Botanical Garden 49: 35-56.

Dussourd, D. E. 1990. The vein drain; or, how insects outsmart plants. Natural History 90: 44-49.

Dussourd, D. E., and A. M. Hoyle. 2000. Poisoned plusiines: toxicity of milkweed latex and cardenolides to some generalist caterpillars. Chemoecology 10: 11-16. **dx.doi.org/10.1007/PL00001810**

Dutech, C., Joly, H. I., and P. Jarne. 2004. Gene flow, historical population dynamics and genetic diversity within French Guianan populations of a rainforest tree species, *Vouacapoua americana*. Heredity 92: 69–77.

Dyer, L. A., D. K. Letourneau, W. Williams, and C. Dodson. 1999. A commensalism between *Piper marginatum* Jacq.(Piperaceae) and a coccinellid beetle at Barro Colorado Island, Panama. Journal of Tropical Ecology 15: 841-846.

Easteal, S. 1981. The history of introductions of *Bufo marinus* (Amphibia : Anura); a natural experiment in evolution. Biological Journal of the Linnean Society 16: 93-113.

Eberhard, W. G. 1975. The ecology and behavior of a subsocial pentatomid bug and two scelionid wasps: Strategy and counterstrategy in a host and its parasites. Smithsonian Contributions to Zoology 205: 1–39.

Edelman, T., and A. D. Kartoredjo. 1978. Verslag van de semi-detail bodemkartering Coropina (concentratiegebied 1). Rapport 61, Dienst Bodemkartering, Paramaribo, Suriname.

Edmond, J. M., M. R. Palmer, C. I. Measures, B. Grant, and R. F. Stallard. 1995. The fluvial geochemistry and denudation rate of the Guayana Shield in Venezuela, Colombia, and Brazil. Geochimica et Cosmochimica Acta 59: 3301-3325.

Edmonds, W. D., and J. Zidek. 2010. A taxonomic review of the neotropical genus *Coprophanaeus* Olscufieff, 1924 (Coleoptera, Scarabaeidae, Scarabaeinae. Insecta Mundi 649. digitalcommons.unl.edu/insectamundi/649

Eidmann, H., and F. Kühlhorn. 1970. Lehrbuch der entomologie. Verlag Paul Parey, Hamburg & Berlin, Germany.

Eisler, R., and S.N. Wiemeyer. 2004. Cyanide hazards to plants and animals from gold mining and related water issues. Reviews of Environmental Contamination and Toxicology 183: 21-54

Eisma, D., P. G. E. F. Augustinus, and C. Alexander. 1991. Recent and subrecent changes in the dispersal of Amazon mud. Netherlands Journal of Sea Research 28: 181-192.

Eisner, T. 2003. For love of insects. The Belknap Press of Harvard University Press, Cambridge, MA, USA, and London, England, UK.

Ellison, A. M, J. S. Denslow, B. A. Loiselle, and M. D. Brenes. 1993. Seed and seedling ecology of Neotropical Melastomataceae. Ecology 74: 1733-1749.

Ellison, A. M., M. S. Bank, B. D. Clinton, E. A. Colburn, K. Elliott, C. R. Ford , D. R. Foster, B. D. Kloeppel, C. M. Lovett, and J. Mohan. 2005. Loss of foundation species: consequences for the structure and dynamics of forested ecosystems. Frontiers in Ecology and the Environment 3: 479–486.

Eltz, T., A. Sager, and K. Lunau. 2005. Juggling with volatiles: exposure of perfume by displaying male orchid bees. Journal Comparative Physiology A 191: 575-581.

Eltz, T., Y. Zimmermann, J. Haftmann, R. Tweke, W. Francke, J. J. G. Quezada-Euan, and K. Lunau. 2007. Enfleurage, lipid recycling and the origin of perfume collection in orchid bees. Proceedings Royal Society B 274: 2843-2848.

Emanuels, J. 1968. Klassificatie der seizoenen. Report series 2, no. 6 of Meteorologische Dienst, Ministerie van Bouwwerk, Verkeer en Waterstaat, Paramaribo, Suriname.

EMMONS, L. H. 1989. Jaguar Predation on Chelonians. Journal of Herpetology 23: 311-314.

EMMONS, L. H., AND F. FEER. 1997. Neotropical rainforest mammals: a field guide (2nd edition). University of Chicago Press, Chicago, IL, USA.

ENGEL, M. S., AND F. DINGEMANS-BAKELS. 1980. Nectar and pollen resources for stingless bees (Meliponinae, Hymenoptera) in Surinam (South America). Apidologie: 341-350. **hal.archives-ouvertes.fr/hal-00890527**

ENGEL, S. 1981. Estuarine and lagoon stocks. Fisheries Department Suriname, Ministry of Agriculture, Paramaribo.

ERFTEMEIJER, P., AND P.A. TEUNISSEN. 2009. ICZM Plan Suriname – mangrove report: analysis of problems and solutions for the management of mangrove forests along Suriname's 'wild coast'. Lievense, Breda, The Netherlands.

ERSPAMER, V., G. F. ERSPAMER, C. SEVERINI, R. L. POTENZA, D. BARRA, G. MIGNOGNA, AND A. BLANCHI. 1993. Pharmacological studies of "Sapo" from the frog *Phyllomedusa bicolor* skin: a drug used by the Peruvian Matses Indians in Shamanic hunting practices. Toxicon 31: 1099-1111.

ERWIN, T. L. 1982. Tropical forests: their richness in Coleoptera and other arthropod species. The Coleopterists Bulletin 36: 74-75.

ERWIN, T. L. 1983. Tropical forest canopies: the last biotic frontier. Bulletin of the Entomological Society of America 29: 14-19.

ERWIN, T. L. 1997. Biodiversity at its utmost: tropical forest beetles. *In* M. L. Reaka-Kudla, D. E. Wilson, and E. O. Wilson (Eds.). Biodiversity II, pp. 27-40. Joseph Henry Press, Washington, DC, USA.

ESPARZA-DÍAZ, G., A. OLGUIN, L. K. CARTA, A. M. SKANTAR, AND R. T. VILLANUEVA. 2013. Detection of *Rhynchophorus palmarum* (Coleoptera: Curculionidae) and identification of associated nematodes in South Texas. Florida Entomologist 96: 1513-1521.

ESTRADA-PEÑA A., A. A. GUGLIELMONE, AND A. J. MANGOLD. 2004. The distribution and ecological 'preferences' of the TICK *Amblyomma cajennense* (Acari: Ixodidae), an ectoparasite of humans and other mammals in the Americas. Annals of Tropical Medicine and Parasitology 98: 283-92.

ESTRELIA, A., E. E. SANCHEZ, J. A. GALAN, W. A. TAO, B. GUERRERO, L. F. NAVARRETE, AND A. RODRIGUEZ-ACOSTA. 2011. Characterization of toxins from the broad-banded water snake *Helicops angulatus* (Linnaeus, 1758): isolation of a cysteine-rich secretory protein, Helicopsin. Archives of Toxicology 85: 305-313.

EVANGELISTA, D. A., K. CHAN, K. L. KAPLAN, M. M. WILSON, AND J. L. WARE. 2015. The Blattodea s.s. (Insecta, Dictyoptera) of the Guiana Shield. Zookeys 475: 37-87.

EVANGELISTA, D.A., E. SYLVAIN, C. M. MENDOZA, AND K. GUZMAN. 2016. New and enigmatic cockroaches (Insecta: Dictyoptera: Blattodea) of Guyana. The Journal of Natural History 1-28. **doi: 10.1080/00222933.2016.1193647**

EVANS, J. L., J. D. FUENTES, X. HU, AND H. HAMILTON. 2011. Earth-Atmosphere Interactions: Tropical Storm and Hurricane Activity in the Caribbean and Consequent Health Impacts. The Journal of Race & Policy 7: 53-74.

EVERT, R. F., S. E. EICHHORN, AND P. H. RAVEN. 2013. Raven Biology of plants. W. H. Freeman and Company Publishers, New York.

FANSHAWE, D. B. 1952. The Vegetation of Guyana. A preliminary review. Imperial Forestry Institute, University of Oxford, Oxford, England, UK.

FAWCETT, H. W. 1981. Het Brokopondo projekt. Suralco, Paramaribo, Suriname.

FEDIGAN, L.M., AND L.M. ROSE. 1995. Interbirth interval variation in three sympatric species of neotropical monkey. American Journal of Primatology 37: 9-24.

FERGUSON, M. W. J. 1985. The reproductive biology and embryology of the crocodilians. *In* C. Gans, F. Billett, and P. F. A. Maderson (Eds.). Biology of the Reptilia, Vol. 14, Development A, pp. 329-491. John Wiley & Sons, New York, NY, USA.

FERMIN, P. 1765. Histoire naturelle de la Hollande equinoxiale: ou déscription des animaux, plantes, fruits, et autres curiosités naturelles, qui se trouvent dans la colonie de Surinam; avec leurs noms différents, tant François, que Latins, Hollandois, Indiens & Négre-Anglois. M. Magerus, Amsterdam [The Netherlands].

FERNANDEZ, A. Z., A. TABLANTE, F. BARTOLI, S. BEGUIN, H. C. HEMKER, AND R. APITZ-CASTRO. 1999. Expression of biological activity of draculin, the anticoagulant factor from vampire bat saliva, is strictly dependent on the appropriate glycosylation of the native molecule. Biochimica et Biophysica Acta 1425: 291-299.

FERNANDEZ, M., J. LOBO, E. CHACO, AND M. QUESADA. 2008. Curculionid beetles in aborted flower buds and immature fruits of *Ceiba pentandra* (Bombacaceae). Plant Ecology 194: 1–4.

FERREIRA, A., B. KIAUTA, AND A. ZAHA. 1979. Male germ cell chromosomes of thirty-two Brazilian dragonflies. Odonatologica 8: 5-22.

FERREIRA, L. V., D. A. CUNHA, AND P. PAROLIN. 2014. Effects of logging on *Virola surinamensis* in an Amazonian floodplain forest. Environment Conservation Journal 15: 1-8.

FERRETTI, N., G. POMPOZZI, S. COPPERI, A. GONZALEZ, AND F. PEREZ-MILES. 2013. Sexual behaviour of mygalomorph spiders: when simplicity becomes complex an update of the last 21 years. Journal of Arachnology 16: 85–93.

FEUILLET, C. 2009. Checklist of the plants of the Guiana shield 1. An update to the angiosperms. Journal of the Botanical Research Institute of Texas 3(2): 799-814.

FFRENCH, R. P., AND F. HAVERSCHMIDT. 1970. The Scarlet Ibis in Surinam and Trinidad. Living Bird 9: 147-165.

FIABOE, K. K. M., A.T. PETERSON, M. T. K. KAIRO, AND A. L. RODA. 2012. Predicting the potential worldwide distribution of the Red Palm Weevil *Rhynchophorus ferrugineus* (Olivier) (Coleoptera: Curculionidae) using ecological niche modeling. Florida Entomologist 95: 659-673.

FINCKE, O. M. 1984. Giant damselflies in a tropical forest: reproductive biology of *Megaloprepus coerulatus* with notes on *Mecistogaster* (Zygoptera: Pseudostigmatidae). Advances in odonatology 2: 13-27.

FISCH-MULLER, S., R. COVAIN, AND J. H. MOL. (in press). An intregrative framework to reevaluate the Neotropical catfish genus *Guyanancistrus* (Siluriformes: Loricariidae) with particular emphasis on the *Guyanancistrus brevispinis* complex. [submitted to PLOS ONE]

FISHBEIN, M. 2001. Evolutionary innovation and diversification in the flowers of Asclepiadaceae. Annals of the Missouri Botanical Garden 88: 603-623.

FITTKAU, E. J. 1967. On the ecology of Amazonian rainforest streams. Atas do Simpósio sôbre a Biota Amazônica 3: 97-108.

FITTKAU, E. J. 1970. Role of caimans in the nutrient regime of mouth-lakes of Amazon affluents (an hypothesis). Biotropica 2: 138-142.

FITTKAU, E. J., AND H. KLINGE. 1973. On biomass and trophic structure of the Central Amazonian rain forest ecosystem. Biotropica 5: 2-14.

FLECKER, A. S. 1996. Ecosystem engineering by a dominant detritivore in a diverse tropical stream. Ecology 77: 1845-1854.

FLECKER, A. S., P. B. MCINTYRE, J. W. MOORE, J.T. ANDERSON, B. W. TAYLOR, AND R. O. HALL. 2010. Migratory fishes as material and process subsidies in riverine ecosystems. American Fisheries Society Symposium 73: 559-592.

FLEMING, T. H. 1988. The short-tailed fruit bat: a study in plant-animal interactions. University of Chicago Press, Chicago, IL, USA.

FLEMING, T. H. 2004. Dispersal ecology of neotropical *Piper* shrubs and treelets. *In* L. A. Dyer and A. D. N. Palmer (Eds.). *Piper*: a model genus for studies of phytochemistry, ecology, and evolution, pp. 58-77. Kluwer Academic Publishers, Dordrecht, the Netherlands.

FLORES-VINDAS, E. M. 1994. *Pentaclethra macroloba, Calophyllum brasiliense, Terminalia amazonia, Terminalia oblonga*. Arboles y Semillas del Neotrópico 3: 1-110.

FOCKE, H. C. 1849. Enumeratio diagnostic quarundam orchidearum surinamensium. Tijdschrift voor Wis- en Natuurkundige Wetenschappen 2: 122-125.

FOCKE, H. C. 1855. Aanteekeningen betreffende de in Suriname voorkoomende soorten van het geslacht *Vanilla*. West Indië 1: 275-280.

FOLGARAIT, P. J. 2013. Leaf-Cutter ant parasitoids: current knowledge. Psyche 2013: 539780. **dx.doi.org/10.1155/2013/539780**

FONSECA, G. A. B., C. M. RODRIGUEZ, G. MIDGLEY, J. BUSCH, L. HANNAH, AND R. A. MITTERMEIER. 2007. No forest left behind. PLoS Biology 5: 1645-1646.

FONTAINE, J. J., Y. ATTEMA, P. J. BENIOT, F. C. BUBBERMAN, S. W. DE GROOT, J. J. JANSSEN, C. DE JONG, B. NELEMANS, J. L. VOLDERS, AND J. B. CH. WEKKER. 1980. Uit Suriname's historie: fragmenten uit een bewogen verleden. Uitgeversmaatschappij Walburg Pers, Zutphen, The Netherlands.

FORGET, P.-M. 1991. Scatterhoarding of *Astrocaryum paramaca* by *Proechimys* in French Guiana: comparison with *Myoprocta exilis*. Tropical Ecology 32: 155-167.

FORGET, P.-M. 1994. Recruitment pattern of *Vouacapoua americana* (Caesalpiniaceae), a rodent-dispersed tree species in French Guiana. Biotropica 26: 408-419.

FORGET, P.-M. 1996. Removal of seeds of *Carapa procera* (Meliaceae) by rodents and their fate in rain forest in French Guiana. Journal of Tropical Ecology 12: 751-761.

FORGET, P.-M. 1997. Ten-year seedling dynamics of *Vouacapoua americana* in French Guiana. Biotropica 29: 124-126.

FORGET, P.-M., AND D. S. HAMMOND. 2005. Rainforest vertebrates and food plant diversity in the Guiana Shield. *In* D. S. Hammond (Ed.). Tropical forests of the Guiana Shield: Ancient forests in a modern world, pp. 233-294. CABI Publishing, Cambridge, MA, USA.

FORGET, P.-M., AND L. CUIJPERS. 2008. Survival and scatterhoarding of frugivore-dispersed seeds as a function of forest disturbance. Biotropica 40: 380-385.

FORGET, P.-M., AND P. A. JANSEN. 2007. Hunting increases dispersal limitation in the tree *Carapa procera*, a nontimber forest product. Conservation Biology 21: 106-113.

FORSBERG, B., C. A. ARAUJO-LIMA, L. A. MARTINELLI, R. L.VICTORIA, AND J. A. BONASSI. 1993. Autotrophic carbon sources for fish of the Central Amazon. Ecology 74: 643-652.

FOSSETTE, S., L. KELLE, M. GIRONDOT, E. GOVERSE, M. L. HILTERMAN, B. VERHAGE, B. DE THOISY, AND J.-Y. GEORGES. 2008. The world's largest leatherback rookeries: A review of conservation-oriented research in French Guiana / Suriname and Gabon. Journal of Experimental Marine Biology and Ecology 356: 69-82.

FOUQUET, A., B. P. NOONAN, M. T. RODRIGUES, N. PECH, A. GILLES, AND N. J. GEMMELL. 2012. Multiple quaternary refugia in the eastern Guiana shield revealed by comparative phylogeography of 12 frog species. Systematic Biology 61: 461–89.

FOUQUET, A., J-P. VACHER, V. KADOSOE, R. JAIRAM, AND P. OUBOTER. 2015. Checklist of the Amphibians of the Sipaliwini area, Suriname. Herpetology Notes 8: 63-68.

FOUQUET, A., M. VENCES, M. D. SALDUCCI, A. MEYER, C. MARTY, M. BLANC, AND A. GILLES. 2007a. Revealing cryptic diversity using molecular phylogenetics and phylogeography in frogs of the *Scinax ruber* and *Rhinella margaritifera* species groups. Molecular Phylogenetics and Evolution 43: 567–582.

FOUQUET, A., P. GAUCHER, M. BLANC, AND C. M. VELEZ-RODRIGUEZ. 2007b. Description of two new species of *Rhinella* (Anura: Bufonidae) from the lowlands of the Guiana shield. Zootaxa 1663: 17–32.

FRAGA, L. M., N. J. REIS, AND R. DALL'AGNOL. 2009. Cauarane - Coeroeni belt– the main tectonic feature of the Central Guyana shield, northern Amazonian Craton. Simpósio de Geologia da Amazônia 11, Manaus, pp. 1-3. Sociedade Brasileira de Geologia Nucleo Norte, Universidade Federal do Pará, Belem, Pará, Brazil.

FRAGASZY, D. M., VISALBERGHI, E., AND L. M. FEDIGAN. 2004. The Complete Capuchin: The biology of the genus *Cebus*. Cambridge University Press, Cambridge, England, UK.

FRAGOSO, J. M. V. 1997. Tapir-generated seed shadows: scale-dependent patchiness in the Amazon rain forest. Journal of Ecology 85: 519-529.

FRAGOSO, J. M. V. 1998. Home range and movement patterns of white-lipped peccary (*Tayassu pecari*) herds in the northern Brazilian Amazon. Biotropica 30: 458-469.

FRAGOSO, J. M. V., AND J. M. HUFFMAN. 2000. Seed-dispersal and seedling recruitment patterns by the last Neotropical megafaunal element in Amazonia, the tapir. Journal of Tropical Ecology 16: 369-385.

FRANK, J. H., J. P. PARKMAN, AND F. D. BENNETT. 1995. *Larra bicolor* (Hymenoptera: Sphecidae) a biological control agent of *Scapteriscus* mole crickets (Orthoptera: Gryllotalpidae) established in Northern Florida. The Florida Entomologist 78: 619-623.

FREDERICKSON, M. E. 2009. Conflict over reproduction in an ant-plant symbiosis: Why *Allomerus octoarticulatus* ants sterilize *Cordia nodosa* trees. The American Naturalist 173: 675-681.

FRENCH, J. A., BALES, K. L., BAKER, A. J., AND J. M. DIETZ. 2003. Endocrine monitoring of wild dominant and subordinate female *Leontopithecus rosalia*. International Journal of Primatology 24: 1281-1300.

FRÉRY, N., R. MAURY-BRACHET, E. MAILLOT, M. DEHEEGER, B. DE MÉRONA, AND A. BOUDOU. 2001. Gold-mining activities and mercury contamination of native Amerindian communities in French Guiana: key role of fish in dietary uptake. Environmental Health Perspectives 109: 449-456.

FRITH, C. B. 1978. The function of display and coloration in the Sunbittern. Avicultural Magazine 84: 150-157.

FROMARD, F., H. PUIG, E. MOUGIN, G. MARTY, J. L. BETOULLE, AND L. CADAMURO. 1998. Structure, above-ground biomass and dynamics of mangrove ecosystems: new data from French Guiana. Oecologia 115: 39-53.

FROST, D. R. 2015. Amphibian Species of the World: an Online Reference. Version 6.0 (October/November 2015). American Museum of Natural History, NY, USA. research.amnh.org/herpetology/amphibia/index.html.

FUKUSHIMA, C. S. 2011.Taxonomical revision & cladistic analysis of *Avicularia* Lamarck 1818 (Araneae, Theraphosidae, Aviculariinae). PhD dissertation, Instituto de Biociências, Universidade de São Paolo, São Paolo, Brazil.

FULY, A. L., O. L. MACHADO, E. W. ALVER, AND C. R. CARLINI. 1997. Mechanism of inhibitory action on platelet activation of a phospholipase A2 isolated from *Lachesis muta* (Bushmaster) snake venom. Thrombosis and Heamostasis 78: 1372-1380.

FUNK, V., T. HOLLOWELL, P. BERRY, C. KELLOFF, AND S. N. ALEXANDER. 2007. Checklist of the Plants of the Guiana Shield (Venezuela: Amazonas, Bolivar, Delta Amacuro; Guyana, Surinam, French Guiana). Contributions from the United States National Herbarium 55: 1-584.

FUTUYMA, D. J., AND G. MORENO. 1988. The evolution of ecological specialization. Annual Review of Ecology and Systematics 19: 207-233.

GABLE, F., AND D. AUBREY. 1990. Potential Impact of Contemporary Changing Climate on Caribbean Coastlines. Ocean and Shoreline Management 13: 35-67.

GAJAPERSAD, K., A. MACINTOSH, A. BENITEZ, AND E. PAYÁN. 2011. A survey of the large mammal fauna of the Kwamalasamutu region, Suriname. In O'Shea, B. J., L. E. Alonso, and T. H. Larsen (Eds.). A Rapid Biological Assessment of the Kwamalasamutu region, Southwestern Suriname, pp. 150-156. RAP Bulletin of Biological Assessment 63, Conservation International, Arlington, VI, USA.

GALATTI, U. 1992. Population Biology of the Frog *Leptodactylus pentadactylus* in a Central Amazonian rainforest. Journal of Herpetology 26(1): 23-31.

GAMBLE T., A. M. BAUER, G. R. COLLI, E. GREENBAUM, T. R. JACKMAN, L. J. VITT, AND A. M. SIMONS. 2011. Coming to America: Multiple Origins of New World Geckos. Journal of Evolutionary Biology 24: 231-244.

GARB J. E., A. GONZALEZ, AND R. G. GILLESPIE. 2004. The black widow spider genus *Latrodectus* (Araneae: Theridiidae): phylogeny, biogeography, and invasion history. Molecular Phylogenetics and Evolution 31: 1127–1142.

GARBER, P. A. 1988. Foraging decisions during nectar feeding by tamarin monkeys (*Saguinus mystax* and *Saguinus fuscicollis*, Callitrichidae, Primates) in Amazonian Peru. Biotropica 20: 100-106.

GARCIA-DAVILA, C., D. CASTRO-RUIZ, J. F. RENNO, W. CHOTA-MACUYAMA, F. M. CARVAJAL-VALLEJOS, H. SANCHEZ, C. ANGULO, C. NOLORBE, J. ALVARADO, G. ESTIVALS, J. NUNEZ-RODRIGUEZ, AND F. DUPONCHELLE. 2015. Using barcoding of larvae for investigating the breeding seasons of Pimelodid catfishes from the Marañon, Napo and Ucayali rivers in the Peruvian Amazon. Journal of Applied Ichthyology 31: 40-51.

GARRETT, R. W., K. A. CARLSON, M. S. GOGGANS, M. H. NESSON, C. A. SHEPARD, AND R. M. S. SCHOFIELD. 2016. Leaf processing behavior in *Atta* leafcutter ants: 90% of leaf cutting takes place inside the nest, and ants select pieces that require less cutting. Royal Society Open Science 3: 150111. **doi: 10.1098/rsos.150111**

GARRISON, J. S. E., AND C. L. GASS. 1999. Response of a traplining hummingbird to changes in nectar availability. Behavioral Ecology 10: 714-725.

GARRISON, R. W., N. VON ELLENRIEDER, AND J. A. LOUTON. 2006. Dragonfly genera of the New World: An illlustrated and annotated key to the Anisoptera. The Johns Hopkins University Press, Baltimore, MD, USA.

GARRISON, R. W., N. VON ELLENRIEDER, AND J. A. LOUTON. 2010. Damselfly genera of the New World. An illustrated and annotated Key to the Zygoptera. The John Hopkins University Press, Baltimore, MD, USA.

GAUTIER, J. Y., P. PLANQUETTE, AND Y. ROUGER. 1988. Etude éthologique de la relation mâle-femelle au cours du cycle de reproduction chez *Hoplosternum littorale*. Revue d'Ecologie (Terre et Vie) 43: 389-398.

GEERINCKX, T., A. HERREL, AND D. ADRIAENS. 2011. Suckermouth armored catfish resolve the paradox of simultaneous respiration and suction attachment: a kinematic study of *Pterygoplichthys disjunctivus*. Journal of Experimental Zoology 315: 121-131.

GEIGER, S. P., J. J. TORRES, AND R. E. CRABTREE. 2000. Air breathing and gill ventilation frequencies in juvenile tarpon, *Megalops atlanticus*: responses to changes in dissolved oxygen, temperature, hydrogen sulfide, and pH. Environmental Biology of Fishes 59: 181-190.

GEIJSKES, D. C. 1942. Observations on temperature in a tropical river. Ecology 23: 106-110.

GEIJSKES, D. C. 1943. De visscherij in het district Nickerie. Ministry of Agriculture, Paramaribo, Suriname.

GEIJSKES, D. C. 1954. Het dierlijk voedsel van de bosnegers aan de Marowijne. Vox Guyane 1: 61-83.

GEIJSKES, D. C. 1984. Notes on Megaloptera from the Guyanas, S. Am. *In* U. Gepp, H. Aspock, and H. Holzel (Eds.). Progress in World's Neuropterology, Proceedings of the 1st International Symposium on Neuropterology, 22-26 September, pp. 79-84. Graz, Austria.

GENOWAYS, H. H., H. A. REICHART, AND S. L. WILLIAMS. 1982. The Suriname small mammal survey: a case study of the cooperation between research and national conservation needs. *In* M.A. Mares, and H.H. Genoways (Eds.). Mammalian biology in South America, pp. 491-594. University of Pittsburgh, Pymatuning Laboratory of Ecology, Special Publication Series, Volume 6, Linesville, PA, USA.

GENTRY, A. H. 1982. Patterns of neotropical plant species diversity. Evolutionary Biology 15: 1-84.

GERMAN, P. P. 2008. Beavers of the fish world: can wood-eating catfishes actually digest wood? PhD dissertation, University of Florida, Gainesville, FL, USA.

GERNAAT, H. B. P. E., B. G. BECKLES, AND T. VAN ANDEL. 2012. Butterflies of Suriname. A natural history. KIT Publishers, Amsterdam, The Netherlands.

GÉRY, J. 1977. Characoids of the world. TFH Publications, Neptune City, NJ, USA.

GHOSH, S. 2010. Wetland macrophytes as toxic metal accumulators. International Journal of Environmental Sciences 1: 523.

GIANNININI, A., Y. KUSHNIR, AND M. CANE. 2000. Interannual variability of Caribbean rainfall, ENSO and the Atlantic Ocean. Journal of Climate 13: 297-311.

GIBBS, A.K. AND C.N. BARRON 1993. Geology of the Guiana shield. Oxford University Press, Oxford, England, UK.

GIBERNAU, M., D. BARABÉ, D. LABAT, P. CERDAN, AND A. DEJEAN. 2003. Reproductive biology of *Montrichardia arborescens* (Araceae) in French Guiana. Journal of Tropical Ecology 19: 103-107.

GIBSON, J. P., AND N. T. WHEELWRIGHT. 1995. Genetic structure in a population of a tropical tree *Ocotea tenera* (Lauraceae): influence of avian seed dispersal. Oecologia 103: 49-54.

GILBERT, J. A., J. K. JACKSON, AND R. KNIGHT. 2014. The Earth Microbiome project: successes and aspirations. BMC Biology 12: 69. **www.biomedcentral.com/1741-7007/12/69**

GILL, F. B. 1988. Trapline foraging by hermit hummingbirds: competition for an undefended, renewable resource. Ecology 69: 1933-1942.

GILLESPIE, L. J., AND W. S. ARMBRUSTER. 1997. A contribution to the Guianan flora: Dalechampia, *Haematostemon*, *Omphalea*, *Pera*, *Plukenetia*, and *Tragia* (Euphorbiaceae) with notes on subfamily Acalyphoideae. Smithsonian Contributions to Botany 86: 14-28.

GILLIARD, E. T. 1962. On the breeding behavior of the Cock-of-the-Rock (Aves, *Rupicola rupicola*). Bulletin of the American Museum of Natural History 124: 31-68.

GIVNISH, T. J., K. C. MILLAM, P. E. BERRY, AND K. J. SYTSMA. 2007. Phylogeny, adaptive radiation, and historical biogeography of bromeliaceae inferred from ndhF sequence data. Aliso 23: 3–26.

GIVNISH, T. J., T. M. EVANS, M. L. ZJHRA, T. B. PATTERSON, P. E. BERRY, AND K. J. SYTSMA. 2000. Molecular evolution, adaptive radiation, and geographic diversification in the amphiatlantic family Rapateaceae: evidence from ndh F sequences and morphology. Evolution 54: 1915-1937.

GLASTRA, R. 1983. Notes on a population of *Caiman crocidilus crocodilus* depleted by hide hunting in Suriname. Biological Conservation 26:149-162.

GLEASON, H. A. 1926. The individualistic concept of the plant association. The Bulletin of the Torrey Botanical Club 53: 7-26.

GLOOR, M., R. J. W. BRIENEN, D. GALBRAITH, T. R. FELDPAUSCH, J. SCHÖNGART, J.-L. GUYOT, J. C. ESPINOZA, J. LLOYD, AND O. L. PHILLIPS. 2013. Intensification of the Amazon hydrological cycle over the last two decades. Geophysical Research Letters 40: 1729–1733.

GMD. 1977. Geological map of Suriname 1: 500.000. Geologisch Mijnbouwkundige Dienst (GMD), Paramaribo, Suriname.

GOMEZ, D., AND M. THÉRY. 2004. Influence of ambient light on the evolution of colour signals: comparative analysis of a Neotropical rainforest bird community. Ecology Letters 7: 279-284.

GONÇALVES JR., J. F., F. A. ESTEVES, AND M. CALLISTO. 2003. Chironomids colonization on *Nymphaea ampla* L. detritus during a degradative ecological succession experiment in a Brazilian coastal lagoon. Acta Limnology Brasiliensia 15: 21-27.

GONZALEZ, F., AND N. PABON-MORA. 2015 Trickery flowers: the extraordinary chemical mimicry of *Aristolochia* to accomplish deception to its pollinators. New Phytologist 206: 10-13.

GONZÁLEZ, N. AND C. VISPO. 2003. Aspects of the diets and feeding ecologies of fish from nine floodplain lakes of the lower Caura, Venezuelan Guayana. Scientia Guaianae 12: 329-366.

GONZALEZ, V. H., AND V. MAHNERT. 2008. A new host record for *Dasychernes inquilinus* (Arachnida, Pseudoscorpiones, Chernetidae), with an overview of pseudoscorpion-bee relationships. Journal of Arachnology 35: 470–474.

GOODLAND, R. J. A. 1979. Environmental optimization in hydrodevelopment of tropical forest regions. *In* R.S. Panday (Ed.). Man-made Lakes and Human Health, pp. 10-20. Proceedings of the Symposium on Man-made Lakes and Human Health, Paramaribo, Suriname.

GÖRNER, K., AND K. DOBAT. 1991. Reise nach Suriname, Pflanzen- und Landschafstbilder der Louise von Panhuys 1763-1844. Helmut Burkhardt, Senkenbergischen Bibliothek, Johann Wolfgang Goethe-Universität, Frankfurt am Main, Germany.

GORZULA, S. J. 1977. Foam nesting in leptodactylids: a possible function. British Journal of Herpetology 5: 657-659.

GOTTSBERGER, G. 1994. As anonaceae do cerrado e a sua polinizaçao (The Annonaceae of the Brazilan cerrado and their pollination). Revista Brasileira de Biologia 54: 391-402.

GOTTSBERGER, G. 1999. Pollination and evolution in neotropical Annonaceae. Plant Species Biology 14: 143-152.

GOUDA, E. J. 1987. Bromeliaceae, subfamily Tillandsioideae. *In* A. R. A. Gorts-Van Rijn (Ed.). Flora of the Guianas, Series A: Phanerogams fasc. 3. Koeltz Scientific Books, Koenigstein, Germany.

GOULDING, M. 1980. The fishes and the forest: explorations in Amazonian natural history. University of California Press, Berkeley and Los Angeles, CA, USA.

GOULDING, M., M. L. CARVALHO, AND E. G. FERREIRA. 1988. Rio Negro rich life in poor water. SPB Publishing, The Hague, The Netherlands.

GOVERSE, E. 2003. The sea turtles of Suriname 2002 project: aerial survey of the coastline of eastern Suriname and nesting beach characteristics. Forest and Environmental Conservation Project, WWF Guianas, Paramaribo, Suriname.

GRANT, T., D. R. FROST, J. P. CALDWELL, R. GAGLIARDO, C. F. B. HADDAD, P. J. R. KOK, D. B. MEANS, B. P. NOONAN, W. E. SCHARGEL, AND W. C. WHEELER. 2006. Phylogenetic systematics of dart-poison frogs and their relatives (Amphibia: Athesphatanura: Dendrobatidae). Bulletin of the American Museum of Natural History 299: 1-262.

GRATTO-TREVOR, C., R. I. G. MORRISON, D. MIZRAHI, D. B. LANK, P. HICKLIN, AND A. L. SPAANS. 2012. Migratory connectivity of Semipalmated Sandpipers: winter distribution and migration routes of breeding populations. Waterbirds 35: 83-95.

Graves, G. R. 1990. Function of crest displays in Royal Flycatchers (*Onychorhynchus*). Condor 92: 522-524.

Grazia, J, A. R. Panizzi, C. Greve, C. F. Schwertner, L. A. Campos, T. A. Garbelotto, and J. A. M. Fernandes. 2015. Stink Bugs (Pentatomidae). *In* A. R. Panizzi, and J. Grazia. (Org.). True Bugs (Heteroptera) of the Neotropics (1st ed.), pp. 681-756. Springer Science, Dordrecht, The Netherlands.

Grazia, J., and C. F. Schwertner. 2017. Stink bug classification, phylogeny, biology and reproductive behavior. *In* A. ockl, and M. Borges (Eds.). Stink bugs biorational control based on communication processes, pp. 1-30. CRC Press, Boca Raton, FL, USA. **doi.org/10.1201/9781315120713-2**

Greene, H. W., and R. W. McDiarmid. 1981. Coral snake mimicry: Does it occur? Science 213: 1207-1212.

Greeney, H. F. 2001. The insects of plant held waters: a review and bibliography. Journal of Tropical Ecology 17: 241-260.

Greeney, H. F., and J. Port. 2010. A nest of Brown Nunlet (*Nonnula brunnea*) and observations on the nesting of other Ecuadorian puffbirds. Ornitología *Colombiana* 9: 31-37.

Gregory L. T., and M. A. Norconk. 2013. Comparative socioecology of sympatric, free-ranging white-faced and bearded sakis in Suriname: preliminary data. *In* L. M. Veiga, A. A. Barnett, S. F. Ferrari & M. A. Norconk (Eds). Evolutionary Biology and Conservation of Titis, Sakis and Uacaris, pp. 285-294. Cambridge University Press, Cambridge, England, UK.

Gregory-Wodzicki, K. M. 2000. Uplift history of the Central and Northern Andes: A review. Geological Society of America Bulletin 112: 1091–1105.

Grenand, P., C. Moretti, H. Jacquemin, and M.-F. Prévost. 2004. Pharmacopées Traditionelles en Guyane – Créoles, Palikur, Wayãpi. Collection Mémoires n°108, Orstom éditions, Paris, France.

Gribel, R., P. E. Gibbs, and A. L. Queiro. 1999. Flowering phenology and pollination biology of *Ceiba pentandra* (Bombacaceae) in central Amazonia. Journal of Tropical Ecology 15: 247–263.

Griffith, G., H. Mac Donald, and G. Smith. 2016. Context analysis: to assist World Wildlife Fund Guianas in the strategic planning process. Attune, Paramaribo, Suriname.

Grimm, E. C., S. Lozano-Garcia, H. Behling, and V. Markgraf. 2001. Holocene Vegetation and Climate Variability in the Americas. *In* V. Markgraf (Ed.). Past Climate Variability from the Last Glacial Maximum to the Holocene in South America, pp. 325-370. Academic Press, University of Colorado, Boulder, CO, USA.

Grishin, E. V. 1998. Black widow spider toxins: the present and the future. Toxicon 36: 1693–1701.

Groen, J. 1998. Hydrogeological investigations in Suriname. *In* T. E. Wong, D. R. De Vletter, L. Krook, J. I. S. Zonneveld, and A. J. Van Loon (Eds.). The history of earth sciences in Suriname, pp. 129-174. Netherlands Institute of Applied Geoscience TNO, and Royal Netherlands Academy of Arts and Sciences, Amsterdam, The Netherlands.

Groen, J. 2002. The effects of transgressions and regressions on coastal and offshore groundwater. PhD dissertation, Vrije Universiteit Amsterdam, Amsterdam, The Netherlands.

Grumbine, R. E.1994. What is ecosystem management? Conservation Biology 8: 27-38.

Guedes, F. L., D. Attili-Angelis, and F. C. Pagnocca. 2012. Selective isolation of dematiaceous fungi from the workers of *Atta laevigata* (Formicidae: Attini). Folia Microbiologia 57: 21-6. **doi: 10.1007/s12223-011-0081-6**

Guglielmone, A.A., L. Beati, D. M. Barros-Battesti, M. B. Labruna, S. Nava, J. M. Venzal, A. J. Mangold, M. P. Szabó, J. R. Martins, D. González-Acuña, and A. Estrada-Peña. 2006. Ticks (Ixodidae) on humans in South America. Experimental and Applied Acarology 40:83-100.

Guitet S., O. Brunaux, J.-J. de Granville, S. Gonzalez, and C. Richard-Hansen. 2015a. Catalogue des habitats forestiers de Guyane. DEAL Guyane, Office National de Forêts, Cayenne, French Guiana.

Guitet, S., R. Pélissier, O. Brunaux, G. Jaouen, and D. Sabatier. 2015b. Geomorphological landscape features explain floristic patterns in French Guiana rainforest. Biodiversity and Conservation 24: 1215-1237.

Gunst, N., S. Boinski, and D. M. Fragaszy. 2010. Development of skilled detection and extraction of embedded prey by wild brown capuchin monkeys (*Cebus apella apella*). Journal of Comparative Psychology 124: 194-204.

Gutierrez, E. 1999. Two New Species of the Cockroach Genus *Eurycotis* Stål (Dictyoptera: Blattidae: Polyzosteriinae) from Cuba and Puerto Rico (West Indies). Proceedings of the Natural Sciences of Philadelphia 149: 65-69.

Gutierrez, E. 2013. El Género Eurycotis (Dictyoptera: Blattaria: Blattidae: Polyzosteriinae) 1, Especies De La Hispaniola, Primera Parte: Seis Especies Nuevas. Novitates Caribaea 6: 16-35.

Haber, W. A., and G. W. Frankie. 1989. A tropical hawkmoth community: Costa Rican dry forest Sphingidae. Biotropica 21: 155-172.

HACAS. 2016. Hydrogeological Assessment of the Coastal Aquifers in Suriname. Surinaamse Waterleiding Maatschappij, Paramaribo, Suriname.

Haddad-Jr, V., D. Garrone-Neto, J. B. Paula-Neto, F. P. L. Marques, and K. C. Barbaro. 2004. Freshwater stingrays: study of epidemiologic, clinic and therapeutic aspects based on 84 envenomings in humans and some enzymatic activities of the venom. Toxicon 43: 287-294.

Haffer, J. 1969. Speciation in Amazonian forest birds. Science 165: 131-137.

Hakim, M. A., S. Yang, and R. Lai. 2015. Centipede venoms and their components: resources for potential therapeutic applications. Toxins 7: 4832-4851.

Hallé, F., and R. A. A, Oldeman. 1975. An essay on the architecture and dynamics of growth of tropical trees. Penerbit Universiti Malaya, Kuala Lumpur, Malaysia.

Hallé, F., R. A. A. Oldeman, and P. B. Tomlinson. 1978. Tropical trees and forests: an architectural analysis. Springer Verlag, Berlin, Germany.

Hamada, N., A. L. S. Gomes, G. Couturier, and B. Ronchi-Teles. 1998. Insectos associados a gravioleira (*Annona muricata* L., Annonaceae) na regiao de Manaus, Amazonas, Brasil. Acta Amazonica 28: 425-431.

HAMMOND, D. S. (ED.). 2005. Tropical Forests of the Guiana Shield: Ancient Forests in a Modern World. CABI publishing, Wallingford, England, UK, and Cambridge, MA, USA.

HAMMOND, D. S., AND V. K. BROWN. 1995. Seed size of woody plants in relation to disturbance, dispersal, soil type in wet neotropical forests. Ecology 76: 2544-2561.

HANSKI, I, AND Y. CAMBEFORT. 1991. Dung Beetle Ecology. Princeton University Press, Princeton, NJ, USA.

HARCOURT, R. 1707. Scheeps-togt van Robert Harcourt na Gujana, gedaan in het jaar 1608. Aanwijsende de gelegentheeden en hoedanigheeden, deses landschaps, des selfs eylanden, rivieren, grens-palen, verscheydenheyd der volkeren en talen, jaar-getijden, tijd-rekeningen, dood-malen, spijse en drank: als mede veelerley soort van dieren, vogelen, vissen fryuten, suyker-riet, katoen, verw-stoffen, kostelijke gommen, balsem, droogeryen, tabak, & c. Door den reysiger selfs in het Engels beschreeven, en nu alder-eerste uyt die spraak vertaalt. Met een Volkoomen Register verrijkt. Pieter Vander Aa, Leyden [Leiden, The Netherlands].

HARDNER, J. J., R. E. GULLISON, AND P. P. LOWRY II. 2015. Offsets: conservation served by flexibility. Nature 524: 415.

HARIPERSAD-MAKHANLAL, A., AND P. E. OUBOTER. 1993. Limnology: physico-chemical parameters and phytoplankton composition. *In* P.E. Ouboter (Ed.). The freshwater ecosystems of Suriname, pp. 53-75. Kluwer Academic Publishers, Dordrecht, The Netherlands.

HARRIS, J. G., AND M. W. HARRIS. 1994. Plant identification terminology: an illustrated glossary. Spring Lake Publishing, Spring Lake, UT, USA

HARTE, J., AND J. KITZES. 2015 Inferring regional-scale species diversity from small-plot censuses. PLOS ONE 10: e0117527. **doi: 10.1371/journal.pone.0117527**

HARTKE, T. R., AND R. B. ROSENGAUS. 2011. Heterospecific pairing and hybridization between *Nasutitermes corniger* and *N. ephratae.* Naturwissenschaften 98: 745-753.

HARTMANN, T. 2009. Pyrrolizidine alkaloids: the successful adoption of a plant chemical defense. *In* W. E. Conner (Ed.). Tiger moths and woolly bears. Behavior, ecology and evolution of the Arctiidae, pp. 55-81. Oxford University Press, Oxford, England, UK.

HARTSINCK, J. J. 1770. Beschryving van Guiana of de Wilde Kust in Zuid-America. Betreffende de aardrykskunde en historie des lands, de zeeden en gewoontes der inwooners, de dieren, vogels, visschen, boomen en gewassen, als mede de eerste ontdekking dier kust, de bezittingen des Spanjaarden, Franschen en Portugeezen en voornamelyk de volksplantingen der Nederlanderen, als Essequebo, Demerary, Berbice, Suriname; en derzelver rivieren, met de noodige kaarten en afbeeldingen der forten. Waarbij komt eene verhandeling over den aart en de gewoontes der neger-slaaven. Gerrit Tielenburg, Amsterdam [The Netherlands; 2 volumes]

HARVEY, B. C., AND T. E. LISLE. 1998. Effects of suction dredging on streams: a review and an evaluation strategy. Fisheries Habitat 23: 8-17.

HARVEY, M. S. 2013a, Order Pseudoscorpiones. *In* Z.-Q. Zhang (Ed.). Animal Biodiversity An Outline of Higher-Level Classification and Survey of Taxonomic Richness (Arachnida). Zootaxa 3703: 1–82.

HARVEY, M. S. 2013b. Pseudoscorpions of the World, version 3.0. Western Australian Museum, Perth, Australia. **www.museum.wa.gov.au/catalogues/pseudoscorpions**

HAVERSCHMIDT, F. 1953. Notes on the life history of the Blood-colored Woodpecker in Surinam. Auk 70: 21-25.

HAVERSCHMIDT, F. 1954a. Evening flights of the Southern Everglade Kite and the Blue and Yellow Macaw in Surinam. Wilson Bulletin 66: 264-265

HAVERSCHMIDT, F. 1954b. The occurrence of *Muscivora tyrannus* in Surinam. Auk 71: 327.

HAVERSCHMIDT, F. 1962. Notes on the feeding habits and food of some hawks of Surinam. Condor 64: 154-158.

HAVERSCHMIDT, F. 1968. Birds of Surinam. Oliver & Boyd, Edinburgh, Scotland, and London, England, UK.

HAVERSCHMIDT, F. 1970. Notes on the Snail Kite in Surinam. Auk 87: 580-584.

HAVERSCHMIDT, F., AND G. F. MEES. 1994. Birds of Suriname. Vaco, Paramaribo, Suriname.

HAYS, P., AND R. VIEIRA. 2005. Mercury contamination, a legacy to handicap a generation. WWF Guianas, Paramaribo, Suriname.

HEBERT, P. D. N., A. CYWINSKA, S. L. BALL AND J. R. DEWAARD. 2003. Biological identifications through DNA barcodes. Proceedings of the Royal Society London B 270: 313-321.

HECKENBERGER, M. J., A. KUIKURO, U. T. KUIKURO, J. C. RUSSELL, M. SCHMIDT, C. FAUSTO, AND B. FRANCHETTO. 2003. Amazonia 1492: pristine forest or cultural parkland? Science 301: 1710-1713.

HECKMAN, C. W. 2006. Encyclopedia of South American Aquatic Insects: Odonata - Anisoptera. Springer, Dordrecht, The Netherlands.

HECKMAN, C. W. 2008. Encyclopedia of South American Aquatic Insects: Odonata - Zygoptera. Springer, Dordrecht, The Netherlands.

HEEMSKERK, M., AND K. DELVOYE. 2007. Trio baseline study: a sustainable livelihoods perspective on the Trio indigenous peoples of South Suriname. Amazon Conservation Team, Paramaribo, Suriname.

HEEMSKERK, M., K. DELVOYE, D. NOORDAM, AND P.A. TEUNISSEN. 2007. Wayana baseline study: a sustainable livelihoods perspective on the Wayana indigenous peoples living in and around Puleowime (Apetina), Palumeu, and Kawemhakan (Anapaike) in Southeast Suriname. Amazon Conservation Team, Paramaribo, Suriname.

HEER K., E. K. V. KALKO, L. ALBRECHT, R. GARCÍA-VILLACORTA, F. C. STAEPS, AND E. A. HERRE. 2015. Spatial scales of genetic structure in free-standing and strangler figs (*Ficus*, Moraceae) inhabiting forests. PLOS ONE 10: e0133581. [**doi:10.1371/journal.pone.0133581**]

HEILBRON, T. 2012. Botanical relics of the plantations of Suriname. MSc Thesis, Universiteit Leiden, the Netherlands. **www.bhsproject.com/wp-content/uploads/2013/10/botanicalrelics_upload.pdf**

HEINRICHS, A. 2010. The Amazon rain forest. Marshall Cavendish Benchmark, New York NY, USA.

HEITHAUS, E. R., T. H. Fleming, and P. A. Opler. 1975. Foraging patterns and resource utilization in seven species of bats in a seasonal tropical forest. Ecology 56: 841-854.

HEKKING, W. H. A. 1988. Violaceae Part I: Rinorea and Rinoreocarpus. Flora Neotropica Monograph 46: 1-207.

HELED, J., AND A. J. DRUMMOND. 2010. Bayesian inference of species tree from multilocus data. Molecular Biology and Evolution 27: 570-580.

Helfman, G. S. 2007. Fish conservation. A guide to understanding and restoring global aquatic biodiversity and fishery resources. Island Press, Washington, DC, USA.

Helgen, K. M., C. M. Pinto, R. Kays, L. E. Helgen, M. T. N. Tsuchiya, A. Quinn, D. E. Wilson, and J. E. Maldonado. 2013. Taxonomic revision of the olingos (*Bassaricyon*), with description of a new species, the Olinguito. Zookeys 324: 1–83.

Helman, A. 1982. Avonturen aan de Wilde Kust, de geschiedenis van Suriname met zijn buurlanden. Verbeek + Verbeek Boekproducties, Eindhoven, The Netherlands.

Hemming, J. 2008. Tree of rivers, the story of the Amazon. Thames & Hudson, London, England, UK.

Henderson, P. A., and H. F. Hamilton. 1995. Standing crop and distribution of fish in drifting and attached floating meadow within an Upper Amazonian varzea lake. Journal of Fish Biology 47: 266–276.

Henderson, P. A., and I. Walker. 1990. Spatial organization and population density of the fish community of the leaf litter banks within a central Amazonian blackwater stream. Journal of Fish Biology 37: 401-411.

Henderson, R. W., A. Nickerson, and S. Ketcham. 1976. Short term movements of the snakes *Chironius carinatus*, *Helicops angulatus* and *Bothrops atrox* in Amazonian Peru. Herpetologica 32: 304-310.

Hendrison, J. 2009. Capacity Needs Assessment in the Forest Sector of Suriname. Ministry of Physical Planning, Land- and Forest Management (RGB) / SBB, ITTO Pre-Project PPD 97/04: Improved Strategies to achieve SFM in Suriname (Part III). RGB and ITTO, Paramaribo, Suriname.

Henkel, T. W., M. C. Aime, M. Chin, S. L. Miller, and R. Vilgalys. 2012. Ectomycorrhizal fungal sporocarp diversity and discovery of new taxa in *Dicymbe corymbosa* monodominant forest of the Guiana Shield. Biodiversity and Conservation 21: 2195-2220.

Henry, O., F. Feer, and D. Sabatier. 2000. Diet of the lowland tapir (*Tapirus terrestris* L.) in French Guiana. Biotropica 32: 364–368.

Hentrich, H. 2008. The reproductive biology of Euglossine-pollinated plants in the natural reserve Nouragues, French Guiana. PhD dissertation, Ulm University, Ulm, Germany. **vts.uni-ulm.de/docs/2008/6618/vts_6618_9053.pdf**

Herlein, J. D. 1718 Beschryvinge van de volks-plantinge Zuriname: vertonende de opkomst dier zelver colonie, de aandbouw en bewerkinge der zuiker-plantagien. Neftens den aard der eigene natuurlijke inwoonders of Indianen; als ook de slaafsche Afrikaansche Mooren; deze beide natien haar levens-manieren, afgoden-dienst, regering, zeden, gewoonten en dagelijksche bezigheden. Mitsgaders een vertoog van de bosch grond, water- en pluim-gediertens; de veel vuldige heerlijke vrugten, melk-agtige zappen, gommen, olyen, en de gehele gesteltheid van de Karaïbaansche kust. Meindert Injema, Leeuwarden [The Netherlands].

Hero, J. M., W. E. Magnusson, C. F. D. Rocha, and C. P. Catterall. 2001. Antipredator defenses influence the distribution of amphibian prey species in the Central Amazon rain forest. Biotropica 33: 131–141.

Hero, J.-M., and U. Galatti. 1990. Characteristics distinguishing *Leptodactylus pentadactylus* and *Leptodactylus knudseni* in the Central Amazon rainforest. Journal of Herpetology 24: 226-228.

Herrera, M. and O. Pellmyr (Eds). 2009. Plant animal interactions: an evolutionary approach. John Wiley & Sons, New York, NY, USA.

Herzog, S., A. Hamel-Leigue, T. Larsen, D. Mann, R. Soria-Auza, B. Gill, W. Edmonds, and S. Spector. 2013. Elevational distribution and conservation biogeography of phanaeine dung beetles (Coleoptera: Scarabaeinae) in Bolivia. PLoS ONE 8, e64963. **doi.org/10.1371/journal.pone.0064963**

Heyborne, W., and S. P. Mckessy. 2013. Identification and characterization of a taxon-specific three-finger toxin from the venom of the Green Vinesnake (*Oxybelis fulgidus*; family Colubridae). Biochimie 95: 1923-1932.

Heyer, W. R. 1969. The adaptive ecology of species groups of the genus *Leptodactylus* (Amphibia, Leptodactylidae). Evolution 23: 421-428.

Heyligers, P. G. 1963. Vegetation and soil of a white sand savanna in Suriname. Van Eedenfonds, Amsterdam, The Netherlands / Verhandelingen der Koninklijke Nederlandse Akademie van Wetenschappen, Afd. Natuurkunde, tweede reeks 54.

Hilbert, F., D. Sabatier, J. Andrivot, C. Scotti-Saintagne, S. Gonzalez, M. F. Prévost, and C. Richard-Hansen. 2011. Botany, genetics and ethnobotany: a crossed investigation on the elusive tapir's diet in French Guiana. Plos One 6: e25850.

Hill, J. 2016. The Environmental Impact of Roads. **www.environmentalscience.org/roads**

Hill, R. L., and D. J. Green. 1971. Surinam turtle notes - 2: investigation of the damage by the crab *Ocypode quadrata* to the eggs of the green turtle *Chelonia mydas*. Stinasu Mededeling 2: 10-13.

Hilterman, M. L., and E. Goverse. 2006. Annual report on the 2005 leatherback turtle research and monitoring project in Suriname. World Wildlife Fund - Guianas Forests and Environmental Conservation Project (WWF-GFECP), Technical Report of the IUCN Netherlands Committee (IUCN NL), Amsterdam, The Netherlands.

Hilterman, M., and E. Goverse. 2007. Nesting and Nest Success of the Leatherback Turtle (*Dermochelys coriacea*) in Suriname, 1999–2005. Chelonian Conservation and Biology 6: 87-100.

Hilty, S. L. 2003. Birds of Venezuela. Princeton University Press, Princeton, NJ, USA.

Hindwood, K. A. 1959. The nesting of birds in the nests of social insects. Emu 59: 1-43.

Hiwat H., and G. Bretas. 2011. Ecology of Anopheles darling Root with respect to vector importance: a review. Parasites and Vectors 4: 177

Hiwat H., J. Issaly, P. Gaborit, A. Somai, A. Samjhawan, P. Sardjoe, T. Soekhoe, and R. Girod. 2009. Behavioral heterogeneity of *Anopheles darlingi* (Diptera, Culicidae) and malaria transmission dynamics along the Maroni River Suriname, French Guiana. Transactions of The Royal Society for Tropical Medicine and Hygiene 104: 207-213

Hiwat H., L. Hardjopawiro, W. Takken, and L. Villegas. 2012. Novel strategies lead to near elimination of malaria in previously high-risk areas in Suriname, South America. Malaria Journal 11: 10. **doi:10.1186/1475-2875-11-10**

Hiwat, H. 2014. Triatominae species of Suriname (Heteroptera: Reduviidae) and their role as vectors of Chagas Disease. Memorias do Instituto Oswaldo Cruz 109: 452-458. **doi: 10.1590/0074-0276130408**

HOFFMAN, B. 2009. Drums and arrows: ethnobotanical classification and use of tropical forest plants by a Maroon and Amerindian community in Suriname, with implications for biocultural conservation. University of Hawai'i, Honolulu, HI, USA.

HOFFMAN, B. 2012. Exploring biocultural contexts: comparative woody plant knowledge of indigenous and Afro-American Maroon communities in Suriname, South America. *In* R. Voeks and J. Ashford (Eds.). African Ethnobotany in the Americas, pp. 335-393. Springer, New York, NY, USA.

HOFFMAN, B., AND S. RUYSSCHAERT. 2017. Lianas of the Guianas – a guide to woody climbers in the tropical forests of Guyana, Suriname, and French Guiana. LM Publishers, Volendam, The Netherlands.

HOFWIJKS, S., AND R. MADHAR. 2006. Assessment of the bushmeat and wildplant trade in Suriname. The Guianas Regional Programme: World Wildlife Fund Guianas, Paramaribo, Suriname.

HOGUE, C. L. 1993. Latin American insects and entomology. University of California Press, Berkeley and Los Angeles, CA, USA, and Oxford, England, UK.

HOLBROOK, K. M. 2011. Home range and movement patterns of toucans: implications for seed dispersal. Biotropica 43: 357-364.

HOLLDOBLER, B., AND E. O. WILSON. 2011. The Leafcutter Ants. W. W. Norton & Company, New York, NY, USA.

HOLLOWELL, T., AND R. P. REYNOLDS. 2005. Introduction. *In* Hollowell, T. and R.P. Reynolds (Eds.). Checklist of the Terrestrial Vertebrates of the Guiana Shield, pp. 1-6. Bulletin of the Biological Society of Washington 13.

HOLMLUND, C. M., AND M. HAMMER. 1999. Ecosystem services generated by fish population. Ecological Economics 29: 253-268.

HOLTHUIS, L. B. 1959. The Crustacea Decapoda of Suriname (Dutch Guiana). Zoologische Verhandelingen Leiden 44: 1-296.

HONDIUS, J. 1599. Nieuwe caerte van het wonderbaer ende goudrijcke landt Guiana, gelegen onder de linie aequinocticael, tusschen Brasilien ende Peru nieuwelick besocht door Sir Walter Raleigh ridder van Engelandt, in het jaer 1594, 95 ende 1596. Amsterdam [The Netherlands].

HOOGMOED, M. S. 1973. Notes on the herpetofauna of Suriname IV. The lizards and amphisbaenians of Suriname. Biogeographica IV. Dr. W. Junk, The Hague, The Netherlands.

HOOGMOED, M. S. 1979. The Herpetofauna of the Guianan Region. *In* W. E. Duellman (Ed.). The South American Herpetofauna: Its origin, evolution and dispersal. University of Kansas Museum of Natural History Monographs 7: 241-279.

HOOGMOED, M. S. 1981. Introduced Species of Reptiles in Surinam. Notes on the Herpetofauna of Surinam VIII. Amphibia-Reptilia 1: 277-285.

HOOGMOED, M. S. 1993. The herpetofauna of floating meadowns. *In* Ouboter, P.E. (Ed.). The Freshwater Ecosystems of Suriname, pp. 199-213. Kluwer Academic Publishers, Dordrecht, The Netherlands.

HOOGMOED, M. S., AND T. C. S. AVILA-PIRES. 1991. Annotated checklist of the herpetofauna of Petit Saut, Sinnamary River, French Guiana. Zoologische Mededelingen Leiden 65: 53-88.

HOORN, C. 1997. Palynology of the Pleistocene glacial/interglacial cycles of the Amazon Fan (Holes 940A, 944A, and 946A). *In* R. D. Flood, D. J. W. Piper, A. Klaus, and L. C. Peterson (Eds.). Proceedings Ocean Drilling Program, Scientific Results 115, pp. 397–409. Ocean Drilling Program, College Station, Texas, USA. **doi: 10.2973/odp.proc.sr.155.226.1997**

HOORN, C., F. P. WESSELINGH, H. TER STEEGE, M. A. BERMUDEZ, A. MORA, J. SEVINK, I. SANMARTÍN, A. SANCHEZ-MESEGUER, C. L. ANDERSON, J. P. FIGUEIREDO, C. JARAMILLO, D. RIFF, F. R. NEGRI, H. HOOGHIEMSTRA, J. LUNDBERG, T. STADLER, T. SÄRKINEN, AND A. ANTONELLI. 2010. Amazonia through time: Andean uplift, climate change, landscape evolution, and biodiversity. Science 330: 927-931.

HOORN, M. C. 1994. Fluvial palaeoenvironments in the intracratonic Amazonas Basin (Early Miocene–early Middle Miocene, Colombia). Palaeogeography, Palaeoclimatology, Palaeoecology 110: 1–54.

HOOVER, T. M., AND J. D. ACKERMAN. 2004. Near-bed hydrodynamic measurements above boulders in shallow torrential streams: implications for stream biota. Journal of Environmental Engineering and Science 3: 365-378.

HOPKINS, C. D., AND W. HEILIGENBERG. 1978. Evolutionary designs for electric signals and electroreceptors in gymnotoid fishes of Surinam. Behavioral Ecology and Sociobiology 3: 113-134.

HOPKINS, C. D., N. C. COMFORT, J. BATIAN, AND A. H. BASS. 1990. Functional analysis of sexual dimorphism in an electric fish, *Hypopomus pinnicaudatus*, order Gymnotiformes. Brain, Behavior and Evolution 35: 350-367.

HOPKINS, H. C. 1986. *Parkia* (Leguminosae: Mimosoideae). Flora Neotropica 43: 1-123.

HOREAU, V., P. CERDAN, A. CHAMPEAU, AND S. RICHARD. 1998. Importance of aquatic invertebrates in the diet of rapid-dwelling fish in the River Sinnamary, French Guiana. Journal of Tropical Ecology 14: 851-864.

HORN, M. H., S. B. CORREA, P. PAROLIN, B. J. A. POLLUX, J. T. ANDERSON, C. LUCAS, AND M. GOULDING. 2011. Seed dispersal by fishes in tropical and temperate fresh waters: the growing evidence. Acta Oecologica 37: 561-577.

HORNBY, C. S. HARPER, J. MACDONALD, AND D. ZELLER. 2015. Reconstruction of Suriname's marine fisheries catches from 1950-2010. Fisheries Centre, University of British Columbia, Vancouver, Canada. **www.seaaroundus.org/doc/publications/wp/2015/Hornby-et-al-Suriname.pdf**

HOSSAERT-MCKEY, M., J. ORIVEL, E. LABEYRIE, L. PASCAL, J. H. C. DELABIE, AND A. DEJEAN. 2001. Differential associations with ants of three co-occurring extrafloral nectary-bearing plants. Ecoscience 8: 325-335.

HOUBEN, L., AND A. MOLENAAR. (undated). Rehabilitation of the East-West Road Connection, causes and repair methods of road cracks in recently rehabilitated road stretch and advice on technical specifications for a planned new road (draft report), pp. 1-16. Delft University of Technology, Delft, The Netherlands.

HOUTTUYN, M. 1764. Natuurlyke historie of uitvoerige beschryving der dieren, planten en mineraalen, volgens het samenstel van den Heer Linnaeus, met naauwkeurige afbeeldingen. Part I. Vol. 7. De Visschen. F. Houttuyn, Amsterdam [The Netherlands].

HOWARD, F.W., R. M. GIBLIN-DAVIS, D. MOORE, AND R. G. ABAD. 2001. Insects on Palms. CAB International, Wallingford, England, UK.

Howe, H. F. 1984. Implications of seed dispersal by animals for tropical reserve management. Biological Conservation 30: 261-281.

Howe, H. F., and G. A. V. Kerckhove. 1980. Nutmeg dispersal by tropical birds. Science 210: 925-927.

Howse, P., and K. Wolfe. 2011. Giant silk moths: colour, mimicry & camouflage. Papadakis Publisher, Berkshire, England, UK.

Hu, C., E.T. Montgomery, R. W. Schmitt, and F. E. Muller-Karger. 2004. The dispersal of the Amazon and Orinoco River water in the tropical Atlantic and Caribbean Sea: observation from space and S-PALACE floats. Deep-Sea Research II 51: 1151-1171.

Hua, M. H., Y. S. Ao, X. E. Yang, and T. Q. Li. 2008. Treating eutrophic water for nutrient reduction using an aquatic macrophyte (*Ipomoea aquatic* Forsskal) in a deep flow technique system. Agricultural Water Management 95: 607-615. **doi: 10.1016/j.agwat.2008.01.001**

Huber, O. 1995. Geographical and Physical Features. *In* P. E. Berry, B. K. Holst, and K. Yatskievych (Eds.) Flora of the Venezuelana Guayana. Volume 1: Introduction, pp. 1–51. Missouri Botanical Gardens Press, St. Louis, MO, USA.

Husar, S. L. 1978. *Trichechus manatus*. Mammalian Species 93: 1-5.

Husson, A. M. 1978. The mammals of Surinam. Zoölogische Monographieën van het Rijksmuseum van Natuurlijke Historie 2, Brill, Leiden, The Netherlands.

IACHR. 2015. Case of the Kaliña and Lokono peoples v. Suriname. Judgment of November 25, 2015, San José, Costa Rica. Inter-American Commission on Human Rights, Organisation of American States, Washington, DC, USA.

IBAMA. 2008. Lista Oficial das Espécies da Flora Brasileira Ameaçadas de Extinção. *In* Instrução Normativa N° 6, de 23 de setembro de 2008. IBAMA, Brasilia, DF, Brazil. **www.ibama.gov.br/documentos/lista-de-especiesameacadas-de-extincao**

IUCN. 2012. IUCN Red List Categories and Criteria: Version 3.1. Second edition. International Union for the Conservation of Nature, Gland, Switzerland, and Cambridge, England, UK.

IUCN. 2015. IUCN Red List of threatened species. International Union for the Conservation of Nature, Gland, Switzerland. **www.iucnredlist.org** [accessed Nov. 2015]

IUCN. 2016. IUCN Red List of threatened species. International Union for the Conservation of Nature, Gland, Switzerland. **www.iucnredlist.org** [accessed Dec. 2016]

Ivanyisky, S. J., and J. S. Albert. 2014. Systematics and biogeography of Sternarchellini (Gymnotiformes:Apteronotidae): diversification of electric fishes in large Amazonian rivers. Neotropical Ichthyology 12: 565-584.

Izar, P., Stone, A., Carnegie, S., and E. S. Nakai. 2009. Sexual selection, female choice and mating systems. *In* P. A. Garber, A. Estrada, J. C. Bicca-Marques, E. W. Heymann, and K. B. Strier (Ed.). South American Primates, pp.157-189. Springer, New York, NY, USA.

Jack, K. M. 2011. The cebines. *In* C.J. Campbell, A. Fuentes, K.C. MacKinnon, S.K. Bearder and R.M. Stumpf. Primates in Perspective, pp. 108-122. Oxford University Press, Oxford, England, UK.

Jacobs, W., and M. Renner. 1974. Taschenlexicon zur biologie der insecten. Gustav Fisher Verlag, Stuttgart, Germany.

Jäger-Zürn, I., M. Spies, C. T. Philbrick, C. P. Bove, and A. Mora-Olivo. 2013. Plant galls (cecidia) in the neotropical water plant family Podostemaceae induced by larvae of Chironomidae. Spixiana 36: 97-112.

Jagessar, R. C., G. Hoolas, and A. R. Maxwell. 2013. Phytochemical screening, isolation of betulinic acid, trigonelline and evaluation of heavy metals ion content of *Doliocarpus dentatus*. Journal of Natural Products 6: 5-16.

Jahn, A. E., D. J. Levey, V. R. Cuety, J. P. Ledezma, D. T. Tuero, J. W. Fox, and D. Masson. 2013. Long-distance bird migration within South America revealed by light-level geolocators. Auk 130: 223-229.

Jahn, A. E., J. I. Giraldo, M. MacPherson, D. T. Tuero, J. Sarasola, J. Cereghetti, D. A. Masson, and M. V. Morales. 2016. Demographic variation in timing and intensity of feather molt in migratory Fork-tailed Flycatchers (*Tyrannus s. savana*). Journal of Field Ornithology 87: 143-154.

Jansen, P. A., and P. M. Forget. 2001. Scatter-hoarding rodents and tree regeneration. *In* F. Bongers, P. Charles-Dominique, P.-M. Forget, and M. Théry (Eds.). Nouragues. Dynamics and plant-animal interactions in a neotropical rainforest, pp. 275–288. Kluwer Academic Publishers, Dordrecht, The Netherlands.

Jansen, P. A., F. Bongers, and L. Hemerik. 2004. Seed mass and mast seeding enhance dispersal by a Neotropical scatter-hoarding rodent. Ecological Monographs 74: 569-589.

Jansen-Jacobs, M. J., A. Gröger, U. P. D. Raghoenandan, O. S. Bánki, and C. I. E. E. van 't Klooster. 2002. Botanical exploration in Suriname, Voltzberg – van Stockumberg. Report, National Herbarium of the Netherlands – Utrecht University, Utrecht, The Netherlands; Botanical Garden München – Nymphenburg, München, Germany; National Herbarium of Suriname – Anton de Kom University Suriname, Paramaribo, Suriname.

Jansma, R. 1994. Ecology of some Northern Suriname savannas. Koeltz Scientific Books, Königstein, Germany.

Janssen, B.H., and J.F. Wienk (Eds.). 1990. Mechanized annual cropping on low fertility acid soils in the humid tropics. A case study of the Zanderij soils in Suriname. Wageningen University Papers 90-5, Wageningen University, Wageningen, The Netherlands.

Januchowski-Hartley, S. R., P. B. McIntyre, M. Diebel, P. J. Doran, D. M. Infante, C. Joseph, and J. D. Allan. 2013. Restoring aquatic ecosystem connectivity requires expanding inventories of both dams and road crossings. Frontiers Ecology Environment 11: 211-217.

Janzen, D. H. 1979. How to be a fig. Annual Reviews of Ecology and Systematics 10: 13-51.

Janzen, D. H. 1985. *Spondias mombin* is culturally deprived in megafauna-free forest. Journal of Tropical Ecology 1: 131-155.

Janzen, D. H. 2011. How does an "All Taxa Biodiversity Inventory (ATBI)" promote and facilitate global biodiversity conservation? Biodiversity 4: 4-10.

Janzen, D. H., and P. S. Martin. 1982. Neotropical anachronisms: the fruits the Gomphotheres ate. Science 215: 19–27.

Janzen, D. H. 1984. Two ways to be a tropical big moth: Santa Rosa saturniids and sphingids. Oxford Surveys in Evolutionary Biology 1: 85-140.

Jarau, S., J. W. van Veen, R. Twele, C. Reichle, E. H. Gonzales, I. Aguilar, W. Francke, and M. Ayasse. 2010. Workers make the queens in *Melipona* bees: identification of geraniol as a caste determining compound from labial glands of nurse bees. Journal Chemical Ecology 36: 565-569.

Johnson, N. F., and C. A. Triplehorn. 2004. Borror and DeLoong's introduction to the study of insects. Cengage Learning, Boston, MA, USA.

Joron, M. 2009. Aposematic coloration. *In* V. H. Resh, and R. T. Cardé (Eds.). Encyclopedia of Insects, pp. 33–37. Elsevier (Academic Press), Oxford, England, UK.

Judd, W. S., C. S. Campbell, E. A. Kellogg, P. F. Stevens, and M. J. Donoghue. 2016. Plant systematics: A phylogenetic approach. Sinauer Associates, MA, USA.

Julen, C. R. 1974. Vegetatiekundig onderzoek op het Matapica-Krofajapasikreek-gebied en het oostelijk gedeelte van het Natuurreservaat Wia Wia. MSc dissertation, Institute for Systematic Botany, Utrecht University, Utrecht, The Netherlands.

Julien-Laferrière, D. 1993. Radio-tracking observations on ranging and foraging patterns by kinkajous (*Potos flavus*) in French Guiana. Journal of Tropical Ecology 9:19-32.

Julien-Laferrière, D. 1999. Fruit consumption, seed dispersal and seed fate in the vine *Strychnos erichsonii* in a French Guianan forest. Revue d'Ecologie-La Terre et La Vie 54: 315-326.

Jullien, M., and J.-M. Thiollay. 1998. Multi-species territoriality and dynamic of Neotropical forest understorey bird flocks. Journal of Animal Ecology 67: 227-252.

Jungfer, K.-H., and P. Weygoldt. 1999. Biparental care in the tadpole-feeding Amazonian treefrog *Osteocephalus oophagus*. Amphibia-Reptilia 20: 235-249.

Juniper, T., and M. Parr. 1998. Parrots: a guide to parrots of the world. Yale University Press, New Haven, CT, USA.

Junk, W. J. 1984. Ecology, fisheries and fish culture in Amazonia. *In* H. Sioli (Ed.). The Amazon: limnology and landscape ecology of a mighty tropical river and its basin, pp. 443-476. Junk, Dordrecht, the Netherlands.

Junk, W. J. 1985. Temporary fat storage, an adaptation of some fish species to the water level fluctuations and related environmental changes of the Amazon River. Amazoniana 9: 315- 351.

Junk, W. J., P. B. Bayley, and R. E. Sparks. 1989. The flood pulse concept in river-floodplain systems. Special Publications of the Canadian Journal of Fisheries and Aquatic Sciences 106: 110-127.

Kaminski, A. C., and M. L. Absy. 2006. Bees visitors of three species of Clusia (Clusiaceae) flowers in Central Amazonia. Acta Amazonica 36: 259-263. **dx.doi.org/10.1590/S0044-59672006000200016**

Kappler, A. 1854. Zes jaren in Suriname. Schetsen en tafereelen uit het maatschappelijke en militaire leven in deze kolonie. W. F. Dannenfelser, Utrecht, The Netherlands.

Kappler, A. 1881. Holländisch-Guiana: Erlebnisse und Erfahrungen während eines 43 jährigen Aufenthalts in der Kolonie Surinam. Commissions-Verlag von W. Kohlhammer, Stuttgart, Germany.

Kappler, A. 1887. Surinam, sein Land, seine Natur, Bevölkerung, und seine Kultur-Verhältnisse met Bezug auf Kolonisation. J. G. Cotta'sche Verlagsbuchhandlung, Stuttgart, Germany.

Kardong, K. V. 1982. The evolution of the venom apparatus in snakes from Colubrids to Viperids & Elapids. Memórias do Instituto de Butantan 46: 106-118.

Karr, J. R., S. K. Robinson, J. G. Blake, and R. O. Bierregaard, Jr. 1990. Birds of four Neotropical forests. *In* A. Gentry (Ed.). Four Neotropical Forests, pp. 237-269. Yale University Press, New Haven, CT, USA.

Karubian, J., J. Fabara, D. Yunes, J. P. Jorgenson, D. Romo, and T. B. Smith. 2005. Temporal and spatial patterns of macaw abundance in the Ecuadorian Amazon. Condor 107: 617-626.

Kay, K. M., and D. W. Schemske. 2003. Pollinator assemblages and visitation rates for eleven species of Neotropical Costus (Costaceae). Biotropica 35: 198-207.

Kay, K. M., P. A. Reeves, R. G. Olmstead, and D. W. Schemske. 2005. Rapid speciation and the evolution of hummingbird pollination in neotropical *Costus* subgenus Costus (Costaceae): evidence from nrDNA ITS and ETS sequences. American Journal of Botany 92: 1899-1910.

Kays, R. W. 1999. Food preferences of kinkajous (*Potos flavus*) a frugivorous carnivore. Journal of Mammalogy 80: 589-599.

Kays, R. W., and J. L. Gittleman. 2001. The social organization of the kinkajou *Potos flavus* (Procyonicae). Journal of Zoology 253: 491-504.

Kays, R., M. C. Crofoot, W. Jetz, and M. Wikelski. 2015. Terrestrial animal tracking as an eye on life and planet. Science 348, aaa2478. **doi: 10.1126/science.aaa2478**

Kays, R., P. A. Jansen, E. M. H. Knecht, R. Vohwinkel, and M. Wikelski. 2011. The effect of feeding time on dispersal of *Virola* seeds by toucans determined by GPS tracking and accelerometers. Acta Oecologica 37: 625-631.

Kearns, C. A., and D. W. Inouye. 1993. Techniques for pollination biologists. University Press of Colorado, Niwot, CO, USA

Keenleyside, M. H. A. 1955. Some aspects of the schooling behaviour of fish. Behaviour 8: 183-248.

Keenleyside, M. H. A. 1979. Diversity and adaptation in fish behaviour. Springer, Berlin and Heidelberg, Germany, and New York, NY, USA.

Keenleyside, M. H. A. and B. F. Bietz. 1981. The reproductive behaviour of *Aequidens vittatus* (Pisces, Cichlidae) in Surinam, South America. Environmental Biology of Fishes 6: 87-94.

Kelleher, M.K. 2005. Discards in the world's marine fisheries: An update. FAO Fisheries Technical Paper 470, Food and Agriculture Organization of the United Nations (FAO), Rome, Italy.

Kelloff, C. L., and V. A. Funk. 2004. Phytogeography of the Kaieteur Falls, Potaro Plateau, Guyana: floral distributions and affinities. Journal of Biogeography 31: 501-513.

Kelly, D. 1994. The evolutionary ecology of mast seeding. Tree 9: 465-470.

Kempraj, V., and S. K. Bhat. 2008. Ovicidal and larvicidal activities of *Cyperus giganteus* Vahl and *Cyperus rotundus* Linn. essential oils against *Aedes albopictus* (Skuse). Natural Product Radiance 7: 416-419.

Kenny, J. S. 1969. Amphibia of Trinidad. Studies of the Fauna of Curacao and other Caribbean Islands 29: 1-78.

Kerkvliet, J. D., and J. G. Beerlink. 1991. Pollen analysis of honeys from the coastal plain of Surinam. Journal of Apicultural Research 30: 25-31.

Kerman, I. 2010. Exploitation of the jaguar, *Panthera onca,* and other large forest cats in Suriname. WWF-Guianas, Paramaribo, Suriname.

Khan, M. W. (Ed.). 1993. Nematode Interactions. Chapman & Hall, London, England, UK.

Killip, E. P. 1937. Passifloraceae. Flora of Suriname 3: 306-327.

Kinzey, W. G. 1992. Dietary and dental adaptations in the Pitheciinae. American Journal of Physical Anthropology 88: 499-514.

Kinzey, W. G., and M. A. Norconk. 1990. Hardness as a basis of fruit choice in two sympatric primates. American Journal of Physical Anthropology 81, 5-15.

Kips, P. A., and A. R. Snel. 1979. Landschap, bodemgesteldheid en landgeschiktheid langs de midden-Suriname rivier (Goejaba-Botopassi). DBK Report 63. Dienst Bodemkartering Suriname, Paramaribo, Suriname.

Klaver, M., E. W. F. de Roever, J. A. M. Nanne, P. R. D. Mason, and G. R. Davies. 2015. Charnockites and UHT metamorphism in the Bakhuis Granulite Belt, western Suriname: Evidence for two separate UHT events. Precambrian Research 262: 1–19.

Kleinfeldt, S. E. 1978. Ant-gardens: the interaction of *Codonanthe crassifolia* (Gesneriaceae) and *Crematogaster longispina* (Formicidae). Ecology 59: 449-456.

Klinge, H. 1965. Podzol soils in the Amazon Basin. Journal of Soil Science 16: 95-103.

Klinge, H. 1967. Podzol soils: a source of black-water rivers in Amazonia. Atas do Simpósio sôbre a Biota Amazônica 3: 117-125.

Knight, F. M., J. Lombardi, J. P. Wourms, and J. R. Burns. 1985. Follicular placenta and embryonic growth of the viviparous four-eyed fish (*Anableps*). Journal of Morphology 185: 131-142.

Knöppel, H. A. 1970. Food of Central Amazonian fishes. Amazoniana 2: 257-352.

Kohda, M., M. Tanimura, M. Kikue-Nakamura, and S. Yamagishi. 1995. Sperm drinking by female catfishes: a novel mode of insemination. Environmental Biology of Fishes 42: 1-6.

Koster, J. M., Hodgen J. J., Venegas, M. D., and T. J. Copeland. 2010. Is meat flavor a fact in hunters' prey choice decisions? Human Nature 21: 219-242.

Kovařík, F. 2009. Illustrated catalog of scorpions. Part I. Introductory remarks; keys to families and genera; subfamily Scorpioninae with keys to *Heterometrus* and *Pandinus* species. Clairon Production, Prague, Czech Republic.

Kovařík, F., and A. A. Ojanguren Affilastro. 2013. Illustrated catalog of scorpions. Part II. Bothriuridae; Chaerilidae; Buthidae I. Genera *Compsobuthus*, *Hottentotta*, *Isometrus*, *Lychas*, and *Sassanidotus*. Clairon Production, Prague, Czech Republic.

Kovařík, F., R. Teruel, M. A. C. Cozijn, and M. Seiter. 2013. *Tityus carolinae* sp. n. from Suriname and Guyana (Scorpiones: Buthidae). Euscorpius 178: 1–9.

Kramer, D. L. and M. McClure. 1980. Aerial respiration in the catfish *Corydoras aeneus* (Callichthyidae). Canadian Journal of Zoology 58: 1984-1991.

Kramer, K. U., and J. van Donselaar .1968. A sketch of the vegetation and flora of the Kappel Savanna near Tafelberg, Suriname. Verhandelingen der Koninklijke Nederlandse Akademie van Wetenschappen, Afd. Natuurkunde, serie C 71: 495-524.

Krekorian, C. O. N. 1976. Field observations in Guyana on the reproductive biology of the spraying characid, *Copeina arnoldi* Regan. American Midland Naturalist 96: 88-97.

Kress, W. J., and D. E. Stone. 1993. Morphology and floral biology of *Phenakospermum* (Strelitziaceae), an arborescent herb of the Neotropics. Biotropica 25: 290-300.

Kress, W. J., L. M. Prince, W. J. Hahn, and E. A. Zimmer. 2001. Unraveling the evolutionary radiation of the families of the Zingiberales using morphological and molecular evidence. Systematic Biology 50: 926–944.

Krijger, C. L., M. Opdam, M. Théry, and F. Bongers. 1997. Courtship behaviour of manakins and seed bank composition in a French Guianan rain forest. Journal of Tropical Ecology 13: 631-636.

Kronmüller, C., and J. G. E. Lewis. 2015. On the function of the ultimate legs of some Scolopendridae (Chilopoda, Scolopendromorpha). ZooKeys 510: 269–278. Advance online publication. **doi.org/10.3897/zookeys.510.8674**

Krook, L. 1970. Climate and sedimentation in the Guianas during the Last Glacial and Holocene. 8th Guiana Geological Conference, 1969. Georgetown, Guyana.

Krook, L. 1979. Sediment petrographical studies in northern Suriname. PhD dissertation, Vrije Universiteit Amsterdam, Amsterdam, The Netherlands.

Krook, L. 1994. De Geologische en geomorfologische ontwikkeling van Noord Suriname. *In* L. J. van der Steen (Ed.). Recente geologische en mijnbouwkundige ontwikkelingen in Suriname, pp. 23-40. Natuurwetenschappelijke Studiekring voor het Caraïbisch gebied, Amsterdam, The Netherlands.

Krook, L., and M. A. Mulders. 1971. Geological and related pedological aspects of the Upper Coesewijne Formation. Mededelingen Geologisch Mijnbouwkundige Dienst Suriname 21: 183-208.

Kroonenberg, S. B. 1976. Amphibolite facies and granulite-facies metamorphism in the Coeroeni-Lucie area, southwestern Surinam. PhD dissertation, University of Amsterdam, Amsterdam, The Netherlands; Mededelingen Geologisch Mijnbouwkundige Dienst Suriname 25: 101-289.

Kroonenberg, S. B. 1978 Precambrian paleosols at the base of the Roraima Formation in Surinam. Geologie en Mijnbouw 57: 445-450.

Kroonenberg, S. B., and E. W. F. de Roever. 2010. Geological Evolution of the Amazonian Craton. *In* C. Hoorn, and F. P. Wesselingh (Eds.). Amazonia, Landscape and Species Evolution, pp. 9-28. Blackwell Publishing, Oxford, England, UK.

Kroonenberg, S. B., and P.J. Melitz. 1983. Summit levels, bedrock control and the etchplain concept in the basement of Suriname. Geologie en Mijnbouw 62: 389-399.

Kroonenberg, S. B., E. W. F. de Roever, L. M. Fraga, N. J. Reis, M. T. Faraco, J.-M. Lafon, U. G. Cordani, and T. E. Wong. 2016. Paleoproterozoic evolution of the Guiana Shield in Suriname: a revised model. Netherlands Journal of Geosciences (Geologie en Mijnbouw) 95: 491–522.

Kubitzki, K., and H. Kurz. 1984. Synchronized dichogamy and dioecy in neotropical Lauraceae. Plant Systematics and Evolution 147: 253-266.

Kunz, T. H, and C. A. Diaz. 1995. Folivory in fruit-eating bats, with new evidence from *Artibeus jamaicensis* (Chiroptera: Phyllostomidae). Biotropica 27: 106-120.

Kushlan, J. A. 2009. Feeding repertoire of the Boat-billed Heron (*Cochlearius cochlearius*). Waterbirds 32: 408-414.

Labruna, M. B., M. Amaku, J. A. Metzner, A. Pinter, and F. Ferreira. 2003. Larval behavioral diapause regulates life cycle of *Amblyomma cajennense* (*Acari*: Ixodidae) in southeast Brazil. Journal Medical Entomology 40: 170-178.

Langeland, K. A., and K. C. Burks. 2000. Identification and biology of non-native plants in Florida's natural areas. Institute of Food and Agricultural Sciences (IFAS), University of Florida, Gainesville, FL, USA.

Langenheim, J. H. 2003. Plant resins: chemistry, evolution, ecology and ethnobotany. Timber Press, Portland, OR, USA.

Lanjouw, J., and H. Uittien. 1935. Surinaamse geneeskruiden in de tijd van Linnaeus. De West-Indische Gids 17: 173-190.

Lannoo, M. J., and S. J. Lannoo. 1993. Why do electric fishes swim backwards? An hypothesis based on gymnotiform foraging behavior interpreted through sensory constraints. Environmental Biology of Fishes 36: 157-165.

Lasso, C. A., R. S. Rosa, P. Sánchez-Duarte, M. A. Morales-Betancourt, and E. Agudelo-Córdoba. 2014. Rayas de aqua dulce (Potamotrygonidae) de Suramérica. Parte I. Instituto de Investigación de Recursos Biológicos Alexander von Humboldt, Bogotá, Colombia.

Laurance, W. F., H. E. M. Nascimento, S. G. Laurance, R. Condit, S. D'angelo, and A. Andrade. 2004. Inferred longevity of Amazonian rainforest trees based on a long-term demographic study. Forest Ecology and Management 190: 131–143.

Laurance, W. F., M. Goosem, and S. G. W. Laurance. 2009. Impacts of roads and linear clearings on tropical forests. Trends in Ecology and Evolution 24: 659-669.

Lavilla, E. O., M. Vaira, and L. Ferrari. 2003. A new species of *Elachistocleis* (Anura: Microhylidae) from the Andean Yungas of Argentina, with comments on the *Elachistocleis ovalis–E. bicolor* controversy. Amphibia-Reptilia 24: 269–284.

Leal, I. R., E. Fischer, C. Kost, M. Tabarelli, and R. Wirth. 2006. Ant protection against herbivores and nectar thieves in *Passiflora coccinea* flowers. Ecoscience 13: 431-438.

Lee, H.B. 1958. Zeevisserij-onderzoek, Coquette Survey Report. Report no. 20, Ministry of Agriculture Paramaribo, Suriname.

Leentvaar, P. 1975. Hydrobiological observations in Surinam with special reference to the man-made Brokopondo Lake. Studies of the Fauna of Suriname and other Guyanas 56: 1-191.

Lehmann, P., H. Lazzarotto, and R. E. Reis. 2014. *Parotocinclus halbothi*, a new species of small armored catfish (Loricariidae: Hypoptopomatinae), from the Trombetas and Marowijne River basins, in Brazil and Suriname. Neotropical Ichthyology 12: 27-33.

Leighton, M., and D. R. Leighton. 1983. Vertebrate responses to fruiting seasonality within a Bornean rain forest. *In* S. L. Sutton, T. C. Whitmore, and A. C. Chadwick (Eds.). Tropical Rain Forests: Ecology and Management, pp. 181-196. Blackwell Scientific Publications, Oxford, UK.

Lemaire, C. 1988. Les Saturniidae américains – the Saturniidae of America – los Saturniidae Americanos (= Attacidae). Ceratocampinae. Museo Nacional de Costa Rica, San José, Costa Rica.

Lenselink, K., and R. van der Weert. 1970. Estimating free water evaporation in Suriname. De Surinaamse Landbouw 21: 70-79.

Lescure, J., and C. Marty. 2000. Atlas des Amphibiens de Guyane. Publication Scientifiques de M.N.H.N. 45, Paris, France.

Leuenberger, B. E. 1997. Cactaceae. Flora of the Guianas, Series A: Phanerogams, no. 18.

Lever, C. 2001. The Cane Toad. The history and ecology of a successful colonist. Westbury Publishing, West Yorkshire, England, UK.

Levey, D. J., and F. G. Stiles. 1992. Evolutionary precursors of long-distance migration: resource availability and movement patterns in Neotropical landbirds. American Naturalist 140: 447-476.

Levi, H. W. 1959. The spider genus *Latrodectus* (Araneae, Theridiidae). Transactions of the American Microscopical Society 78: 7-43.

Levin, S.A., H. C. Muller-Landau, R. Nathan, and J. Chave. 2003. The ecology and evolution of seed dispersal: a theoretical perspective. Annual Review of Ecology, Evolution, and Systematics 34: 575-604.

Levis C., F. R. C. Costa, F. Bongers, M. Peña-Claros, C. R. Clement, A. B. Junqueira, E. G. Neves, E. K. Tamanaha, F. O. G. Figueiredo, R. P. Salomão, C. V. Castilho, W. E. Magnusson, O. L. Phillips, J. E. Guevara, D. Sabatier, J.-F. Molino, D. Cárdenas López, A. M. Mendoza, N. C. A. Pitman, A. Duque, P. Núñez Vargas, C. E. Zartman, R. Vasquez, A. Andrade, J. L. Camargo, T. R. Feldpausch, S. G. W. Laurance, W. F. Laurance, T. J. Killeen, H. E. Mendonça Nascimento, J. C. Montero, B. Mostacedo, I. L. Amaral, I. C. Guimarães Vieira, R. Brienen, H. Castellanos, J. Terborgh, M. de Jesus Veiga Carim, J. R. da Silva Guimarães, L. de Souza Coelho, F. D. de Almeida Matos, F. Wittmann, H. F. Mogollón, G. Damasco, N. Dávila, R. García-Villacorta, E. N. H. Coronado, T. Emilio, D. de Andrade Lima Filho, J. Schietti, P. Souza, N. Targhetta, J. A. Comiskey, B. S. Marimon, B.-H. Marimon Jr., D. Neill, A. Alonso, L. Arroyo, F. A. Carvalho, F. C. de Souza, F. Dallmeier, M. P. Pansonato, J. F. Duivenvoorden, P. V. A. Fine, P. R. Stevenson, A. Araujo-Murakami, G. A. Aymard C., C. Baraloto, D. D. do Amaral, J. Engel, T. W. Henkel, P. Maas, P. Petronelli, J. D. Cardenas Revilla, J. Stropp, D. Daly, R. Gribel, M. Ríos Paredes, M. Silveira, R. Thomas-Caesar, T. R. Baker, N. F. da Silva, L. V. Ferreira, C. A. Peres, M. R. Silman, C. Cerón, F. C. Valverde, A. Di Fiore, E. M. Jimenez, M. C. Peñuela Mora, M. Toledo,

E. M. Barbosa, L. C. de Matos Bonates, N. C. Arboleda, E. de Sousa Farias, A. Fuentes, J.-L. Guillaumet, P. Møller Jørgensen, Y. Malhi, I. P. de Andrade Miranda, J. F. Phillips, A. Prieto, A. Rudas, A. R. Ruschel, N. Silva, P. von Hildebrand, V. A. Vos, E. L. Zent, S. Zent, B. B. L. Cintra, M. T. Nascimento, A. A. Oliveira, H. Ramirez-Angulo, J. F. Ramos, G. Rivas, J. Schöngart, R. Sierra, M. Tirado, G. van der Heijden, E. V. Torre, O. Wang, K. R. Young, C. Baider, A. Cano, W. Farfan-Rios, C. Ferreira, B. Hoffman, C. Mendoza, I. Mesones, A. Torres-Lezama, M. N. U. Medina, T. R. van Andel, D. Villarroel, R. Zagt, M. N. Alexiades, H. Balslev, K. Garcia-Cabrera, T. Gonzales, L. Hernandez, I. Huamantupa-Chuquimaco, A. G. Manzatto, W. Milliken, W. P. Cuenca, S. Pansini, D. Pauletto, F. R. Arevalo, N. F. Costa Reis, A. F. Sampaio, L. E. Urrego Giraldo, E. H. Valderrama Sandoval, L. Valenzuela Gamarra, C. I. A. Vela, and H. ter Steege. 2017. Persistent effects of pre-Columbian plant domestication on Amazonian forest composition. Science 355: 925-931. **doi: 10.1126/science.aal0157**

Lewis, S. E., and D. E. Wilson. 1987. *Vampyressa pusilla*. Mammalian Species 292: 1-5.

Lijding, H. W. 1958. Proeven met tilapia in Suriname / Experimental breeding of tilapia in Suriname. De Surinaamse Landbouw 6: 183-194.

Lill, A. 1974. Behavior of the Grassland Sparrow and two species of seed-finches. Auk 91: 35-43.

Lim, B. K. 2012a. Biogeography of mammals from the Guianas of South America. *In* B. D. Patterson, and L. P. Costa (Eds.). Bones, Clones, and Biomes: The History and Geography of Recent Neotropical mammals. pp. 230–258, University of Chicago Press, Chicago, IL, USA.

Lim, B. K. 2012b. Preliminary assessment of neotropical mammal DNA barcodes: an underestimation of biodiversity. The Open Zoology Journal 5 (Supplement 1-M3): 10-17.

Lim, B. K. 2016. Review of mammalogical research in the Guianas of northern South America. Integrative Zoology 11: 151–61.

Lim, B. K., and H. Banda. 2013. Rapid assessment program (RAP) survey of small mammals in the Grensgebergte and Kasikasima region of southeastern Suriname. *In* L. E. Alonso, and T.H. Larsen (Eds.). A rapid biological assessment of the Upper Palumeu River watershed (Grensgebergte and Kasikasima) of southeastern Suriname, pp. 161-169. RAP Bulletin of Biological Assessment 67, Conservation International, Arlington, VA, USA.

Lim, B. K., and S. Joemratie. 2011. Rapid Assessment Program (RAP) survey of small mammals in the Kwamalasamutu region of Suriname. *In* B. J. O'Shea, L. E. Alonso, and T. H. Larsen (Eds.). A rapid biological assessment of the Kwamalasamutu Region, Suriname, pp, 144-149. RAP Bulletin of Biological Assessment 63, Conservation International, Arlington, VA, USA.

Lim, B. K., M. D. Engstrom, and J. Ochoa G. 2005a. Mammals. *In* T. Hollowell, and R. P. Reynolds (Eds.). Checklist of the terrestrial vertebrates of the Guiana Shield, pp. 77-92. Bulletin of the Biological Society of Washington 13, Washington, DC, USA.

Lim, B. K., M. D. Engstrom, H. H. Genoways, F. M. Catzeflis, K. A. Fitzgerald, S. L. Peters, M. Djosetro, S. Brandon, and S. Mitro. 2005b. Results of the Alcoa Foundation-Suriname expeditions. XIV. Mammals of Brownsberg Nature Park, Suriname. Annals of Carnegie Museum 74: 225-274.

Lim. B. K., and F. M. Catzeflis. 2014. History of mammalogy in the Guianas: historia de la mastozoología en las Guayanas. *In* J. Ortega, J. L. Martínez, and D. G. Tirira (Eds.). Historia de la mastozoología en Latinoamérica, las Guayanas y el Caribe, pp. 265-292. Editorial Murciélago Blanco y Asociación Ecuatoriana de Mastozoología, Quito, Equador, and Mexico City, Mexico.

Lima, A., W. E. Magnusson, M. Menin, L. K. Erdtmann, D. J. Rodrigues, C. Keller, and W. Hödl. 2006. Guia de Sapos da Reserva Adolpho Ducke-Amazonia Central. Guide to the Frogs of Reserva Adolpho Ducke-Central Amazonia. Attema, Manaus, Brazil.

Lima, C. C., P. L. L. Rosangela, and L. M. Conserva. 2014. Dilleniaceae family: an overview of its ethnomedicinal uses, biological and phytochemical profile. Journal of Pharmacognosy and Phytochemistry 3: 181-204.

Lin, C. P. 2006. Social behavior and life history of Membracine treehoppers. Journal of Natural History 40: 1887–1907.

Lindeman, J. C. 1953. The vegetation of the coastal region of Suriname. Van Eedenfonds, Amsterdam, The Netherlands / Kemink en Zoon N.V., Utrecht, The Netherlands.

Lindeman, J. C., and A. M. W. Mennega. 1963. Bomenboek voor Suriname. Dienst 's Lands Bosbeheer, Paramaribo, Suriname.

Lindeman, J. C., and S. P. Moolenaar. 1959. Preliminary survey of the vegetation types of Northern Suriname. Van Eedenfonds, Amsterdam, The Netherlands / Kemink en Zoon N.V., Utrecht, The Netherlands.

Lindenmayer, D. B., J. Fischer, A. Felton, R. Montague-Drake, A. D. Manning, D. Simberloff, K. Youngentob, D. Saunders, D. Wilson, A. M. Felton, C. Blackmore, A. Lowe, S. Bond, N. Munro, and C. P. Elliott. 2007. The complementarity of single-species and ecosystem-oriented research in conservation research. Oikos 116: 1220-1226.

Link, A. and A. DiFiore. 2006. Seed dispersal by spider monkeys and its importance in the maintenance of neotropical rain-forest diversity. Journal of Tropical Ecology 22: 235-246.

Linnaeus, C. 1753. Species Plantarum, exhibentes plantas rite cognitas, ad genera relatas, cum differentiis specificis, nominibus trivialibus, synonymis selectis, locis natalibus, secundum systema sexuale digestas. Tomus I et II. Impensis Laurentii Salvii, Holmiæ [Stockholm, Sweden; 2 volumes].

Linnaeus, C. 1758. Systema naturae per regna tria naturae, secundum classes, ordines, genera, species, cum characteribus, differentiis, synonymis, locis. Tomus I., Editio Decima, Reformata. Impensis Direct. Laurentii Salvii, Holmiæ [10th ed., first volume; Stockholm, Sweden].

Linnaeus, C. 1763. D. A. G. dissertatio botanico-medica, sistens lignum quassiae. Upsaliæ [Uppsala, Sweden].

Linnaeus, C. 1775. D. D. Plantæ Surinamenses. Typis Edmannianis, Upsaliæ [Uppsala, Sweden].

Lira G., E. Torres, R. Soares, S. Mendonça, M. Costa, K. Silva, S. Simon, and K. Veras. 2007. Nutritional value of crustaceans from Estuary-Lagoon Complex Mundaú/Manguaba-Alagoas. Revista do Instituto Adolfo Lutz, São Paulo 66: 261-267.

Lobova, T. A., C. K. Geiselman, and S. A. Mori. 2009. Seed dispersal by bats in the Neotropics. Memoirs of the New York Botanical Garden 101: 1-471.

Lobova, T. A., S. A. Mori, F. Blanchard, H. Peckham, and P. Charles-Dominique. 2003. Cecropia as a food resource for bats in French Guiana and the significance of fruit structure in seed dispersal and longevity. American Journal of Botany 90: 388-403.

Locey, K. J., and J. T. Lennon. 2016. Scaling laws predict global microbial diversity. PNAS 113: 5970-5975. **www.pnas.org/cgi/doi/10.1073/pnas.1521291113**

Lockwood, J. L., M. F. Hoopes, and M. P. Marchetti. 2013. Invasion ecology (2nd ed.). Wiley-Blackwell, Chichester, England, UK.

Lokvam, J., and J. F. Braddock. 1999. Anti-bacterial function in the sexually dimorphic pollinator rewards of *Clusia grandiflora* (Clusiaceae). Oecologia 119: 534-540

Lomáscolo, S. B., P. Speranza, and R. T. Kimball. 2008. Correlated evolution of fig size and color supports the dispersal syndromes hypothesis. Oecologia 156: 783-796.

López-Hernández, D. 2001. Nutrient dynamics (C, N and P) in termite mounds of *Nasutitermes ephratae* from savannas of the Orinoco Llanos (Venezuela). Soil Biology and Biochemistry 33: 747-753.

Lord, C., Y. Fermon, F. J. Meunier, M. Jégu, and P. Keith. 2007. Croissance et longévité du Watau yaike, *Tometes lebaili* (Serrasalminae), dans le bassin du Haut Maroni (Guyane française). Cybium 31: 359-367.

Löttker P., Huck, M., Heymann, E. W., and M. Heistermann. 2004. Endocrine correlates of reproductive status in breeding and nonbreeding wild female moustached tamarins. International Journal of Primatology 25: 919-937.

Lougheed, S. C., C. Gascon, D. A. Jones, J. P. Bogart, and P. T. Boag. 1999. Ridges and rivers: a test of competing hypotheses of Amazonian diversification using a dart-poison frog (*Epipedobates femoralis*). Proceedings of the Royal Society, Biological Sciences 266: 1829–1835.

Lowe-McConnell, R. H. 1962. The fishes of the British Guiana continental shelf, Atlantic coast of South America, with notes on their natural history. Zoological Journal of the Linnean Society 44: 669-700.

Lowe-McConnell, R. H. 1964. The fishes of the Rupununi savanna district of British Guiana, South America. Part 1. Ecological groupings of fish species and effects of the seasonal cycle on the fish. Zoological Journal of the Linnean Society 45: 103-144.

Lowe-McConnell, R. H. 1969. The cichlid fishes of Guyana, South America, with notes on their ecology and breeding behaviour. Zoological Journal of the Linnean Society 48: 255-302.

Lowe-McConnell, R. H. 1984. The status of studies on South Amertican freshwater food fishes. *In* T. M. Zaret (Ed.), Evolutionary ecology of Neotropical freshwater fishes, pp. 139-156. Dr. W. Junk Publishers, The Hague, The Netherlands.

Lowe-McConnell, R. H. 1987. Ecological studies in tropical fish communities. Cambridge University Press, Cambridge, England, UK.

Lowe-McConnell, R. H. 2000. Land of waters: explorations in the natural history of Guyana, South America. The Book Guild, Lewes, England, UK.

Lowery, G. H., and W. W. Dalquest. 1951. Birds from the state of Veracruz, Mexico. Museum of Natural History. University of Kansas Publications, Lawrence, KS, USA.

Lucanus, O. 2009. The Amazon below water. Panta Rhei, Hannover, Germany.

Lucas, Y., R. Boulet, and A. Domeny 1982. Acid soils of French Guyana. *In* Wienk, J. F. and H. A. De Wit (Eds.). Workshop on low fertility acid soils of the American humid tropics, Paramaribo, Suriname, November 1981. IICA, San Jose, Costa Rica.

Lücking, R., J. Mata-Lorenzen, and L. G. Dauphin. 2010. Epizoic liverworts, lichens and fungi growing on Costa Rican shield mantis (Mantodea: *Choeradodis*). Studies on Neotropical Fauna and Environment 45: 175–186

Luger, M., T. W. J. Garner, R. Ernst, W. Hödl, and S. Lötters. 2008. No evidence for precipitous declines of harlequin frogs (*Atelopus*) in the Guyanas. Studies on Neotropical. Fauna and Environment 43: 177-180

Luger, M., W. Hödl, and S. Lötters. 2009. Site fidelity, home range behaviour and habitat utilization of male harlequin toads (Amphibia: *Atelopus hoogmoedi*) from Suriname: relevant aspects for conservation breeding. Salamandra 45: 211-218.

Lujan, N. K., and K. W. Conway. 2015. Life in the fast lane: a review of rheophily in freshwater fishes. *In* R. Riesch, M. Tobler, and M. Plath (Eds.). Extremophile fishes: ecology, evolution, and physiology of teleosts in extreme environments, pp 107-136. Springer, Cham, Heidelberg, New York, Dordrecht, London.

Lujan, N. K., K. A. Roach, D. Jacobsen, K.O. Winemiller, V. M. Vargas, V. R. Ching, and J. A. Maestre. 2013. Aquatic community structure across an Andes-to-Amazon fluvial gradient. Journal of Biogeography 40: 1715-1728.

Lujan, N. K., and J. W. Armbruster. 2011. Geological and hydrological history of the Guyana Shield and historical biogeography of its fishes. *In* J. A. Albert, and R. E. Reis (Eds.). Historical Biogeography of Neotropical Freshwater Fishes, pp. 211-224. University of California Press, Berkeley, A, USA.

Lundberg, J. G., L. G. Marshall, J. Guerrero, B. Horton, M. C. S. L. Malabarba and F. Wesselingh. 1998. The stage for Neotropical fish diversification: a history of tropical South American rivers. *In* L. R. Malabarba, R. E Reis, R. P. Vari, Z. M. S. Lucena and C. A. S. Lucena (Eds.). Phylogeny and classification of Neotropical fishes, pp 13-48. EDIPUCRS, Porto Alegre, Brazil.

Lundberg, J. G., M. Kottelat, G. R. Smith, M. Stiassny and T. Gill. 2000. So many fishes, so little time: an overview of recent ichthyological discoveries in fresh waters. Annals of the Missouri Botanical Garden 87: 26-62.

Luquet, P., T. Boujard, P. Planquette, Y. Moreau and G. Hostache. 1989. The culture of *Hoplosternum littorale*: state of the art and perspectives. In Advances in tropical aquaculture, pp 511-516. Aquacop-Ifremer, Paris, France.

Lynch Alfaro, J. W., Boubli, J. P., Olson, L. E., Di Fiore, A., Wilson, B., Gutiérrez-Espeleta, G. A., Chiou, K. L., Schulte, M., Neitzel, S., Ross, V., Schwochow, D., Nguyen, M. T. T., Farias, I., Janson, C. H., and M. E. Alfaro. 2012. Explosive Pleistocene range expansion leads to widespread Amazonian sympatry between robust and gracile capuchin monkeys. Journal of Biogeography 39: 272-288.

Lynch Alfaro, J. W., Boubli, J. P., Paim, F. P., Ribas, C. C., da Silva, M. N. F., Messias M. R., Röhe, F., Mercês M. P., Silva Júnior, J. S., Silva, C. R., Pinho, G. M., Koshkarian, G., Nguyen, M. T. T., Harada, M. L., Rabelo, R. M., Queiroz, H. L., Alfaro, M. E., and I. P. Farias. 2015. Biogeography of squirrel monkeys (genus *Saimiri*): South-central Amazon origin and rapid pan-Amazonian diversification of a low and primate. Molecular Phylogenetics and Evolution 82: 436-454.

Lynch Alfaro, J. W., J. Silva de Sousa Jr., A. B. Rylands. 2012. How different are robust and gracile capuchin monkeys? An argument for the use of *Sapajus* and *Cebus*. American Journal of Primatology 74: 273-286.

Maas, P. J. M. 1971. Floristic observation on forest types in Western Suriname. I. Mededelingen van het Botanisch Museum en Herbarium van de Rijksuniversiteit te Utrecht 350: 269-284 / Proceedings Koninklijke Nederlandse Akademie van Wetenschappen (Amsterdam), Series C 74: 269-302.

Maas, P. J. M. 1972. Costoideae (Zingiberaceae). Flora Neotropica Monographs, 8. Hafner, New York, NY, USA.

Maas, P. W. T. 1970. Contamination of the palm weevil (*Rhynchophorus palmarum*) with the red ring nematode (*Rhadinaphelenchus cocophilus*) in Surinam. Nematologica 16: 429-433.

Mabberley, D. J. 1992. Tropical Rain Forest Ecology, 2nd edition. Blackie Academic & Professional, Chapman & Hall, Glasgow, Scotland, UK.

MacLean, S. A. 2010. Sunbittern (*Eurypyga helias*), Neotropical Birds Online (T. S. Schulenberg, Ed.). Cornell Lab of Ornithology. **neotropical.birds.cornell.edu/portal/species/overview?p_p_spp=142356**

Madrid, V. M., J. Y. Aller, R. C. Aller, and A. Y. Christoserdov. 2001. High prokaryote diversity and analysis of community structure in mobile mud deposits off French Guiana: identification of two new bacterial candidate divisions. FEMS Microbiology Ecology 37: 197-209.

Magalhães, J. A. S., A. H. A. de Moraes Neto, and F. C. Miguens. 2008. Nematodes of *Rhynchophorus palmarum*, L. (Coleoptera: Curculionidae), vector of the Red Ring disease in coconut plantations from the north of the Rio de Janeiro State. Parasitology Research 102: 1281. **doi: 10.1007/s00436-008-0906-7**

Magnusson, W. E. 1985. Habitat selection, parasites and injuries in Amazonian crocodiles. Amazoniana 9: 193-204.

Maguire, B. 1945. The first botanical exploration of the Table Mountain in Suriname – part I and II. Journal of the New York Botanical Garden 46: 253-272 and 277-287.

Maguire, B. 1970. On the flora of the Guayana Highland. Biotropica 2: 85-100.

Maguire, B. et al. 1948. Plant explorations in Guiana in 1944, chiefly to the Tafelberg and Kaieteur Plateau – parts I through VI. Bulletin Torrey Botanical Club 75: 56-115, 189-230, 286-323, 374-438, 523-580 and 633-671.

Mahabali, s., and P. Spanoghe. 2013. Mitigation of two insecticides by wetland plants: feasibility study for the treatment of agricultural runoff in Suriname (South America). Water, Air & Soil Pollution, 225: 1771.

Mahajan, M., V. Manon, C. W. Spaander and J. J. B. van Lanschot. 2016. Esophageal fishbone impaction as a result of ingestion of "trapoen" (Surinamese dried fish). Endoscopy 48: E3.

Malavasi, A, D. Midgarden, A. van Sauers-Muller, and M. J. Godoy. 2013. Manual for the control of the carambola fruit fly in South America, operations manual, revised. Carambola Fruit Fly Programma, Ministry of Agriculture and Animal Husbandry, Paramaribo, Suriname.

Mallet, J. 1995. A species definition for the modern synthesis. Trends in Ecology and Evolution 10: 294-299.

Mallet, J., and L. E. Gilbert, L. E. 1995. Why are there so many mimicry rings? Correlations between habitat, behaviour and mimicry in *Heliconius* butterflies. Biological Journal of the Linnean Society 55: 159-180.

Maloof, J. E., and D. W. Inouye. 2000. Are nectar robbers cheaters or mutualists? Ecology 81: 2651-2661.

Manzi, M., and O. T. Coomes. 2009. Managing Amazonian palms for community use: a case of aguaje palm (*Mauritia flexuosa*) in Peru. Forest Ecology and Management 257: 510-517.

Mariano, J. R., M. C. Makrakis, E. A. L. Kashiwaqui, E. F. Celestino, and S. Makrakis. 2012. Longitudinal habitat disruption in Neotropical streams: fish assemblages under the influence of culverts. Neotropical Ichthyology 10: 771-784.

Marini-Filho, O. J. 1999. Distribution, composition, and dispersal of ant gardens and tending ants in three kinds of central Amazonian habitats. Tropical Zoology 12: 289-296.

Markham, C. L. (Ed.). 2010. The Letters of Amerigo Vespucci and other documents illustrative of his career. Cambridge University Press, New York, NY, USA.

Maros, A., A. Louveaux, C. Lelarge, and M. Girondot. 2006. Evidence of the exploitation of marine resource by the terrestrial insect *Scapteriscus* didactylus through stable isotope analyzes of its cuticle. BMC Ecol 6:6. **doi.org/10.1186/1472-6785-6-6**

Maros, A., A. Louveaux, E. Liot, J. Marmet J., and M. Girondot. 2005. Identifying characteristics of *Scapteriscus* spp. (Orthoptera: Gryllotalpidae) apparent predators of marine turtle eggs. Environmental Entomology 34: 1063-1070.

Maros, A., A. Louveaux, M. H. Godfrey, and M. Girondot. 2003. *Scapteriscus didactylus* (Orthoptera, Gryllotalpidae), predator of leatherback turtle eggs in French Guiana. Marine Ecology Progress Series 249: 289-296.

Marrenga, M., and C. Ruleman. 2008. Rapids of the Suriname River. Ralicon/Buanda, Paramaribo, Suriname.

Marshall, J., Y. Kushnir, D. Battisti, P. Change, A. Czaja, R. Dickson, J. Hurrel, M. McCartney, R. Saravanan, and M. Visbeck. 2001. North Atlantic Climate Variability: phenomena, impacts and mechanism, International Journal of Climatology 21: 1863-1898.

Marshall, S. D. 2001. Tarantulas and other Arachnids. Barron's Educational Series, Inc., Hauppauge, NY, USA.

Martín, J., M. B. Raymúndez, J. Vallès, T. Garnatje, and E. Raimúndez. 2012. Palynological study of the Venezuelan species of the genus *Hymenocallis* (Amaryllidaceae). Plant Systematics and Evolution 298: 695-701.

Martínez, A. E., and J. P. Gomez. 2013. Are mixed-species bird flocks stable through two decades? American Naturalist 181: E53-E59. **doi: 10.1086/669152**

Martis, A., G. J. van Oldenburg, and G. Burgers. 2002. Predicting rainfall in the Dutch Caribbean – more than El Ni o? International Journal of Climatology 22: 1219-1234.

Massemin, D., D. Lamy, J. P. Pointier and O. Gargominy. 2009. Coquillages et escargots de Guyane. Muséum National d'Histoire Naturelle, Paris, France.

MATHEWS, A. A. 1977. Studies on termites from the Mato Grosso state, Brazil. PhD dissertation. Academia brasileira de ciências, Rio de Janeiro, Brazil.

MATTIG, S., AND H. GREVEN. 1994. Beobachtungen zur Balz und Paarung des Vieraugenfisches *Anableps anableps* (Linné, 1758) (Anablepidae, Cyprinodontiformes). Zeitschrift für Fischkunde 2: 167-192.

MAUGIET, Q., D. OLIVIER, N. VANDEWALLE, AND P. VANDEWALLE. 2010. Ontogeny of swmming movements in bronze corydoras (*Corydoras aeneus*). Canadian Journal of Zoology 88: 378–389.

MAURY-LECHON, G. 1991. Comparative dynamics of tropical rain forest regeneration in French Guyana. *In* T. C. Whitmore, and A. Gomez-Pompa (Eds.). Rainforest Regeneration and Management, pp. 295-302. Man and the Biosphere Series, Smithsonian Institution, Washington, DC, USA.

MAYER, A., AND S. VASCONCELOS. 2013. Necrophagous beetles associated with carcasses in a semi-arid environment in Northeastern Brazil: Implications for forensic entomology. Forensic Science International 226: 41-45.

MAZZOLDI, C., V. LORENZI, AND M. B. RASOTTO. 2007. Variation of male reproductive apparatus in relation to fertilization modalities in the catfish families Auchenipteridae and Callichthyidae (Teleostei: Siluriformes). Journal of Fish Biology 70: 243-256.

MCCANN, S., O. MOERI, T. JONES, C. SCOTT, G. KHASKIN, R. GRIES, S. O'DONNELL, AND G. GRIES. 2013. Strike fast, strike hard: the Red-throated Caracara exploits absconding behavior of social wasps during nest predation. PLoS One 8(12): e84114. **doi: 10.1371/journal.pone.0084114**

MCCAULEY, R.D., J. FEWTRELL, A.J. DUNCAN, C. JENNER, M-N. JENNER, J.D. PENROSE, R.I.T. PRINCE, A. ADHITYA, J. MURDOCH, AND K. MCCABE. 2000. Marine seismic surveys: a study of environmental implications. Appea Journal 2000: 692-708.

MCCRONE J. D., AND H. W. LEVI. 1964 North American widow spiders of the *Latrodectus curacaviensis* group (Araneae: Theridiidae). Psyche 71: 12-27. **doi: 10.1155/1964/86469**

MCDONALD, F. J. D. 1966. The genitalia of North American Pentatomoidea (Hemiptera: Heteroptera). Quaestiones Entomologicae 2: 7-150.

MCGRANAHAN, G., D. BALK, AND B. ANDERSON. 2007. The rising tide: assessing risks of climate change and human settlements in low elevation coastal zones. Environment & Urbanization 19: 17-37.

MCGREW, W. C., L. F. MARCHANT, R. W. WRANGHAM, AND H. KLEIN. 1999. Manual laterality in anvil use: wild chimpanzees cracking *Strychnos* fruits. Laterality 4: 79-87.

MCINNIS, D. O., P. RENDON, E. JANG, A. VAN SAUERS-MULLER, R. SUGAYAMA, AND A. MALAVASI. 1999. Interspecific Mating of introduced, sterile *Bactrocera dorsalis* with wild *B. carambolae* (Diptera: Tephritidae) in Suriname: A potential case for cross-species Sterile Insect Technique. Annals of the Entomological Society of America 92: 758-765.

MCLELLAN. E., AND ALLAN, C. 2015. Wildlife Crime Initiative Annual Update 2015. WWF and TRAFFIC, Gland, Switzerland.

MCNULTY, K. P. 2016. Hominin taxonomy and phylogeny: what's in a name? Nature Education Knowledge 7:2.

MCPHADEN, M. 2004. Evolution of the 2002/03 El Niño. Bulletin of the American Meteorological Society 85: 677-695

MEBS, D., M. JANSEN, G. KÖHLER, W. POGODA, AND G. KAUERT. 2010. Myrmecophagy and alkaloid sequestration in amphibians: a study on *Ameerega picta* (Dendrobatidae) and *Elachistocleis* sp. (Microhylidae) frogs. Salamandra 46: 11-15.

MEDEM, F. M. 1958. The crocodilian genus Paleosuchus. Fieldiana (Zoology) 39(21): 227-247.

MEDEM, F. M. 1981. Los Crocodylia de Sur America. Vol. 1. Los Crocodylia de Colombia. Colciencias, Bogota, Colombia.

MEDEM, F. M. 1983. Reproductive data on *Platemys platycephala* (Testudines: Chelidae) in Colombia. *In* Rhodi, A, and K. Miyata (Eds.). Advances in Herpetology and Evolutionary Biology; Essays in Honor of Ernest E. Williams, pp 429-434. Museum of Comparative Zoology, Harvard University.

MEDEM, F. M., O. F. CASTANO, AND M. LUGO-R. 1979. Contribucion al conoscimiento sobra la reproduccion y el crecimiento de los "morrocoyes" (*Geochelone carbonaria* y *G. denticulata*; Testudines, Testudinidae). Caldasia 12: 497-511.

MEDENDORP, C. 1999. Gerrit Schouten (1779-1839). Botanische tekeningen en diorama's uit Suriname. Koninklijk Instituut voor de Tropen, Amsterdam, The Netherlands.

MEDENDORP, C. 2008. Kijkkasten uit Suriname. De diorama's van Gerrit Schouten. KIT Publishers, Amsterdam, The Netherlands.

MEES, G. F. 1974. The Auchenipteridae and Pimelodidae of Suriname (Pisces, Nematognathi). Zoologische Verhandelingen Leiden 132: 1-256.

MEES, G. F. 1988. In Memoriam: François Haverschmidt (1906-1987). Ardea 76: 211-221.

MELITZ, P. J. 1975. Legendarapport bij de landvormkaart van Suriname tussen 4° en 5° graden Noorderbreedte. DBK Report 54, Dienst Bodemkartering, Paramaribo, Suriname.

MENNEGA, A. M. W. (ED.). 1976. *Fa joe kan tak' mi no moi*, Deel I, Flora. Natuurgids Serie B 4, Stinasu, Paramaribo, Suriname.

MERIAN, M. S. 1705. Metamorphosis insectorum surinamensium. ofte verandering der Surinaamsche insecten. Waar in de Surinaamsche rupsen en wormen met alle des zelfs veranderingen na het leven afgebeeld en beschreeven worden, zynde elk geplaast op die gewassen, bloemen en vruchten, daar sy op gevonden zyn; waar in ook de generatie der kikvorschen, wonderbaare padden, hagedissen, slangen, spinnen en mieren werden ver toond en beschreeven, alles in America na het leven en levensgroote geschildert en beschreeven. Maria Sibylla Merian, Amsterdam. [The Netherlands]

MERIAN, M. S. 1719. Over de voortteeling en wonderbaerlyke veranderingen der Surinaemsche insecten, waer in de Surinaemsche rupsen en wormen, met alle derzelver veranderingen, naer het leeven afgebeeldt, en beschreeven worden; zijnde elk geplaest op dezelfde gewassen, bloemen, en vruchten, daer ze op gevonden zijn; beneffens de beschrijving dier gewassen. Waer in ook de wonderbare padden, hagedissen, slangen, spinnen, en andere zeltzame gediertens worden vertoont, en beschreeven. Alles in Amerika door den zelve M. S. Meiraen near het leeven, en leevensgrootte geschildert, en nu in't koper overgebracht. Benevens een Aenhangsel van de verandering van visschen in kikvorschen, en van kikvorschen in visschen. Johannes Oosterwyk, Amsterdam. [The Netherlands]

Mesquita, A. 2000. Sea level variations along the Brazilian coast: A short review. *In* A. Klein (Ed.). Proceedings of the Brazilian symposium on sandy beaches: morphodynamics, ecology, uses, hazards and management, pp. 21-31. Itajai, Santa Catarina, Brazil.

Metcalf, J L., C. Turney, R. Barnett, F. Martin, S. C. Bay, J. T. Vilstrup, L. Orlando, R. Salas-Gismondi, D. Loponte, M. Medina, M. De Nigris, T. Civalero, P. M. Fernández, A. Gasco, V. Duran, K.L.Seymour, C. Otaola, A. Gil, R. Paunero, F. J. Prevosti, C. J. A. Bradshaw, J. C. Wheeler, L. Borrero, J. J. Austin, and A. Cooper. 2016. Synergistic roles of climate warming and human occupation in Patagonian megafaunal extinctions during the last Deglaciation. Science Advances 2. **doi: 10.1126/sciadv.1501682**

Meunier, F. J., N. Journiac, S. Lavoué and N. Rabet. 2002. Caracteristiques histologiques des marques de croissance squelettique chez l'atipa *Hoplosternum littorale* (Hancock, 1828) (Teleostei, Siluriformes), dans le Marais de Kaw (Guyane francaise). Bulletin Français de la Pêche et de la Pisciculture 364: 71-86.

Meyer de Schauensee, R. 1970. A Guide to the Birds of South America. Published for the Academy of the Natural Sciences of Philadelphia by InterCollegiate Press Inc, Wynnewood, PA, USA.

Miglino, L., L. Corral, A. Killmer, C. Elias, M. da Cunha, W. Grisley, and G. Barreda. 2005. Country Environmental Assessment. Inter-American Development Bank, Paramaribo, Suriname.

Mikolji, I. 2013. Freshwater natural aquarium (video). **www.youtube.com/watch?v=TzzJU810AIk**

Miles, M. A., A. A. de Souza, and M. Povoa. 1981. Chagas Disease in the Amazon Basin III. Ecotopes of ten triatomine bug species (Hemiptera: Reduviidae) from the vicinity of Belem, Para State, Brazil. Journal of Medical Entomology 18: 266-278.

Miller, G. T., and S. E. Spoolman. 2008. Environmental science (12th ed.). Brooks/Cole Thomson Learning Inc., Belmont, CA, USA.

Miller, G. T., and S. E. Spoolman. 2016. Environmental science (15th ed.). Cengage Learning, Boston, MA, USA.

Miller, M. J. 2009. Ecology of anguilliform leptocephali: remarkable transparent fish larvae of the ocean surface layer. Aqua-BioScience Monographs 2: 1-94.

Milliken, W., B. Klitgård, and A. Baracat (Eds.). 2009 onwards. Neotropikey - Interactive key and information resources for flowering plants of the Neotropics. Kew Botanic Gardens, Richmond, England, UK. **www.kew.org/neotropikey**

Milton, P. Y. 2012. Plant genetic resources of Suriname (Country Report). P. Y. Milton, Paramaribo, Suriname

Ministerie van Landbouw, Veeteelt en Visserij. 2013. Visserij management plan voor Suriname 2014-2018. Onderdirectoraat Visserij, Paramaribo, Suriname.

Ministry of Labour, Technological Development and Environment. 2013. Suriname Second National Communication to the United Nations Framework Convention on Climate Change. Ministry of Labour, Technological Development and Environment, Paramaribo, Suriname.

Ministry of Labour, Technological Development, and Environment. 2006. National biodiversity strategy. National Institute for Environment and Development in Suriname, Paramaribo, Suriname.

Ministry of Labour, Technological Development, and Environment. 2013. National biodiversity action plan (NBAP) 2012-2016. Ministry of Labour, Technological Development, and Environment, Paramaribo, Suriname.

Ministry of Natural resources. 1999. Nomination dossier Central Suriname Nature Reserve for World Heritage List. The Government of Suriname, Paramaribo, Suriname.

Mittermeier, J. C., K. Zyskowski, E. S. Stowe, and J. E. Lai. 2010. Avifauna of the Sipaliwini savanna (Suriname) with insights into its biogeographic affinities. Bulletin of the Peabody Museum of Natural History 51: 97-122.

Mittermeier, R. A. 1991. Hunting and its effect on wild primate populations in Suriname. *In* J. Robinson, and K. Redford (Eds.). Neotropical wildlife use and conservation, pp. 93-107. University of Chicago Press, Chicago, IL, USA.

Mittermeier, R. A., and M. G. M. van Roosmalen. 1981. Preliminary observations on habitat utilization and diet in eight Surinam monkeys. Folia Primatologia 36: 1-39.

Mittermeier, R. A., Rylands, A. B., and J. P. Boubli. 2008. *Ateles paniscus*. The IUCN Red List of Threatened Species. Version 2015.2. **www.iucnredlist.org** [accessed Jul. 2015]

Mittermeier, R. A., Rylands, A., van Roosmalen, M. G. M., Norconk, M. A., Konstant, W. R., and Famolare, L. (Eds.). 2008. Monkeys of the Guianas: Guyana, Suriname, French Guiana. Tropical Pocket Guide Series, Conservation International, Washington, DC, USA.

Mittermeijer, R. A. 1977. Distribution, synecology and conservation of Suriname monkeys. PhD dissertation, Biological Anthropology Department, Harvard University, Cambridge, MA, USA.

Mittermeijer, R. A., S. A. J. Malone, M. J. Plotkin, F. Baal, K. Mohadin, J. MacKnight, M. Werkhoven, and T. B. Werner. 1990. Conservation Action Plan for Suriname. Conservation International, Suriname Forest Service, World Wildlife Fund, Stinasu, and University of Suriname.

Moberly, W. R. 1968. The metabolic responses of the Common Iguana, *Iguana iguana*, to walking and diving. Compendium of Biochemy and Physiology 27: 21-32.

Moermond, T. C., and J. S. Denslow. 1985. Neotropical avian frugivores: patterns of behavior, morphology, and nutrition, with consequences for fruit selection. Ornithological Monographs 36: 856-897.

Mohadin, K. 1980. Hydrobiologische verkenning van Noord Suriname. Instituut voor de Opleiding van Leraren. Thesis, Paramaribo, Suriname.

Mohan, S., M. Tiller, G. van der Voet, and H. Kanhai. 2005. Mercury exposure of mothers and newborns in Surinam: a pilot study. Clinical Toxicology 43: 101-104.

Mokross, K., T. B. Ryder, M. Corrêa Côrtes, J. D. Wolfe, and P. C. Stouffer. 2014. Decay of interspecific avian flock networks along a disturbance gradient in Amazonia. Proceedings of the Royal Society B 281: 20132599.

Mol, J. H. A. 1993. Structure and function of floating bubble nests of three armoured catfishes (Callichthyidae) in relation to the aquatic environment. *In* P.E. Ouboter (Ed.). The freshwater ecosystems of Suriname, pp. 167-197. Kluwer Academic Publishers, Dordrecht, The Netherlands.

MOL, J. H. A. 1994. Effects of salinity on distribution, growth and survival of three neotropical armoured catfishes (Siluriformes - Callichthyidae). Journal of Fish Biology 45: 763-776.

MOL, J. H. A. 1995. Ontogenetic diet shifts and diet overlap among three closely related neotropical armoured catfishes. Journal of Fish Biology 47: 788-807.

MOL, J. H. A. 1996a. Reproductive seasonality and nest site differentiation in three closely related armoured catfishes (Siluriformes: Callichthyidae). Environmental Biology of Fishes 45: 363-381.

MOL, J. H. A. 1996b. Impact of predation on early stages of the armoured catfish *Hoplosternum thoracatum* (Siluriformes - Callichthyidae) and implications for the syntopic occurrence with other related catfishes in a Neotropical multi-predator swamp. Oecologia 107: 395-410.

MOL, J. H. A. 2006. Attacks on humans by the piranha *Serrasalmus rhombeus* in Suriname. Studies on Neotropical Fauna and Environment 41: 189-195.

MOL, J. H. A. 2012a. Occurrence of a freshwater pipefish *Pseudophallus* cf. *brasiliensis* (Syngnathidae) in Corantijn River, Suriname, with notes on its distribution, habitat, and reproduction. Cybium 36: 45-53.

MOL, J. H. A. 2012b. The freshwater fishes of Suriname. Brill, Leiden, the Netherlands.

MOL, J. H. A., AND D. PONTON. 2003. Growth of young armoured catfish *Megalechis thoracata* in neotropical swamps and a rain-forest creek as revealed by daily micro-increments in otoliths. Journal of Tropical Ecology 19: 301-313.

MOL, J. H. A., AND F. L. VAN DER LUGT. 1996. Distribution and feeding ecology of the African tilapia *Oreochromis mossambicus* (Teleostei, Perciformes, Cichlidae) in Suriname (South America) with comments on the tilapia –kwikwi (*Hoplosternum littorale*) (Teleostei, Siluriformes, Callichthyidae) interaction. Acta Amazonica 25: 101-116

MOL, J. H. A., AND K. WAN TONG YOU. 2013. Fishes of the Palumeu River, Suriname. *In* L. E Alonso and T. Larsen (Eds.), A rapid biological assessment of the upper Palumeu River Watershed (Grensgebergte and Kasikasima) of southeastern Suriname, pp 119-130. RAP Bulletin of Biological Assessment 67. Conservation International, Arlington, USA.

MOL, J. H. A., AND P. E. OUBOTER. 2004. Downstream effects of erosion from small-scale gold mining on the instream habitat and fish community of a small Neotropical rainforest stream. Conservation Biology 18: 201-214.

MOL, J. H. A., B. DE MÉRONA II, P. E. OUBOTER, AND S. SAHDEW. 2007a. The fish fauna of Brokopondo Reservoir, Suriname, during 40 years of impoundment. Neotropical Ichthyology 5: 351-368.

MOL, J. H. A., D. RESIDA, J. RAMLAL, AND C. BECKER. 2000. Effects of El Niño-related drought on freshwater and brackish water fishes in Suriname, South America. Environmental Biology of Fishes 59: 429-440.

MOL, J. H. A., J. S. RAMLAL, C. LIETAR, AND M. VERLOO. 2001. Mercury contamination in freshwater, estuarine and marine fishes in relation to small-scale gold mining in Suriname, South America. Environmental Research 86: 183-197.

MOL, J. H. A., K. WAN TONG YOU, I. VREDE, A. FLYNN, P. E. OUBOTER, AND F. VAN DER LUGT. 2007b. Fishes of Lely and Nassau Mountains, Suriname. *In* L. A. Alonso and J. H. Mol (Eds.), A rapid biological assessment of the Lely and Nassau Plateaus Suriname (with additional information on the Brownsberg Plateau), pp 107-118. Conservation International, Arlington, USA.

MOL, J. H. A., P. WILLINK, B. CHERNOFF, AND M. COOPERMAN. 2006. Fishes of the Coppename River, Central Suriname Nature Reserve, Suriname. *In* L. E. Alonso, and H. J. Berrenstein (Eds.), A rapid biological assessment of the aquatic ecosystems of the Coppename River Basin, Suriname, pp 67-79. Conservation International, Washington, DC, USA.

MOLGO, I., AND B. P. E. DE DIJN. 2007. Orchids and orchid bees of the Brownsberg, Nassau and Lely ranges. *In* L. E. Alonso, and J. H. Mol (Eds.). A rapid biological assessment of the Lely and Nassau Plateaus (with additional information on the Brownsberg Plateau), pp. 86-91. RAP Bulletin of Biological Assessments 43, Conservation International, Arlington, VA, USA.

MOLINARI J., E. E. GUTIÉRREZ, A. A. DE ASCENÇÃO, J. M. NASSAR, A. ARENDS, AND R. J. MÁRQUEZ 2005. Predation by giant centipedes, *Scolopendra gigantea*, on three species of bats in a Venezuelan cave. Caribbean Journal of Science 41: 340–346.

MONSELS, D. M. 2016. Bauxite deposits in Suriname: Geological context and resource development. Netherlands Journal of Geosciences (Geologie en Mijnbouw) 95: 405–418.

MORA, C., D. P. TITTENSOR, S. ADL, A. G. B. SIMPSON, AND B. WORM. 2011. How many species are there on Earth and in the Ocean? PLoS Biology 9: e100127.

MORALES, J. F. 2002. Asclepiadaceae. *In* S. A. Mori, G. Cremers, C. A. Gracie, J. D. Granville, S. Heald, M. Hoff, and J. D. Mitchell (Eds.). Guide to the vascular plants of Central French Guiana: part 2. Dicotyledons. Memoirs of the New York Botanical Garden 76: 89-95.

MORALES, K., AND T. VINICIUS. 2003. Amazon rainforest: biodiversity and biopiracy Student BMJ 13: 386-7

MORAN, V. C., AND T. R. E. SOUTHWOOD. 1982. The guild composition of arthropod communities in trees. Journal of Animal Ecology 51: 289-306.

MOREIRA, J. R., M. R. ALVAREZ, T. TARIFA, V. PACHECO, A. TABER, D. G. TIRIRA, E. A. HERRERA, K. M. P. M. B. FERRAZ, J. ALDANA-DOMÍNGUEZ, AND D. W. MACDONALD. 2013. Taxonomy, natural history and distribution of the capybara. *In* J. R. Moreira (Ed.). Capybara: biology, use and conservation of an exceptional neotropical species, pp. 3-37. Springer, New York, NY, USA.

MOREL, F. M. M., A. M. L. KRAEPIEL, AND M. AMYOT. 1998. The chemical cycle and bioaccumulation of mercury. Annual Review of Ecology and Systematics 29: 543-566.

MORENO, J.,E., Y. RUBIO-PALIS, E. PAEZ, AND V. SANCHEZ. 2007. Abundance, biting behavior and parous rates of anopheline mosquito species in relation to malaria incidence in gold mining areas of southern Venezuela. Medical and Veterinary Entomology 21: 339-349

MORI, S. A., AND G. T. PRANCE. 1993. Lecythidaceae. Flora of the Guianas Series A, Phanerogams, 12: 1-44.

MORI, S. A., AND J. D. BOEKE. 1987. Chapter XII. Pollination. *In* S. A. Mori & Collaborators (Eds.). The Lecythidaceae of La Fumée Mountain, French Guiana. Memoirs of the New York Botanical Garden 44: 137-155.

Mori, S. A., G. Cremers, C. A. Gracie, J. D. Granville, M. Hoff, and J. D. Mitchell (Eds.). 1997. Guide to the vascular plants of central French Guiana: part 1. Pteridophytes, gymnosperms, and monocotyledons. Memoirs of the New York Botanical Garden 76, Bronx, New York, NY, USA.

Mori, S. A., G. Cremers, C. A. Gracie, J. D. Granville, S. Heald, M. Hoff, and J. D. Mitchell (Eds.). 2002. Guide to the vascular plants of central French Guiana: part 2. Dicotyledons. Memoirs of the New York Botanical Garden 76, Bronx, New York, NY, USA.

Mori, S. A., J. E. Orchard, and G. T. Prance. 1980. Intrafloral pollen differentiation in the New World Lecythidaceae, subfamily Lecythidoideae. Science 209: 400-403.

Mori, S. A., N. P. Smith, X. Cornejo, and G. T. Prance. 2010 onward. The Lecythidaceae Pages. The New York Botanical Garden, Bronx, New York, NY, USA. **sweetgum.nybg.org/lp/index.php**

Morrison, R. I. G., D. S. Mizrahi, R. K. Ross, O. H. Ottema, N. de Pracontal, and A. Narine. 2012. Dramatic declines of Semipalmated Sandpipers on their major wintering areas in the Guianas, northern South America. Waterbirds 35: 120-134.

Morton, E. S. 1971. Food and migration habits of the Eastern Kingbird in Panama. Auk 88: 925-926.

Mota de Oliveira, S. (Ed.). 2012 onwards. Flora of the Guianas. Royal Botanic Garden, Kew, Richmond, England, UK. **portal.cybertaxonomy.org/flora-guianas/**

Moure, J. S., G. A. R. Melo, and L. R. R. Faria Jr. 2012. Euglossini Latreille, 1802. *In* J.S. Moure, D. Urban, and G. A. R. Melo (Eds.). Catalogue of Bees (Hymenoptera, Apoidea) in the Neotropical Region - online version. **www.moure.cria.org.br/catalogue** [accessed Oct. 2015]

Mrosovsky, N. and, C. L. Yntema. 1980. Temperature dependence of sexual differentiation in sea turtles: Implications for conservation practices. Biological Conservation 18: 271-280.

Muedeking, M. H., and W. R. Heyer. 1976. Description of eggs and reproductive patterns of *Leptodactytylus pentadactylus* (Amphibia: Leptodactylidae). Herpetologica 32: 137-139.

Mueller, U. G., and N. Gerardo. 2002. Fungus-farming insects: Multiple origins and diverse evolutionary histories. PNAS 99: 15247-15249. **www.pnas.org/cgi/doi/10.1073/pnas.242594799**

Müller, U. K., E. J. Stamhuis, and J. J. Videler. 2000. Hydrodynamics of unsteady fish swimming and the effects of body size: comparing the flow fields of fish larvae to audults. Journal of Experimental Biology 203: 193-206.

Munn, C. A. 1986. Birds that "cry wolf". Nature 319: 143-145.

Munn, C. A., and J. W. Terborgh. 1979. Multi-species territoriality in Neotropical foraging flocks. Condor 81: 338-347.

Murphy, J. C. 1997. Amphibians and Reptiles of Trinidad and Tobago. Krieger Publishers, Malabar.

Murphy, R. C. 1955. Feeding habits of the Everglade Kite (*Rostrhamus sociabilis*). Auk 72: 204-205.

Myers, C. W., and M. A. Donnelly. 1997. A tepui herpetofauna on a granitic mountain (Tamacuari) in the borderland between Venezuela and Brazil: Report from the Phipps Tapirapec Expedition. American Museum Novitates 3213: 1-71.

Myers, J. and M. D. Loveless. 1976. Nesting aggregations of the Euglossine bee *Euplusia surinamensis* (Hymenoptera: Apidae): individual interactions and the advantage of living together. Canadian Entomologist 108: 1-6.

Nagelkerken, I., S. J. M. Blaber, S. Bouillon, P. Green, M. Haywood, L. G. Kirton, J.-O. Meynecke, J. Pawlik, H. M. Penrose, A. Sasekumar, and P. J. Somerfield. 2008. The habitat function of mangroves for terrestrial and marine fauna: a review. Aquatic Botany 89: 155–185.

Naipal, S., and M. Amatali. 1999a. Project Country Study Climate Change Suriname Water Resources (Technical report 4). Ministry of Public Works, Paramaribo, Suriname, and Institute for Environmental Studies, Amsterdam, The Netherlands.

Naipal, S., and M. Amatali. 1999b. Project Country Study Climate Change Suriname and First Steps Towards Coastal Zone Management. Ministry of Public Works, Paramaribo, Suriname, and Institute for Environmental Studies, Amsterdam, The Netherlands.

Naka, L. N. 2011. Avian distribution patterns in the Guiana Shield: implications for the delimitation of Amazonian areas of endemism. Journal of Biogeography 38: 681-696.

Naples, V. L. 1999. Morphology, evolution and function of feeding in the giant anteater (*Myrmecophaga tridactyla*). Journal of Zoology 249: 19-41.

National Planning Office of Suriname. 1988. Suriname planatlas. National Planning Office, Paramaribo, Suriname, and Organization of American States, Washington, DC, USA.

NEDECO. 1968. Suriname Transportation Study. Report of the hydraulic investigation. NEDECO, Den Haag, The Netherlands.

Neild, A. F. E. 2008. The butterflies of Venezuela, Part 2: Nymphalidae II (Acraeinae, Libytheinae, Nymphalinae, Ithomiinae, Morphinae). Meridian Publications, London, England, UK.

Nelson, A. 2016. Staatsolie's VISION 2030: the contributions of petroleum geology to Surinamese society. Netherlands Journal of Geosciences (Geologie en Mijnbouw) 95: 375-392,.

Nelson, J. S., T. C. Grande, and M. V. H. Wilson. 2016. Fishes of the world (5th ed.). Wiley, New York, NY, USA.

Nelson, M. E., and M. A. MacIver. 1999. Prey capture in the weakly electric fish *Apteronotus albifrons*: sensory acquisition strategies and electrosensory consequences. Journal of Experimental Biology 202: 1195-1203.

Neme, L. 2015. Journey to oblivion: unraveling Latin America's illegal wildlife trade. **news.mongabay.com/2015/11/journey-to-oblivion-unraveling-latin-americas-illegal-wildlife-trade/**

Netto-Ferreira, A. L., M. P. Albrecht, J. L. Nessimian, and É. P. Caramaschi. 2007. Feeding habits of *Thoracocharax stellatus* (Characiformes: Gasteropelecidae) in the upper rio Tocantins, Brazil. Neotropical Ichthyology 5: 69-74.

Nevis, A. B., J. M. Martinelli, A. S. S. Carvalho, and V. J. I. Nahum. 2009. Abundance and spatial-temporal distribution of the family Portunidae (Crustacean, Decapoda) in the Curuçá estuary on the northern coast of Brazil Brazilian. Journal of Aquatic Science and Technology 13: 71-79.

Nichols, D. K., E. W. Lamirande, A. P. Pessier, and J. E. Longcore. 2001. Experimental transmission of Cutaneous Chytridiomycosis in Dendrobatid Frogs. Journal of Wildlife Diseases 37: 1-11.

Nichols, J. D. 1994. *Terminalia amazonia* (Gmel.) Exell.: development of a native species for reforestation and agroforestry. Commonwealth Forestry Review 73: 9-13.

Nicholson, K. E., B. I. Crother, C. Guyer, and J. M. Savage. 2012 .It is time for a new classification of anoles (Squamata: Dactyloidae). Zootaxa 3477: 1–108.

Nickle, D. A. 2003. A Revision of the mole cricket genus *Scapteriscus* with the description of a morphologically similar new genus (Orthoptera: Gryllotalpidae: Scapteriscinae). Transactions of the American Entomological Society 129: 411-485.

Nico, L. G., H. L. Jelks, and T. Tuten. 2009. Non-native suckermouth armored catfishes in Florida: description of nest burrows and burrow colonies with assessment of shoreline conditions. ANSRP Bulletin 9: 1-30.

Niewiarowski, S., E. P. Kirby, T. M. Bruszynski, and K. Stocker. 1979. Thrombocytin, a serine protease from Bothrops atrox venom. 2. Interaction with platelets and plasma-clotting factors. Biochemistry 18: 3570-3577.

Nijssen, H. 1970. Revision of the Surinam catfishes of the genus *Corydoras* Lacépède, 1803 (Pisces, Siluriformes, Callichthyidae). Beaufortia 18: 1-75.

NIMOS. 2003. Forest sector environmental assessment and action plan. Volume 1: Forest resource use and environmental assessment. National Institute for Environment and Development (NIMOS), Paramaribo, Suriname.

NIMOS. 2005. Environmental assessment guidelines Volume 1: Generic. NIMOS Office of Environmental and Social Assessment, Paramaribo, Suriname.

Noonan, B. P. 2000. Does the phylogeny of pelomedusoid turtles reflect vicariance due to continental drift? Journal of Biogeography 27 1245–1249.

Noonan, B. P., and K. P. Wray. 2006. Neotropical diversification: the effects of a complex history on diversity within the poison frog genus *Dendrobates*. Journal of Biogeography 33: 1007-1020.

Noonan, B. P., and P. Gaucher. 2005. Phylogeography and demography of Guianan harlequin toads (*Atelopus*): diversification within a refuge. Molecular Ecology 14: 3017–3031.

Noonan, B. P., and P. Gaucher. 2006. Refugial isolation and secondary contact in the dyeing poison frog *Dendrobates tinctorius*. Molecular Ecology 15: 4425–4435.

Noonan, B. P., and P. T. Chippindale. 2006. Vicariant origin of malagasy reptiles supports late cretaceous antarctic land bridge. American Naturalist 168: 730–741.

Noordam, D. 1986. Verslag van de semi-detail bodemkartering t.b.v. de aquacultuur te Saronbrug e.o. Rapport 134. Dienst Bodemkartering, Paramaribo, Suriname.

Noordam, D. 1993. The Geographical Outline. *In* P. E. Ouboter (Ed.) The Freshwater Ecosystems of Suriname. pp. 13–28, Kluwer Academic Publishers, Dordrecht, The Netherlands.

Noordam, D., and M. W. P. M. Schellekens. 1988. Rapport van de semi-detail bodemkartering van kaartblad 13b. Rapport 129. Dienst Bodemkartering, Paramaribo, Suriname.

Noordam, D., and P. A. Teunissen. 2008. Oil exploration activities in the Coronie Swamp in Suriname 2. Report of the heli-sample flight of November 22, 2007. Consultancy report prepared for Paradise Oil Company N.V., Paramaribo, Suriname.

Noordam, D., and S. T. Carilho. 1998. The soil survey department (DBK) of Suriname. *In* T. E. Wong, D. R. de Vletter, L. Krook, J. I. S. Zonneveld, and A. J. van Loon (Eds.). The history of earth sciences in Suriname, pp. 431-442. Royal Netherlands Academy of Arts and Sciences & Netherlands Institute of Applied Geoscience TNO, Amsterdam, The Netherlands.

Norconk, M. A. 1996. Seasonal variation in the diets of white-faced and bearded sakis (*Pithecia pithecia* and *Chiropotes satanas*) in Guri Lake, Venezuela. Pp. 403-423 *In* M. A. Norconk, A. L. Rosenberger, & P. A. Garber (Eds). Adaptive radiations in Neotropical Primates. Plenum Publishers, New York, NY, USA.

Norconk, M. A. 2006. A long-term study of group dynamics and female reproduction in Venezuelan *Pithecia pithecia*. International Journal of Primatology 27: 653-674.

Norconk, M. A. 2011. Sakis, uacaris, and titi monkeys. In C. J. Campbell, A. Fuentes, K. C. MacKinnon, S. K. Bearder and R. M. Stumpf (Eds.). Primates in perspective, pp. 122-139. Oxford University Press, Oxford, England, UK.

Norconk, M. A., and M. Veres. 2011. Physical properties of fruit and seeds ingested by primate seed predators with emphasis on sakis and bearded sakis. Anatomical Record 294: 202-2011.

Norconk, M. A., and N .L. Conklin-Brittain. 2016. Bearded saki feeding strategies on an island in Lago Guri, Venezuela. American Journal of Primatology 78: 07-522.

Norconk, M. A., Oftedal, O. T., Power, M. L., Jakubasz, M., and A. Savage. 2002. Digesta passage and a preliminary study of fiber digestibility in captive white-faced sakis, *Pithecia pithecia*. American Journal of Primatology 58: 23-34.

Norconk, M. A., Raghanti, M. A. Martin, S. K., Grafton, B. W., Gregory, L. T., and B. P. E. De Dijn. 2003. Primates of Brownsberg Natuurpark, Suriname, with particular attention to the pitheciins. Neotropical Primates 11: 94-100.

Norde, R., and F. H. F. Oldenburger. 1975. The Vegetation of the Sipaliwini Savanna in Southern Suriname. **sipaliwinisavanna.com/docs/vegetation_sipaliwini_savanna.pdf**

Norden, N., J. Chave, P. Belbenoit, A. Caubère, P. Châtelet, P.-M. Forget, and C. Thébaud. 2007. Mast fruiting is a frequent strategy in woody species of eastern South America. PLoS ONE 2(10): e1079. **doi: 10.1371/journal.pone.0001079**

Nota, D. J. G. 1958. Sediments of the western Guiana shelf. PhD dissertation / Mededelingen van de Landbouwhogeschool te Wageningen 58: 1-98.

Nota, D. J. G. 1967. Geomorphology and sediments of the western Surinam shelf, a preliminary note. Geologie en Mijnbouw 48: 1-4.

Nota, D. J. G. 1971. Morphology and sediments off the Marowijne River, eastern Surinam shelf. Hydrographic Newsletter 6: 31-36.

Novakowski, G. C., R. Fugi, and N. S. Hahn. 2004. Diet and dental development of three species of *Roeboides* (Characiformes: Characidae). Neotropical Ichthyology 2: 157-162.

Nurmohamed, R. 2008. The impact of climate change and climate variability on water resources in Suriname. PhD dissertation, Anton de Kom Universiteit van Suriname, Paramaribo, Suriname.

Nurmohamed, R., and S. Naipal. 2005. Effects of global climate change and climate variability on Suriname's climate and water resources. Interactie (Journal of the University of Suriname): 47-60.

Nurmohamed, R., S. Naipal, and C. Becker. 2007. Changes and variation in the discharge regime of the Upper Suriname River Basin and its relationship with the tropical Pacific and Atlantic SST anomalies. Journal of Hydrological Processes 22: 1650-1659.

O'Dea, A., H. A. Lessios, A. G. Coates, R. I. Eytan, S. A. Restrepo-Moreno, A. L. Cione, L. S. Collins, A. de Queiroz, D. W. Farris, R. D. Norris, R. F. Stallard, M. O. Woodburne, O. Aguilera, M.-P. Aubry, W. A. Berggren, A. F. Budd, M. A. Cozzuol, S. E. Coppard, H. Duque-Caro, S. Finnegan, G. M. Gasparini, E. L. Grossman, K. G. Johnson, L. D. Keigwin, N. Knowlton, E. G. Leigh, J. S. Leonard-Pingel, P. B. Marko, N. D. Pyenson, P. G. Rachello-Dolmen, E. Soibelzon, L. Soibelzon, J. A. Todd, G. J. Vermeij, and J. B. C. Jackson. 2016. Formation of the Isthmus of Panama. Science Advances 2 (8), e1600883. **DOI: 10.1126/sciadv.1600883**

O'Shea, B. J. 2005. Notes on birds of the Sipaliwini savanna and other localities in southern Suriname, with six new species for the country. Ornitología Neotropical 16: 361-370.

O'shea, B. J., and O. Ottema. 2007. Environmental and social impact assessment for the proposed Bakhuis bauxite mining project. Specialist study: Ornithology. SRK Consulting, Cape Town, South Africa.

O'shea, B. J., and S. Ramcharan. 2013. A rapid assessment of the avifauna of the upper Palumeu watershed, southeastern Suriname. *In* L. E. Alonso, and T. H. Larsen (Eds.). A rapid biological assessment of the upper Palumeu River watershed (Grensgebergte and Kasikasima) of southeastern Suriname, pp. 145-160. Conservation International, Arlington, VA, USA.

O'Shea, B. J., L. E. Alonso, and T. H. Larsen (Eds.). 2011. A rapid biological assessment of the Kwamalasamutu Region, Suriname. RAP Bulletin of Biological Assessment 63, Conservation International, Arlington, VA, USA.

Ohara, W. M., I. D. Da Costa and M. L. Fonseca. 2016. Behaviour, feeding habits and ecology of the blind catfish *Phreatobius sanguijuela* (Ostariophysi: Siluriformes). Journal of Fish Biology 89: 1285-1301

O'Herne, L. 1969. A photogeological study of the basal complex of Surinam. Mededelingen Geologisch Mijnbouwkundige Dienst Suriname 20: 53-149.

Olamendi-Portugal, T, C. F. V. Batista, R. Restano-Cassulini, V. Pando, O. Villa-Hernandez, A. Zavaleta-Martinez-Vargas, M. C. Salas-Arruz, R. C. Rodriguez de la Vega, B. Becerril, and L. D. Possani. 2008. Proteomic analysis of the venom from the fish eating coral snake *Micrurus surinamensis*: Novel toxins, their function and phylogeny. Proteomics 8: 1919–1932.

Oldeman, R. A. A. 1974. L'Architecture de la forêt guyanaise. Mémoires ORSTOM 73, ORSTOM, Paris, France.

Oldenburger, F. H. F, R. Norde, and H. T. Riezebos. 1973. Ecological investigations on the vegetation of the Sipaliwini-savanna area (Southern Suirname). Institute of Systematic Botany & Laboratory of Physical Geography, Utrecht University, Utrecht, The Netherlands.

Oldenburger, F. H. F., and R. Norde. 2009. The desert and rock pavement vegetation on the inselberg slopes of the 4-Gebroeders Mountains, the Great Granite Plate and the isolated boulders within the Sipaliwini Savanna Area. **sipaliwinisavanna.com/docs/desert_and_rock_pavement_vegetation_inselberg_slopes.pdf**

Oliveira, A. C. M., and S. F. Ferrari. 2000. Seed dispersal by black-handed tamarins, *Saguinus midas niger* (Callitrichinae, Primates): implications for the regeneration of degraded forest habitats in eastern Amazonia. Journal of Tropical Ecology 16: 709-716.

Oliveira, C., G. S. Avelino, K. T. Abe, T. C. Mariguela, R. C Benine, G. Ortí, R. P Vari, and R. M. Corrêa e Castro. 2011. Phylogenetic relationships within the speciose family Characidae (Teleostei: Ostariophysi: Characiformes) based on multilocus analysis and extensive ingroup sampling. BMC Evolutionary Biology 11: 275. **doi: 10.1186/1471-2148-11-275**

Oliveira, E. C. S., R. H. Tardin, F. R. Poletto, and S. M. Simáo. 2013. Coordinated feeding behavior of the Guiana dolphin, *Sotalia guianensis* (Cetacea: Delphinidae), in southeastern Brazil: a comparison between populations. Zoologia 30: 585-591.

Oliveira, M. E., and M. Martins. 2001. When and where to find a pitviper: Activity patterns and habitat use of the Lancehead, *Bothrops atrox*, in Central Amazonia, Brazil. Herpetological Natural History 8: 101–110.

Oliver, J. R. 2008. The archaeology of agriculture in ancient Amazonia. *In* H. Silverman, and W. H. Isbell (Eds.). The handbook of South American archaeology, pp. 185-216. Springer, New York, NY, USA.

Ollerton, J. and S. Liede. 1997. Pollination systems in the Asclepiadaceae: a survey and preliminary analysis. Biological Journal of the Linnean Society 62: 593-610.

Olsder, K. 2004. Sustainable ecotourism in the Guiana Shield region. Netherlands Committee for IUCN, Amsterdam, The Netherlands.

Oostburg, B. F. J, and H. Hiwat. 1998. Araneism in Suriname. Surinaams Medisch Bulletin 13: 37-47.

Oostburg, B. F., J. E. Anijs, G. P. Oehlers, H. Hiwat, and S. M. Burke-Hermelijn. 2003. Case Report: the first parasitologically confirmed autochthonous case of acute Chagas disease in Suriname. Royal Society of Tropical Medicine and Hygiene 97: 166-167.

Organ, J. M. 2010. Structure and function of platyrrhine caudal vertebrae. The Anatomical Record 293: 730-745.

Ostendorf, F. W. 1962. Nuttige planten en sierplanten in Suriname. Bulletin Landbouwproefstation in Suriname 73. Landbouwproefstation, Paramaribo, Suriname.

Ottema, O. H., and A. L. Spaans. 2008. Challenges and advances in shorebird conservation in the Guianas, with a focus on Suriname. Ornitología Neotropical 19 (Suppl.): 339-346.

Ottema, O. H., J. H. J. M. Ribot, and A. L. Spaans. 2009. Annotated Checklist of the Birds of Suriname. WWF Guianas, Paramaribo.

Ouboter, P. E. (Ed.). 1993. The Freshwater Ecosystems of Suriname. Kluwer Academic Publishers, Dordrecht, The Netherlands.

Ouboter, P. E. 1989. The impact of an Indian village on caimans. Area reports: Suriname. IUCN Specialist Group Newsletter 8: 28.

Ouboter, P. E. 1996. Ecological studies on crocodilians in Suriname. Niche segregation and competition in three predators. PhD dissertation, SPB-Academic Publishing, Amsterdam, The Netherlands.

Ouboter, P. E. 2000. Wildlife management in Suriname. Bioconsult, Paramaribo, Suriname.

Ouboter, P. E. 2001. Assessment of traded wildlife species. Report #GFECP07 to WWF-Guianas, September 2001. WWF Guianas, Paramaribo, Suriname.

Ouboter, P. E. 2002. Directory of protected areas of Suriname. IBER, on behalf of Conservation International Suriname, Paramaribo, Suriname.

Ouboter, P. E., and J. H. A. Mol. 1993 The fish fauna of Suriname. *In* P.E. Ouboter (Ed.). The freshwater ecosystems of Suriname, pp 133-154. Kluwer Academic Publishers, Dordrecht, The Netherlands.

Ouboter, P. E., and L. M. R. Nanhoe. 1987. Notes on nesting and parental care in *Caiman crocodilus crocodilus* in northern Suriname and an analysis of crocodilian nesting habits. Amphibia-Reptilia 8: 331-348.

Ouboter, P. E., and L. M. R. Nanhoe. 1988. Habitat selection and migration of *Caiman crocodilus crocodilus* in a swamp and swamp-forest habitat in northern Suriname. Journal of Herpetology 22: 283-294.

Ouboter, P. E., and L. M. R. Nanhoe. 1989. Notes on the Dynamics of a population of *Caiman crocodilus crocodilus* in Northern Suriname and its Implications for Management. Biological Conservation 48: 243-264.

Ouboter, P. E., and R. Jairam. 2012. Amphibians of Suriname. Brill Academic Publishers, Leiden, The Netherlands.

Ouboter, P. E., and V. S. Kadosoe. 2016. Three years of continuous monitoring of the large terrestrial mammals of Brownsberg Nature Park, Suriname. Academic Journal of Suriname 7:643-660.

Ouboter, P. E., G. Landburg, J. Quick, J. Mol, and F. van der Lugt. 2012. Mercury levels in pristine and gold mining impacted aquatic ecosystems of Suriname, South America. Ambio 41: 873-882.

Ouboter, P. E., R. Jairam, and C. Kasanpawiro. 2011. A rapid assessment of the amphibians and reptiles of the Kwamalasamutu region (Kutari/Lower Sipaliwini Rivers), Suriname. *In* O'Shea, B.J., L.E. Alonso & T.H. Larsen (Eds.). A Rapid Biological Assessment of the Kwamalasamutu region, Southwestern Suriname, pp. 124-127. RAP Bulletin of Biological Assessment 63, Conservation International, Arlington, VA, USA.

Ouboter, P. E., R. Jairam, and K. Wan Tong You. 2007. Additional records of amphibians and reptiles from Nassau Mountain, Suriname. *In* Alonso, L. E. and J. H. Mol (Eds.). A Rapid Biological Assessment of the Lely and Nassau Plateaus, Suriname (with additional information on the Brownsberg Plateau), pp. 126-129. RAP Bulletin of Biological Assessment 43, Conservation International, Arlington, VA, USA.

Ouboter, P. E., S. Starke, and U. Satnarain. 2010. Trade in bush meat and medicinal plants in Suriname. World Wildlife Fund Guianas – Wildlife Management Conservation Component of the Guiana Sustainable Natural Resource Management Project, Paramaribo, Suriname.

Ouboter, P. E., V. Kadosoe, and R. Jairam. 2016. Dry season survival strategy of amphibians in a rainforest at Brownsberg Nature Park, Suriname (South America). Herpetological Review 47: 17-19.

Oudschans Dentz, F. 1941. De fortuinlijke loopbaan in Suriname van den Zweed C.G. Dahlberg. De West-Indische Gids 23: 269-279.

Pagnocca F.C., Masiulionis V.E., and A. Rodrigues. 2012. Selective specialized fungal parasites and opportunistic fungi in gardens of Attine ants. Psyche 2012: 905109. **dx.doi.org/10.1155/2012/905109**

Panday-Verheuvel, M. P. 1982. Factors influencing water quality and life in the Suriname River. Rapport OW&V/WLA no P83-1, Ministerie Openbare Werken, Paramaribo, Suriname.

Panday-Verheuvel, M. P. 1983. Factors influencing water quality in the Suriname River. *In* D. Bekker and H. Ehrenburg (Eds.). Future of roads and rivers in Suriname and neighbouring countries. Proceedings FURORIS congress Paramaribo, Suriname, December 2-7, 1982, pp. 473-502.

Panday-Verheuvel, M. P., and J. Buys. 1976. De invloed van het Brokopondo Stuwmeer op de waterkwaliteit van de Suriname Rivier benedenstrooms van de dam. Rapport OW&V/WLA no 81-6, Ministerie Openbare Werken, Paramaribo, Suriname.

Panfili, J., H. de Pontual, H. Troadec, and P. J. Wright (Eds.). 2002. Manual of fish sclerochronology. Ifremer-IRD, Brest, France.

Panizzi, A. R. 1997. Wild hosts of pentatomids: Ecological significance and role in their pest status on crops. Annual Review of Entomology 42: 99–122.

Panizzi, A. R. 2000. Suboptimal nutrition and feeding behavior of hemipterans on less preferred plant food sources. Anais da Sociedade Entomológica do Brasil 29: 1–12.

Panizzi, A. R., J. E. McPherson, D. G. James, M. Javahery, and R. M. McPherson. 2000. Stink bugs (Pentatomidae). *In* C. W. Schaefer, and A. R. Panizzi (Eds.). Heteroptera of economic importance (1st ed.), pp. 421–474. CRC Press, Boca Raton, FL, USA.

Parahoe, M., A. Soetosenojo, S. Naipal, G. Wesenhagen, V. Wortel, K. Tjon, J. Wirjosentono, R. Sabajo, M. Sewotaroeno, J. Mol, Y. Babb, G. Evans, C. Gangadien, and Y. Jadhav. 2008. Final report on biodiversity and economic valuation of Bigi Pan Multiple Use Management Area. CELOS, and FTeW (Center for Agricultural Research in Suriname, and Faculty of Technology), Anton de Kom University of Suriname, Paramaribo, Suriname. [in four parts]

Paredes, J. Chave, D. de Andrade Lima Filho, P. Møller Jørgensen, A. Fuentes, J. Schöngart, F. C. Valverde, A. Di Fiore, E. M. Jimenez, M. C. Peñuela Mora, J. F. Phillips, G. Rivas, T. R. van Andel, P. von Hildebrand, B. Hoffman, E. L. Zent, Y. Malhi, A. Prieto,

Parker, M. 2015. Willoughbyland: England's lost colony. Hutchinson, London, England, UK.

Parker, T. A. III, and E. O. Willis. 1997. Notes on three tiny grassland flycatchers, with comments on the disappearance of South American fire-diversified savannas. Ornithological Monographs 48: 549-555.

Parkman J. P., J. H. Frank, T. J. Walker, and D.J. Schuster. 1996. Classical biological control of *Scapteriscus* spp. (Orthoptera: Gryllotalpidae) in Florida. Environmental Entomology 25: 1415-1420.

Parolin, P., F. Wittmann, and L. V. Ferreira. 2013. Fruit and seed dispersal in Amazonian floodplain trees – a review. Ecotropica 19: 19-36.

Pazin, V. F. V., E. E. Magnusson, J. Zuanon, and F. P. Mendonça. 2006. Fish assemblages in temporary ponds adjacent to 'terra firme' streams in Central Amazonia. Freshwater Biology 51: 1025-1037.

Peel, M. C., B. Finlayson, and T. A. McMahon. 2007. Updated world map of the Koppen-Geiger climate classification. Hydrology and Earth System Sciences 11: 1633–1644.

Pejchar, L., R. M. Pringle, J. Ranganathan, J. R. Zook, G. Duran, F. Oviedo, and G. C. Daily. 2008. Birds as agents of seed dispersal in a human-dominated landscape in southern Costa Rica. Biological Conservation 141: 536-544.

Pelizza S., L. A. Elíades, M. C. Saparrat, M. N. Cabello, A. C. Scorsetti, and C. Lange. 2012. Screening of Argentine native fungal strains for biocontrol of the grasshopper *Tropidacris collaris*: relationship between fungal pathogenicity and chitinolytic enzyme activity. World Journal of Microbiology and Biotechnology. 28: 1359-1366.

Penard, F. P., and A. P. Penard. 1908-1910. De vogels van Guyana. Penard, Paramaribo, Suriname [2 volumes].

Péneau, J., A. Nguyen, A. Flores-Ferrer, D. Blanchet, and S. Gourbière. 2016. Amazonian Triatomine Biodiversity and the Transmission of Chagas Disease in French Guiana: In Medio Stat Sanitas. PLoS Neglected Tropical Diseases 10: e0004427. **doi: 10.1371/journal.pntd.0004427**

Pennington, T. D. 2006. Flora da Reserva Ducke, Amazonas, Brasil: Sapotaceae. Rodriguésia 57: 251-366.

Pereira, A. C., J. B Silva, R. Goldenber, G. A. R. Melo, and I. G. Varassin. 2011. Flower color change accelerated by bee pollination in *Tibouchina* (Melastomataceae) Flora 206: 491-497.

Pereira, L. H. G., R. Hanner, F. Foresti and C. Oliveira. 2013 Can DNA barcoding accurately discriminate megadiverse Neotropical freshwater fish fauna? BMC Genetics 14: 20.

Peres, C. A. 1990. A Harpy Eagle successfully captures an adult male Red Howler Monkey. Wilson Bulletin 102: 560-561.

Peres, C. A. 2000. Identifying keystone plant resources in tropical forests: the case of gums from *Parkia* pods. Journal of Tropical Ecology 16: 287-317.

Peres, C. A., L. C. Schiesari, and C. L. Dias-Leme. 1997. Vertebrate predation of Brazil-nuts (*Bertholletia excelsa*, Lecythidaceae), an agouti-dispersed Amazonian seed crop: a test of the escape hypothesis. Journal of Tropical Ecology 13: 69-79.

Perreijn, K. 2002. Symbiotic nitrogen fixation by leguminous trees in tropical rain forest in Guyana. Tropenbos-Guyana Series 11, Tropenbos-Guyana Programme, Georgetown, Guyana.

Peterson, G. D., and M. Heemskerk. 2001. Deforestation and forest regeneration following small-scale gold mining in the Amazon: the case of Suriname. Environmental Conservation 28: 117-126.

Petrunkevitch, A. 1926. Tarantula versus tarantula-hawk: a study in instinct. Journal of Experimental Zoology 42: 367–397.

Piechowski D., S. Dötterl, and G. Gottsberger. 2010. Pollination biology and floral scent chemistry of the Neotropical chiropterophilous *Parkia pendula*. Plant Biology 12: 172-182.

Pine, R. H. 1982. Current status of South American mammalogy. *In* M. A. Mares and H. H. Genoways (Eds.). Mammalian biology in South America, pp. 27–37, University of Pittsburgh, Pymatuning Laboratory of Ecology, Special Publication Series, Volume 6, Linesville, PA, USA.

Piso, W., G. Marcgrav von Liebstad, and J. de Laet. 1648. Historia naturalis Brasiliae : auspicio et beneficio illustriss. I. Mauriti Com. Nassau illius provinciae et maris summi praefecti adornata: in qua non tantum plantae et animalia, sed et indigenarum morbi, ingenia et mores describuntur et iconibus supra quingentas illustrantur. Franciscus Hackius, Leiden, and Louis Elzevir, Amsterdam [The Netherlands].

Playfair, M. 2007. Law compliance, and prevention and control of illegal activities in the forest sector in Suriname: Country assessment preliminary version. The World Bank, Washington, DC, USA.

Ploeg, A. 1987. Review of the cichlid genus *Crenicichla* Heckel, 1840 from Surinam, with descriptions of three new species (Pisces, Perciformes, Cichlidae). Beaufortia 37: 73-98.

Plotkin, M. J. 1986. Ethnobotany and the conservation of the tropical forest with special reference to the Indians of southern Suriname. PhD dissertation, Department of Biology, Tufts University, Medford, MS, USA.

Plotkin, M. J. 1993. Tales of a shaman's apprentice. Penguin books, New York, NY, USA.

Plumpton, D. L., and J. K. Jones, Jr. 1992. *Rhynchonycteris naso*. Mammalian Species 413: 1-5.

Poda, N. 1761. Insecta musei graecensis, quae in ordines, genera et species juxta systema naturae Caroli Linnaei. Graz, Austria.

Poderoso, J. C. M., M. K. M Da Costa, M. E. Correia-Oliveira, P. C. Dantas, J. C. Zanuncio, and G. T. Riberio. 2013. Occurrence of *Tropidacris collaris* (Orthoptera: Acridoidea: Romaleidae) damaging *Casuarina glauca* (Casuarinaceae) plants in the municipality of Central Bahia, Brazil. Florida Entomologist 96: 268-269.

Podolsky, R. D. 1990. Effects of mixed-species association on research use by *Saimiri sciureus* and *Cebus apella*. American Journal of Primatology 21:147-158.

POELS, R. L. H. 1987. Soils, water and nutrients in a forest ecosystems in Suriname. PhD dissertation, Agricultural University Wageningen, Wageningen, The Netherlands.

POHLMANN, K., J. ATEMA, AND T. BREITHAUPT. 2004. The importance of the lateral line in nocturnal predation of piscivorous catfish. Journal of Experimental Biology 207: 2971-2978.

PONDER, W., AND D. LUNNEY. 1999. The Other 99%.The conservation and biodiversity of invertebrates. Transactions of the Royal Society of New South Wales, Mosman, New South Wales, Australia.

PONS, T. L. 1972. Een vegetatiekundig onderzoek in het westelijk gedeelte van het Wia Wia - reservaat, Suriname. PhD dissertation, Institute for Systematic Botany, Utrecht University, Utrecht, The Netherlands.

PONTON, D., AND S. MÉRIGOUX. 2000. Comparative morphology and diet of young cichlids in the dammed Sinnamary river, French Guiana, South America Journal of Fish Biology 56: 87–102.

PONTON, D., AND S. MÉRIGOUX. 2001. Description and ecology of some early life stages of fishes in the river Sinnamary (French Guiana, South America). Folia Zoologica Monograph 50: 1-116.

PONTON, D., J. H. MOL, AND J. PANFILI. 2001. Use of otolith microincrements for estimating the age and growth rate of young armoured catfish *Hoplosternum littorale*. Journal of Fish Biology 10: 1274-1285.

POOL, M. 2013. Incorrect use of the IUCN classification may lead to bad management decisions: the case of the *Trichechus manatus manatus* in Suriname. Academic Journal of Suriname 2013: 375-377.

POTTER, A. B. 2011. Gray-winged Trumpeter (*Psophia crepitans*), Neotropical Birds Online (T. S. Schulenberg, Ed.). Cornell Lab of Ornithology. **neotropical.birds.cornell.edu/portal/species/overview?p_p_spp=142996**

PRADO, C. P. A., L. M. GOMIERO, AND O. FROEHLICH. 2006. Spawning and parental care in *Hoplias malabaricus* (Teleostei, Characiformes, Erythrinidae) in the southern Pantanal, Brazil. Brazilian Journal of Biology 66: 697-702.

PRANCE, G. T. 1972. Monograph of Chrysobalanaceae. Flora Neotropica Monographs 9: 1-410.

PRANCE, G. T. 1980. A note on the pollination of *Nymphaea amazonum* Mart. & Zucc. (Nymphaeaceae). Brittonia 32: 505-507.

PRANCE, G. T. 1992. Ant association with *Parinari excelsa* (Chrysobalanaceae) in Marajo, Brazil. Biotropica 24: 102-104.

PRANG, G. 2008. An industry analysis of the freshwater ornamental fishery with particular reference to the supply of Brazilian freshwater ornamentals to the UK market. Uakari 3: 7-51.

PRÉVOST, M.-F. 1982. Comportement de *Cecropia obtusa* et de *Cecropia sciadophylla* dans les premiers stades de la régénération forestière, piste de Saint-Elie, en Guyane. Ecosystème Forestier Guyanais: Bulletin de Liaison du Groupe de Travail, ECEREX 6: 231-250.

PRICE, R. 1991. Subsistence on the plantation periphery: crops, cooking, and labour among eighteenth-century Suriname maroons. Slavery & Abolition 12: 107-127.

PRICE, R. 2011. Rainforest warriors: human rights on trial. University of Pennsylvania Press, Philadelphia, PA, USA.

PRIEM, H. N. A., N. A. I. M. BOELRIJK, E. H. HEBEDA, E. A.T. VERDURMEN, AND R. H. VERSCHURE. 1971. Isotopic ages of the Trans-Amazonian acidic magmatism and the Nickerie Episode in the Precambrian basement of Surinam, South America. Geological Society of America Bulletin 82: 1667-1680.

PRIEM, H. N. A., N. A. I. M. BOELRIJK, E. H. HEBEDA, E. A.T. VERDURMEN, AND R. H. VERSCHURE. 1973. Age of the Precambrian Roraima Formation in northeastern South America: Evidence from isotopic dating of Roraima pyroclastic volcanic rocks in Suriname. Geological Society of America Bulletin 84: 1677–1684.

PRITCHARD, P. C. H., AND P. TREBBAU. 1984. The Turtles of Venezuela. Contributions to Herpetology 2, Society for the Study of Amphibians and Reptiles, Ithaca, NY, USA.

PROISY, C., N. GRATIOT, E. J. ANTHONY, A. GARDEL, F. FROMARD, AND P. HEURET. 2009. Mud bank colonization by opportunistic mangroves: A case study from French Guiana using lidar data. 2009. Continental Shelf Research 29:632-641. **doi: 10.1016/j.csr.2008.09.017**

PULLE, A., J. LANJOUW, A. L. STOFFERS, AND J. C. LINDEMAN (EDS.). 1928 onwards [series of publications]. Flora of Suriname. Vereniging Koloniaal Instituut / Koninklijk Instuut voor de Tropen, Amsterdam, and E. J. Brill, Leiden, The Netherlands.

QUETIN-LECLERCQ, J., L. ANGENOT, AND N. G. BISSET. 1990. South American Strychnos species. Ethnobotany (except curare) and alkaloid screening. Journal of Ethnopharmacology 28: 1-52.

RABASSA, J. 2014. Some concepts on Gondwana landscapes: long-term landscape evolution, genesis, distribution and age. *In* J. Rabassa, and C. Ollier (Eds.). Gondwana Landscapes in southern South America, pp 9-46. Springer, Heidelberg, Germany.

RABB, G. B., AND M. S. RABB 1960. On the mating and egg laying behavior of the Suriname toad, *Pipa pipa*. Copeia 1960(4): 271-276.

RABINOWITZ, A. R., AND B. G. NOTTINGHAM. 1986. Ecology and behaviour of the jaguar (*Panthera onca*) in Belize, Central America. Journal of Zoology 210: 149-159.

RABINOWITZ, D. 1978. Dispersal properties of mangrove propagules. Biotropica 10: 47–57.

RAGHAVAN, V. 2012. Developmental biology of flowering plants. Springer Science & Business Media, New York, NY, USA.

RAGHOENANDAN, U.P.D. 2000. The Guianas, (Guiana, Suriname, French Guaiana). *In* S. Porembski, and W. Barthlott (Eds.). Inselbergs, ecological studies 134, pp 316-338. Springer Verlag, Berlin-Heidelberg, Germany

RAHM, M., B. JULIAN, A. LAUGER, R. DE CARVALHO, L. VALE, J. TOTATRAM, K. A. CORT, M. DJOJODIKROMO, M. HARDJOPRAJITNO, S. NERI, R. VIEIRA, E. WATANABE, M. DO CARMO BRITO, P. MIRANDA, C. PALOENG, V. MOE SOE LET, S. CRABBE, AND M. CALMEL. 2015. Monitoring the impact of gold mining on the forest cover and freshwater in the Guiana Shield. Reference year 2014. REDD+ for the Guiana Shield Project, and WWF Guianas, Paramaribo, Suriname.

RAJAGOPOLAN, B., KUSHIR, Y. TOURRE, AND M. CANE. 1997. Temporal Variability of North Atlantic Oscillation and Tropical Atlantic SST. Lamont-Doherty Earth Observatory of Columbia University, New York, NY, USA.

RAMCHARAN, S. 2011. Het vakblad op veldbezoek: Peperpot Natuurpark. Vakblad Bos en Natuur 6: 10-11.

RAMDARSHAN, A., ALLOING-SÉGUIER, T., MERCERON, G., AND L. MARIVAUX. 2011. The primate community of Cachoeira (Brazilian Amazonia): A model to decipher ecological partitioning among extinct species. Plos One 6: e27392.

RAMÍREZ, S., R. L. DRESSLER, AND M. OSPINA. 2002. Abejas euglosinas (Hymenoptera: Apidae) de la Región Neotropical: Listado de especies con notas sobre su biología. Biota Colombiana 3: 7-118.

RAMIREZ-GOMEZ, S. O. I., G. BROWN, P. A. VERWEIJ, AND R. BOOT. 2016. Participatory mapping to identify indigenous community use zones: implications for conservation planning in southern Suriname. Journal for Nature Conservation 29: 69-78.

RAMJAI, R. 2001. The Omai disaster in Guyana. The Geographical Bulletin 43: 83-90.

RANNALA, B. 2015. The art and science of species delimitation. Current Zoology 61: 846-853.

RÄSÄNEN, M. E., A. M. LINNA, J. C. SANTOS, AND F. R. NEGRI. 1995. Late miocene tidal deposits in the amazonian foreland basin. Science 269: 386–390.

RASMUSSEN, J., AND J. WALLACE. 1983. Meteorological aspects of the 1982-83 El Niño/Southern Oscillation Episode. Science 222: 104-127.

RATNASINGHAM, S., AND P.D.N. HEBERT. 2007. The Barcode Of Life Data System (**www.barcodinglife.org**). Molecular Ecology Notes 7: 355-364.

RAYNER, N. A., D. E. PARKER, E. B. HORTON, C. K. FOLLAND, L.V. ALEXANDER, D. P. ROWELL, E. C. KENT, AND A. KAPLAN. 2003. Global analyses of sea surface temperature, sea ice, and night marine air temperature since the late nineteenth century. Journal of Geophysical Research 108, 4407. **doi: 10.1029/2002JD002670**, D14

RECH, A. R., F. C. L. MANENTE-BALESTIERI, AND M. L. ABSY. 2011. Reproductive biology of *Davilla kunthii* A. St-Hil. (Dilleniaceae) in Central Amazonia. Acta Botanica Brasilica 25: 487-496.

REEBS, S. 2001. Fish behavior in the aquarium and in the wild. Comstock Publishing Associates, Ithaca, NY, USA.

REECE, J. B., L. A. URRY, M. L. CAIN, S. A. WASSERMAN, P. V. MINORSKY, R. B. JACKSON. 2009. Biology (10th ed.). Pearson Benjamin Cummings, San Francisco, CA, USA.

REEMER, M. 2010. A second survey of Surinam Syrphidae (Diptera): introduction and Syrphinae. Tijdschrift voor Entomologie 153: 163-196.

REEMER, M. 2013. Taxonomic exploration of Neotropical Microdontinae (Diptera: Syrphidae) mimicking stingless bees. Zootaxa 3697: 1-88.

REEMER, M. 2014. A review of Microdontinae (Diptera: Syrphidae) of Surinam, with a key to the Neotropical genera. Tijdschrift voor Entomologie 157: 27-57.

REEMER, M. 2016. Syrphidae of Surinam: Eristalinae and synthesis. Tijdschrift voor Entomologie 159: 97-142.

REGAN, B. C., C. JULLIOT, B. SIMMEN, F. VIE, P. CHARLES-DOMINIQUE, AND J. D. MOLLON. 1998. Frugivory and colour vision in *Alouatta seniculus*, a trichromatic platyrrhine monkey. Vision Research 38: 3321-3327.

REGO, F. N. A. A., C. A. RHEIMS, AND E. M. VENTICINQUE. 2004. Notes on the predation of an aviculariine spider (Araneae: Theraphosidae, *Avicularia* sp.) by *Pepsis frivaldszkyi* (Hymenoptera: Pompilidae) in Brazilian Amazonia. British Tarantula Society Journal 20:18-25.

REID, F. A. 1997. A field guide to the mammals of Central America and Southeast Mexico. Oxford University Press, New York, USA.

REIS, N. J., W., TEIXEIRA, M. A. HAMILTON, F. BISPO-SANTOS, M. E. ALMEIDA, AND M. S. D'AGRELLA-FILHO. 2013. Avanavero mafic magmatism, a late Paleoproterozoic LIP in the Guiana Shield, Amazonian craton: U–Pb ID-TIMS baddeleyite,geochemical and paleomagnetic evidence. Lithos 174: 175-195.

REIS, R. E. 2013. Conserving the freshwater fishes of South America. International Zoo Yearbook 47: 65-70.4

REIS, R. E., S. O. KULLANDER, AND C. J. FERRARIS JR. (EDS.). 2003. Check list of the freshwater fishes of South and Central America. Edipucrs, Porto Alegre, Brazil.

REITSMA, E. 2008. Maria Sibylla Merian & dochters, Vrouwenlevens tussen kunst en wetenschap. Waanders Uitgeverij, Zwolle, The Netherlands.

REMSEN, J. V. JR., J. I. ARETA, C. D. CADENA, S. CLARAMUNT, A. JARAMILLO, J. F. PACHECO, J. PÉREZ-EMÁN, M. B. ROBBINS, F. G. STILES, D. F. STOTZ, AND K. J. ZIMMER. 2017. A classification of the bird species of South America (version 22 April 2017). American Ornithologists' Union. **www.museum.lsu.edu/~Remsen/SACCBaseline.html**

REMSEN, J. V. JR., M. A. HYDE, AND A. CHAPMAN. 1993. The diets of Neotropical trogons, motmots, barbets, and toucans. Condor 95: 178-192.

RENNER, S. S. 1989. Survey of reproductive biology in Neotropical Melastomataceae and Memecylaceae. Annals of the Missouri Botanical Garden 76: 496-518.

RENSSEN, T. A. 1974. Twelve bird species new for Suriname. Ardea 62: 118-122.

RENTON, K., AND D. J. BRIGHTSMITH. 2009. Cavity use and reproductive success of nesting macaws in lowland forest of southeast Peru. Journal of Field Ornithology 80: 1-8.

RENTZ, D. C. F. 2007. The changa mole cricket, *Scapteriscus didactylus* (Latreille), a New World pest established in Australia (Orthoptera: Gryllotalpidae). Australian Journal of Entomology 34: 303 – 306.

RETTIG, N. 1978. Breeding behavior of the Harpy Eagle (*Harpia harpyja*). Auk 95: 629-643.

REYENGA, T. W. 1971. Verspredingsecologie van *Biomphalaria glabrata* (Say, 1818) in relatie tot bilharziasis in Suriname. PhD dissertation, Leiden University, Leiden, The Netherlands.

RIBAS, C. C., C. Y. MIYAKI, AND J. CRACRAFT. 2009. Phylogenetic relationships, diversification and biogeography in Neotropical *Brotogeris* parakeets. Journal of Biogeography 36: 1712–1729.

RIEHL, C. 2008. Communal calling and prospecting by Black-headed Trogons (*Trogon melanocephalus*). Wilson Journal of Ornithology 120: 248-255.

RIÉRA, B., H. PUIG, AND J.-P. LESCURE. 1989. La dynamique de la forêt naturelle. Bois et Forêts des Tropiques 219: 69-78.

Riéra, B., R. Pélissier, and F. Houllier. 1998. Caractérisation d'une mosaïque forestière et sa dynamique en en forêt tropicale humide sempervirente. Biotropica 30: 251-260.

Riezebos, H. T. 1979. Geomorphology and soils of Sipaliwini savanna, southern Suriname. PhD dissertation, Utrechtse Geografische Studies 12, Geografisch Instituut, Universiteit Utrecht, Utrecht, The Netherlands.

Riskin, D. K., and J. W. Hermanson. 2005. Independent evolution of running in vampire bats. Nature 434: 292.

Rivas, J. A. 2000. The life history of the green anaconda (*Eunectes murinus*), with emphasis on its reproductive biology. PhD dissertation, University of Tenessee, Knoxville, TN, USA.

Rivas, L. H., L. D. Giustina, L. N. Luz, I. V. Karsburg, T. N. S. Pereira, and A. A. B. Rossi. 2013. Genetic diversity in natural populations of *Theobroma subincanum* Mart. in the Brazilian Amazon. Genetics and Molecular Research 12: 4998-5006.

Robbins, M. B., M. J. Braun, and D. W. Finch. 2004. Avifauna of the Guyana southern Rupununi, with comparisons to other savannas of northern South America. Ornitología Neotropical 15: 173-200.

Robbins, M. B., M. J. Braun, C. M. Milensky, B. K. Schmidt, W. Prince, N. H. Rice, D. W. Finch, and B. J. O'Shea. 2007. Avifauna of the upper Essequibo River and Acary Mountains, southern Guyana. Ornitología Neotropical 18: 339-368.

Roberts, T. R. 1970. Scale-eating American characoid fishes, with special reference to *Probolodus heterostomus*. Proceedings of the California Academy of Sciences 38: 383-390.

Roberts, T. R. 1973. Osteology and relationships of the Prochilodontidae, a South American family of characoid fishes. Bulletin Museum of Comparative Zoology 145: 213-235.

Robertson, A., and C. Mechoso. 2002. Links between the American Ocean and the South American Climate Variability. CLIVAR Exchanges 25: 1-4.

Robinson, M. H. 1969.The defensive behaviour of some orthopteroid insects from Panama. Transactions of the Entomological Society of London 121: 281–303.

Robinson, S. K. 1988. Foraging ecology and host relationships of Giant Cowbirds in southeastern Peru. Wilson Bulletin 100: 224-235.

Rocha M., C. C. Santos Júnior, G. A. Damasceno-Júnior, V. J. Pott, and A. Pott. 2015. Effect of fire on a monodominant floating mat of *Cyperus giganteus* Vahl in a neotropical wetland. Brazilian Journal of Biology 75: 114-124. **doi: 10.1590/1519-6984.08613**

Rocha, R. G., E. Ferreira, Y. L. R. Leite, C. Fonseca, and L. P. Costa. 2011. Small mammals in the diet of barn owls, *Tyto alba* (Aves: Strigiformes) along the mid-Araguaia River in central Brazil. Zoologia 28: 709-716.

Rodrigues, A. A., and G. A. Venturieri. 1997. Floral biology of "cupui" (*Theobroma subincanum*) - Sterculiaceae. Anais do I Encontro de Botânica Econômica de Espécies Nativas da América Latina; CATI - Campinas/SP/Brasil (21 e 22 de agôsto).

Rodríguez-Flores, C., F. G. Stiles, and M. Del Coro Arizmendi. 2012. Pollination network of a hermit hummingbird community (Trochilidae, Phaethornithinae) and their nectar resources in the Columbian Amazon. Ornitologia Neotropical 23: 85-100.

Roe, D. 2008. Trading nature. A report, with case studies, on the contribution of wildlife trade management to sustainable livelihoods and the Millennium Development Goals. TRAFFIC International and WWF International, Cambridge, England, UK.

Roeleveld, W. 1969. Pollen analysis of two sections in the Young Coastal Plain of Surinam. Geologie en Mijnbouw 48: 215-224.

Roeleveld, W., and A. J. Van Loon. 1979. The Holocene development of the Young Coastal Plain of Surinam. Geologie en Mijnbouw 48: 215-224.

Rojas R., M. A. Marini, and M. T. Z. Coutinho. 1999. Wild birds as hosts of *Amblyomma cajennense* (Fabricius, 1787) (Acari: Ixodidae). Memorias do Instituto Oswaldo Cruz 94: 315-322.

Rojas, B., and J. A. Endler. 2013. Sexual dimorphism and intra-populational colour pattern variation in the aposematic frog *Dendrobates tinctorius*. Evolution and Ecology 27: 739-753.

Rojas-Sandoval, J., K. Budde, M. Fernández, E. Chacón, M. Queseda, and J. A. Lobo. 2008. Phenology and pollination biology of *Ceiba pentandra* (Bombacaceae) in the wet forest of south eastern Costa Rica. Stapfia 88: 539-545.

Rolander, D. 2008. Daniel Rolander's journal, the Suriname journal: composed during an exotic journey (translated by J. Dobreff, C. Dahlman, D. Morgan & J. Tipton). *In* L. Hansen (Ed.). The Linnaeus apostles: global science & adventure, Vol. 3, Book 3, pp. 1215–1576. IK Foundation & Company, London & Whitby, England, UK.

Rolston, L. H., and F. J. D. McDonald. 1979. Keys and diagnoses for the families of Western Hemisphere Pentatomoidea, subfamilies of Pentatomidae and tribes of Pentatominae (Hemiptera). Journal of the New York Entomological Society 87:189-207.

Rolston, L. H., and F. J. D. McDonald. 1981. Conspectus of Pentatomini genera of the Western Hemisphere - Part 2 (Hemiptera: Pentatomidae). Journal of the New York Entomological Society 88: 257-272.

Rolston, L. H., and F. J. D. McDonald. 1984. A conspectus of Pentatomini of the Western Hemisphere - Part 3 (Hemiptera: Pentatomidae). Journal of the New York Entomological Society 92: 69-86.

Rolston, L. H., F. J. D. McDonald, and D. B. Thomas, Jr. 1980. A conspectus of Pentatomini genera of the Western Hemisphere - Part I (Hemiptera: Pentatomidae). Journal of the New York Entomological Society 88: 120-132.

Romero, G. A., and C. E. Nelson. 1986. Sexual dimorphism in *Catasetum* orchids: forcible pollen emplacement and male flower competition. Science 232: 1538-1540.

Romero, R. M., and L. Casatti. 2012. Identification of key microhabitats for fish assemblages in tropical Brazilian savanna streams. Internationale Revue Hydrobiologie 97: 526–541.

Rondeel, A. J. 1965. The Surinam fish protection legislation. De Surinaamse Landbouw 4: 63-64.

Root, R. B. 1967. The niche exploitation pattern of the Blue-Gray Gnatcatcher. Ecological Monographs 37: 317-350.

ROOT, R. B. 1973. Organization of a plant-arthropod association in simple and diverse habitats: the fauna of collards (*Brassica oleracea*). Ecological Monographs 43: 95-124.

ROSA, R. K., R. I. BARBOSA, AND S. KOPTUR. 2014. Which factors explain reproductive output of *Mauritia flexuosa* (Arecaceae) in forest and savanna habitats of northern Amazonia? International Journal of Plant Sciences 175: 307-318.

ROSSMAN, D. A. 1973. Missellaneous notes on the South American water snake genus *Helicops*. HISS-New Journal 1: 189-191.

ROSSMAN, D. A. 1984. Life history. *Helicops angulatus*. Reproduction. Herpetological Review 15: 20.

ROSTAIN S. 2010. Pre-Columbian earthworks in coastal Amazonia. Diversity 2: 331-352.

ROTTBØLL, C. F. 1778 Descriptiones plantarum qvarundam Surinamensium cum fragmento materiæ medicæ & oeconomicæ Surinamensis. Acta Literaria Universitatis Hafniensis 1: 267–304. [Copenhagen, Denmark]

ROUBIK, D. W. 1979. Nest and colony characteristics of stingless bees from French Guiana (Hymenoptera: Apidae). Journal Kansas Entomological Society 52: 443-470.

ROUBIK, D. W. 1989. Ecology and natural history of tropical bees. Cambridge University Press, Cambridge, NY, USA.

ROUBIK, D. W. 1993. Direct costs of forest reproduction, bee-cycling and the efficiency of pollination modes. Journal of Biosciences 18: 537-552.

ROUBIK, D. W. 2006. Stingless bee nesting biology. Apidologie 37: 124-143.

ROUBIK, D. W., AND P. E. HANSON. 2004. Abejas de orquídeas de la América tropical: Bíologia y guía de campo / Orchid bees of tropical America: Biology and field guide. Instituto Nacional de Biodiversidad (INBio), Santo Domingo de Heredia, Costa Rica.

ROUBIK, D. W., N. M. HOLBROOK, AND G. PARRA V. 1985. Roles of nectar robbers in reproduction of the tropical treelet *Quassia amara* (Simaroubaceae). Oecologia 66: 161–167.

ROY, R. 2004. Revision et phylogenie des Choeradodini Kirky, 1904 (Dictyoptera, Mantodea). Bulletin de la Société Entomologique de France 109: 113–128.

ROZE, J. A. 1996. Coral snakes of the Americas: biology, identification and venoms. Krieger Publishers, Malabar, FL, USA.

ROZENDAAL, J. A. 1990. Epidemiology and control of malaria in Suriname with special reference to *Anopheles darlingi*. PhD dissertation, Leiden University, Leiden, The Netherlands.

RÜDIGERA, A. L., A. C. SIANIB, AND V. F. VEIGA. 2007. The chemistry and pharmacology of the South America genus Protium Burm. f. (Burseraceae). Pharmacognosy Reviews 1: 93-104.

RUIZ, A., M. SANTOS, AND J. CAVELIER. 2000. Estudio fenologico de Cactaceae en el enclave seco de la Tatacoa, Colombia. Biotropica 32: 397-407.

RULL, V. 2008. Speciation timing and neotropical biodiversity: the Tertiary-Quaternary debate in the light of molecular phylogenetic evidence. Molecular Ecology 17: 2722–2729.

RUMPHIUS, G. E. 1705. D'Amboinsche rariteitkamer, behelzende een beschryvinge van allerhande zoo weeke als harde schaalvissen, te weeten raare krabben, kreeften, en diergelijke zeedieren, als mede allerhande hoorntjes en schulpen die men in d'Amboinsche Zee vindt; Daar benenven zomminge mineraalen, gesteenten, en soorten van aarde, die in d'Amboinsche, en zommige omleggende eilanden gevonden worden. Francois Halma, Amsterdam [The Netherlands].

RUSSO, S. E., C. J. CAMPBELL, J. L. DEW, P. R. STEVENSON, AND S. A. SUAREZ. 2005. A multi-forest comparison of dietary preferences and seed dispersal by *Ateles* spp. International Journal of Primatology 26: 1017-1037.

RUTISHAUSER, R., AND M. GRUBERT. 1999. The architecture of *Mourera fluviatilis* (Podostemaceae): developmental morphology of inflorescences, flowers, and seedlings. American Journal of Botany 86: 907-922.

RUYSSCHAERT, S. 2017. Non-timber forest products in Suriname: diversity, knowledge and use in an Amerindian and Maroon community. Scientific report, University of Gent, Gent, Belgium.

SABAJ, M. H., J. W. ARMBRUSTER, AND L. M. PAGE. 1999. Spawning in *Ancistrus* (Siluriformes: Loricariidae) with comments on the evolution of snout tentacles as a novel reproductive strategy : larval mimicry. Ichthyological Exploration of. Freshwaters 10: 217-229.

SABATIER, D., AND H. PUIG. 1985. Phénologie et saisonnalité de la floraison et de la fructification en forêt dense guyanaise. Mémoires de Muséum Nacional d'Histoire Naturelle, Paris, Série A 132: 173-184.

SABINO, J., AND J. ZUANON. 1998. A stream fish assemblage in Central Amazônia: distribution, activity pattern and feeding behaviour. Ichthyological Exploration of Freshwaters 8: 201–210.

SAGEBAKKEN, G., I. AHNESJO, K. B. MOBLEY, I. B. GONCALVES, AND C. KVARNEMO. 2010. Brooding fathers, not siblings, take up nutrients from embryos. Proceedings Royal Society B 277: 971–977.

SAKAKIBARA, A. M., AND O. EVANGELISTA. 2010. *Membracis foliata* (Linnaeus) (Hemiptera; Membracidae: Membracinae) and allied species: an effort towards their correct determination. Journal of Natural History 44: 2131–2148.

SANDOVAL M. C., M. M. VEIGA, J. HINTON, AND S. SANDNER. 2006. Application of sustainable development concepts to an alluvial mineral extraction project in Lower Caroni River, Venezuela. Journal of Cleaner Production 14: 415-426.

SANMARTÍN, I., F. RONQUIST, AND I. SANMARTIN. 2004. Southern Hemisphere Biogeography Inferred by Event-Based Models: Plant versus Animal Patterns. Systematic Biology 53: 216–243.

SANTOS, J. C., L. A. COLOMA, AND D. C. CANNATELLA. 2003a. Multiple, recurring origins of aposematism and diet specialization in poison frogs. Proceedings of the National Academy of Sciences USA 22: 12792-12797.

SANTOS, J. C., L. A. COLOMA, K. SUMMERS, J. P. CALDWELL, R. REE, AND D. C. CANNATELLA. 2009. Amazonian amphibian diversity is primarily derived from late Miocene Andean lineages. PLoS Biology 7: e56.

Santos, J. O. S., P. E. Potter, N. J. Reis, L. A. Hartmann, I. R. Fletcher, and N. J. McNaughton. 2003b. Age, source and regional stratigraphy of the Roraima Supergroup and Roraima-like outliers in northern South America based on U-Pb geochronology. Geological Society of America Bulletin 115: 331-348.

Sarthou, C., J. F. Villiers, and J. F. Ponge. 2003. Shrub vegetation on tropical granitic inselbergs in French Guiana. Journal Vegetation Science 14: 645–652.

Sarthou, C., S. Samadi, and M.-C. Boisselier-Dubayle. 2001. Genetic structure of the saxicole *Pitcairnia geyskesii* (Bromeliaceae) on inselbergs in French Guiana. American Journal of Botany 88: 861–868.

Saunders, D. L., J. J. Meeuwig, and C. J. Vincent. 2002. Freshwater protected areas: strategies for conservation. Conservation Biology 16: 30-41.

Sazima, I. 1980. Behavior of two Brazilian species of parodontid fishes *Apareidon piracicabae* and *A. ibitensis*. Copeia 1980: 166-169

Sazima, I. 1983. Scale-eating in characoids and other fishes. Environmental Biology of Fishes 9: 87-101.

Sazima, I., and F. A. Machado.1990. Underwater observations of piranhas in western Brazil. Environmental Biology of Fishes 28: 17-31.

Sazima, I., L. N. Carvalho, F. P. Mendonça, and J. Zuanon. 2006. Fallen leaves on the water-bed: diurnal camouflage of three night-active fish species in an Amazonian streamlet. Neotropical Ichthyology 4: 119-122.

SBB. 2016. Forestry sector analysis 2015. Stichting voor Bosbeheer en Bostoezicht, Paramaribo, Suriname.

Scartozzoni, R. R., M. da Graça Salomão, and Selma M. de Almeida-Santos. 2009. Natural history of the vine snake *Oxybelis fulgidus* (Serpentes, Colubridae) from Brazil. South American Journal of Herpetology 4: 81-89.

Schaefer C. W. *2009. Prosorrhyncha* (Heteroptera and Coleorrhyncha). *In* V. H. Resh, and R. T. Cardé (Eds.). Encyclopedia of insects (2nd ed.), pp. 839-855. Academic Press, Amsterdam, The Netherlands.

Schaefer, S. A., and D. J. Stewart. 1993. Systematics of the *Panaque dentex* species group (Siluriformes, Loricariidae), wood-eating armored catfishes from tropical South America. Ichthyological Exploration of Freshwaters 4: 39-56.

Schaefer, S. A., F. Provenzano, M. De Pinna, and J. N. Baskin. 2005. New and noteworthy Venezuelan glanapterygine catfishes (Siluriformes, Trichomycteridae), with discussion of their biogeography and psammophily. American Museum Novitates 3496: 1-27.

Scheltema, W. 1964. Het kawfoetoe verschijnsel in Suriname (een vorm van microreliëf). DBK Rapport 16, Dienst Bodemkartering, Paramaribo, Suriname.

Schemske, D. W. 1980. The evolutionary significance of extrafloral nectar production by *Costus woodsonii* (Zingiberaceae): an experimental analysis of ant protection. The Journal of Ecology: 959-967.

Schemske, D. W. 1982. Ecological correlates of a neotropical mutualism: ant assemblages at *Costus* extrafloral nectaries. Ecology: 932-941.

Schiesari, L., J. Zuanon, C. Azevedo-Ramos, M. Garcia, M. Gordo, M. Messias, and E. M. Vieira. 2003. Macrophyte rafts as dispersal vectors for fishes and amphibians in the Lower Solimo s River, Central Amazon. Journal of Tropical Ecology 19: 333–336.

Schindler, D. E., M. D Scheuerell, J. W Moore, S. M Gende, T. B Francis, and W. J Palen. 2003. Pacific salmon and the ecology of coastal ecosystems. Frontiers Ecology Environment 1: 31–37.

Schlesinger, W. H., and E. S. Bernhardt. 2013. Biogeochemistry, An analysis of global change. Academic Press, Oxford, England, UK.

Schlosser, I. J. 1987. The role of predation in age- and size-related habitat use by stream fishes. Ecology 68: 651-659.

Schmidt, P., R. L. H. Poels, and W. B. J. Jonkers. 2011. Tree growth, recruitment and mortality after logging and refinement. *In* M. J. A. Werger (Ed.). Sustainable management of tropical rainforests: the CELOS management system, pp. 46-73 Tropenbos International, Paramaribo, Suriname.

Schofield C.J., and W.R. Dolling. 1993. Bedbugs and Kissing bugs (Bloodsucking Hemiptera). *In* R.P. Lane, and R. Crosskey (Ed). Medical Insects and Arachnids, pp. 78-119. Chapman and Hall, London, England, UK.

Schuh, R. T., and J. A. Slater. 1995. True bugs of the world (Hemiptera: Heteroptera): classification and natural history (1st ed.). Cornell University Press, New York, NY, USA.

Schulenberg, T. S. 2009. Harpy Eagle (*Harpia harpyja*), Neotropical Birds Online (T. S. Schulenberg, Ed.). Cornell Lab of Ornithology. neotropical.birds.cornell.edu/portal/species/overview?p_p_spp=142356

Schulz, J. P. 1960. Ecological studies on rain forest in northern Suriname. Noord-Hollandsche Uitgevers Maatschappij, Amsterdam, The Netherlands / Verhandelingen der Koninklijke Nederlandse Akademie van Wetenschappen, Afd. Natuurkunde, tweede reeks 53: 1-267.

Schulz, J. P. 1975. Sea Turtles Nesting in Surinam. Stichting Natuurbehoud Suriname, Paramaribo, Suriname / Zoologische Verhandelingen Leiden 143: 3-143.

Schulz, J. P. 1980. Zeeschildpadden die in Suriname leggen. Stichting Natuurbehoud Suriname, Paramaribo, Suriname.

Schulze, E.-D., E. Beck, and K. Müller-Hohenstein. 2005. Plant Ecology. Springer, New York, NY, USA.

Schwassmann, H. O. 1992. Seasonality of reproduction in Amazonian fishes. *In* W.C. Hamlett (Ed.), Reproductive biology of South American vertebrates, pp 71-81. Springer-Verlag, New York, NY, USA.

Scoble, M. J. 1995. The Lepidoptera: form, function and diversity. Oxford University Press, Oxford, England, UK.

Seaforth, C. E., G. S., Williams, and W. F. Tinto. 1998. Alkaloids from *Hymenocallis tubiflora*. Fitoterapia 69-79.

Sealy, J. R. 1954. Review of the genus *Hymenocallis*. Kew Bulletin 9: 201–240.

Sedano, R. E., and K. J. Burns. 2010. Are the northern Andes a species pump for Neotropical birds? Phylogenetics and biogeography of a clade of Neotropical tanagers (Aves: Thraupini). Journal of Biogeography 37: 325-343.

Sedgwick, C. W. 2010. White-winged Swallow (*Tachycineta albiventer*), Neotropical Birds Online (T. S Schulenberg, Editor). Cornell Lab of Ornithology. **neotropical.birds.cornell.edu/portal/species/overview?p_p_spp=523116**

Sefer, E. 1957. A formiga *Dolichoderas mucronifer* Roger. Norte Agronômico 3: 55-57.

Seymour, K. L. *1989. Panthera onca*. Mammalian Species 340: 1-9.

Seymour, R. S., and P. G. Matthews. 2006. The role of thermogenesis in the pollination biology of the Amazon waterlily *Victoria amazonica*. Annals of Botany 98: 1129-1135.

Shaffer, C. A. 2013. Feeding ecology of northern bearded sakis (*Chiropotes sagulatus*) in Guyana. Amercian Journal of Primatology 75: 68-580.

Shang, Y., Feng, P., & Wang, C. 2015. Fungi That Infect Insects: Altering Host Behavior and Beyond. PLoS Pathogens 11: e1005037. **doi.org/10.1371/journal.ppat.1005037**

Sheldon, A. L. 1988. Conservation of stream fishes: patterns of diversity, rarity, and risk. Conservation Biology 2: 149-156.

Shelley R. M., G. B. Edwards, and A. Chagas Jr. 2005. Introduction of the Centipede *Scolopendra morsitans* L., 1758, into northeastern Florida, the first authentic North American record, and a review of its global occurrences (Scolopendramorpha: Scolopendridae: Scolopendrinae). Entomological News 116: 39-58.

Shelley, R. M., and S. B. Kiser. 2000. Neotype designation and a diagnostic account for the centipede, *Scolopendra gigantea* L. 1758, with an account of *S. galapagoensis* Bollman 1889 (Chilopoda, Scolopendromorpha, Scolopendridae). Tropical Zoology 13:159-170

Shibuya, A., M. L. G. Araujo, and A. S. Zuanon. 2009. Analysis of stomach contents of freshwater stingrays (Elasmobranchii, Potamotrygonidae) from the middle Negro River, Amazonas, Brazil. Pan-American Journal of Aquatic Sciences 4: 466-475.

Shu, D-G., H-L. Luo, S. Conway Morris, X-L. Zhang, S-X. Hu, L. Chen, J. Han, M. Zhu, Y. Li, and L-Z. Chen. 1999. Lower Cambrian vertebrates from south China. Nature 402: 42-46.

Sick, H. 1993. Birds in Brazil. Princeton University Press, Princeton, NJ, USA.

Sigrist, M. R., and M. Sazima. 2004. Pollination and reproductive biology of twelve species of neotropical Malpighiaceae: stigma morphology and its implications for the breeding system. Annals of Botany 94: 33-41.

Silva, C. A. M., L. A. Simeoni, And D. Silveira. 2009. Genus *Pouteria*: Chemistry and biological activity. Brazilian Journal of Pharmacognosy 19: 501-509.

Simberloff, D. 1998. Flagships, umbrellas, and keystones: is single-species management passé in the landscape era? Biological Conservation 83: 247-257.

Simberloff, D., and T. Dayan. 1991. The guild concept and the structure of ecological communities. Annual Review of Ecology and Systematics 22: 115-143.

Simons, S. M., P. L. Júnior, F. Faria, I. F. Batista, D. M. Barros-Battesti, M. B. Labruna, and A. M. Chudzinski-Tavassi. 2011. The action of *Amblyomma cajennense* tick saliva in compounds of the hemostatic system and cytotoxicity in tumor cell lines. Biomedicine and Pharmacotherapy 65: 443-450.

Sinka M.E., Y. Rubio-Palis, S. Manguin, A. P. Patil, W. H. Temperley, P. W. Gething, T. Van Boeckel, C. W. Kabaria, R. E. Harbach, and S. I. Hay. 2010. The dominant *Anopheles* vectors of human malaria in the Americas; occurrence data, distribution maps and bionomics. Parasites and Vectors 3: 72.

Sioli, H. 1950. Das Wasser im Amazonasgebiet. Forschungen und Fortschritte 26: 274-280.

Sizer, N. and R. Rice. 1995. Backs to the wall in Suriname. World Resources Institute, Washington, DC, USA.

Skutch, A. F. 1960. Life histories of Central American birds II. Pacific Coast Avifauna 34.

Skutch, A. F. 1962. Life history of the white-tailed trogon *Trogon viridis.* Ibis 104: 301-313.

Skutch, A. F. 1971. Life history of the Keel-billed Toucan. Auk 88: 381-424.

Skutch, A. F. 1972. Studies of tropical American birds. Publications of the Nuttall Ornithological Club 10: 1-228.

Skutch, A. F. 1980. Arils as food of tropical American birds. Condor 82: 3142.

Small Grants Programme. 2016. Conservation of agro-biodiversity and providing communities key crop planting material. UNDP Small Grants Programme, Paramaribo, Suriname. **www.facebook.com/sgpsuriname** [accessed Dec. 2016]

Small, G. E., P. J. Torres, L. M. Schweitzer, J. H. Duff, and C. M. Pringle. 2013. Importance of terrestrial arthropods as subsidies in lowland Neotropical rain forest stream ecosystems. Biotropica 45: 80-87.

Smit, N., A. Niel, L. Bruce, and K. A. Hadfield. 2014. Global diversity of fish parasitic isopod crustaceans of the family Cymothoidae. Parasites and Wildlife 3: 188–197.

Smith, D. G. 1979. Guide to the leptocephali. NOAA Technical Report NMFS Circular 424. U.S. Department of Commerce, National Oceanic and Atmospheric Administration and National Marine Fisheries Service, USA.

Smith, L. C. Jr., M. Smith, and P. Ashcroft. 2011. Analysis of environmental damages from British Petroleum's Deepwater Horizon oil spill. Albany Law Review 74: 563-585.

Smith, M. E., T. W. Henkel, J. K. Uehling, A. K. Fremier, H. D. Clarke, and R. Vilgalys. 2013. The ectomycorrhizal fungal community in a Neotropical forest dominated by *Pakaraimaea dipterocarpacea*. PLoS ONE 8: e55160.

Smith, N. G. 1968. The advantage of being parasitized. Nature 219: 690-694.

Smith, R. J., and W. L. Jungers. 1997. Body mass in comparative primatology. Journal Human Evolution 32: 523-559.

Smythe, N. 1970. Relationships between fruiting seasons and seed dispersal methods in a neotropical forest. American Naturalist 104: 25-35.

Smythe, N., and O. Brown de Guanti. 1995. Domestication and husbandry of the paca (*Agouti paca*). Food and Agriculture Organization of the United Nations 26: 1-93.

Snodgrass, R. E. 1935. Principles of insect morphology. McGraw-Hill Book Co., New York, NY, USA, and London, England, UK.

SNOW, B. K. 1961. Notes on the behavior of three Cotingidae. Auk 78: 150-161.

SNOW, B. K. 1972. A field study of the calfbird, *Perissocephalus tricolor*. Ibis 114: 139-162.

SNOW, D. W. 1971. Notes on the biology of the Cock-of-the-rock (*Rupicola rupicola*). Journal für Ornithologie 112: 323-333.

SNOW, D. W. 1981. Tropical frugivorous birds and their food plants: a world survey. Biotropica 13: 1-14.

SNYDER, N. F. R., AND H. W. KALE II. 1983. Mollusk predation by Snail Kites in Colombia. Auk 100: 93-97.

SOBRAL-LEITE, M., J. A. DE SIQUEIRA FILHO, C. ERBAR, AND I. C. MACHADO. 2011. Anthecology and reproductive system of *Mourera fluviatilis* (Podostemaceae): Pollination by bees and xenogamy in a predominantly anemophilous and autogamous family. Aquatic Botany 95: 77-87.

SOBREVILLA, C. 2008. The role of indigenous peoples in biodiversity conservation: the natural but often forgotten partners. The World Bank, Washington, DC, USA.

SOLANO, P. J., S. DUROU, B. CORBARA, A. QUILICHINI, P. CERDAN, M. BELIN-DEPOUX, J. H. C. DELABIE, AND A. DEJEAN. 2003. Myrmecophytes of the understory of French Guianian rainforests: their distribution and their associated ants. Sociobiology 41: 605–614.

SOLOMON, S. E. 2007. Biography and evolution of widespread leaf-cutting ants *Atta* spp. (Formicida, Attini). Thesis, University of Texas, Austin, TX, USA.

SOLOMON, S., D. QIN, M. MANNING, Z. CHEN, M. MARQUIS, K.B. AVERYT, M. TIGNOR, AND H.L. MILLER (EDS.). 2007. Contribution of Working Group I to the Fourth Assessment Report of the Intergovernmental Panel on Climate Change. Cambridge University Press, Cambridge, England, United Kingdom, and New York, NY, USA.

SOMMEIJER, M. J. 1984. Distribution of labour among workers of *Melipona favosa* F.: age-polyethism and worker oviposition. Insectes Sociaux 31: 171-184.

SOMMEIJER, M. J. 1985. The social behavior of *Melipona favosa* (F): some aspects of the activity of the queen in the nest. Journal Kansas Entomological Society 58: 386-396.

SOMMEIJER, M. J., L. L. M. DE BRUIJN, AND F. J. A. J. MEEUWSEN. 2004. Behaviour of males, gynes and workers at drone congregations sites of the stingless bee *Melipona favosa* (Apidae: Meliponini). Entomologische Berichten 64: 10-15.

SOMMEIJER, M. J., L. L. M. DE BRUIJN, F. J. A. J. MEEUWSEN, AND E. J. SLAA. 2003. Reproductive behaviour of stingless bees: nest departures of non-accepted gynes and nuptial flight in *Melipona favosa* (Hymenoptera: Apidae, Meliponini). Entomologische Berichten 63: 7-13.

SOMMEIJER, M. J., T. X. CHINH, AND F. J. A. J. MEEUWSEN. 1999. Behavioural data on the production of males by workers in the stingless bee *Melipona favosa* (Apidae, Meliponinae). Insectes Sociaux 46: 92-93.

SOUSA, J. C., R. F. BERTA, E. A. GOIS., N. C. FONTENELE-CARDI, J. E. R. HONORIO JR., K. KONNO, M. RICHARDSON, M. F. G. ROCHA, A. A. C. M. CAMARGO D. C. PIMENTA, B. A. CARDI, AND K. M. CARVALHO. 2009. Leptoglycin: A new Glycine/Leucine-rich antimicrobial peptide isolated from the skin secretion of the South American frog *Leptodactylus pentadactylus* (Leptodactylidae). Toxicon 2009: 1-10.

SOUTHALL, T., N. PFEIFFER, S. SINGH-RENTON, M. GILL. 2011. MSC sustainable fisheries certification: Suriname Atlantic seabob shrimp. Food Certification International, Inverness, Scotland, UK.

SOUZA, G. R., A. H. KRAHL, AND A. G. SILVA. 2012. Population structure and spatial pattern of *Catasetum discolor* (Lindl.) Lindl. (Orchidaceae) in a sandy coastal plain of southeastern Brazil. Natureza 10: 105-110.

SPAANS, A. L. 1978. Status and Numerical Fluctuations of Some North American Waders along the Surinam Coast. The Wilson Bulletin 90: 60-83.

SPAANS, A. L. 1990. Problems in assessing trends in breeding populations of Scarlet Ibis (*Eudocimus ruber*) and other ciconiiform birds. *In* P. C. Frederick, L. Gonzalo Morales, A. L. Spaans & C. S. Luthin (Eds.). The Scarlet Ibis (*Eudocimus ruber*): Status, Conservation and Recent Research, pp. 1-6. Special Publication 11, International Waterfowl and Wetlands Research Bureau, Slimbridge, England, UK.

SPAANS, A. L. 2003. Kustvogels van Suriname / Coastal Birds of Suriname. Foundation for Nature Conservation in Suriname (STINASU), Paramaribo, Suriname.

SPAANS, A. L., O. H. OTTEMA, AND J.-H. J. M. RIBOT. 2016. Field guide to the birds of Suriname. Brill, Leiden, The Netherlands.

SPARKS, J. S., AND W. L. SMITH. 2005. Freshwater fishes, dispersal ability, and nonevidence: "Gondwana life rafts" to the rescue. Systematic Biology 54: 158–165.

SPLITGERBER, F. L. 1840. Observationes de *Voyria*. Tijdschrift Natuurlijke Geschiedenis en Physiologie 7: 129-139.

SPLITGERBER, F. L. 1842. De plantis novis surinamensibus. Tijdschrift Natuurlijke Geschiedenis en Physiologie 9: 95-114.

SPOTTE, S., P. PETRY, AND J. A. S. ZUANON. 2001. Experiments on the feeding behavior of the hematophagous candiru, *Vandellia* cf. *plazaii*. Environmental Biology of Fishes 60: 459-464.

SPS/OAS. 1988. Suriname Planatlas. National Planning Office of Suriname, Paramaribo, Suriname, and Organization of American States, Washington, DC, USA.

ST. JOHN, T. V. 1980. A survey of micorrhizal infection in an amazonian rain forest. Acta Amazonica 10: 527-533.

STARACE, F. 2013. Serpents et Amphisbenes de Guyane francaise (second edition). Ibis Rouge, Matoury, French Guiana.

STATZNER, B., AND T. F. HOLM. 1982. Morphological adaptations of benthic invertebrates to stream flow - an old question studied by means of a new technique (Laser Doppler Anemometry). Oecologia 53: 290-292.

STEDMAN, J. G. 1806. Narrative of a five years' expedition, against the revolted negroes of Surinam, in Guiana, on the wild coast of South America; from the year 1772, to 1777: elucidating the history of that country, and describing its productions ... with an account of the Indians of Guiana, & negroes of Guinea. J. Johnson, and T. Payne, London, England, UK.

STEDMAN, J. G. 1972. Narrative, of a five years' expedition, against the revolted negroes of Surinam (edited by R. A. J. van Lier). University of Massachusetts Press, Amherst, MA, USA.

STERRINGA, J. T. 1971. Een vegetatiekundig en natuurbehoudsonderzoek in het Wia Wia natuurreservaat, Suriname. Dissertation, Natuurbeheer en Natuurbehoud, Landbouw Hogeschool Wageningen, Wageningen, The Netherlands.

STEVENS, P. F. 2001 onwards. Angiosperm Phylogeny Website. Version 12, July 2012. **www.mobot.org/MOBOT/research/APweb/**

STICHTING VOLKSLECTUUR SURINAME. 2004. Wordlist English-Sranam with a list of plant and animals names (4th ed.). Vaco N.V. Uitgeversmaatschappij, Paramaribo, Suriname.

STILES, F. G., AND A. F. SKUTCH. 1989. A guide to the birds of Costa Rica. Comstock Publishing Associates, Cornell University Press, Ithaca, NY, USA.

STILES, F. G., AND L. ROSSELLI. 1993. Consumption of fruits of the Melastomataceae by birds - How diffuse is coevolution? Vegetatio 108: 57-73.

STORK, N. E. 1987. Guild structure of arthropods from Bornean rain forest trees. Ecological Entomology 12: 69-80.

STORK, N. E. 1997. Measuring global biodiversity and its decline. *In* M. L. Reaka-Kudla, D. E. Wilson, and E. O. Wilson (Eds.). Biodiversity II, pp. 41-68. Joseph Henry Press, Washington, DC, USA.

STORK, N. E., J. MCBROOM, C. GELY, AND A. J. HAMILTON. 2015 New approaches narrow global species estimates for beetles, insects, and terrestrial arthropods. PNAS 112: 7519-7523.

STORTI, E. F. 2002. Biologia da polinização e sistema reproductive de *Passiflora coccinea* Aubl. em Manaus, Amazonas, Brasil. Acta Amazônica 32: 421-429.

STOUFFER, P. C., AND R. O. BIERREGAARD, JR. 1995. Use of Amazonian forest fragments by understory insectivorous birds. Ecology 76: 2429-2445.

STRADLING, D. J. 1978. The growth and maturation of the "tarantula", *Avicularia avicularia* L. Zoological Journal of the Linnean Society 62: 291-303.

STRONZA A., AND W. H. DURHAM. 2008. Ecotourism and Conservation in the Americas. CAB International, Wallingford, England, UK.

STROPP, J., L. L. DO AMARAL, G. AYMARD, O. BANKI, C. V. DE CASTILLO, C. FEREIRA, T. W. HENKEL, D. C. LOPEZ, W. MAGNUSON, F. D. DE ALMEIDA MATOS, W. MILLIKEN, A. OLIVEIRA, D. PAULETTO, O. P. PHILLIPS, R. THOMAS, AND H. TER STEEGE. 2011. Comparing tree communities of white-sand and terra-firme forests across three Amazonian regions. *In* Stropp (Ed.). Towards an understanding of tree diversity in Amazonian forests, pp. 31-46. PhD dissertation, Institute of Environmental Science, Utrecht University, Utrecht, the Netherlands.

STUTCHBURY, B. J. M., AND E. S. MORTON. 2001. Behavioral Ecology of Tropical Birds. Academic Press, San Diego, CA, USA.

STUYFZAND, P. J. 1989. A new hydrochemical classification of water types. Regional Characterization of Water quality. *In* S. Ragone (Ed.). Water Quality Proceedings of the Baltimore Symposium, Maryland, pp. 89-98. IAHS Publication 182, IAHS Press, Wallingford, England, UK.

STYRSKY, J. D., AND M. D. EUBANKS. 2007 Ecological consequences of interactions between ants and honeydew-producing insects. Proceedings Royal Society B: Biological Sciences 274: 151–164.

SUSSMAN, R. W., AND W. G. KINZEY. 1984. The ecological role of the Callitrichidae: A review. American Journal of Physical Anthropology 64: 419-449.

SUTTON M.1999. Aboriginal Ownership of National Parks and Tourism. Cultural Survival Quarterly 23: 55-56.

SWAN, C. 2002. From blowfish to flower still life paintings, classification and its images, circa 1600. *In* P. H. Smith, and P. Findlen (Eds.). Merchants & marvels, commerce, science, and art in early modern Europe, pp. 109-136. Routledge, New York, NY, USA.

SWAN, C. 2004. Collecting naturalia in the shadow of early modern Dutch trade. *In* L. Schieringer, and C. Swan (Eds.). Colonial botany, science, commerce, and politics in the early modern world, pp. 223-236. University of Pennsylvania Press, Philadelphia, PA, USA.

SWENNEN, C., P. DUIVEN, AND A. L. SPAANS. 1982. Numerical density and biomass of macrobenthic animals living in the intertidal zone of Suriname, South America. Netherlands Journal of Sea Research 15: 406-418.

SYMBIONT CONSULTING. 2007. Capaciteitsopbouw en Trainingsplan voor de Afdeling Natuurbeheer van de Dienst Landsbosbeheer (NB/LBB). Symbiont Consulting, Paramaribo, Suriname.

TABERELLI, M., AND C. PERES. 2002. Abiotic and vertebrate seed dispersal in the Brazilian Atlantic forest: implications for forest regeneration. Biological Conservation 106: 165-176. **doi: 10.1016/S0006-3207(01)00243-9**

TARDIN, R. H. O., M. A. ESPÉCIE, L. LODI, AND S. M. SIMÃO. 2013. Parental care behavior in the Guiana dolphin, *Sotalia guianensis* (Cetacea: Delphindae), in Ilha Grande Bay, southeastern Brazil. Zoologia 30: 15-23.

TAVARES-DIAS M., C. S. O. ARAÚJO, M. S. BARROS, AND G. M. VIANA. 2014. New hosts and distribution records of, a parasite Cymothoidae of fishes from the Amazon. *Braga patagonica*. Brazilian Journal of Aquatic Science and Technology 18: 91-97.

TAYLOR, D. S. 2012. Twenty-four years in the mud: what have we learned about the natural history and ecology of the mangrove rivulus, *Kryptolebias marmoratus*? Integrative and Comparative Biology 52: 724–736.

TAYLOR, D. S., B. J. TURNER, W. P. DAVIS, AND B. B. CHAPMAN. 2008. A novel terrestrial fish habitat inside emergent logs. American Naturalist 171: 263-266.

TAYLOR, G. J., W. RIBI, M. BECH, A. J. BODEY, C. RUA, A. STEUWER, E. J. WARRANT, AND E. BAIRD. 2016 The dual function of orchid bee ocelli as revealed by X-ray microtomography. Current Biology 26: 1-6.

TAYLOR, K. E, R. J. STOUFFER, AND G. A. MEEHL. 2012. An Overview of CMIP5 and the Experiment Design. Bulletin of the American Meteorological Society 93: 485–498.

TER STEEGE, H. 1998. The use of forest inventory data for a National Protected Area Strategy in Guyana. Biodiversity and Conservation 7: 1457-1483.

TER STEEGE, H. 2000. Plant diversity in Guyana, with recommendations for a national protected area strategy. Tropenbos Series 18, Tropenbos Foundation, Wageningen, the Netherlands.

ter Steege, H., and D. S. Hammond. 2001. Character convergence, diversity, and disturbances in tropical rain forest in Guyana. Ecology 82: 3197-3212.

ter Steege, H., D. Sabatier, H. Castellanos, T. van Andel, J. Duivenvoorden, A. Adalardo de Oliviera, R. Ek, R. Lilwah, P. Maas, and S. Mori. 2000. An analysis of the floristic composition and diversity of Amazonian forests including those of the Guiana Shield. Journal of Tropical Ecology 16: 801-828.

ter Steege, H., N. C. Pitman, D. Sabatier, C. Baraloto, R. P. Salomão, J. E. Guevara, O. L. Phillips, C. V. Castilho, W. E. Magnusson, J. F. Molino, A. Monteagudo, P. Núñez Vargas, J. C. Montero, T. R. Feldpausch, E. N. Honorio Coronado, T. J. Killeen, B. Mostacedo, R. Vasquez, R. L. Assis, J. Terborgh, F. Wittmann, A. Andrade, W. F. Laurance, S. G. W. Laurance, B. S. Marimon, B.-H. Marimon Jr., I. C. Guimarães Vieira, I. L. Amaral, R. Brienen, H. Castellanos, D. Cárdenas López, J. F. Duivenvoorden, H. F. Mogollón, F. D. de Almeida Matos, N. Dávila., R. García-Villacorta, P. R. Stevenson Diaz, F. Costa, T. Emilio, C. Levis, J. Schietti, P. Souza, A. Alonso, F. Dallmeier, A. J. Duque Montoya, M. T. Fernandez Piedade, A. Araujo-Murakami, L. Arroyo, R. Gribel, P. V. A. Fine, C. A. Peres, M. Toledo, G. A. Aymard C.,T. R. Baker, C. Cerón, J. Engel, T. W. Henkel, P. Maas, P. Petronelli, J. Stropp, C. E. Zartman, D. Daly, D,. Neill, M. Silveira, M. Ríos Paredes, J. Chave, D. de Andrade Lima Filho, P. Møller Jørgensen, A. Fuentes, J. Schöngart, F. do Cornejo Valverde, A. Di Fiore, E. M. Jimenez, M. C,. Peñuela Mora, J. Fernando Phillips, G. Rivas, T. R. van Andel, P. von Hildebrand, B. Hoffman, E. L. Zent, Y. Malhi, A. Prieto, A. Rudas, A. R. Ruschell, N. Silva, V. Vos, S. Zent, A. A. Oliveira, A. Cano Schutz, T. Gonzales, M. T. Nascimento, H. Ramirez-Angulo, R. Sierra, M. Tirado, M. N. Umaña Medina, G. van der Heijden, C. I. A. Vela, E. Vilanova Torre, C. Vriesendorp, O. Wang, K. R. Young, C. Baider, H. Balslev, C. Ferreira, I. Mesones, A. Torres-Lezama, L. E. Urrego Giraldo, R. Zagt, M. N. Alexiades, L. Hernandez, I. Huamantupa-Chuquimaco, W. Milliken, W. Palacios Cuenca, D. Pauletto, E. Valderrama Sandoval, L. Valenzuela Gamarra, K. G. Dexter, K. Feeley, G. Lopez-Gonzalez, and M. R. Silman. 2013. Hyperdominance in the Amazonian tree flora. Science 342 (6156) **doi:10.1126/science.1243092**

ter Steege, H., N. C. Pitman, D. Sabatier, H. Castellanos, P. van Der Hout, D. C. Daly, M. Silveira, O. Phillips, R. Vasquez, T. van Andel, and J. Duivenvoorden. 2003. A spatial model of tree α-diversity and tree density for the Amazon. Biodiversity and Conservation 12: 2255–2277.

ter Steege, H., N. C. Pitman, O. Phillips, J. Chave, D. Sabatier, A. Duque, J.-F. Molino, M.-F. Prevost, R. Spichiger, H. Castellanos, P. von Hildebrand, and R. Vasquez. 2006. Continental-scale patterns of canopy tree composition and function across Amazonia. Nature 443: 444–447.

ter Steege, H., O. S. Banki, and P. Haripersaud. 2007. Plant diversity of the bauxite plateaus of North East Suriname. *In* L. E. Alonso, and J. H. Mol (Eds.). A rapid biological assessment of the Lely and Nassau Plateaus (with additional information on the Brownsberg Plateau), pp. 76-85. RAP Bulletin of Biological Assessments 43, Conservation International, Arlington, VA, USA.

ter Steege, H., O. S. Banki, M. Jansen-Jacobs, G. Ramharakh, and K. Tjon. 2005. Plant diversity of the Lely Mountains, Suriname: Report of the Nov-Dec 2004 expedition. Nationaal Herbarium Nederland, Utrecht Branch, Utrecht University, Utrecht, The Netherlands. **www.researchgate.net/publication/46668375_Plant_diversity_of_the_Lely_Mountains_Suriname_Report_of_the_Nov-Dec_Expedition**

ter Steege, H., R. W. Vaessen, D. Cárdenas-López, D. Sabatier, A. Antonelli, S. Mota de Oliveira, N. C. A. Pitman, P. Møller Jørgensen, and R. P. Salomão. 2016. The discovery of the Amazonian tree flora with an updated checklist of all known tree taxa. Scientific Reports 6: 29549.

ter Steege, H., V. G. Jetten, A. M. Polak, and M. J. A. Weger. 1993. Tropical rain forest types and soil factors in a watershed area in Guyana. Journal of Vegetation Science 4(5): 705-716.

Terborgh, J. 1983. Five New World Primates. Princeton University Press, Princeton, NJ, USA.

Terborgh, J. 1992. Diversity and the Tropical Rain Forest. Scientific American Library, New York, NY, USA.

Teunissen, P. A. 1976. Vegetation changes in a dammed up fresh water swamp in NW Suriname. Acta Amazonica 6: 117-150.

Teunissen, P. A. 1978a. Ecosystemen van Suriname: Biogeografie van Zuid-Amerika. Instituut voor de Opleiding van Leraren, Paramaribo, Suriname.

Teunissen, P. A. 1978b. Ecosystemen van Suriname: Bovenrivieren en soela-complexen Instituut voor de Opleiding van Leraren, Paramaribo, Suriname.

Teunissen, P. A. 1978c. Ecosystemen van Suriname: Rotssavannes. Instituut voor de Opleiding van Leraren, Paramaribo, Suriname.

Teunissen, P. A. 1978d. Ecosystemen van Suriname: Tropisch regenbos. Instituut voor de Opleiding van Leraren, Paramaribo, Suriname.

Teunissen, P. A. 1978e. Inleiding Surinaamse ecosystemen: Biogeografie van Suriname. Instituut voor de Opleiding van Leraren, Paramaribo, Suriname.

Teunissen, P. A. 1978f. Inleiding Surinaamse ecosystemen: De Surinaamse landschappen. Instituut voor de Opleiding van Leraren, Paramaribo, Suriname.

Teunissen, P. A. 1978g. Reconnaissance Map Surinam Lowland Ecosystems (Coastal Region and Savanna Belt), scale 1:200 000. STINASU, Paramaribo, Suriname, and De Walburg Pers, Zutphen, The Netherlands.

Teunissen, P. A. 1979. Geschiedenis van het biologisch onderzoek in Suriname. Instituut voor de Opleiding van Leraren (IOL), Paramaribo, Suriname.

Teunissen, P. A. 1980a. Ecosystemen van Suriname: Savannes. Instituut voor de Opleiding van Leraren, Paramaribo, Suriname.

Teunissen, P. A. 1980b. Overzicht van Surinaamse Laagland Ecosystemen met vegetatietabellen. Report, Lands Bosbeheer (LBB) / Stichting Natuurbehoud Suriname, Paramaribo, Suriname.

Teunissen, P. A. 1984. De vegetatie van Suriname. Instituut voor de Opleiding van Leraren, Paramaribo, Suriname.

Teunissen, P. A. 1987a. Ecosystemen van Suriname: Graszwampen en zwampbossen. Instituut voor de Opleiding van Leraren, Paramaribo, Suriname.

Teunissen, P. A. 1987b. Ecosystemen van Suriname: Het brakwater-kustgebied. Instituut voor de Opleiding van Leraren, Paramaribo, Suriname.

Teunissen, P. A. 1987c. Ecosystemen van Suriname: Open zoetwater-voorkomens. Instituut voor de Opleiding van Leraren, Paramaribo, Suriname.

Teunissen, P. A. 1987d. Ecosystemen van Suriname: Zoetwaterzwampen. Instituut voor de Opleiding van Leraren, Paramaribo, Suriname.

Teunissen, P. A. 1988a. Ecosystemen van Suriname: Cultuurgebieden. Instituut voor de Opleiding van Leraren, Paramaribo, Suriname.

Teunissen, P. A. 1988b. Ecosystemen van Suriname: De Oceaan. Instituut voor de Opleiding van Leraren, Paramaribo, Suriname.

Teunissen, P. A. 1988c. Vegetation. *In* E. R. Nahar, C.A.F. Pigot, J. R. Pinas, and P. A. Teunissen (Eds.). Suriname Planatlas, pp. 15 and B7. Stichting Planbureau Suriname, Paramaribo, Suriname, and Organization of American States, Washington, DC, USA

Teunissen, P. A. 1993 Vegetation and vegetation succession of the freshwater wetlands. *In* P. E. Ouboter (Ed.). The freshwater ecosystems of Suriname, pp. 77-98. Kluwer Academic Publishers, Dordrecht, The Netherlands.

Teunissen, P. A. 2000. Milieu problemen en milieu-zorg in Suriname. Bureau Openbare Gezondheidszorg, Paramaribo, Suriname.

Teunissen, P. A. 2003. Position paper for the development of a framework policy and strategic plan for the sustainable management of the non-urban environmental sub-sector in Suriname: Protected areas and wildlife management. Ministry of Planning and Development Cooperation, Paramaribo, Suriname.

Teunissen, P. A., and J. T. Wildschut. 1970. Vegetation and flora of the savannas in the Brinckheuvel Nature Reserve, Northern Suriname. Verhandelingen der Koninklijke Nederlandse Akademie van Wetenschappen, Afd. Natuurkunde, tweede reeks 59: 1-60.

Thatcher, V. E. 2006. Amazon Fish Parasites (2nd ed.). Pensoft Publishers, Sofia, Bulgaria.

Théry, M. 1992. The evolution of leks through female choice: differential clustering and space utilization in six sympatric manakins. Behavioral Ecology and Sociobiology 30: 227-237.

Théry, M., S. Pincebourde, and F. Feer. 2008. Dusk light environment optimizes visual perception of conspecifics in a crepuscular horned beetle. Behavioral Ecology 19: 627-634. **doi: 10.1093/beheco/arn024**

Théveniaut, H., and P. Freyssinet. 2002. Timing of lateritization on the Gulana Shield: synthesis of paleomagnetic results from French Guiana and Suriname. Palaeogeography, Palaeoclimatology, Palaeoecology 178: 91-117.

Thiollay, J.-M. 1991. Foraging, home range use and social behaviour of a group-living rainforest raptor, the Red-throated Caracara *Daptrius americanus*. Ibis 133: 382-393.

Thiollay, J.-M. 1992. Influence of selective logging on bird species diversity in a Guianan rain forest. Conservation Biology 6: 47-63.

Thiollay, J.-M. 2007. Raptor communities in French Guiana: distribution, habitat selection, and conservation. Journal of Raptor Research 41: 90-105.

Thiollay, J.-M., and M. Jullien. 1998. Flocking behaviour of foraging birds in a Neotropical rain forest and the antipredator defense hypothesis. Ibis 140: 382-394.

Thomas, B. T., and S. D. Strahl. 1990. Nesting behavior of the Sunbittern (*Eurypyga helias*) in Venezuela. Condor 92: 576-581.

Thomas, L., and J. Middleton. 2003. Guidelines for management planning of protected areas. Best Practice Protected Area Guidelines Series No.10. IUCN, Gland, Switzerland.

Thomas, M. F. 1975. Tropical geomorphology. Macmillan, London, England, UK.

Thompson, C. L., and M. A. Norconk. 2011. Within-group social bonds in white-faced saki monkeys (*Pithecia pithecia*) display male-female pair preferences. American Journal of Primatology 73: 1051-1061.

Thompson, C. L., M. A. Norconk, and P. L. Whitten. 2012. Why fight? Selective forces favoring between-group aggression in a variably pair-living primate, the white-faced saki (*Pithecia pithecia*). Behaviour 149: 795-820.

Thompson, F. C. 1999. A key to the genera of flower flies (Diptera: Syrphidae) of the Neotropical Region including descriptions of new genera and species and a glossary of taxonomic terms. Contributions on Entomology, International 3: 322-378.

Thompson, F. C., G. E. Rotheray, and M. A. Zumbado. 2010. Syrphidae (flower flies). *In* B. V. Brown, A. Borkent, J. M. Cumming, D. M. Wood, N. E. Woodley, and M.A. Zumbado (Eds.). Manual of Central American Diptera, Volume 2, pp. 763-792. NRC Research Press, Ottawa, Canada.

Thorne, B. 1985. Numerical and biomass caste proportions in colonies of the termites *Nasutitermes corniger* and *N. ephratae* (Isoptera; Termitidae). Insectes Sociaux 32: 411-426.

Thorne, B. L. 1980. Differences in nest architecture between the neotropical arboreal termites *Nasutitermes corniger* and *Nasutitermes ephratae* (Isoptera: Termitidae). Psyche 87: 235-244.

Thorne, B. L. 1982. Polygyny in termites: multiple primary queens in colonies of *Nasutitermes corniger* (Motschuls.) (Isoptera: Termitidae). Insectes Sociaux 29: 102-117.

Thorson, T. B., D. R. Brooks, and M. A. Mayes. 1983a. The evolution of freshwater adaptation in stingrays. National Geographic Research Reports 15: 663-694.

Thorson, T. B., J. K. Langhammer, and M. I. Oetinger. 1983b. Reproduction and development of the South American freshwater stingrays *Potamotrygon circularis* and *P. motoro*. Environmental Biology of Fishes 9: 3-24.

Tilles, J. P. 1979. Les Cymothoidae (Isopoda, Flabellifera; parasites de poisson) du Rijksmuseum van Natuurlijke Historie de Leiden II. Afrique, Amerique et Regions Indo-Ouest-Pacifiques. Zoologische Mededelingen 54: 245-275.

Tillett, S. S. 2003. Passifloraceae. Flora of the Venezuelan Guayana 7: 625-667.

Tippery, N.P., C. T. Philbrick, C. P. Bove, and D. H. Les. 2011. Systematics and phylogeny of neotropical riverweeds (Podostemaceae: Podostemoideae). Systematic Botany 36: 105–118.

Tito de Morais, A., and L. Tito de Morais. 1994. The abundance and diversity of larval and juvenile fish in a tropical estuary. Estuaries 17: 216-225.

Tito de Morais, L., and J. Raffray. 1999. Movements of *Hoplias aimara* during the filling phase of the Petit-Saut dam, French Guiana. Journal of Fish Biology 54: 627-635.

Toledo, L. F., and C. F. B. Haddad. 2009. Colors and Some Morphological Traits as Defensive Mechanisms in Anurans. International Journal of Zoology 2009: 1-12.

Torreias, S. R. D. S., R. L. Ferreira-Keppler, B. S. Godoy, and N. Hamada. 2010. Mosquitoes (Diptera, Culicidae) inhabiting foliar tanks of *Guzmania brasiliensis* Ule (Bromeliaceae) in central Amazonia, Brazil. Revista Brasileira de Entomologia 54: 618-623.

Trail, P. W. 1983. Cock-of-the-rock, jungle dandy. National Geographic 146: 830-839.

Trail, P. W. 1987. Predation and antipredator behavior at Guianan Cock-of-the-Rock leks. Auk 104: 496-507.

Trail, P. W. 1990. Why should lek-breeders be monomorphic? Evolution 44: 1837-1852.

Trigo, J. R. 2000. The chemistry of antipredator defense by secondary compounds in neotropical Lepidoptera: facts, perspectives and caveats. Journal of the Brazilian Chemical Society 6: 551-561. **dx.doi.org/10.1590/S0103-50532000000600002**

Tucker, G. C., and K. Gandhi. 2013. Nomenclatural notes on Neotropical Cyperus (Cyperaceae). Harvard Papers in Botany 18: 149-154.

Turienzo, P., O. di Iorio, and V. Mahnert. 2010. Global checklist of pseudoscorpions (Arachnida) found in birds' nests. Revue Suisse de Zoologie 117: 557–598.

Turnbull, M., and N. J. Fashing. 2002. Efficacy of the ventral abdominal secretion of the cockroach *Eurycotis floridana* (Blattaria: Blatticae) as a defense allomone. Journal of Insect Behavior 15: 369-384.

Turner, I. M. 2001. The ecology of trees in the tropical rain forest. Cambridge University Press, Cambridge, England, UK.

Uetz, P., and J. Hošek. 2015. The Reptile Database. **www.reptile-database.org** [accessed Nov. 2015]

Uribe S., B. V. Brown, G. Correa, and A. Ortiz. 2016. Phorids associated with nests of *Atta cephalotes* (Hymenoptera: Formicidae) in a forest and a plantation. Revista Colombiana de Entomología 42: 48-53.

Ururahy-Rodrigues, A., J. A. Rafael, R. Ferreira Wanderley, H. Marques, and J. R. Pujol-Luz. 2008. *Coprophanaeus lancifer* (Linnaeus, 1767) (Coleoptera, Scarabaeidae) activity moves a man-size pig carcass: Relevant data for forensic taphonomy. Forensic Science International 182: e19–e22. **doi.org/10.1016/j.forsciint.2008.09.009**

US Geological Survey. 2012. Assessment of undiscovered conventional oil and gas resources of South America and the Caribbean, 2012. Fact Sheet 2012-3046, World Petroleum Assessment Project. US Geological Survey, US Department of the Interior, Reston, VA, USA.

Valente, V. C., S. A. Valente, F. Noireau, H. J. Carrasco, and M. A. Miles. 1998. Chagas disease in the Amazon Basin: association of *Panstrongylus geniculatus* (Hemiptera: Reduviidae) with domestic pigs. Journal of Medical Entomology 35: 99-103

van Andel, T. 2010. Suriname door de ogen van een achttiende-eeuwse biologiestudent. OSO (Tijdschrift voor Surinamistiek en het Caraibisch gebied) 29: 375-384.

van Andel, T. and P. Westers. 2010. Why Surinamese migrants in the Netherlands continue to use medicinal herbs from their home country. Journal of Ethnopharmacology 127: 694-701.

van Andel, T. and R. Havinga. 2008. Sustainability aspects of commercial medicinal plant harvesting in Suriname. Forest Ecology and Management 256: 1540-1545.

van Andel, T., A. MacKinven, and O. Banki. 2003. Commercial non-timber forest products of the Guiana Shield: an inventory of commercial NTFP extraction and possibilities for sustainable harvesting. The Netherlands Committee for IUCN, Amsterdam, The Netherlands.

van Andel, T., and S. Ruysschaert. 2011. Medicinale en rituele planten van Suriname. KIT Publishers, Amsterdam, The Netherlands.

van Andel, T., J. Behari-Ramdas, R. Havinga, and S. Groenendijk. 2007. The medicinal plant trade in Suriname. Ethnobotany Research & Applications 5: 351-372.

van Andel, T., P. Maas, and J. Dobreff. 2012a. Ethnobotanical notes from Daniel Rolander's Diarium Surinamicum (1754-1756): Are these plants still used in Suriname today? Taxon 61: 852-863 + supplement (12 pp.).

Van Andel, T., S. Mitchell, G. Volpato, I. Vandebroek, J. Swier, S. Ruysschaert, and N. Raes. 2012b. In search of the perfect aphrodisiac: Parallel use of bitter tonics in West Africa and the Caribbean. Journal of Ethnopharmacology 143: 840-850.

Van Andel, T., S. Ruysschaert, K. Van de Putte, and S. Groenendijk. 2013. What makes a plant magical? Symbolism and sacred herbs in Afro-Surinamese winti rituals. *In* R. Voeks, and J. Rashford (Eds.). African Ethnobotany in the Americas, pp. 247-284. Springer, New York, NY, USA.

van Andel, T., S. Veldman, P. Maas, G. Thijsse, and M. Eurlings. 2012c. The forgotten Hermann Herbarium: a 17th century collection of useful plants from Suriname. Taxon 61: 1296-1304.

van Benthem, M., and S. Van Dijk. 2015. Plenty of FSC timber in Suriname; but the market must adapt to the forest. Bosberichten 2105 #6, Probos, Wageningen, The Netherlands.

van Berkel, A. 1695. Amerikaansche voyagien, behelzende een reis na Rio de Berbice, gelegen op het vaste land van Guiana, aande Wilde-kust van America, mitsgaders een andere na de colonie van Suriname, gelegen in het noorder deel van het gemelde landschap Guiana, ondermengd met alle de byzonderheden noopende de zeden, gewoonten, en levenswijs der inboorlingen, boom- en aardgewasser, waaren en koopmanschappen, en andere aanmerkelijke zaaken. Johan ten Hoorn, Amsterdam [The Netherlands].

Van der Eyk, J. J. 1957. Reconnaissance soil survey in northern Surinam. PhD dissertation, Landbouwuniversiteit Wageningen, Wageningen, The Netherlands.

van der , T. 1963. A Palynological Study on the Quaternary of British Guyana. Leidse Geologische Mededelingen 29: 126–168.

van der Hammen, T. 1966. De palynologie van het Guiana-bekken. Tijdschrift van het Koninklijk Nederlands Aardrijkskundig Genootschap, 2nd Series 83: 128-138.

van der Hammen, T. 1974. The Pleistocene changes of vegetation and climate in tropical South America. Journal of Biogeography 1: 3–26.

van der Hammen, T., and H. Hooghiemstra. 1996. The history of the forests and the climate of tropical South América. *In* C. S. Harcourt, and J. A. Sayer (Eds.). The conservation atlas of tropical forests, America, pp. 17-26. Simon & Schuster, London, England, UK.

van der Hammen, T., and M. Absy. 1994. Amazonia during the last glacial. Palaeogeography, Palaeoclimatology, Palaeoecology 109: 247-261.

van der Heide, J. 1982. Lake Brokopondo: filling phase limnology of a man-made lake in the humid tropics. Kanters, Alblasserdam, The Netherlands.

van der Steege, J. G. 1986. Ecologische aspecten van de reconditionering van de wegverbinding tussen Burnside en Wageningen in Suriname. Dienst Lands Bosbeheer, Paramaribo, Suriname.

van der Steege, J. G. 1990.Bestrijdingsmiddelen van het Bigi Pan gebied nabij Nieuw Nickerie (Distict Nickerie, Suriname) augustus 1984 t/m december 1987. Dienst Lands Bosbeheer, Paramaribo, Suriname.

van der Werff, H., and J. G. Richter. 1996. Toward an improved classification of Lauraceae. Annals of the Missouri Botanical Garden 83: 409–418.

van Dijck, P. 2013. The IIRSA Guianese Shield hub: the case of Suriname. *In* P. van Dijck (Ed.). The impact of the IIRSA road infrastructure programme on Amazonia, pp. 151-180. Routledge, Abingdon, England, UK.

van Dinther, J. M. 1960. Insect pests of cultivated plants in Surinam. Bulletin Landbouwproefstation Suriname 69. Landbouwproefstation, Paramaribo, Suriname.

van Donselaar, J. 1965. An ecological and phytogeographic study of northern Surinam savannas. Wentia 14: 1-163.

van Donselaar, J. 1968. Phytogeogaphic notes on the savanna flora of southern Surinam (South America). Acta Botanica Neerlandica 17: 393-404.

van Donselaar, J. 1969. Observations on savanna vegetation-types in the Guianas. Vegetatio 27: 271-312.

van Donselaar, J. 1970. Floristic and ecological data on the lianes of the Brokopondo District, Surinam. Acta Botanica Neerlandica 19: 287-296.

van Donselaar, J., and J. P. Schulz. (undated). On the flora and vegetation of granite exposures in the Voltzberg region (Suriname), with notes on some similar areas in Suriname. [unpublished report]

van Donselaar, J., P. A. Teunissen, M. C. M. Werkhoven, and J. T Wildschut. 1964-1975. Successie-onderzoek in de Perica-zwamp (parts I through IV). Lands Bosbeheer, Paramaribo, Suriname.

van Donselaar-Ten Bokkel Huinink, J. 1966. Structure, root systems and periodicity of savanna plants and vegetations in Suriname. Wentia 17: 1-162.

van Gelder, R. 1993. Noordnederlandse verzamelingen in de zeventiende eeuw. *In* E. Bergvelt, D. Meijers, and M. Rijnders (Eds.). Verzamelen, van rariteitenkabinet tot kunstmuseum, pp. 123-44. Unieboek / Het Spectrum, Heerlen, The Netherlands.

van Lier, R. A. J. 1971. Frontier society, a social analysis of the history of Surinam. Martinus Nijhoff, Den Haag, The Netherlands. [Translation of 'Samenleving in een grensgebied', 2nd ed., 1971]

van Regteren Altena, C. O. 1960. On a small collection of land Mollusca from Suriname. Basteria 24: 48-51.

van Regteren Altena, C. O. 1975. The marine Mollusca of Suriname (Dutch Guiana) Holocene and recent, part III, Gastropoda and Cephalopoda. Zoölogische Verhandelingen (Leiden) 139: 1-104.

van Roosmalen, M. G. M. 1985a. Fruits of the Guianan flora. Institute of Systematic Botany, Utrecht University, Utrecht, The Netherlands.

van Roosmalen, M. G. M. 1985b. Habitat preferences, diet, feeding strategy and social organization of the black spider monkey (*Ateles paniscus paniscus* Linnaeus 1758) in Surinam. Acta Amazonica 15, supplement: 1-238.

van Sauers-Muller, A. 2008. Carambola fruit fly situation in Latin America and the Caribbean. Proceedings of the Caribbean Food Crops Society 44: 135-144.

van 't Klooster, C. I. E. A., J. C. Lindeman, and M. J. Jansen-Jacobs. 2003. Index of vernacular plant names of Suriname. Blumea Supplement 15: 1-322.

van Traa, A. 1946. Suriname 1900-1940. Uitgeverij W. van Hoeve, Deventer, the Netherlands.

Vanderhoff, E. N., and B. Grafton. 1996. Behavior of tamarins, tanagers and manakins foraging in a strangler fig (*Ficus* sp.) in Suriname, South America: implications for seed dispersal. Biota Neotropica 9: 419-423.

Vanni, M. J. 2002. Nutrient cycling by animals in freshwater ecosystems. Annual Review of Ecology and Systematics 33: 341-370.

Vannote, R. L., G. W. Minshall, K. W. Cummins, J. R. Sedell, and C. E. Cushing. 1980. The river continuum concept. Canadian Journal of fisheries and Aquatic Sciences 37: 130-137.

Vanzolini, P. E., and E. E. Williams. 1970. South American Anoles: the Geographic Differentiation and Evolution of the *Anolis chrysolepis* species group (Sauria, Iguanidae). Arquivos de Zoologia, Sao Paulo 19: 1-124.

Vari, R. P. 1982. Inventory, biology and ecology of the fishes in the Corantijn River System, Suriname. Environmental Impact of the Kabalebo Project, Ministry of Development, Paramaribo, Suriname.

Vari, R. P., C. J. Ferraris, and M. C. C. de Pinna. 2005. The Neotropical whale catfishes (Siluriformes: Cetopsidae: Cetopsinae), a revisionary study. Neotropical Ichthyology 3: 127-238.

Vari, R. P., C. J. Ferraris, and P. Keith. 2003. A new *Pseudocetopsis* species (Siluriformes: Cetopsidae) from Suriname and French Guiana. Proceedings of the Biological Society of Washington 116: 692-698.

Veen, A. W. L. 1970. On geogenesis and pedogenesis in the Old Coastal Plain of Surinam (South America). PhD dissertation, Amsterdam, The Netherlands.

Velazco, P. M., and B. K. Lim. 2014. A new species of broad-nosed bat *Platyrrhinus* Saussure, 1860 (Chiroptera: Phyllostomidae) from the Guianan Shield. Zootaxa 3796: 175-193.

Verberk, W. C. E. P., D. T. Bilton, P. Calosi, and J. I. Spicer. 2011. Oxygen supply in aquatic ectotherms: partial pressure and solubility together explain biodiversity and size patterns Ecology 92: 1565–1572.

Vercauteren, D. R., and J.-P. Gautier. 1993. On the occurrence of nocturnal and diurnal loud calls, differing in structure and duration, in red howlers (*Alouatta seniculus*) of French Guiana. Folia Primatologica 60: 195-209.

Vermeer, K., R. W. Risebrough, A. L. Spaans, and L. M. Reynolds. 1974. Pesticide effects on fishes and birds in the ricefields of Surinam, South America. Environmental Pollution 7: 227-236.

Vermeulen, F. 2015. The Killi's of the lost world. Pastelaria Studios Publishers, Papel, Portugal.

Versteeg, A. H. 1985. The Prehistory of the Young Coastal Plain of West Suriname. PhD dissertation, Rijksuniversiteit Leiden, Leiden, The Netherlands / Berichten van de Rijksdienst van Oudheidkundig Bodemonderzoek 35: 653-750.

Versteeg, A. H. 2003. Suriname voor Columbus / Suriname before Columbus. Libri Musei Surinamensis 1, Stichting Surinaams Museum, Paramaribo, Suriname.

Versteeg, A. H. 2007. Werehpai: an archeological site in SW-Suriname. Stichting Surinaams Museum, and Conservation International Suriname, Paramaribo, Suriname.

Vilhena, K. S. S., G. M. S. P. Guilhon, A. P. S. Filho, A. P. S. Souza. M. G. B. Zoghbi, L. S. Santos, M. S. P. Arruda, and A. C. Arruda. 2009. Inhibitory activity of essential oil of *Cyperus giganteus* Vahl. on weed species of Amazon. Allelopathy Journal 23: 221-227.

Villiers, A. 1979. Initiation a l'entomologie, Tome I, anatomie, biologie et classification Société Nouvelle des Editions Boubeé & Cie., Paris, France.

Vink, A. T. 1965. Surinam timbers. Surinam Forest Service, Ministry of Development, Paramaribo, Suriname.

Vital M. J., A. L. Carneiro, L. F. Rocha, R. C. Das Neves Amorim, M. R. Camargo, and A. M. Pohlit. 2015. Chemical composition, ethnopharmacology and biological activity of *Geissospermum* allemão species (Apocynaceae Juss.). Revista Fitos Eletrônica 8: 137-146.

Vitt, L. J., and J. P. Caldwell. 2013. Herpetology (4th ed.): An Introductory Biology of Amphibians and Reptiles. Academic Press, New York, NY, USA.

Vogel, S. 1990. History of the Malpighiaceae in the light of pollination ecology. Memoirs of the New York Botanical Garden 55: 130-142.

Voigt, C. C., and J. R. Speakman. 2007. Nectar-feeding bats fuel their high metabolism directly with exogenous carbohydrates. Functional Ecology 21: 913–921.

von Ellenrieder, N., and J. Muzón. 1999. The Argentinean species of the genus *Perithemis* Hagen (Anisoptera: Libellulidae). Odonatologica 28: 385-398.

Von Humboldt, A. 1814-1825. Personal narrative of a journey to the equinoctial regions of the new continent (Penguin Classics edition 1995). Penguin Books, London, England, UK.

Voss, R. S. 2013. Opossums (Mammalia: Didelphidae) in the diets of Neotropical pitvipers (Serpentes: Crotalinae): Evidence for alternative coevolutionary outcomes? Toxicon 66: 1–6.

Voss, R. S., D. P. Lunde, and N. B. Simmons. 2001. The mammals of Paracou, French Guiana: a neotropical lowland rainforest fauna. Part 2 Nonvolant species. Bulletin of the American Museum of Natural History 263: 1-236.

Vulinec, K. 2000. Dung beetles, monkeys and conservation. Florida Entomologist 83: 229-241.

Wainwright, P. C., W. L. Smith, S. A. Price, K. L. Tang, J. S. Sparks, L. A. Ferry, K. L. Kuhn, R. I. Eytan and T. J. Near. 2012. The evolution of pharyngognathy: a phylogenetic and functional appraisal of the pharyngeal jaw key innovation in labroid fishes and beyond. Systematic Biology 61: 1001-27.

Walker T. J., and T. E. Moore. 2013. Singing insects of North America (SINA). Department of Entomology and Nematology, Institute of Food and Agricultural Sciences University of Florida, Gainesville, FL, USA.

Walker, L. 1995. Amazonian streams and small rivers. *In* J.G. Tundisi, C.E.M. Bicudo, and T. Matsamura-Tundisi (Eds.). Limnology in Brazil, pp. 167-194. Brazilian Academy of Sciences, Sao Paulo, Brazil.

Walls, J. G. 1994. Jewels of the Rainforest. Poison Frogs of the Family Dendrobatidae. T.F.H. Publishers, Boston, MS, USA.

Walsh, J., and R. Gannon. 1967. Time is short and the water rises. E.P. Dutton, New York, NY, USA.

Walsh, S. J. 1990. A systematic revision of the Neotropical catfish family Ageneiosidae (Teleostei: Ostariophysi: Silurifomes). PhD dissertation, University of Florida, Gainesville, FL, USA.

Wan Tong You, K. 2007. Observations on the behavior of *Harttiella crassicauda* and *Guyanancistrus* n. sp. ('big mouth') of Nassau Mountains in the aquarium. *In* L.A. Alonso, and J.H. Mol (Eds.). A Rapid Biological Assessment of the Lely and Nassau Plateaus, Suriname (with additional information on the Brownsberg Plateau), p 249. Conservation International, Arlington, VA, USA.

Wang, C. 2001. Atlantic climate variability and its associated atmospheric circulation cells. International Journal of Climate 15: 1516-1536.

Wang, C. 2005. ENSO, Atlantic climate variability, and the Walker and Hadley circulations. *In* H. F. Diaz, and R. S. Bradley (Eds.). The Hadley Circulation: present, past and future, pp. 173-202. Springer Netherlands, Haarlem, The Netherlands.

Wang, Y., H. Cheng, R. L. Edwards, X. Kong, X. Shao, S. Chen, J. Wu, X. Jiang, X. Wang, and Z. An. 2008. Millennial- and orbital-scale changes in the East Asian monsoon over the past 224,000 years. Nature 451: 1090–1093.

Ward, P. S. 1999. Systematics, biogeography and host plant as associations of the *Pseudomyrmex viduus* group (Hymenoptera: Formicidae), Triplaris- and Tachigali-inhabiting ants. Zoological Journal of the Linnean Society 126: 451-540.

Warren, G. 1667. An impartial description of Surinam upon the continent of Guiana in America: with a history of several strange beasts, birds, fishes, serpents, insects, and customs of that colony, etc. William Godbid for Nathaniel Brooke, London, UK.

WARREN, R. J., AND GILADI, I. 2014. Ant-mediated seed dispersal: a few ant species (Hymenoptera: Formicidae) benefit many plants. Myrmecological News 20: 129-140.

WATERS, C. N., J. ZALASIEWICZ. C. SUMMERHAYES, A. D. BARNOSKY, C. POIRIER, A. GAŁUSZKA, A. CEARRETA, M. EDGEWORTH, E. C. ELLIS, M. ELLIS, C. JEANDEL, R. LEINFELDER, J. R. MCNEILL, D. DEB. RICHTER, W. STEFFEN, J. SYVITSKI, D. VIDAS, M. WAGREICH, M. WILLIAMS, A. ZHISHENG, J. GRINEVALD, E. ODADA, N. ORESKES, AND A. P. WOLFE. 2016. The Anthropocene is functionally and stratigraphically different from the Holocene. Science 351: 137. **doi: 10.1126/science.aad2622**

WEBB, S. D. 1985. The interrelationships of tree sloths and ground sloths. *In* G. G. Montgomery (Ed.). The ecology and evolution of armadillos, sloths, and vermilinguas, pp. 105–112. Washington, DC: Smithsonian Institution Press.

WEBER, N. A. 1943. Parabiosis in neotropical ant gardens. Ecology 24: 400-404.

WEBSTER, M. S. 1994. Interspecific brood parasitism of Montezuma Oropendolas by Giant Cowbirds: parasitism or mutualism? Condor 96: 794-798.

WEBSTER, T., AND L. ROEBUCK. 2001. Water Resources Assessment of Suriname. US Army Corps of Engineers Mobile District, Mobile, AL, and US Army Corps of Engineers Topographic Engineering Center, Alexandria, VA, USA.

WECKERLE, C. S., AND R. RUTISHAUSER. 2005. Gynoecium, fruit and seed structure of Paullinieae (Sapindaceae). Botanical Journal of the Linnean Society 147: 159–189.

WEI, F.2014. Compendium of Best Practices in Sustainable Tourism. Prepared for United Nations Department of Economic and Social Affairs, Uited Nations, Washington, DC, USA.

WEITZMAN, S. H., AND R. P. VARI. 1988. Miniaturization in South American freshwater fishes: an overview and discussion. Proceedings of the Biological Society of Washington 101: 444-465.

WELDON, P. J., AND J. H. RAPPOLE. 1997. A survey of birds odorous or unpalatable to humans: possible indications of chemical defense. Journal of Chemical Ecology 23: 2609-2633.

WERGER, M. J. A. (ED.). 2011. Sustainable management of tropical rainforests, The CELOS Management System. Tropenbos Series 25, Tropenbos International Suriname, Paramaribo, Suriname.

WERKGROEP ESTUARIENE KUSTSTROOK. 1976. Instelling Bijzonder Beheersgebied Estuariene Kuststrook. Nota over de bestemming van de zoute en brakke kuststrook in Suriname op ecologische grondslag. Report Inter-departmentale Werkgroep, Lands Bosbeheer, Paramaribo, Suriname.

WERKHOVEN, M. 1986. Orchideën van Suriname. Vaco Press, Paramaribo, Suriname.

WERKHOVEN, M. C. M., AND A. H. VERSTEEG. 1980. The vegetation of four mounds in the coastal plain of Suriname. Mededelingen Surinaams Museum 32: 8-37.

WERKHOVEN, M. C. M., AND G. M. T. PEETERS. 1993. Aquatic macrophytes. *In* P.E. Ouboter (Ed.). The freshwater ecosystems of Suriname, pp. 99-112. Kluwer Academic Publishers, Dordrecht, The Netherlands.

WESSELS BOER, J. G. 1965. The indigenous palms of Suriname. E.J. Brill, Leiden, the Netherlands.

WESTBY, G. W. M. 1988. The ecology, discharge diversity and predatory behaviour of gymnotiforme electric fish in the coastal streams of French Guiana. Behavioral Ecology and Sociobiology 22: 341-354.

WETTENGL, K. (ED.). 1998. Maria Sibylla Merian 1647-1717, kunstenares en natuuronderzoekster. Becht / Teylers Museum, Haarlem, The Netherlands.

WETTERER J. K., A. L. WETTERER, C. RUMBAITIS-DEL RIO, C. CHANG, G. VEGA, L. L. MANNE, J. AUKEMA, J. KARBIAN, A. SLOAN, AND M. DESAI. 2000. Diel shifts in treehopper tending by ants and wasps in Costa Rica (Hymenoptera). Sociobiology 36: 123-131.

WEYGOLDT, P. 1969. The biology of pseudoscorpions. Harvard University Press, Cambridge, MA, USA.

WHEELER, W. M. 1921. A new case of parabiosis and the 'ant-gardens' of British Guiana. Ecology 2: 89-103.

WHITE, C. 2012. Brownsberg Nature Park situation analysis 2012. WWF-Guianas, Paramaribo, Suriname.

WHITEHEAD, N. L. (ED.). 1997. The discoverie of the large, rich, and bewtiful empyre of Guiana by Sir Walter Ralegh. University of Oklahoma Press and Manchester University Press, Manchester, England, UK.

WHITFIELD, A. K., M. ELLIOTT, A. BASSET, S. J. M. BLABER, AND R. J. WEST. 2012. Paradigms in estuarine ecology: a review of the Remane diagram with a suggested revised model for estuaries. Estuarine, Coastal and Shelf Science 97: 78-90.

WHITNEY, B. M. 2007. "Kleptoptily": how the Fork-tailed Palm-Swift feathers its nest. Auk 124: 712-715.

WICHERTS, E., AND J. VELTKAMP. 2012. Geschiedenis van de Landsspoorweg. Suriname Informatie 9. Veka Productions, Wageningen, The Netherlands.

WICKLER, W. 1968. Mimicry in plants and animals. McGraw-Hill, New York, NY, USA.

WIERSEMA, J. H., A. R. NOVELO, AND J. R. BONILLA-BARBOSA. 2008. Taxonomy and typification of *Nymphaea ampla* (Salisb.) DC. sensu lato (Nymphaeaceae). Taxon 57: 967-974.

WIESNER-HANKS, M. E. 2006. Early Modern Europe 1450-1789. Cambridge History of Europe, Cambridge University Press, Cambridge, England, UK.

WIEST, F. C. 1995. The specialized locomotory apparatus of the freshwater hatchetfish family Gasteropelecidae. Journal of Zoology 236: 571-592.

WIGGLESWORTH, V. B. 1950. The principles of insect physiology. Methuen & Co., London, UK, and Dutton & Co., New York, NY, USA.

WIJMSTRA, T. A. 1971. The palynology of the Guiana coastal basin. PhD dissertation, Leiden, The Netherlands.

WIJMSTRA, T. A., AND T. VAN DER HAMMEN. 1966. Palynological data on the history of tropical savannas in northern South America. Leidse Geologische Mededelingen 38: 71–90.

WIJNSTRA, T. A. 1969. Palynology of the Alliance well. Geologie en Mijnbouw 48: 125-133.

WILLE, A., AND C. D. MICHENER. 1973. The nest architecture of stingless bees with special reference to Costa Rica. Revista Biologia Tropical 21, Supplement 1: 1-278.

WILLEMS T., A. DE BACKER, M. VINCX, AND K. HOSTENS. 2016a. Catch composition in seabob shrimp *Xiphopenaeus kroyeri* trawl fisheries off Suriname. *In* T. Willems (Ed.). An ecosystem approach to fisheries management: the Atlantic seabob shrimp (*Xiphopenaeus kroyeri*) in Suriname, pp. 172-199. PhD dissertation, Marine Biology Research Group, Biology Department, Faculty of Sciences, Gent University, Gent, Belgium.

WILLEMS, T. 2016. An ecosystem approach to fisheries management: the Atlantic seabob shrimp (*Xiphopenaeus kroyeri*) in Suriname. PhD dissertation, Marine Biology Research Group, Biology Department, Faculty of Sciences, Gent University, Gent, Belgium.

WILLEMS, T., A. DE BACKER, J. H. MOL, M. VINCX, AND K. HOSTENS. 2015a. Distribution patterns of the demersal fish fauna on the inner continental shelf of Suriname. Regional Studies in Marine Science 2: 177–188.

WILLEMS, T., A. DE BACKER, K. WAN TONG YOU, M. VINCX, AND K. HOSTENS. 2015b. Spatio-temporal distribution patterns of the epibenthic community in the coastal waters of Suriname. Continental Shelf Research 108: 25-40.

WILLEMS, T., A. DE BACKER, T. KERKHOVE, N. N. DAKRIET, M. DE TROCH, M. VINCX, AND K. HOSTENS. 2016b. Trophic ecology of Atlantic seabob shrimp *Xiphopenaeus kroyeri*: intertidal benthic microalgae support the subtidal food web off Suriname. Estuarine, Coastal and Shelf Science 182: 146-157.

WILLEMS, T., J. DEPESTELE, A. DE BACKER AND K. HOSTENS. 2016c. By-catch in a tropical shrimp fishery: do Bycatch Reduction Devices and Turtle Exclusion Devices effectively exclude rays. Fisheries Research 175: 35-42.

WILLIAMS, A. B. 1974. The swimming crabs of the genus *Callinectes* (Decapoda: Portunidae). Fishery Bulletin 72: 685- 798.

WILLINK, P. W., K. WAN TONG YOU, AND M. PIQUE. 2011. Fishes of the Sipaliwini and Kutari rivers. *In* B.J. O'Shea, L.E. Alonso, and T.H. Larsen (Eds.). A Rapid Biological assessment of the Kwamalasumutu Region, southwestern Suriname, pp. 189-123. RAP Bulletin of Biological Assessment 63, Conservation International, Arlington, VA, USA.

WILLIS, E. O., AND Y. ONIKI. 1978. Birds and army ants. Annual Review of Ecology and Systematics 9: 243-263.

WILSON, D. E., C. F. ASCORRA, AND S. SOLARI. 1996. Bats as indicators of habitat disturbance. *In* D. E. Wilson, and A. Sandoval (Eds.). Manu: the Biodiversity of Southeastern Peru, pp. 613-25. Smithsonian Institution, Washington, DC, and Editorial Horizonte, Lima, Peru.

WILSON, E. O. 1971. The insect societies. The Belknap Press of Harvard University Press, Cambridge, MA, USA, and London, England, UK.

WILSON, E. O. 1984. Clockwork lives of the Amazonian leafcutter army. Smithsonian 15: 92-101.

WILSON, E. O. 1985. The biological diversity crisis. BioScience 35: 700-706.

WILSON, E. O. 1988. The current state of biological diversity. *In* E. O. Wilson, and F. M. Peter (Eds.). Biodiversity, pp. 3-18. National Academy Press, Washington, DC, USA.

WILSON, E. O. 1992. The diversity of life. Allen Lane The Penguin Press, London, England, UK.

WILSON, K. 2009. Dragonfly Giants. Agrion 13(1): 29-31.

WINEMILLER, K. O. 1987. Feeding and reproductive biology of the currito, *Hoplosternum littorale*, in the Venezuelan llanos with comments on the possible function of the enlarged male pectoral spines. Environmental Biology of Fishes 20: 219-227.

WINEMILLER, K. O. 1989a. Development of dermal lip protuberances for aquatic surface respiration in South American characid fishes. Copeia 1989: 382-390.

WINEMILLER, K. O. 1989b. Ontogenetic diet shifts and resource partitioning among piscivorous fishes in the Venezuelan llanos. Environmental Biology of Fishes 26: 177-199.

WINEMILLER, K. O., AND A. ADITE. 1997. Convergent evolution of weakly electric fishes from floodplain habitats in Africa and South America. Environmental Biology of Fishes 49: 175-186.

WIP, D., T. WARNEKE, A. K. PETERSEN, J. NOTHOLT, C. TEMME, H. KOCK, AND P. CORDY. 2013. Urban mercury pollution in the city of Paramaribo, Suriname. Air Quality Atmosphere and Health 6: 205–213.

WMO/UNDP/GOVERNMENT OF SURINAME. 1972. Public Water Supplies and Sewerage Project – Water Resources – Vol. III (hydrogeological and hydrological studies). Surinaamse Waterleiding Maatschappij, Ministerie van Openbare Werken, Paramaribo, Suriname.

WOLDA, H., AND C. W. SABROSKY. 1986. Insect visitors to two forms of *Aristolochia pilosa* in Las Cumbres, Panama. Biotropica 18: 295-299.

WONG T. E., R. DE KRAMER, P. L. DE BOER, C. LANGEREIS, AND J. SEW-A-TJON. 2009. The influence of sea-level changes on tropical coastal lowlands; the Pleistocene Coropina Formation, Suriname. Sedimentary Geology 216: 125–137.

WONG, T. E. 1989. Revision of the stratigraphy of the coastal plain of Suriname. Natuurwetenschappelijke Studiekring voor Suriname en de Nederlandse Antillen 125: 1-64.

WONG, T. E. 2014. Geological development of the coastal plain of Suriname. Memorias Geological Map of South America Workshop, Villa de Leiva, Colombia, pp. 22-23 and 163-177. Servicio Geologico Combiana / Ministerio de Minas y Energia, Bogota, Colombia.

WONG, T. E., L. KROOK, AND J. I. S. ZONNEVELD. 1998. Investigations in the coastal plain and offshore area of Suriname. *In* T. E. Wong, D. R. de Vletter, L. Krook, J. I. S. Zonneveld, and A. J. van Loon (Eds.). The history of earth sciences in Suriname, pp. 73-100. Royal Netherlands Academy of Arts and Sciences & Netherlands Institute of Applied Geoscience TNO, Amsterdam, The Netherlands.

WORLD TEAM. 2016. Trio & Wayana tribes of Suriname. **diaryofaforeigncouple.files.wordpress.com/2016/03/suriname_tribal.pdf**

WOSIACKI, W. B., D. P. COUTINHO AND L. F. DE ASSIS MONTAG. 2011. Description of a new species of sand-dwelling catfish of the genus *Stenolicmus* (Siluriformes: Trichomycteridae). Zootaxa 2752: 62-86.

WOURMS, J. P. 1981. Viviparity: the maternal-fetal relationship in fishes. American Zoologist 21: 473-515.

Wuster, W., J. E. Ferguson, J. A. Quijada-Mascarenas, C. E. Pook, M. Da Graca Salomao, and R. S. Thorpe. 2005. Tracing an invasion: landbridges, refugia, and the phylogeography of the Neotropical rattlesnake (Serpentes: Viperidae: *Crotalus durissus*). Molecular Ecology 14: 1095–1108.

WWF. 2011. Marine Turtle Bycatch Research WWF Guianas 2010-2011. WWF Guianas, Paramaribo, Suriname. **www.widecast.org/What/AnnualMeeting/2011/2011_Felix_Marie_Louise_Bycatch_Guianas.pdf**

WWF. 2012. Living Guianas Report 2012. WWF Guianas, Paramaribo, Suriname.

WWF. 2015. Living Blue Planet Report: species, habitats and human well-being. WWF International, Gland, Switzerland.

WWF. 2016a. Living Planet Report 2016. Risk and resilience in a new era. WWF International, Gland, Switzerland.

WWF. 2016b. Farming: Pollution. WWF International, Gland, Switzerland. **wwf.panda.org/what_we_do/footprint/agriculture/impacts/pollution/** [accessed Dec. 2016]

Yalcindag E., E. Elguero, C. Arnathau, P. Durand, J. Akiana, T. J. Anderson, A. Aubouy, F. Balloux, P. Besnard, H. Bogreau, P. Carnevale, U. D'Alessandro, D. Fontenille, D. Gamboa, T. Jombart, J. Le Mire, E. Leroy, A. Maestre, M. Mayxay, D. Ménard, L. Musset, P.N. Newton, D. Nkoghé, O. Noya, B. Ollomo, C. Rogier, V. Veron, A. Wide, S. Zakeri, B. Carme, E. Legrand, C. Chevillon, F.J. Ayala, F. Renaud, and F. Prugnolle. 2012. Multiple independent introductions of *Plasmodium falciparum* in South America. Proceedings of the National Academy of Sciences of the United States of America. 109:511–516.

Yonke, T. R. 1990. Order Hemiptera. *In* F. W. Stehr (Ed.). Immature insects, Vol. 2, pp. 22-65. Kendall/Hunt, Dubuque, IA, USA.

Youlatos, D., Couette S., and L. B. Halenar. 2015. Morphology of howler monkeys: a review and quantitative analysis. *In* M. M. Kowalewski, P. A. Garber, L. Cortés-Ortiz, B. Urbani, and D. Youlatos (Eds.). Howler monkeys: Behavior, Ecology and Conservation, pp. 133-176. Springer, New York, NY, USA.

Young, A. M., E. H. Erickson Jr., M. A. Strand, and B. J. Erickson. 1987. Pollination biology of *Theobroma* and *Herrania* (Sterculiaceae). I. Floral biology. Insect Science and Its Application 8: 151-164.

Zahl, P. A. 1975. Hidden worlds in the heart of a plant. National Geographic 147: 388-397.

Zahl, P. A., J. J. A. McLaughlin, and R. J. Gomprecht, 1977, Visual versatility and feeding of the four-eyed fishes, *Anableps*. Copeia 1977: 791-793.

Zeh, D. W., and. J. A. Zeh. 1992. On the function of Harlequin Beetle-riding in the pseudoscorpion *Cordylochernes scorpioides* (Pseudoscorpionida; Chernetidae). Journal of Arachnology 20: 47–51.

Zeh, D. W., Zeh J. A., and G. Tavakilian. 1992. Sexual selection and sexual dimorphism in the Harlequin Beetle *Acrocinus longimanus*. Biotropica 24: 86-96.

Zeh, J. A., and D. W. Zeh. 1990. Cooperative foraging for large prey by *Paratemnus elongatus* (Pseudoscorpionida, Atemnidae). Journal of Arachnology 18: 307–311.

Zeil, J., W. A. Ribi, and A. Narendra. 2014. Polarisation vision in ants, bees and wasps. *In* G. Horváth (Ed.). Polarized light and polarization vision in animal sciences, pp. 41-60. Springer Verlag, Berlin and Heidelberg, Germany.

Zimmerman Y., S. R. Ramírez, and T. Eltz. 2009. Chemical niche differentiation among sympatric species of orchid bees. Ecology 90: 2994-3008.

Zonneveld, J. I. S. 1972. Sulas and sula complexes. Göttinger Geographische Abhandlungen 1972: 93-101.

Zonneveld, J. I. S. 1993. Planation and summit levels in Suriname (S. America). Zeitschrift für Geomorphologie, Supplement Band 93: 29-46.

Zonneveld, J. I. S., and L. Krook. 1998. Geomorphological investigations in the interior of Suriname. *In* T. E. Wong, D. R. de Vletter, L. Krook, J. I. S. Zonneveld, and A. J. van Loon (Eds.). The history of earth sciences in Suriname, pp. 101-127. Royal Netherlands Academy of Arts and Sciences & Netherlands Institute of Applied Geoscience TNO, Amsterdam, The Netherlands.

Zook, D. 2010. Tropical rainforests as dynamic symbiospheres of life. Symbiosis 51: 27–36.

Zuanon, J., and I. Sazima. 2004. Vampire catfishes seek the aorta not the jugular: candirus of the genus *Vandellia* (Trichomycteridae) feed on major gill arteries of host fishes. Aqua Journal of Ichthyology and Aquatic Biology 8: 31-36.

Zuanon, J., F. A. Bockmann, and I. Sazima. 2006. A remarkable sand-dwelling fish assemblage from central Amazonia, with comments on the evolution of psammophily in South American freshwater fishes. Neotropical Ichthyology 4: 107-118.

Zug, G. R., and P. B. Zug. 1979. The Marine Toad, *Bufo marinus*: A natural history resumé of native populations. Smithsonian Contributions to Zoology 284: 1-58.

Zyskowski, K., J. C. Mittermeier, O. Ottema, M. Rakovic, B. J. O'Shea, J. E. Lai, S. B. Hochgraf, J. de León, and K. Au. 2011. Avifauna of the easternmost tepui, Tafelberg in central Suriname. Bulletin of the Peabody Museum of Natural History 52: 153-180.

Index

Authors

Chapter authors

Bart P.E. De Dijn
1) Environmental Services & Support, Leiding 20 #19, Wanica, Suriname, email: **bartdedijn@ess-environment.com**, and
2) Bart De Dyn Environmental Consultancy, Bakamoejoeweg #26, Paramaribo, Suriname, email: **dedijn@yahoo.com**

Jacobus ('Koos') Groen
Acacia Water, Gouda, The Netherlands, email: **koos.groen@acaciawater.com**

Hélène Hiwat
Datrakondrestraat 122, Paramaribo, Suriname, email: **helenehiwat@gmail.com**

Bruce Hoffman
Amazon Conservation Team - Suriname, Doekhieweg Oost #24, Zorg en Hoop, Paramaribo, Suriname, email: **b.hoffman@act-suriname.org**

Gisla Jairam-Doerga
National Herbarium of Suriname (BBS), Anton de Kom Universiteit Suriname, Leysweg, Paramaribo, Suriname, email: **gislaram@gmail.com**

Salomon B. Kroonenberg
1) Delft University of Technology, Stevinweg 1, 2628CN Delft, The Netherlands, email: **s.b.kroonenberg@tudelft.nl**, and
2) Anton de Kom University of Suriname, Leysweg, Paramaribo, Suriname, email: **salomonkroonenberg@gmail.com**

Chantal J. Landburg
1) Environmental Services & Support, Leiding 20 #19, Wanica, Suriname, email: **chantallandburg@ess-environment.com**, and
2) Reeberg project 39, Wanica, Suriname.

Burton K. Lim
Department of Natural History, Royal Ontario Museum, 100 Queen's Park, Toronto, ON M5S2C6, Canada, email: **burtonl@rom.on.ca**

Jan H.A. Mol
Department of Biology, Faculty of Mathematics and Natural Sciences, Anton de Kom University of Suriname, CELOS building, Prof. Dr. Ir. J. Ruinardlaan, Paramaribo, email: **fisheco@celos.sr.org**

Sieuwnath Naipal
Department Infrastructure, Anton de Kom University of Suriname, Leysweg, Paramaribo, Suriname, email: **s.naipal@uvs.edu**

Brice P. Noonan
The University of Mississippi, Department of Biology, Box 1848, University, MS 38677, USA, email: **bnoonan@olemiss.edu**

Dirk Noordam
Edmundstraat 55, Paramaribo, Suriname, email: **dirknoordam@yahoo.com**

Marilyn A. Norconk
Department of Anthropology, Kent State University, 226 Lowry Hall, Kent OH 44242, USA, email: **mnorconk@kent.edu**

Riad Nurmohamed
Department Infrastructure, Anton de Kom University of Suriname, Leysweg, Paramaribo, Suriname, email: **r.nurmohamed@uvs.edu**

Brian J. O'Shea
Research and Collections, North Carolina Museum of Natural Sciences, 11 W. Jones St., Raleigh NC 27601, USA, email: **brian.oshea@naturalsciences.org**

Paul E. Ouboter
National Zoological Collection of Suriname (NZCS), Anton de Kom University of Suriname, Leysweg, Paramaribo, Suriname, email: **p.ouboter@uvs.edu**

Sofie Ruysschaert
World Wildlife Fund Belgium, Emile Jacqmainlaan 90, 1000 Brussel, Belgium, email: **sofie.ruysschaert@wwf.be**

Pieter A. Teunissen
Laan van Louiseshof 47, Paramaribo, Suriname, email: **pieterteunissen@hotmail.com**

Kenneth Wan Tong You
Department of Biology, Faculty of Mathematics and Natural Sciences, Anton de Kom University of Suriname, Leysweg, Paramaribo, Suriname, email: **kgl_wan@hotmail.com**

Additional species account contributors

Antonio A. Agudelo
Instituto Nacional de Pesquisas da Amazônia, Avenida André Araujo 2936, Aleixo, 69067-375, Manaus, AM, Brazil, email: **antonioarnovis@yahoo.com**

Simon Clavier
HYDRECO, BP 823 97388 Kourou cedex, French Guiana, email: **simon.clavier@hydrecolab.com**

Atilano Contreras-Ramos
Instituto de Biología-UNAM, 04510 México, D.F., Mexico, email: **acontreras@ib.unam.mx**

Dominic Evangelista
Muséum National d'Histoire Naturelle - Institut de Systématique, Evolution, Biodiversité, 45 Rue Buffon, 75005 Paris, France, email: **dominicev@gmail.com**

Nicolas Dedieu
INP UPS EcoLab (UMR-CNRS 5245), Université de Toulouse, 31062 Toulouse, France, email: **nicolas.dedieu@hydrecolab.com**

Hajo B. P. E. Gernaat
Department of Biodiversity Discovery, Naturalis Biodiversity Center, P.O. Box 9517, 2300 RA Leiden, The Netherlands, email: **hajo.gernaat@naturalis.nl**

Eric J. Gouda
Botanic Garden, Department of Biology, University of Utrecht (U), Utrecht, The Netherlands; email: **E.J.Gouda@uu.nl**

Jocelia Grazia
Departamento de Zoologia, Universidade Federal Rio Grade do Sul, Av. Bento Gonçalves 9500, Prédio 43435, Sala 223, 91501-970 Porto Alegre, RS, Brazil, email: **jocelia@ufrgs.br**

František Kovařík
P.O.Box 27, CZ – 145 01 Praha 45, Czech Republic, email: **kovarik.scorpio@gmail.com**

Volker Mahnert
Muséum d'Histoire Naturelle, 1, route de Malagnou, 1208 Geneva, Switzerland, email: **volker.mahnert@wanadoo.fr**

Scott Mori
Institute of Systematic Botany, The New York Botanical Garden, Bronx, New York 10458-5126 USA ; email: **smori@nybg.org**

Otte Ottema
Komeetstraat 2, Ephraimszegen, Paramaribo, Suriname, email: **otteottema@gmail.com**

Serano Ramcharan
Cayottestraat 43, Paramaribo, Suriname, email:
sbirdwatching@gmail.com

Menno Reemer
EIS Kenniscentrum Insecten en andere ongewervelden, Naturalis Biodiversity Center, Vondellaan 55, Leiden, The Netherlands, email: **menno.reemer@naturalis.nl**

Julio Rivera
Universidad Particular San Ignacio de Loyola, Av. La Fontana 550, La Molina 00012, Lima, Peru, email: **julio.riverac@usil.pe**

Virginie Roy
University Paris-Est Créteil, iEES-Paris, Créteil, France, email: **roy@u-pec.fr**

Cristiano Schwertner
Departamento de Ciências Biológicas, Universidade Federal de São Paulo, Campus Diadema, Rua Artur Riedel 275, Eldorado, CEP 09972-270, Diadema, SP, Brazil, email: **schwertner@unifesp.br**

David Sillam-Dussès
1) IRD – Sorbonne Universités, iEES-Paris, U 242, Bondy, France, and
2) University Paris 13, Sorbonne Paris Cité, LEEC, EA 4443,
Villetaneuse, France; email: **sillamdusses@univ-paris13.fr**

Arie L. Spaans
Sylvalaan 12, 6816 RB Arnhem, The Netherlands, email: **a.l.spaans@planet.nl**

Alies van Sauers-Muller
Damboentong 282, Saramacca, Suriname, email: **aliesmuller@yahoo.com**

Johan van't Bosch
Newtonplein 62, 2562 JX Den Haag, The Netherlands, email: **johanvantbosch@yahoo.co.uk**

Marcel Wasscher
Minstraat 15 bis, 3582 CA Utrecht, The Netherlands, email: **marcel.hilair@12move.nl**